Fragmentation

Fragmentation

Toward Accurate Calculations on Complex Molecular Systems

Edited by

Mark S. Gordon
Iowa State University, USA

WILEY

This edition first published 2017

Registered Offices
John Wiley & Sons, Inc., 111 River Street, Hoboken, NJ 07030, USA
John Wiley & Sons Ltd, The Atrium, Southern Gate, Chichester, West Sussex, PO19 8SQ, UK

Editorial Offices
9600 Garsington Road, Oxford, OX4 2DQ, UK
The Atrium, Southern Gate, Chichester, West Sussex, PO19 8SQ, UK
111 River Street, Hoboken, NJ 07030-5774, USA

For details of our global editorial offices, for customer services and for information about how to apply for permission to reuse the copyright material in this book please see our website at www.wiley.com/wiley-blackwell

Library of Congress Cataloging-in-Publication Data

Names: Gordon, M. S. (Mark S.), editor.
Title: Fragmentation : toward accurate calculations on complex molecular systems / edited by Professor Mark S. Gordon, Iowa State University, USA.
Description: Chichester, UK ; Hoboken, NJ : John Wiley & Sons, Inc., 2017. | Includes bibliographical references and index.
Identifiers: LCCN 2016057161 (print) | LCCN 2016058050 (ebook) | ISBN 9781119129240 (cloth) | ISBN 9781119129257 (pdf) | ISBN 9781119129264 (epub)
Subjects: LCSH: Fragmentation reactions. | Electron configuration.
Classification: LCC QD281.F7 F738 2017 (print) | LCC QD281.F7 (ebook) | DDC 547/.128–dc23
LC record available at https://lccn.loc.gov/2016057161

A catalogue record for this book is available from the British Library.

Wiley also publishes its books in a variety of electronic formats. Some content that appears in print may not be available in electronic books.

Cover Design: Wiley
Cover Images: (Background) © Esebene/Gettyimages; (Inset Images) Courtesy of the editor

Set in 10/12pt WarnockPro by Aptara Inc., New Delhi, India

10 9 8 7 6 5 4 3 2 1

Contents

List of Contributors

Emily A. Carter
School of Engineering and Applied Science, Princeton University, USA

Garnet K.L. Chan
Frick Chemistry Laboratory, Department of Chemistry, Princeton University, USA

Ajitha Devarajan
Office of University Development, University of Michigan, USA

Johannes M. Dieterich
Department of Mechanical and Aerospace Engineering, Princeton University, USA

Dmitri G. Fedorov
Research Center for Computational Design of Advanced Functional Materials (CD-FMat), National Institute of Advanced Industrial Science and Technology (AIST), Tsukuba, Japan

Alexander Gaenko
Advanced Research Computing, University of Michigan, USA

Kandis Gilliard
Department of Chemistry, University of Illinois at Urbana–Champaign, USA

Mark S. Gordon
Ames Laboratory of United States Department of Energy, USA

Department of Chemistry, Iowa State University, USA

Pradeep K. Gurunathan
Department of Chemistry, Purdue University, USA

Xiao He
School of Chemistry and Molecular Engineering, East China Normal University, China

NYU-ECNU Center for Computational Chemistry, NYU Shanghai, China

So Hirata
Department of Chemistry, University of Illinois at Urbana–Champaign, USA

Jan H. Jensen
Department of Chemistry, University of Copenhagen, Denmark

Carlos A. Jiménez-Hoyos
Frick Chemistry Laboratory, Department of Chemistry, Princeton University, USA

K. V. Jovan Jose*
Department of Chemistry, Indiana University, USA

**Current address: School of Chemistry, University of Hyderabad, India*

Murat Keçeli
Department of Chemistry, University of Illinois at Urbana–Champaign, USA
Argonne National Laboratory, USA

Kazuo Kitaura
Graduate School of System Informatics, Kobe University, Japan

Christoph Köppl
Institute for Theoretical Chemistry, University of Stuttgart, Germany

Caroline M. Krauter
Department of Mechanical and Aerospace Engineering, Princeton University, USA

Jinjin Li
Department of Chemistry, University of Illinois at Urbana–Champaign, USA
National Key Laboratory of Science and Technology on Micro/Nano Fabrication, Department of Micro/Nano Electronics, Shanghai Jiao Tong University, China

Jinfeng Liu
School of Chemistry and Molecular Engineering, East China Normal University, China

Qianli Ma
Institute for Theoretical Chemistry, University of Stuttgart, Germany

Hiromi Nakai
Department of Chemistry and Biochemistry, School of Advanced Science and Engineering, Waseda University, Japan

Krishnan Raghavachari
Department of Chemistry, Indiana University, USA

Michael A. Salim
Department of Chemistry, University of Illinois at Urbana–Champaign, USA

Max Schwilk
Institute for Theoretical Chemistry, University of Stuttgart, Germany

Lyudmila V. Slipchenko
Department of Chemistry, Purdue University, USA

Olaseni Sode
Department of Chemistry, University of Illinois at Urbana–Champaign, USA
Department of Chemistry, Biochemistry, and Physics, The University of Tampa, USA

Casper Steinmann
Department of Physics, Chemistry and Pharmacy, University of Southern Denmark, Denmark

Hans-Joachim Werner
Institute for Theoretical Chemistry, University of Stuttgart, Germany

Theresa L. Windus
Ames Laboratory of United States Department of Energy, USA
Department of Chemistry, Iowa State University, USA

Sebastian Wouters
Center for Molecular Modelling, Ghent University, Belgium
Frick Chemistry Laboratory, Department of Chemistry, Princeton University, USA

Kiyoshi Yagi
Department of Chemistry, University of Illinois at Urbana–Champaign, USA
Theoretical Molecular Science Laboratory, RIKEN, Japan

Takeshi Yoshikawa
Department of Chemistry and Biochemistry, School of Advanced Science and Engineering, Waseda University, Japan

Kuang Yu
Department of Mechanical and Aerospace Engineering, Princeton University, USA

John Z. H. Zhang
School of Chemistry and Molecular Engineering, East China Normal University, China

NYU-ECNU Center for Computational Chemistry, NYU Shanghai, China

Department of Chemistry, New York University, USA

Tong Zhu
School of Chemistry and Molecular Engineering, East China Normal University, China

NYU-ECNU Center for Computational Chemistry, NYU Shanghai, China

Preface

Electronic structure theory, also referred to as *ab initio* quantum chemistry (QC), has attained a high level of maturity and reliability for gas-phase molecules of modest size. Unfortunately, the formal scaling of these methods such as Hartree–Fock (HF), density functional theory (DFT), second-order perturbation theory (MP2), coupled cluster theory (CC), and multi-reference (MR) methods hinder their application to large molecules, to condensed phase systems or to excited electronic state potential energy surfaces. These limitations are especially severe for methods that account for electron correlation, such as MP2, CC, and MR methods, since their scaling with system size is steeper than for the simpler HF and DFT methods. There is therefore a need for computational strategies that nearly retain the accuracy of the most reliable methods while greatly reducing the scaling of these methods as a function of system size. While researchers who are interested in simulations of large molecular systems have often turned to classical molecular mechanics (MM) force fields, MM methods are limited in their applicability. While there are a few exceptions, classical MM cannot realistically treat bond making/bond breaking (the essence of chemistry) or excited state phenomena.

One effective QC approach that has become increasingly popular is referred to as fragmentation (broadly defined) or embedding theory. Fragmentation commonly refers to the physical subdivision of a large molecule into fragments, each of whose energy can be computed on a different compute node, thereby making the overall computation highly parallel. Fragmentation methods of this type scale nearly linearly with system size and can take advantage of massively parallel computers. Fragmentation methods of this type are discussed in Chapters 3, 5, 6, 7, 10, and 11. An alternative approach to physical fragmentation of a molecule is to fragment the wave function, by employing localized molecular orbitals to separate the wave function into domains that can be separately correlated. This approach is based on the fact that electron correlation is short-range. Chapter 1 provides an excellent discussion of local electron correlation methods by one of the leaders in the field.

Embedding methods are similar to fragmentation methods in that a total system is partitioned into multiple subsystems, in a manner that allows the incorporation of interactions among the subsystems. Like fragmentation and local orbital approaches, embedding methods reduce the steep scaling of traditional electronic structure methods. Embedding methods frequently involve multiple levels of theory. Approaches to embedding methods are discussed in Chapters 2, 4, 8, and 9.

The methods that are discussed in this book provide an exciting path forward to the accurate study of large molecules and condensed phase phenomena.

1

Explicitly Correlated Local Electron Correlation Methods

Hans-Joachim Werner, Christoph Köppl, Qianli Ma, and Max Schwilk

Institute for Theoretical Chemistry, University of Stuttgart, Germany

1.1 Introduction

Accurate wave function methods for treating the electron correlation problem are indispensable in quantum chemistry. A well-defined hierarchy of such methods exists, and in principle, these methods allow to approach the exact solution of the non-relativistic electronic Schrödinger equation to any desired accuracy. A much simpler alternative is density functional theory (DFT), which is probably most often used in computational chemistry. However, its failures and uncertainties are well known, and there is no way for systematically improving or checking the results other than comparing with experiment or with the results of accurate wave function methods.

Due to the steep scaling of the computational resources (CPU-time, memory, disk space) with the molecular size, conventional wave function methods such as CCSD(T) (coupled-cluster with single and double excitations and a perturbative treatment of triple excitations) can only be applied to rather small molecular systems. For example, the CPU-time of CCSD(T) scales as $\mathcal{O}(\mathcal{N}^7)$, where $\mathcal{N}$ is a measure of the molecular size (e.g., the number of correlated electrons) and even the simplest electron correlation method, MP2 (second-order Møller-Plesset perturbation theory) scales as $\mathcal{O}(\mathcal{N}^5)$. This causes a "scaling wall" that cannot be overcome. Even with massive parallelization and using the largest supercomputers, this wall can only be slightly shifted to larger systems. However, it is well known that electron correlation in insulators is a short-range effect. The pair correlation energies decay at long-range with R^{-6}, where R is the distance between two localized spin orbitals. Therefore, the steep scaling is unphysical. It results mainly from the use of canonical molecular orbitals, which are usually delocalized over larger parts of the molecule.

The scaling problem can be much alleviated by exploiting the short-range character of electron correlation using local orbitals and by introducing local approximations. This was first proposed in the pioneering work of Pulay *et al.* [1–6], and in the last 20 years enormous progress has been made in developing accurate local correlation methods. There are two different approaches, both of which are based on the use of local orbitals. The traditional one is to treat the whole molecule in one calculation and to apply various approximations that are based on the fast decay of the correlation energy. We will denote

Fragmentation: Toward Accurate Calculations on Complex Molecular Systems, First Edition.
Edited by Mark S. Gordon.

such methods "local correlation methods." A large variety of such approaches has been published in the past [7–59].

The second approach is the so-called "fragmentation methods" [60–87], in which the system is split into smaller pieces. These pieces are treated independently, mostly using conventional methods (although the use of local correlation methods is also possible). The total correlation energy of the system is then assembled from the results of the fragment calculations. Various methods differ in the way in which the fragments are chosen and the energy is assembled. A special way of assembling the energy using a many-body expansion is used in the so-called incremental methods [88–96], but these also belong to the group of fragmentation methods. Fragmentation methods will be described in other chapters of this volume and are therefore not the subject of this chapter. However, in Section 1.13, we will comment on the relation of local correlation and fragmentation methods.

Another problem of the CCSD(T) method is the slow convergence of the correlation energy with the basis set size. Very large basis sets are needed to obtain converged results, and this makes conventional high-accuracy electron correlation calculations extremely expensive. This problem is due to the fact that the wave function has a cusp for $r_{12} \rightarrow 0$, where r_{12} is the distance between two electrons. The cusp is due to the singularity of the Coulomb operator $\frac{1}{r_{12}}$, and cannot be represented by expanding the wave function in antisymmetrized products of molecular orbitals (Slater determinants). This leads to the very slow convergence of the correlation energy with the size of the basis set, and in particular with the highest angular momentum of the basis functions. This problem can be solved by including terms into the wave function that depend explicitly on the distance r_{12}, and these methods are known as "explicit correlation methods" [97–155].

The combination of explicit correlation methods with local approximations has been particularly successful [140–153]. As will be explained and demonstrated later in this chapter, this does not only drastically reduce the basis set incompleteness errors, but also strongly reduces the errors caused by local approximations. Local correlation methods employ two basic approximations. The first is based on writing the total correlation energy as a sum of pair energies. Each pair describes the correlation of an electron pair (in a spin-orbital formulation), or, more generally, the correlation of the electrons in a pair of occupied local molecular orbitals (LMOs). Depending on the magnitude of the pair energies, it is possible to introduce a hierarchy of "strong," "close," "weak," or "distant" pairs [7, 18, 31, 32]. Different approximations can be introduced for each class, ranging from a full local coupled-cluster (LCCSD) treatment for strong pairs to a non-iterative perturbation correction for distant pairs, which can be evaluated very efficiently using multipole approximations [12, 13]. We will denote such approximations as "pair approximations." The second type of local approximations is the "domain approximation," which is applied to each individual pair. A domain is a subset of local virtual orbitals which is spatially close to the LMO pair under consideration. Asymptotically, the number of orbitals in each pair domain (the "domain sizes") become independent of the molecular size. Also the number of pairs in each class (except for the distant pairs) becomes independent of the molecular size. This leads to linear scaling of the computational effort as a function of molecular size, as has already been demonstrated for LMP2 and up to the LCCSD(T) level of theory more than 25 years ago [12–18].

The critical question is, of course, how quickly the correlation energy as well as relative energies (e.g., reaction energies, activation energies, intermolecular interaction energies, and electronic excitation energies) converge with the domain sizes and how they depend on the pair approximations. The domain sizes which are necessary to reach a certain accuracy (e.g., 99.9% of the canonical correlation energy) depends sensitively on the choice of the virtual orbitals. As is known since the 1960s, fastest convergence is obtained with pair natural orbitals (PNOs) [156], and this has first been fully exploited in the seminal PNO-CI and PNO-CEPA methods of Meyer [157, 158], and somewhat later also by others [159–163]. The problem with this approach is that the PNOs are different for each pair and non-orthogonal between different pairs. This leads to complicated integral transformations and prevented the application of PNO methods to large molecules for a long time. The method was revived by Neese and coworkers in 2009 and taken up also by others (including us) later on [32, 33, 48–57, 146–150]. The problem of evaluating the integrals was overcome by using local density-fitting approximations [22]. Furthermore, the integrals are first computed in a basis of projected atomic orbitals (PAOs), which are common to all pairs, and subsequently transformed to the pair-specific PNO domains [54, 146, 147, 150]. Also, hybrid methods, in which so-called orbital-specific virtuals (OSVs) [164–167] are used at an intermediate stage, have been proposed [53, 146, 147]. Later sections of this chapter will explain these approaches in some detail.

Local approximations have also been developed for multi-reference wave functions [168–176]. The description of these methods is beyond the scope of the current article, but we mention that recently very efficient and accurate PNO-NEVPT2 [175] (*N*-electron valence state perturbation-theory) and PNO-CASPT2 [176] (complete active space second-order perturbation theory) methods have been described.

In the current article, we will focus on new developments of well-parallelized PNO-LMP2-F12 and PNO-LCCSD-F12 methods recently developed in our laboratory. These methods also have a close relation to the methods of Neese *et al.* [54–57, 153]. After introducing some benchmark systems, which will be used later on, we will first outline the principles of local correlation and describe the choice of the local occupied and virtual orbitals as well as of the domains. The convergence of the correlation energy as a function of the domain sizes will be demonstrated for various types of virtual orbitals for LMP2. Subsequently, based on these foundations, we will discuss more advanced approximations for distant pairs and close/weak pair approximations used in local coupled cluster methods. Next, we will present an introduction to local explicit correlation methods, and demonstrate the improvements achieved by the F12 approach both for LMP2-F12 and LCCSD-F12. Finally, we will describe some important technical details, such as local density fitting and parallelization. A summary concludes the chapter.

1.2 Benchmark Systems

Some large molecules and reactions, which we have used extensively to benchmark our methods [32, 145, 147, 148], are shown in Figures 1.1 and 1.2. For easy reference, we have given short names to some of the molecules, which are shown in the figure and will be used throughout this article. Reaction **I** is the last step in the synthesis of androstendione. In reaction **II**, testosterone is esterified to make it more lipophilic for a longer

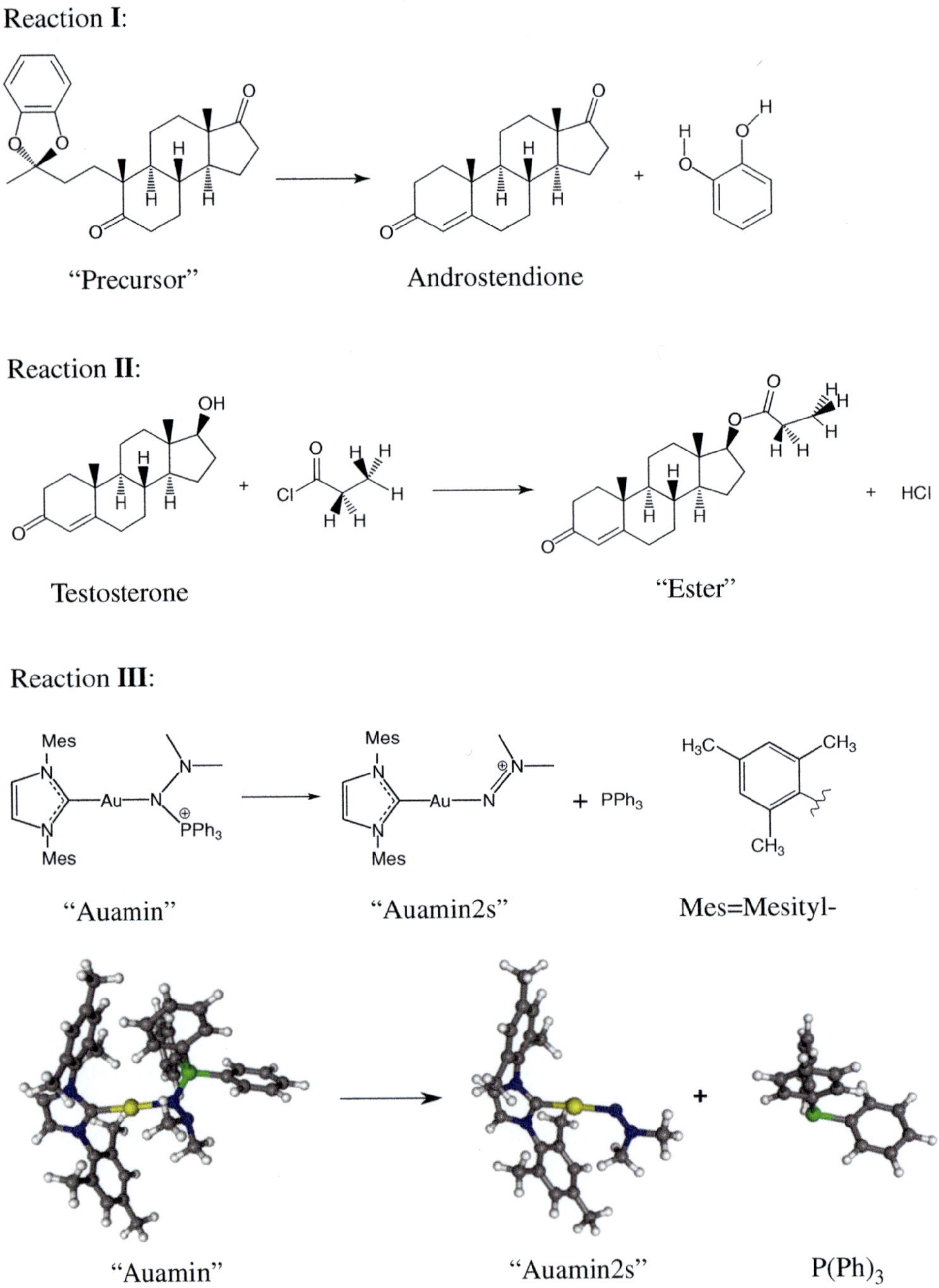

Figure 1.1 Benchmark molecules and reactions.

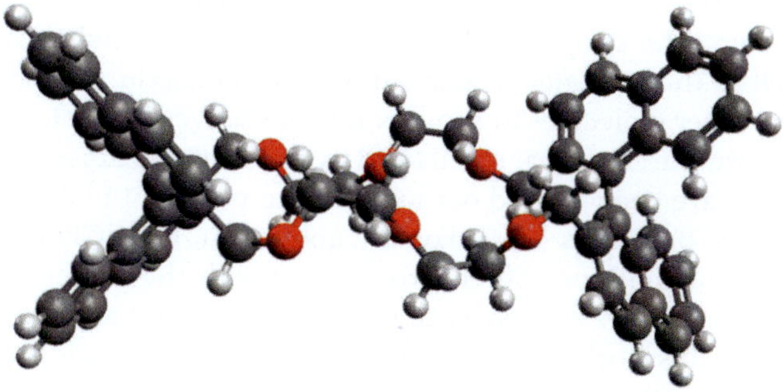

$C_{52}O_6H_{48}$ (106 atoms)

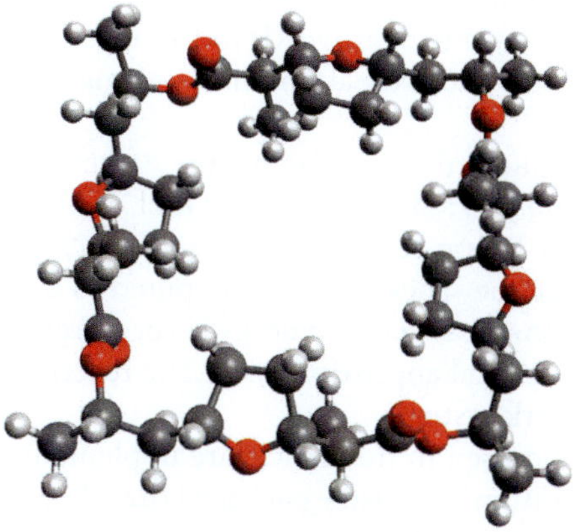

Nonactin (116 atoms)

Elaiophylin (160 atoms)

Figure 1.2 Visualization of a selection of large molecules mentioned in Tables 1.3 and 1.4.

retention time in body tissues. Reaction **III** is the dissociation of a gold(I)-aminonitrene complex ($AuC_{41}H_{45}N_4P$, for simplicity denoted Auamin, see Figure 1.1). This reaction is taken from Ref. [177] and plays an important role in catalytic aziridination and insertion reactions. The Auamin molecule has three phenyl and three mesityl groups and therefore strong long-range dispersion interactions are expected.

In Figure 1.2, some representative medium-sized organic molecules with about 100–150 atoms are shown. The selection of molecules contains a modified crown-ether ($C_{52}O_6H_{48}$) as well as the biologically active molecules nonactin and elaiophylin. We use these and some other molecules to demonstrate the convergence of our local correlation methods toward the canonical results and to perform benchmark calculations.

Most calculations have been carried out using the cc-pVDZ-F12 (VDZ-F12) and cc-pVTZ-F12 (VTZ-F12) basis sets [154]. For the gold atom, the ECP60MDF effective core potential [178] for the inner 60 electrons along with the aug-cc-pVDZ-PP and aug-cc-pVTZ-PP basis sets [179] were used. Density-fitting (DF) approximations were employed throughout this work for the evaluation of 2-electron integrals, and the aug-cc-pVTZ/MP2FIT basis set [180] was used as the DF auxiliary basis. We also used the aug-cc-pVTZ/JKFIT basis, which was derived from the cc-pVTZ/JKFIT basis set [181] by adding for each angular momentum another shell of diffuse functions, as the auxiliary basis for the resolution of the identity (RI) in F12 calculations. Unless otherwise stated, the parameters for local approximations are our program defaults, as described in the following sections.

For reaction **III**, the computed dissociation energy can be directly compared to an experimental gas-phase value [177] of 196.5 ± 11.2 kJ mol^{-1}, which has been obtained by subtracting the PW91/cc-pVTZ-pp zero-point correction [177] of -8.2 kJ mol^{-1} from the measured value. The Hartree–Fock value is computed to be 92 kJ mol^{-1}. Thus, the correlation contribution to the dissociation energy is very large, and, as will be demonstrated later, very sensitive to local approximations. The reaction therefore provides an excellent "difficult" benchmark system.

All methods described and used in this article are implemented in the development version of the MOLPRO quantum chemistry package [182, 183], and will be made available for general use in the next release.

1.3 Orbital-Invariant MP2 Theory

The basic approximations in local correlation methods can be well understood by considering the closed-shell MP2 method. We will therefore discuss local MP2 (LMP2) methods first; more advanced coupled cluster methods will be presented in Section 1.9.

Since we intend to use localized orbitals, the theory must be formulated in a way that is invariant with respect to unitary transformations within the occupied valence and virtual orbital spaces. This means that the Fock matrix is non-diagonal in each of these subspaces. The choice of local occupied and virtual orbitals will be discussed in the subsequent sections.

We will introduce quite a number of different orbital spaces and, for the sake of clarity, the index notation is summarized in Table 1.1.

The first-order MP2 wave function is written as

$$\Psi^{(1)} = \frac{1}{2} \sum_{i,j}^{\text{val}} \sum_{a,b}^{\text{virt}} |\Phi_{ij}^{ab}\rangle T_{ab}^{ij}, \quad T_{ab}^{ij} = T_{ba}^{ji}, \tag{1.1}$$

where i, j are occupied and a, b are virtual orbitals. The summation limit "val" over the first summation indicates that the sum runs only over the correlated valence orbitals

Table 1.1 Index notation for various orbital spaces.

Indices	Orbital space
μ, ν, ρ, σ	Orbital basis functions (AOs)
α, β	Complete space or non-orthogonal resolution of the identity (RI) basis functions
$\bar{\alpha}, \bar{\beta}$	Orthonormal RI functions
A, B	Density-fitting (DF) basis functions
General MOs and PAOs	
i, j, k, l	Localized occupied molecular orbitals (LMOs)
a, b, c, d	Canonical virtual molecular orbitals (VMOs)
$\tilde{r}, \tilde{s}, \tilde{t}, \tilde{u}$	Non-orthogonal projected atomic orbitals (PAOs)
Pair-specific virtuals for domain $[ij]$	
$\tilde{r}^{ij}, \tilde{s}^{ij}, \tilde{t}^{ij}, \tilde{u}^{ij}$	Non-orthogonal pair-specific virtuals (PSVs)
$r^{ij}, s^{ij}, t^{ij}, u^{ij}$	Pseudo-canonical pair-specific virtuals (PC-PSVs)
$\tilde{a}^{ij}, \tilde{b}^{ij}, \tilde{c}^{ij}, \tilde{d}^{ij}$	Pair natural orbitals (PNOs)
$a^{ij}, b^{ij}, c^{ij}, d^{ij}$	Pseudo-canonical pair natural orbitals (PC-PNOs)

(excluding uncorrelated core orbitals). $|\Phi_{ij}^{ab}\rangle$ are doubly excited configurations, which are obtained by applying spin-summed excitation operators $\hat{E}_i^a$ to the Hartree–Fock reference wave function $|0\rangle \equiv |\Phi_{\rm HF}\rangle$:

$$|\Phi_{ij}^{ab}\rangle = \hat{E}_i^a \hat{E}_j^b |0\rangle. \tag{1.2}$$

The MP2 amplitudes T_{ab}^{ij} can be determined by minimizing the Hylleraas functional

$$E_2 = 2\langle \Psi^{(1)}|\hat{H}|0\rangle + \langle \Psi^{(1)}|\hat{H}^{(0)} - E^{(0)}|\Psi^{(1)}\rangle, \tag{1.3}$$

which leads to the first-order amplitude equations

$$R_{ab}^{ij} = \langle \tilde{\Phi}_{ij}^{ab}|\hat{H}|0\rangle + \langle \tilde{\Phi}_{ij}^{ab}|\hat{H}^{(0)} - E^{(0)}|\Psi^{(1)}\rangle \stackrel{!}{=} 0. \tag{1.4}$$

The quantities R_{ab}^{ij} are *residuals*, which must vanish for the optimized amplitudes. Since the configurations $|\Phi_{ij}^{ab}\rangle$ are pairwise non-orthogonal for ij and ji, it is convenient to use in equation (1.4) contravariant configurations

$$|\tilde{\Phi}_{ij}^{ab}\rangle = \frac{1}{6}\left(2|\Phi_{ij}^{ab}\rangle + |\Phi_{ij}^{ba}\rangle\right), \tag{1.5}$$

which have the properties

$$\langle \tilde{\Phi}_{ij}^{ab}|\Phi_{kl}^{cd}\rangle = \delta_{ac}\delta_{bd}\delta_{ik}\delta_{jl} + \delta_{ad}\delta_{bc}\delta_{il}\delta_{jk}, \tag{1.6}$$

$$\langle \tilde{\Phi}_{ij}^{ab}|\Psi^{(1)}\rangle = T_{ab}^{ij}. \tag{1.7}$$

Expanding the first-order wave function as

$$|\Psi^{(1)}\rangle = \sum_{ij}\sum_{ab} |\tilde{\Phi}_{ij}^{ab}\rangle \tilde{T}_{ab}^{ij} \tag{1.8}$$

and equating this to equation (1.1) yields the corresponding contravariant amplitudes

$$\tilde{T}_{ab}^{ij} = 2T_{ab}^{ij} - T_{ba}^{ij}. \tag{1.9}$$

It is then straightforward to derive that

$$\left\langle \tilde{\Phi}_{ij}^{ab} |\hat{H}| 0 \right\rangle = (ai|bj), \tag{1.10}$$

where

$$(ai|bj) = \int_{\mathbb{R}^3} \mathrm{d}\mathbf{r}_1 \int_{\mathbb{R}^3} \mathrm{d}\mathbf{r}_2 \;\; \rho_{ai}(\mathbf{r}_1) \frac{1}{r_{12}} \rho_{bj}(\mathbf{r}_2) \tag{1.11}$$

are the electron repulsion integrals (ERIs) in Mulliken notation and $\rho_{ai}(\mathbf{r}) = \phi_a(\mathbf{r})\phi_i(\mathbf{r})$ are 1-electron charge distributions. In the following, we will often consider the integrals for a fixed pair ij as a matrix

$$[\mathbf{K}^{ij}]_{ab} = (ai|bj). \tag{1.12}$$

Similarly, the amplitudes $T_{ab}^{ij} = [\mathbf{T}^{ij}]_{ab}$ and residuals $R_{ab}^{ij} = [\mathbf{R}^{ij}]_{ab}$ are matrices, where superscripts denote different matrices, and subscripts their indices. Note that the matrices are all defined in the virtual orbital space, that is, their indices always refer to virtual orbitals. The Hylleraas functional (equation 1.3) can then be written as

$$E_2 = \sum_{i,j}^{\text{val}} \sum_{a,b}^{\text{virt}} \left(K_{ab}^{ij} + R_{ab}^{ij} \right) \tilde{T}_{ab}^{ij} = \sum_{i,j}^{\text{val}} \mathrm{tr}[(\mathbf{K}^{ij} + \mathbf{R}^{ij})\tilde{\mathbf{T}}^{ji}]. \tag{1.13}$$

For optimized amplitudes the residuals R_{ab}^{ij} vanish, and the Hylleraas energy E_2 becomes equal to the standard second-order energy expression

$$E^{(2)} = \langle \Psi^{(1)} |\hat{H}| 0 \rangle = \sum_{i,j}^{\text{val}} \mathrm{tr}[\mathbf{K}^{ij}\tilde{\mathbf{T}}^{ji}]. \tag{1.14}$$

The orbital-invariant form of the MP2 residuals is

$$[\mathbf{R}^{ij}]_{ab} = \left[\mathbf{K}^{ij} + \mathbf{F}\mathbf{T}^{ij} + \mathbf{T}^{ij}\mathbf{F} - \sum_{k}^{\text{val}} \left(f_{ik}\mathbf{T}^{kj} + \mathbf{T}^{ik} f_{kj} \right) \right]_{ab}, \tag{1.15}$$

where $[\mathbf{F}]_{ab}$ and f_{kl} are the external (virtual) and internal (occupied) blocks of the closed-shell Fock matrix. In general, the linear equation system $R_{ab}^{ij} = 0$ (cf. equation 1.4) must be solved iteratively. In the special case where canonical HF orbitals are used (i.e., $f_{ij} = \delta_{ij}\epsilon_i$, $F_{ab} = \epsilon_a \delta_{ab}$), the solution reduces to

$$T_{ab}^{ij} = -\frac{K_{ab}^{ij}}{\epsilon_a + \epsilon_b - \epsilon_i - \epsilon_j}. \tag{1.16}$$

If the occupied orbitals are local but the virtual orbitals are still canonical, one can write

$$R_{ab}^{ij} = K_{ab}^{ij} + (\epsilon_a + \epsilon_b - f_{ii} - f_{jj})T_{ab}^{ij} + G_{ab}^{ij} + G_{ba}^{ji},$$
$$G_{ab}^{ij} = -\sum_{k\neq i} f_{ik} T_{ab}^{kj}. \tag{1.17}$$

Approximate amplitudes can then be obtained by neglecting the coupling terms G_{ab}^{ij}

$$T_{ab}^{ij} \approx -\frac{K_{ab}^{ij}}{\epsilon_a + \epsilon_b - f_{ii} - f_{jj}}. \tag{1.18}$$

This is denoted a *semi-canonical* amplitude approximation. The energies obtained with semi-canonical amplitudes are usually too inaccurate for practical applications, but these amplitudes will be useful for generating PNOs, cf. Section 1.6.3, or for distant pair approximations, cf. Section 1.8.

Similarly, the linear amplitude equations can be solved iteratively by computing in each iteration the residuals and from these, by applying first-order perturbation theory, the amplitude updates

$$\Delta T_{ab}^{ij} \approx -\frac{R_{ab}^{ij}}{\epsilon_a + \epsilon_b - f_{ii} - f_{jj}}. \tag{1.19}$$

The semi-canonical amplitudes are the first approximation in this iterative algorithm.

1.4 Principles of Local Correlation

Let us now assume that $|i\rangle$ is a local orbital with charge center at position $\mathbf{R}_i$. We also assume that $|i\rangle$ can be expanded in a local subset of atomic orbitals (AOs) which are spatially close to $\mathbf{R}_i$. Then the charge distributions $\rho_{ai}(\mathbf{r})$ will be local around $\mathbf{R}_i$. This is true even for canonical (delocalized) virtual orbitals $|a\rangle$, since only those parts of them will contribute to the charge distribution ρ_{ai} which overlap with $|i\rangle$. If the virtual orbitals are also local, the locality of the ρ_{ai} is further improved. If the occupied and virtual orbitals are both expanded in AOs that are close to $\mathbf{R}_i$ and $\mathbf{R}_a$, respectively, it follows from the exponential decay of the AOs that the integral $(ai|bj)$ is a sum of contributions that decay exponentially with the distances between the basis functions of $|i\rangle$ and $|a\rangle$.

Similar considerations hold for b and j, and so the integral $(ai|bj)$ will decay exponentially with the distances $R_{ai} = |\mathbf{R}_a - \mathbf{R}_i|$ and $R_{bj} = |\mathbf{R}_b - \mathbf{R}_j|$ (if these are sufficiently large). This forms the basis for *domain approximations*, which means that to each LMO $|i\rangle$, a subset of local virtual orbitals a can be assigned. We will denote such *orbital domains* as $a \in [i]$. Similarly, to each pair ij of LMOs, a pair domain $[ij] \equiv [i] \cup [j]$ can be assigned. In the LMP2 wave function, the excitations $|\Phi_{ab}^{ij}\rangle$ are restricted to the domain $a, b \in [ij]$. The sizes of these domains that are necessary to reach a certain accuracy depend crucially on the choice and construction of the virtual orbitals. This will be discussed in more detail in Section 1.6.

If LMO $|i\rangle$ is far from $|j\rangle$, the local charge distribution ρ_{ai} will also be far from ρ_{bj}. If the virtual orbitals $|a\rangle$, $|b\rangle$ are local with $|a\rangle$ being close to $|i\rangle$ and $|b\rangle$ close to $|j\rangle$, the exchange-type integrals $(bi|aj)$ will decay exponentially and can be neglected at large distances of $|i\rangle$ and $|j\rangle$. Using the semi-canonical amplitude approximation (cf. equation 1.18) the distant pair energies E_{ij}^{dist} can then be approximated as

$$E_{ij}^{\text{dist}} \approx -4 \sum_{a\in[i]} \sum_{b\in[j]} T_{ab}^{ij}(ai|bj) = -4 \sum_{a\in[i]} \sum_{b\in[j]} \frac{(ai|bj)^2}{\epsilon_a + \epsilon_b - f_{ii} - f_{jj}}. \tag{1.20}$$

The integrals $(ai|bj)$ can be efficiently computed using local density-fitting techniques (cf. Section 1.12.1) or approximated by a *multipole expansion* [12, 13, 54] (cf. Section 1.8). Since the densities $\rho_{ai}(\mathbf{r})$ carry no charge ($\int \rho_{ai}(\mathbf{r})dr = 0$), the lowest-order contribution is the dipole–dipole interaction, which decays with R^{-3}, where R_{ij} is the distance between the charge centers of the LMOs i and j. Consequently, in a local orbital basis, the amplitudes and pair energies decay with R_{ij}^{-3} and R_{ij}^{-6}, respectively. This is the basis of *pair approximations*, which simplify the treatment of distant pairs ij, cf. Section 1.8.

1.5 Orbital Localization

Localized occupied molecular orbitals (LMOs) $|i\rangle \equiv |\phi_i^{\text{loc}}\rangle$ can be obtained from the canonical molecular orbitals (CMOs) $|\phi_k^{\text{can}}\rangle$ by a unitary transformation

$$|i\rangle \equiv |\phi_i^{\text{loc}}\rangle = \sum_k^{\text{val}} |\phi_k^{\text{can}}\rangle U_{ki}^{\text{loc}}. \tag{1.21}$$

Uncorrelated core orbitals are excluded from the localization, since mixing of core and valence orbitals would affect the correlation energy even without any local approximations. The localized valence orbitals $|i\rangle$ can be expressed in the atomic orbital (AO) basis $|\mu\rangle \equiv |\chi_\mu\rangle$ by a rectangular transformation matrix $\mathbf{L}$ ($N_{\text{AO}} \times N_{\text{val}}$), which is related to the corresponding canonical MO-coefficient matrix $\mathbf{C}$ by the unitary transformation $\mathbf{U}^{\text{loc}}$:

$$|i\rangle = \sum_{\mu=1}^{N_{\text{AO}}} |\mu\rangle L_{\mu i}, \qquad L_{\mu i} = \sum_k^{\text{val}} C_{\mu k} U_{ki}^{\text{loc}}. \tag{1.22}$$

In the following, the indices i, j, k, l will always refer to LMOs.

There are various criteria that can be used to determine the transformation matrix $\mathbf{U}^{\text{loc}}$. For a recent review, see Ref. [184]. Perhaps the most well-known ones are the methods of Foster and Boys (FB), Pipek-Mezey (PM) [185], and Edmiston and Ruedenberg (ER) [186, 187]. ER localization minimizes the overall inter-orbital electron repulsion. This requires the 2-electron integrals, and is therefore quite expensive and rarely used. In FB localization, the sum of the orbital variances

$$P^{\text{FB}} = \sum_p^{\{x,y,z\}} \sum_{i=1}^{\text{val}} \langle i|(\hat{p} - \langle i|\hat{p}|i\rangle)^2|i\rangle. \tag{1.23}$$

is minimized. PM localization maximizes the sum of the squared Mulliken partial charges q_{iA}:

$$P^{\mathrm{PM}} = \sum_{A}^{\mathrm{atoms}} \sum_{i}^{\mathrm{val}} q_{iA}^2, \tag{1.24}$$

$$q_{iA} = 2 \sum_{\mu \in A} L_{\mu i} \sum_{\nu} S_{\mu\nu} L_{\nu i}, \tag{1.25}$$

where the first sum in the latter equation runs over all basis functions μ centered at atom A. Effectively, this means that the number of centers at which the orbitals are localized is minimized. Pipek-Mezey (PM) localization often suffers from the problem that poor localization may result in calculations with large basis sets containing diffuse functions.

The latter problem can be avoided by using intrinsic bond orbitals (IBOs) [188]. In this case, a molecule-intrinsic minimal basis of polarized atomic orbitals (IAOs) $|\rho\rangle$ [188–191] is created, in which the occupied Hartree–Fock orbitals can be expanded exactly. The IAOs are generated from a suitable minimal basis of free-atom AOs $|\tilde{\rho}\rangle$ using projection operators [188] and are subsequently symmetrically orthogonalized. The free atom AOs $|\tilde{\rho}\rangle$ are stored in a basis set library (we use a subset of the generally contracted cc-pVTZ basis where available). The partial charges are then defined as

$$q_{iA} = 2 \sum_{\rho \in A} \langle i|\rho\rangle\langle\rho|i\rangle \tag{1.26}$$

and used in the PM localization. Alternatively, Knizia [188] has proposed to replace the exponent of two in equation (1.24) by four, which helps to converge the iterative localization procedure in highly symmetric cases where several equivalent solutions exist. We will denote the IBO localizations with exponents 2 and 4 as IBO2 and IBO4, respectively. The resulting IBOs closely resemble PM orbitals, but are very insensitive to basis set variations; in particular, the commonly encountered artifacts of PM orbitals with diffuse basis sets are absent. Additionally, the IBO construction yields stable partial orbital charges which can be used for defining domains (see Section 1.6.2). Closely related localization schemes were also put forward by Lehtola and Jónsson [192] and West *et al.* [191]. The relation between these schemes has been investigated by Janowski [189].

The construction of IBOs is simple and efficient, and since the iterative localization is carried out in the minimal IAO basis, it is much faster than the standard PM localization. Even though it scales cubically with molecular size, the time to generate IBOs is found to be short as compared to a PNO-LMP2 calculation, even for cases with about 10,000 basis functions.

A fourth localization scheme, which has been used in the past in local correlation methods [28], is based on the natural bond orbital (NBO) analysis of Weinhold and coworkers [193–196] and yields so-called natural localized molecular orbitals (NLMOs) [194]. Somewhat similar to IBOs, the NLMOs are obtained by a sequence of transformations which first yields natural atomic orbitals (NAOs), then NBOs, and finally the NLMOs. However, these transformations are much more complicated and less uniquely defined than in the IBO scheme. Based on these transformations partial charges can also be defined [28], which are quite stable with respect to basis set variations and are rather similar to IBO partial charges.

FB localization yields qualitatively different localized orbitals for double bonds than the other schemes. Usually, PM, IBO, and NLMO localization keeps σ- and π-orbitals separate (even in non-planar molecules). In contrary, Boys localization often generates two equivalent "banana-bond"-like orbitals. Some examples are shown in Figure 1.3 for naphtalene. The PM, IBO2, and NBO orbitals are qualitatively similar but not identical. The differences are rather small and therefore only the IBO2 orbitals are shown in Figure 1.3.

Table 1.2 shows a comparison of domain sizes and correlation energies using various localization schemes for a number of aromatic molecules. Despite the qualitative differences of the various localized orbitals, the localization method has only a minor effect on the domain sizes and correlation energies. Exceptions are special cases in which the LMOs are neither symmetry-adapted nor symmetry-equivalent. It may then happen that the domains are neither symmetry-adapted nor symmetry-equivalent, and this can cause spurious dipole moments perpendicular to a molecular plane and other artifacts. An example is the IBO4 localization of naphtalene, cf. Figure 1.3. In longer polycyclic aromatic hydrocarbons such as tetracene or pentacene, none of the localization methods creates symmetry-adapted or equivalent LMOs. In such cases, the non-equivalent orbital domains should be merged so that, overall, the symmetry properties are recovered.

1.6 Local Virtual Orbitals

In this section, we will discuss various types of virtual orbitals and the corresponding domain constructions. The choice of virtual orbitals is most essential for the accuracy and efficiency of local correlation methods. There are schemes which localize virtual orbitals in a similar way as the occupied orbitals, yielding orthonormal localized virtual orbitals (LVOs) [184]. Even though these orbitals have good locality properties and have been successfully used in fragmentation approaches, we have found that, in local correlation methods, much smaller domains are sufficient for a certain accuracy if projected atomic orbitals (PAOs), orbital-specific virtuals (OSVs), or pair natural orbitals (PNOs) are used. PAOs are obtained by projecting atomic orbitals or basis functions to the virtual space, yielding a single set of non-orthogonal functions. In contrast, in OSV and PNO methods, different orthonormal orbital sets are used for each orbital or orbital pair, respectively. These sets are mutually non-orthogonal. In contrast to LVOs, PAOs, OSVs, and PNOs are not obtained by applying a localization criterion, but they are inherently local by construction. The domain sizes that are necessary to obtain a certain fraction of the canonical correlation energy decrease in the order LVO > PAO > OSV > PNO. As will be shown later, the PNO domain sizes are typically one order of magnitude smaller than the PAO ones, which reduces the number of amplitudes to be optimized by two orders of magnitude. LVOs will not be further considered in this article. Before defining PAOs, OSVs, and PNOs in detail, some general considerations and definitions will be presented.

1.6.1 Pseudo-Canonical Pair-Specific Orbitals

In the most general case, one can use completely independent sets of virtual orbitals $\{\tilde{r}^{ij}\}$ for each pair ij. We will denote these as *pair-specific virtuals* (PSVs). PNOs are a special

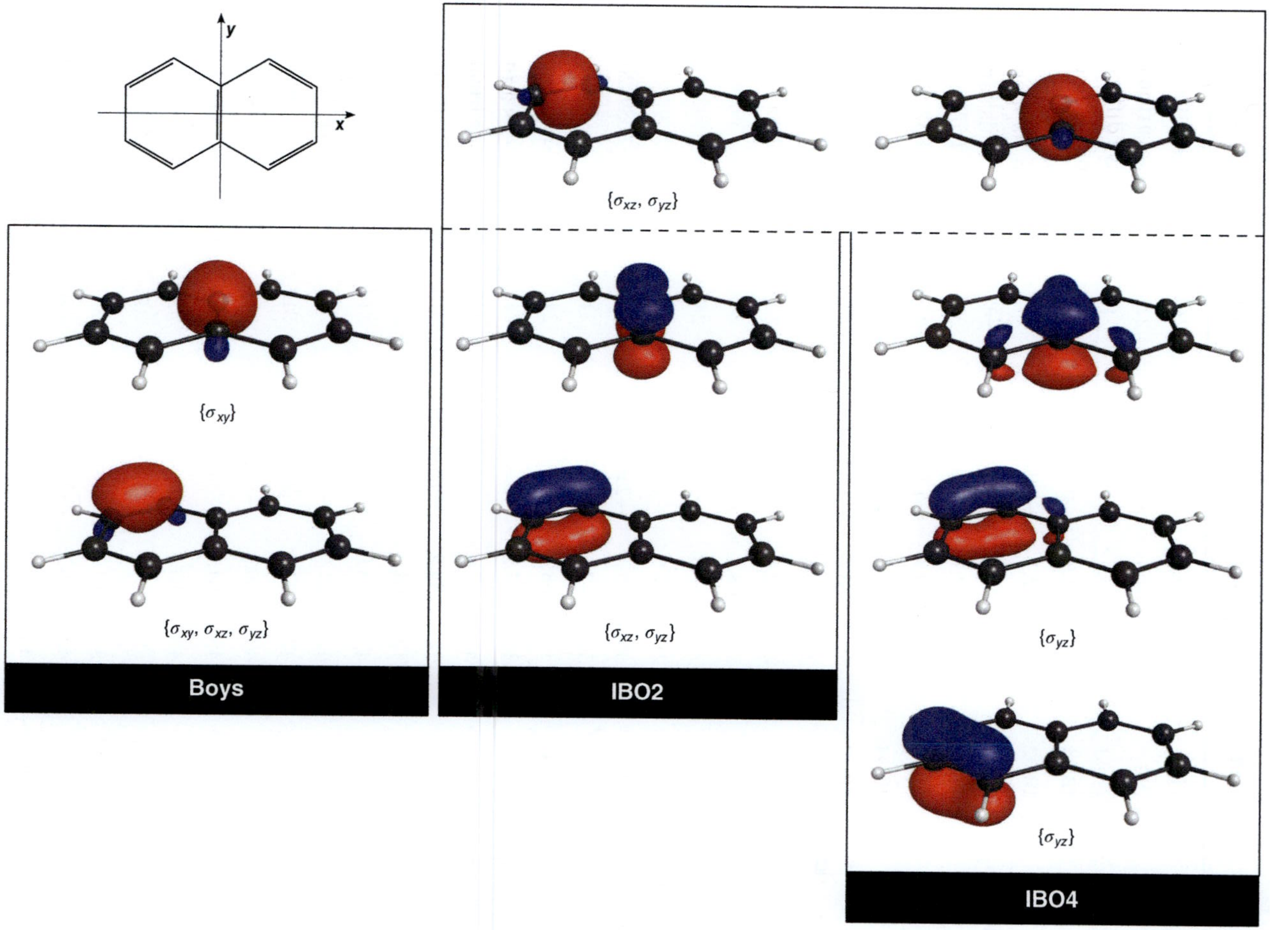

Figure 1.3 Localized C–C bonding orbitals of naphtalene. FB localization yields pairs of equivalent banana bonds, while the other localization schemes keep σ and π orbitals separate. Only the symmetry unique orbitals are shown. The specifications in curly brackets denote the symmetry operations (mirror planes) which generate equivalent orbitals. The IBO4 orbitals are not fully symmetry-adapted or equivalent. Basis set: VTZ.

Table 1.2 Domain sizes and PNO-LMP2 correlation energies (in E_h) using various localization schemes.

		IEXT=1			IEXT=2		
Molecule	**Method**	**PAO**	**PNO**	E_{corr}	**PAO**	**PNO**	E_{corr}
Benzene	FB	130	89.9	−0.947419	218	94.1	−0.949220
	PM	139	81.5	−0.947837	221	83.2	−0.949249
	NBO	130	83.6	−0.947445	218	86.5	−0.949243
	IBO4	139	81.9	−0.947804	221	83.8	−0.949237
	IBO2	139	82.1	−0.947820	221	83.8	−0.949245
Naphtalene	FB	139	85.3	−1.567535	254	91.1	−1.571672
	PM	139	77.1	−1.567584	254	81.3	−1.571771
	NBO	139	79.9	−1.567418	254	84.6	−1.571772
	IBO4	151	79.3	−1.568599	263	82.3	−1.571838
	IBO2	139	78.7	−1.567555	254	83.3	−1.571786
Adrenaline	FB	124	70.4	−2.390946	239	75.0	−2.397254
	PM	128	69.6	−2.391577	243	73.4	−2.397287
	NBO	124	68.4	−2.390728	240	73.1	−2.397304
	IBO4	128	67.5	−2.391431	243	71.5	−2.397313
	IBO2	128	67.9	−2.391531	243	71.8	−2.397338
Tryptophan	FB	130	69.3	−2.632653	247	73.7	−2.639520
	PM	130	72.1	−2.632561	247	76.7	−2.639579
	NBO	130	68.1	−2.632092	247	72.7	−2.639622
	IBO4	133	67.9	−2.633124	249	71.7	−2.639646
	IBO2	131	67.6	−2.632757	247	71.7	−2.639643
Androstenedione	FB	142	59.9	−3.618745	310	63.3	−3.630961
	PM	142	64.6	−3.618731	310	68.4	−3.631017
	NBO	142	63.8	−3.618596	310	67.9	−3.631058
	IBO4	142	61.9	−3.618602	310	65.7	−3.631016
	IBO2	142	61.8	−3.618642	310	65.7	−3.631036
Testosterone	FB	143	59.5	−3.655972	315	63.0	−3.668717
	PM	143	64.2	−3.655940	315	68.1	−3.668737
	NBO	143	63.8	−3.655829	315	69.0	−3.668776
	IBO4	143	61.7	−3.655852	315	65.6	−3.668740
	IBO2	143	61.7	−3.655877	315	65.6	−3.668752
Progesterone	FB	142	59.1	−3.973803	310	62.5	−3.987470
	PM	142	64.2	−3.973799	310	68.0	−3.987538
	NBO	142	63.1	−3.973658	310	67.1	−3.987574
	IBO4	142	61.1	−3.973689	310	64.9	−3.987538
	IBO2	142	61.0	−3.973726	310	64.7	−3.987552

Basis set: VTZ, $T_{\mathrm{LMO}} = 0.2$, $T_{\mathrm{PNO}}^{\mathrm{occ}} = 10^{-8}$. The parameter IEXT and $T_{\mathrm{PNO}}^{\mathrm{occ}}$ determine the PAO and PNO domain sizes, respectively (see Section 1.7).

choice of PSVs, but for the time being, we consider the problem in general. The tilde over the index indicates that the orbitals may be non-orthogonal (but they must always be orthogonal to the occupied orbital space). The PSVs are related to the orthonormal canonical virtuals (CMOs) $\{b\}$ by transformation matrices $\tilde{\mathbf{Q}}^{ij}$:

$$|\tilde{r}^{ij}\rangle = \sum_b |b\rangle \tilde{Q}^{ij}_{b\tilde{r}}. \tag{1.27}$$

These transformations are of course entirely useless unless they produce more local virtual orbitals so that one can restrict the PSV excitation subspaces to domains $[ij]_{\text{PSV}}$. This means that we approximate the pair functions as

$$|\Psi_{ij}\rangle = \sum_{a,b} |\Phi^{ab}_{ij}\rangle T^{ij}_{ab} \approx \sum_{\tilde{r},\tilde{s}\in[ij]_{\text{PSV}}} |\Phi^{\tilde{r}\tilde{s}}_{ij}\rangle T^{ij}_{\tilde{r}\tilde{s}}. \tag{1.28}$$

The transformation matrices $\tilde{\mathbf{Q}}^{ij}$ then become rectangular, with more rows than columns. To keep the formulation compact and to avoid superscripts on subscripts, we will always use the convention that the subscripts of matrices such as $\tilde{\mathbf{Q}}^{ij}$, $\mathbf{T}^{ij}$, or $\mathbf{K}^{ij}$ correspond to the pair ij given by the superscript of the matrix (here pair ij). The CMOs can be considered as a special case in which the domain $[ij]_{\text{CMO}}$ is identical and complete for all pairs.

The PSVs may in general be non-orthogonal with overlap

$$S^{ij,kl}_{\tilde{r}\tilde{s}} = \langle \tilde{r}^{ij}|\tilde{s}^{kl}\rangle = \sum_b \tilde{Q}^{ij}_{b\tilde{r}} \tilde{Q}^{kl}_{b\tilde{s}} = [\tilde{\mathbf{Q}}^{ij\dagger}\tilde{\mathbf{Q}}^{kl}]_{\tilde{r}\tilde{s}}. \tag{1.29}$$

Note that here the indices $\tilde{r}$ and $\tilde{s}$ belong to different pairs, as indicated by the superscripts of the overlap matrix. Inserting equation (1.27) into equation (1.28) yields for the transformation of the amplitudes

$$T^{ij}_{ab} = \sum_{\tilde{r},\tilde{s}} \tilde{Q}^{ij}_{a\tilde{r}} T^{ij}_{\tilde{r}\tilde{s}} \tilde{Q}^{ij}_{b\tilde{s}}. \tag{1.30}$$

Integral and Fock matrices transform in the opposite way, for example,

$$(\tilde{r}^{ij} i|\tilde{s}^{ij} j) = \sum_{a,b} \tilde{Q}^{ij}_{a\tilde{r}} (ai|bj) \tilde{Q}^{ij}_{b\tilde{s}}. \tag{1.31}$$

In matrix form, these transformations can be written as

$$\mathbf{T}^{ij}_{\text{CMO}} = \tilde{\mathbf{Q}}^{ij}\mathbf{T}^{ij}_{\text{PSV}}\tilde{\mathbf{Q}}^{ij\dagger}, \tag{1.32}$$

$$\mathbf{K}^{ij}_{\text{PSV}} = \tilde{\mathbf{Q}}^{ij\dagger}\mathbf{K}^{ij}_{\text{CMO}}\tilde{\mathbf{Q}}^{ij}. \tag{1.33}$$

Since the number of PSVs per pair is in general smaller than the number of CMOs, the inverse transformations are not possible. Equation (1.32) shows that one can always transform the amplitudes from the PSV domain back to the CMO basis, but not vice versa. In contrary, integrals can only be transformed from the larger basis to the smaller one, cf. equation (1.33).

Within each domain $[ij]_{\text{PSV}}$, the energy is still invariant with respect to transformations among the orbitals $|\tilde{r}^{ij}\rangle$. It is useful to apply another transformation within each domain which makes the orbitals *pseudo-canonical* (PC). This means that in the subspace of the domain $[ij]_{\text{PSV}}$, the PSVs become orthonormal and diagonalize the Fock matrix. The pseudo-canonical pair-specific orbitals (PC-PSVs) are denoted $|r^{ij}\rangle$ and are related to the original PSVs by the transformation

$$|r^{ij}\rangle = \sum_{\tilde{s}\in[ij]} |\tilde{s}^{ij}\rangle W^{ij}_{\tilde{s}r}, \quad r^{ij} \in [ij]. \tag{1.34}$$

so that

$$f^{ij}_{rs} = \langle r^{ij}|\hat{f}|s^{ij}\rangle = \epsilon^{ij}_r \delta_{rs}, \quad \langle r^{ij}|s^{ij}\rangle = \delta_{rs}. \tag{1.35}$$

In practice, this is achieved by first orthogonalizing the orbitals (by singular value decomposition if linear dependencies exist), and then diagonalizing the Fock matrix in the basis of the orthogonal PSVs. Note that the PC-PSVs of different pairs are still non-orthogonal. The total transformation from CMOs to PC-PSVs is then

$$|r^{ij}\rangle = \sum_a |a\rangle Q^{ij}_{ar}, \quad r^{ij} \in [ij], \tag{1.36}$$

$$\mathbf{Q}^{ij} = \tilde{\mathbf{Q}}^{ij}\mathbf{W}^{ij}. \tag{1.37}$$

If the domain spans the whole virtual space, the PC-PSVs become equal to the CMOs.

It is straightforward to transform the amplitude equations to the PC-PSV basis by forming $\mathbf{R}^{ij}_{\text{PC-PSV}} = \mathbf{Q}^{ij\dagger}\mathbf{R}^{ij}_{\text{CMO}}\mathbf{Q}^{ij}$ and inserting $\mathbf{T}^{ij}_{\text{CMO}} = \mathbf{Q}^{ij}\mathbf{T}^{ij}_{\text{PC-PSV}}\mathbf{Q}^{ij\dagger}$. Finally, the products $[\mathbf{Q}^{ij\dagger}\mathbf{Q}^{kl}]_{rs}$ are replaced by $[\mathbf{S}^{ij,kl}]_{rs}$ (cf. equation 1.29). This always occurs if an amplitude index is not connected to an integral index. For example, the LMP2 amplitude equations in equation (1.17) become

$$\begin{aligned} R^{ij}_{rs} &= K^{ij}_{rs} + \left(\epsilon^{ij}_r + \epsilon^{ij}_s - f_{ii} - f_{jj}\right) T^{ij}_{rs} + G^{ij}_{rs} + G^{ji}_{sr} = 0, \\ G^{ij}_{rs} &= -\sum_{k\neq i} f_{ik} \sum_{t,u\in[kj]_{\text{PC-PSV}}} S^{ij,kj}_{rt} T^{kj}_{tu} S^{kj,ij}_{us}. \end{aligned} \tag{1.38}$$

The optimized amplitudes are now defined by the conditions $R^{ij}_{rs} = 0$, $r, s \in [ij]_{\text{PC-PSV}}$. Due to the domain approximation, the number of amplitudes and corresponding equations is much smaller than in the canonical case. The Hylleraas functional in the PC-PSV basis is computed as

$$E_{\text{corr}} = \sum_{i\geq j}(2 - \delta_{ij}) \sum_{r,s\,\in\,[ij]} \left(K^{ij}_{rs} + R^{ij}_{rs}\right)\left(2T^{ij}_{rs} - T^{ij}_{sr}\right). \tag{1.39}$$

In the following subsections, we will address the question how to define the PSVs and which are the most suitable ones.

1.6.2 Projected Atomic Orbitals

In a seminal series of papers on local correlation methods, Pulay has proposed to use projected atomic orbitals (PAOs) to span the virtual space [1], and this has been widely used later in local correlation methods. The PAOs are labeled by indices $\tilde{r}$ and are defined as:

$$|\tilde{r}\rangle \equiv |\phi^{\text{PAO}}_{\tilde{r}}\rangle = \hat{P}|\phi^{\text{AO}}_{\tilde{r}}\rangle, \tag{1.40}$$

where $|\phi_{\tilde{r}}^{\mathrm{AO}}\rangle$ are suitable atomic orbitals and $\hat{P}$ projects out the contributions of the occupied space,

$$\hat{P} = 1 - \sum_{i=1}^{N_{\mathrm{occ}}} |\phi_i\rangle\langle\phi_i|, \tag{1.41}$$

so that $\langle\phi_{\tilde{r}}^{\mathrm{PAO}}|\phi_i\rangle = 0$ for all $i, \tilde{r}$. In general, the atomic orbitals $|\phi_{\tilde{r}}^{\mathrm{AO}}\rangle$ can be represented in the AO basis by a block-diagonal matrix $\mathbf{C}^{\mathrm{AO}}$

$$|\phi_{\tilde{r}}^{\mathrm{AO}}\rangle = \sum_{\mu\in A} |\chi_\mu\rangle C_{\mu\tilde{r}}^{\mathrm{AO}}. \tag{1.42}$$

Each non-zero block of the coefficient matrix $\mathbf{C}^{\mathrm{AO}}$ corresponds to one center. In most cases, the coefficient matrix $C_{\mu\tilde{r}}^{\mathrm{AO}}$ is assumed to be diagonal, that is, the AO basis functions themselves are projected.

In the AO basis, the PAOs are represented by a coefficient matrix $\mathbf{P}$,

$$|\tilde{r}\rangle = \sum_{\mu=1}^{N_{\mathrm{AO}}} |\chi_\mu\rangle P_{\mu\tilde{r}}, \tag{1.43}$$

with

$$\mathbf{P} = (\mathbf{1} - \mathbf{L}\mathbf{L}^\dagger\mathbf{S})\mathbf{C}^{\mathrm{AO}} = \mathbf{C}_v\mathbf{C}_v^\dagger\mathbf{S}\mathbf{C}^{\mathrm{AO}}, \tag{1.44}$$

The rectangular ($N_{\mathrm{AO}} \times N_{\mathrm{virt}}$) coefficient matrix $\mathbf{C}_v$ represents the canonical virtual orbitals, and $[\mathbf{S}]_{\mu\nu} = \langle\mu|\nu\rangle$ is the overlap matrix of the basis functions. Thus, the PAOs can be obtained from the canonical virtual orbitals by a transformation

$$|\tilde{r}\rangle = \sum_a |\phi_a\rangle\tilde{Q}_{a\tilde{r}}, \tag{1.45}$$

$$\tilde{\mathbf{Q}} = \mathbf{C}_v^\dagger\mathbf{S}\mathbf{C}^{\mathrm{AO}}. \tag{1.46}$$

Note that the matrix $\tilde{\mathbf{Q}}$ is rectangular ($N_{\mathrm{virt}} \times N_{\mathrm{AO}}$), and the N_{AO} PAOs are non-orthogonal and linearly dependent. Their overlap matrix

$$\mathbf{S}^{\mathrm{PAO}} = \mathbf{P}^\dagger\mathbf{S}\mathbf{P} = \tilde{\mathbf{Q}}^\dagger\tilde{\mathbf{Q}} \tag{1.47}$$

has N_{occ} zero eigenvalues. The linear dependencies in the PAO set are removed by singular value decomposition for individual domains.

The PAOs are neither orbital- nor pair-specific, and each one can be related to an AO (or a basis function) at one atom. In most cases, PAO domains are therefore center-based, that is, all PAOs that originate from basis functions at a selection of centers comprise the orbital domains $[i]_{\mathrm{PAO}}$. The domains are then uniquely defined by a list of centers, independent of the basis set. For each LMO i, the centers can, for example, be selected based on their partial IBO charges Q_{iA} (cf. equation 1.26). PAO pair domains are then defined as the union of the two related orbital domains, $[ij]_{\mathrm{PAO}} = [i]_{\mathrm{PAO}} \cup [j]_{\mathrm{PAO}}$. These are orthonormalized and made pseudo-canonical, as explained in Section 1.6.1. More details about the choice of PAO domains will be given in Section 1.7.

An advantage of the non-orthogonal PAOs is that they are pair-independent, and the overlap or Fock matrices for a domain $[ij]_{\mathrm{PAO}}$ can always be extracted from the full $N_{\mathrm{AO}} \times N_{\mathrm{AO}}$ matrices $\mathbf{S}_{\mathrm{PAO}}$ or $\mathbf{F}_{\mathrm{PAO}}$, respectively. Also the integral transformation is simpler than for truly pair-specific orbitals, in particular for local CCSD methods. In PAO-LMP2 or PAO-LCCSD methods, it is therefore advantageous to compute the residuals first in the non-orthogonal PAO basis, and transform them to the pair-specific PC-PAO basis just for the update of the amplitudes. Subsequently, the amplitude update is transformed back to the non-orthogonal PAO basis:

$$\mathbf{R}^{ij}_{\text{PC-PAO}} = \mathbf{W}^{ij\dagger}\mathbf{R}^{ij}_{\text{PAO}}\mathbf{W}^{ij}, \tag{1.48}$$

$$\Delta\mathbf{T}^{ij}_{\text{PAO}} = \mathbf{W}^{ij}\Delta\mathbf{T}^{ij}_{\text{PC-PAO}}\mathbf{W}^{ij\dagger}. \tag{1.49}$$

These transformations are similar to those in equations (1.32) and (1.33), but with $\mathbf{Q}^{ij}$ replaced by $\mathbf{W}^{ij}_{\text{PAO}}$.

Unfortunately, rather large PAO domains (400-600 PAOs per pair for triple-ζ basis sets) are necessary to recover about 99.9% of the canonical correlation energy in large molecules. Thus, the memory and CPU requirements of such calculations can be very large, despite the formally linear scaling (if distant pairs are approximated). The average domain sizes can be reduced by an order of magnitude if PNOs are used, which will be discussed next.

1.6.3 Pair Natural Orbitals

Pair natural orbitals (PNOs) are obtained by diagonalizing appropriate external pair density matrices. They were first discussed by Edmiston and Krauss [156] and used very successfully in the early 70s in the PNO-CI and PNO-CEPA methods by Meyer [157, 158] and somewhat later also by Ahlrichs, Kutzelnigg, Staemmler, and others [159–163]. In these methods, PNOs were defined separately for singlet ($p = 1$) and triplet pairs ($p = -1$). In terms of the configurations and amplitudes in equation (1.2) these can be defined as

$$|\Phi^{ab}_{ij,p}\rangle = \frac{1}{2}\left(|\Phi^{ab}_{ij}\rangle + p|\Phi^{ba}_{ij}\rangle\right), \tag{1.50}$$

and the corresponding amplitudes are

$$T^{ij,p}_{ab} = \frac{1}{2}\left(T^{ij}_{ab} + pT^{ij}_{ba}\right). \tag{1.51}$$

These configurations and amplitudes are symmetric or antisymmetric with respect to exchange of the indices i, j or a, b and the configurations are orthogonal for $i \geq j$ and $a \geq b$. Note that singlet and triplet refers to the coupling of the two electrons in the virtual orbitals a, b. The configurations $|\Phi^{ab}_{ij,p}\rangle$ have the same total spin eigenvalues as the reference function.

In terms of these configurations and amplitudes, the correlation energy is

$$E^{(2)} = \sum_{i\geq j}\sum_{p=\pm1} N_{ij,p}\sum_{a,b} K^{ij,p}_{ab}T^{ij,p}_{ab}, \tag{1.52}$$

where $N_{ij,1} = (2 - \delta_{ij})$, $N_{ij,-1} = 6(1 - \delta_{ij})$, and $K^{ij,p}_{ab} = \frac{1}{2}(K^{ij}_{ab} + pK^{ij}_{ba})$. The external pair densities are defined as

$$[\mathbf{D}^{ij,p}]_{ab} = N_{ij,p}[\mathbf{T}^{ij,p\dagger}\mathbf{T}^{ij,p}]_{ab}. \tag{1.53}$$

For MP2, the external part of the total unrelaxed one-particle reduced density matrix (1-RDM) is the sum of the pair density matrices

$$\mathbf{D} = \sum_{i\geq j} \sum_{p=\pm 1} \mathbf{D}^{ij,p}. \tag{1.54}$$

The PNOs $|\tilde{a}^{ij,p}\rangle$ for a pair ij, p are obtained by diagonalizing the corresponding pair density $\mathbf{D}^{ij,p}$

$$|\tilde{a}^{ij,p}\rangle = \sum_{b} |b\rangle \tilde{Q}^{ij,p}_{b\tilde{a}}, \tag{1.55}$$

$$[\tilde{\mathbf{Q}}^{ij,p\dagger}\mathbf{D}^{ij,p}\tilde{\mathbf{Q}}^{ij,p}]_{\tilde{a}\tilde{b}} = \delta_{\tilde{a}\tilde{b}} n^{ij,p}_{\tilde{a}}. \tag{1.56}$$

The eigenvalues $n^{ij,p}_{\tilde{a}}$ are denoted *natural pair occupation numbers.* Since the transformation $\tilde{\mathbf{Q}}^{ij,p}$ is orthogonal and the singlet amplitude matrices are symmetric in the indices a, b, the matrices $\mathbf{Q}^{ij,p}$ also diagonalize the singlet amplitude matrices

$$[\tilde{\mathbf{Q}}^{ij,1\dagger}\mathbf{T}^{ij,1}\tilde{\mathbf{Q}}^{ij,1}]_{\tilde{a}\tilde{b}} = \delta_{\tilde{a}\tilde{b}} t^{ij,1}_{\tilde{a}}. \tag{1.57}$$

The non-zero elements $t^{ij,1}_{\tilde{a}}$ are related to the natural pair occupation numbers by the relation

$$n^{ij,1}_{\tilde{a}} = N_{ij,1} \left| t^{ij,1}_{\tilde{a}} \right|^2. \tag{1.58}$$

For the triplet pairs, all natural occupation numbers are doubly degenerate (if the matrix dimension is odd, the last one is zero), and the transformation of the amplitude matrix yields a block-diagonal matrix of the form

$$[\tilde{\mathbf{Q}}^{ij,-1\dagger}\mathbf{T}^{ij,-1}\tilde{\mathbf{Q}}^{ij,-1}] = \begin{pmatrix} 0 & -t^{ij,-1}_1 & 0 & 0 & \dots \\ t^{ij,-1}_1 & 0 & 0 & 0 & \dots \\ 0 & 0 & 0 & -t^{ij,-1}_2 & \dots \\ 0 & & t^{ij,-1}_2 & 0 & \dots \\ \dots & & & & \end{pmatrix}. \tag{1.59}$$

In this case, the natural occupation numbers are

$$n^{ij,-1}_{2\tilde{a}-1} = n^{ij,-1}_{2\tilde{a}} = N_{ij,-1} \left| t^{ij,p}_{\tilde{a}} \right|^2, \quad \tilde{a} = 1, 2, \dots, \mathrm{INT}(N_{\mathrm{virt}}/2), \tag{1.60}$$

where $\mathrm{INT}(N/2)$ equals $N/2$ if N is even and $(N-1)/2$ if N is odd. In the following, we will denote approximations which keep the diagonal or block-diagonal structure of the amplitude matrices as *diagonal PNO approximation.*

In total, the diagonal approximation reduces the number of non-zero amplitudes roughly by a factor $\frac{2}{3}N_{\mathrm{virt}}$, that is, depending on the number of virtual orbitals, by 2–3 orders of magnitude — an enormous reduction. An even further reduction is possible by neglecting amplitudes that correspond to small natural pair occupation numbers. This means that domains of PNOs are selected using an occupation number threshold $T^{\mathrm{occ}}_{\mathrm{PNO}}$, keeping only PNOs for which $n^{ij,p}_{\tilde{a}} \geq T^{\mathrm{occ}}_{\mathrm{PNO}}$. Unfortunately, the amplitudes must be known in advance to determine the PNOs, and so effectively nothing is gained except for a very compact representation of the wave function. The question is, however, how good an approximation can be obtained if the initial amplitudes are approximated at

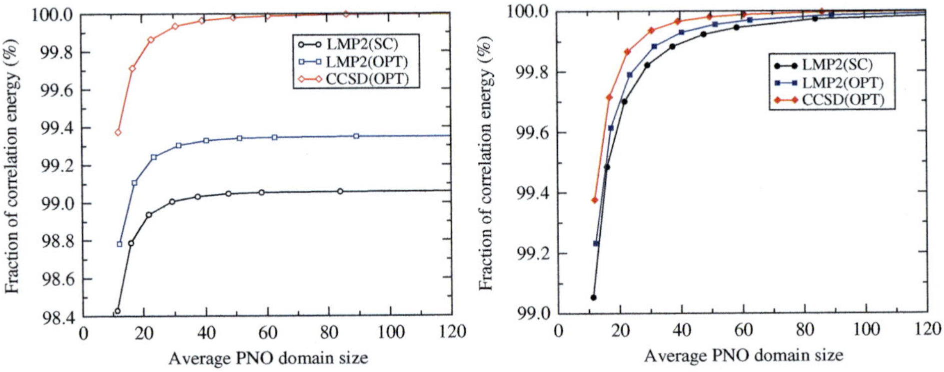

Figure 1.4 PNO-LCCSD correlation energies of propionyl chloride relative to the canonical values (in percent) as a function of the PNO domain sizes using separate PNOs for singlet and triplet pairs for three choices of amplitudes to generate the PNOs. *Left panel*: diagonal approximation (see text). *Right panel*: all amplitudes. Basis set: VDZ-F12.

a lower level. For example, one could either use semi-canonical or optimized LMP2 amplitudes to determine the PNOs, and then use these in a CCSD calculation, still keeping the (block) diagonal form of the amplitudes. The effect of these approximations is demonstrated in the left panel of Figure 1.4 for propionyl chloride (CH_3CH_2COCl) using the VDZ-F12 basis set. The figure shows the percentage of LCCSD correlation energy as a function of the average pair domain size relative to the full canonical calculation. The PNOs have been determined using either semi-canonical LMP2 amplitudes, fully optimized LMP2 amplitudes, or fully optimized LCCSD amplitudes, and the threshold $T_{\rm PNO}^{\rm occ}$ has been varied. If LCCSD amplitudes are used to define the PNOs, the correlation energy converges quickly to the canonical limit. With on average 30 PNOs per pair, more than 99.9% of the canonical correlation energy is already recovered. However, if approximate amplitudes are used, only a smaller fraction of the correlation energy is obtained, since the assumed diagonal form of the amplitudes becomes an approximation. With semi-canonical and optimized MP2 amplitudes, only 99.06% and 99.35%, respectively, of the canonical limit is recovered, even if all PNOs are used. Note that pseudo-canonical orbitals cannot be used in the diagonal approximation, which leads to slower convergence of the LCCSD iterations.

Essentially, the semi-canonical diagonal approximation corresponds to what was used by W. Meyer in his seminal PNO-CI calculations in the early 70s (in fact, he used a somewhat better amplitude approximation). Reaching $\geq 90\%$ of the correlation energy for a given basis set was a big breakthrough at that time. However, even recovering 99% of the correlation energy is insufficient for computing relative energies (e.g., reaction energies) of large molecules, since 1% of the correlation energy can be a huge amount of absolute energy. Today, we therefore aim at approximations which recover about 99.9% of the canonical correlation energy for a given basis set.

The error of the LCCSD correlation energy obtained with full PNO domains (i.e., using all PNOs) is removed if the diagonal amplitude approximation is given up, and all amplitudes $T_{\tilde{a}\tilde{b}}^{ij,p}$ are used. Still, domain approximations can be applied by only including PNOs with natural occupation numbers greater than $T_{\rm PNO}^{\rm occ}$. The convergence of the correlation energy with the domain sizes is shown in the right panel of Figure 1.4. Now the energy smoothly converges to the canonical limit for all 3 types of PNOs; as expected, convergence is fastest (and virtually identical to that of the diagonal amplitude approximation) if the PNOs are generated from LCCSD, but also the errors caused by using approximate amplitudes to generate the PNOs are small. In order to reach 99.92–99.93% of the canonical correlation energy one needs about 30, 40, and 50 PNOs per orbital pair with LCCSD, LMP2(opt), and LMP2(SC) initial amplitudes, respectively (the domain sizes depend on the basis set).

When full amplitude matrices are used, there is no particular advantage of using singlet and triplet configurations. In fact, using the original configurations $|\Phi_{ij}^{ab}\rangle$ and their contravariant counterparts leads to a simpler and more efficient formulation of the CCSD equations [197–199]. One can define symmetric spin-summed pair densities as

$$\mathbf{D}^{ij} = \sum_{p=\pm 1} N_{ij,p} \mathbf{D}^{ij,p} = \frac{1}{1+\delta_{ij}} \left(\tilde{\mathbf{T}}^{ij\dagger} \mathbf{T}^{ij} + \tilde{\mathbf{T}}^{ij} \mathbf{T}^{ij\dagger} \right). \tag{1.61}$$

and diagonalize these to obtain one set of PNOs for each pair $i \geq j$;

$$|\tilde{a}^{ij}\rangle = \sum_{b} |b\rangle \tilde{Q}_{b\tilde{a}}^{ij}, \tag{1.62}$$

$$\left[\mathbf{Q}^{ij\dagger} \mathbf{D}^{ij} \mathbf{Q}^{ij}\right]_{\tilde{a}\tilde{b}} = \delta_{\tilde{a}\tilde{b}} n_{\tilde{a}}^{ij}. \tag{1.63}$$

Again, PNO domains can be determined using the condition $n_a^{ij} > T_{\text{PNO}}^{\text{occ}}$. Subsequently, the PNOs in the domain are made pseudo-canonical, cf. equations (1.34)–(1.37).

The convergence of the correlation energy with respect to the average PNO domain size is shown in Figure 1.5 for spin-independent PNOs (left panel) and singlet and triplet PNOs (right panel). Somewhat unexpectedly it is found that the spin-independent PNOs generated from semi-canonical or optimized LMP2 amplitudes yield faster convergence to the canonical limit than the singlet and triplet ones. Overall, this requires a smaller

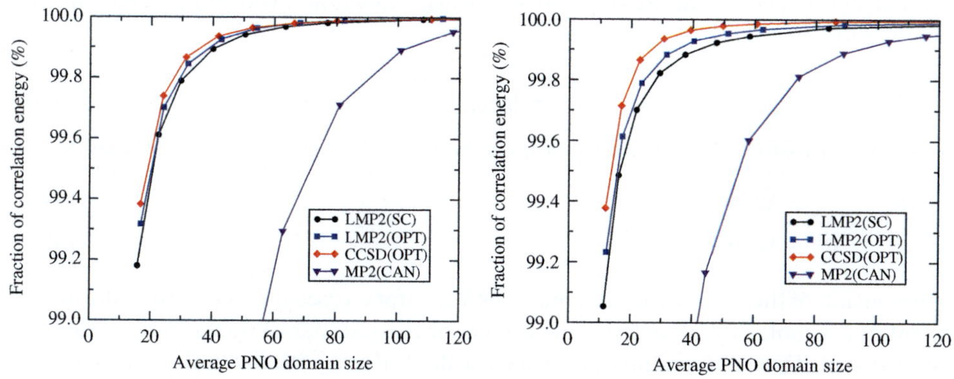

Figure 1.5 PNO-LCCSD correlation energies of propionyl chloride relative to the canonical values (in percent) as a function of the PNO domain sizes using various PNO types. *Left panel*: spin-averaged PNOs. *Right panel*: separate PNOs for singlet and triplet pairs. Basis set: VDZ-F12.

number of PNOs to reach a certain accuracy, in particular if a high percentage of correlation energy is desired. This behavior was also found for some other test molecules. Also shown in the figures is the case that canonical occupied orbitals are used, and the PNOs are generated from canonical MP2 amplitudes. This leads to much larger domains and clearly demonstrates the advantage of using LMOs, even for this relatively small molecule.

1.6.4 Linear Scaling PNO Generation

In the calculations presented in the previous section, the PNOs were generated from LMP2 amplitudes in the full virtual space. The computation of these amplitudes scales asymptotically as $\mathcal{O}(N^5)$, where N is a measure of the molecular size. Furthermore, the calculation of the pair density matrices and their diagonalization also scales as $\mathcal{O}(N^5)$. If distant pairs are neglected or treated by multipole approximations (cf. Section 1.8) this is reduced to $\mathcal{O}(N^4)$. Another problem is the calculation of the integrals $\mathbf{K}^{ij}_{\text{PNO}}$. Since the PNOs are different for each pair, their generation would be very expensive and require to keep all matrices $\mathbf{Q}^{ij}$ (or the AO → PNO transformation matrices $\mathbf{V}^{ij} = \mathbf{C}_v\mathbf{Q}^{ij}$) in memory.

These problems can be overcome by combining PAO and PNO approximations [54, 146, 150]. In this case, the initial LMP2(SC) amplitudes are computed in the PC-PAO basis, using large PAO domains (one can also fully optimize the PAO-LMP2 amplitudes, but this is much more expensive). The integrals $(\tilde{r}i|\tilde{s}j)$, $\tilde{r},\tilde{s} \in [ij]_{\text{PAO}}$ are then first computed in the non-orthogonal PAO basis. This can be done very efficiently using local density fitting, cf. Section 1.12.1. Next, the integrals are transformed to the pair-specific PC-PAO domains

$$\mathbf{K}^{ij}_{\text{PC-PAO}} = \mathbf{W}^{ij\dagger}\mathbf{K}^{ij}_{\text{PAO}}\mathbf{W}^{ij} \tag{1.64}$$

and used to determine the PC-PAO amplitudes T^{ij}_{rs}, $r,s \in [ij]_{\text{PC-PAO}}$. The PNO generation then proceeds exactly as explained in the previous section, except that the transformation matrices $\tilde{\mathbf{Q}}^{ij}$ are replaced by transformation matrices $\mathbf{V}^{ij}$, which transform from the PC-PAO basis to the PNO basis. A further transformation $\mathbf{U}^{ij}$ transforms from the initial PNO basis to the PC-PNO basis. Thus, the whole process proceeds in the following sequence

$$\text{AO} \xrightarrow{\mathbf{P}} [ij]_{\text{PAO}} \xrightarrow{\mathbf{W}^{ij}} [ij]_{\text{PC-PAO}} \xrightarrow{\mathbf{V}^{ij}} [ij]_{\text{PNO}} \xrightarrow{\mathbf{U}^{ij}} [ij]_{\text{PC-PNO}} \tag{1.65}$$

and the total transformation from the non-orthogonal PAO to the PC-PNO basis is

$$\mathbf{K}^{ij}_{\text{PC-PNO}} = \tilde{\mathbf{V}}^{ij\dagger}\mathbf{K}^{ij}_{\text{PAO}}\tilde{\mathbf{V}}^{ij}, \tag{1.66}$$

$$\tilde{\mathbf{V}}^{ij} = \mathbf{W}^{ij}\mathbf{V}^{ij}\mathbf{U}^{ij}. \tag{1.67}$$

The dimension of the transformation matrices $\tilde{\mathbf{V}}^{ij}_{\text{PNO}}$ for a given ij is asymptotically independent of the molecular size; the same is true for the number of operations needed to compute them. Thus, if the number of pairs ij scales linearly with the molecular size, the total memory and CPU time for the PNO generation also scales linearly with molecular size, provided we can compute the integrals $\mathbf{K}^{ij}_{\text{PAO}}$ in a linear scaling fashion. This problem will be addressed in Section 1.12.1.

1.6.5 Orbital-Specific Virtuals (OSVs)

Orbital-specific virtuals (OSVs) [164–167] are identical to the PNOs for diagonal pairs ii and can be obtained by diagonalizing either the amplitude matrices $\mathbf{T}^{ii}$ or the pair density matrices $\mathbf{D}^{ii}$. This yields one set of virtual orbitals for each valence orbital. Orbital domains $[i]_{\text{OSV}}$ can be based on the natural pair occupation numbers using a threshold $T_{\text{OSV}}^{\text{occ}}$, exactly as for PNOs. Pair domains $[ij]_{\text{OSV}}$ are then formed by taking the union of the related orbital domains, $[ij]_{\text{OSV}} = [i]_{\text{OSV}} \cup [j]_{\text{OSV}}$. Since the OSVs in $[i]_{\text{OSV}}$ and $[j]_{\text{OSV}}$ are mutually non-orthogonal and may be linearly dependent, the orbitals in the domain $[ij]_{\text{OSV}}$ must be orthogonalized and thus become pair-specific. Subsequently they are made pseudo-canonical, yielding a domain $[ij]_{\text{PC-OSV}}$.

Although a (near) linear scaling OSV-LCCSD(T) method has been implemented [166, 167], this has not been very successful for two reasons: First, the integral transformations are quite complicated and often more expensive than with PAOs. Second, even though the OSVs are near optimal for describing the electron correlation in a single orbital, they are much less well suited for describing inter-orbital correlations and long-range correlations. Therefore, the threshold $T_{\text{OSV}}^{\text{occ}}$ must be chosen rather small (10^{-9}) in order to obtain accurate results. This yields pair domains that are two to three times larger than the PNO domains, and consequently the computational effort for solving the LCCSD equations is at least one order of magnitude larger than with PNOs. Perhaps even more important, the number of 3-external and 4-external integrals is much larger than in the PNO case.

However, OSVs can be successfully used at an intermediate stage to generate the PNOs [146, 147]. In this approach semi-canonical amplitudes T_{rs}^{ii} are computed in the PC-PAO orbital domains $[i]_{\text{PC-PAO}}$ and used to generate the PC-OSVs. This leads to a significant reduction of the orbital domain sizes. For each LMO, a transformation matrix $\mathbf{Q}_{\tilde{r}a}^{i}$ (PAO → PC-OSV) is stored. Next, OSV pair domains are generated and transformed to pseudo-canonical domains $[ij]_{\text{PC-OSV}}$ (transformation matrices $\mathbf{W}^{ij}$). The integrals $\mathbf{K}^{ij}$ are first computed in the PAO basis and subsequently transformed to the PC-OSV basis. This can be done on the fly using the transformation matrices $\mathbf{Q}^{i}$, $\mathbf{Q}^{j}$, and $\mathbf{W}^{ij}$. Finally, semi-canonical amplitudes $\mathbf{T}_{\text{PC-OSV}}^{ij}$ are calculated and from these the PNOs, just as in the PAO-PNO method outlined in the previous section. The whole process can be summarized as follows:

$$\text{AO} \xrightarrow{\mathbf{P}} [i]_{\text{PAO}} \xrightarrow{\mathbf{Q}^{i}} [i]_{\text{PC-OSV}}, \tag{1.68}$$

$$[ij]_{\text{OSV}} = [i]_{\text{PC-OSV}} \cup [j]_{\text{PC-OSV}}, \tag{1.69}$$

$$[ij]_{\text{OSV}} \xrightarrow{\mathbf{W}^{ij}} [ij]_{\text{PC-OSV}} \xrightarrow{\mathbf{V}^{ij}} [ij]_{\text{PNO}} \xrightarrow{\mathbf{U}^{ij}} [ij]_{\text{PC-PNO}}. \tag{1.70}$$

The last three steps are exactly as in equation (1.65), in which the OSV intermediates are skipped. The advantage of this stepwise approach is that the OSV pair domains are significantly smaller than the PAO pair domains, and therefore the computational effort for the transformations in equation (1.70) (which is dominated by the diagonalizations and depends cubically on the domain sizes) is significantly reduced as compared to equation (1.65). Furthermore, the memory requirements for the storage of the transformation matrices $\mathbf{V}^{ij}$ (which in this case transform from the OSV to the PC-PNO pair domains) are much smaller. This is important for the parallel implementation, in which all data are kept in distributed high-speed memory (cf. Section 1.12.2).

1.7 Choice of Domains

For the accuracy of local correlation methods, the careful choice of domains is essential. In this section, we will summarize how in our current program the PAO, OSV, and PNO domains are chosen.

The PAO orbital domains are selected in a 2-step procedure: first, relatively small *primary domains* are chosen based on IBO partial charges with a threshold T_{LMO} (default 0.2). Typically, this includes only one center for a lone pair, two centers for σ and isolated π bonds, and two to three centers for conjugated or aromatic π bonds. However, these domains are too small for high-accuracy calculations. Therefore, in a second step, *extended domains* are generated which include all atoms that are bound to any atom in the primary domain. Centers are considered to be bound if the bond length is smaller than 1.2 × the sum of the atomic radii. This can be repeated by including all centers that are bound to any atom in the previous extension shell. The advantage of the connectivity criterion IEXT is its independence of the atomic types, since the atomic radii are directly related to the bond lengths. In addition, a distance criterion REXT is used and all centers that are within a radius REXT of any center in the primary domain are included. This ensures that atoms which happen to be close to an atom in the primary domain without being directly bound (e.g., at transition states or in molecular complexes) are also included in the extended domains. The default is to include two shells of neighbors (`IEXT=2`), and a radius of `REXT=5` (distances in a_0). In order to avoid distance criteria, Neese *et al.* [56] have recently proposed a criterion based on the differential overlap of LMOs and PAOs, but so far there is no evidence that this works better than our simple scheme. Alternatively, a completeness criterion has been described by Boughton and Pulay [200], but this also requires a ranking of the centers according to partial charges of the LMOs and is not well suited to define extended domains. Energy related criteria to define PAO pair domains have also been proposed [26, 201], but have rarely been used so far.

Extensive tests [147] have shown that an OSV threshold of 10^{-9} (or smaller) is necessary in order not to spoil the accuracy of the subsequent PNO calculation. But even with this small threshold, the OSV pair domains are typically three times smaller than the default PAO domains. This significantly speeds up the PNO generation and reduces the memory requirements.

The convergence of the correlation energy with the PNO threshold has already been demonstrated in Section 1.6.3. We have chosen a default threshold $T_{\mathrm{PNO}}^{\mathrm{occ}} = 10^{-8}$. It turned out, however, that this is insufficient for rather distant pairs. It may then happen that no PNOs are selected at all for such pairs, and this in turn affects the long-range dispersion energy. We have therefore introduced an additional energy criterion. After diagonalizing the pair density matrix, the orthogonal transformation matrix $\mathbf{V}^{ij}$ is used to transform the amplitudes $\mathbf{T}^{ij}$ and integrals $\mathbf{K}^{ij}$ into the PNO basis (without a further domain approximation, i.e., $\mathbf{V}^{ij}$ is a square matrix). Then, the PNO pair correlation energies $E_{\mathrm{corr}}^{\mathrm{PNO}}$ can be computed by adding one PNO after the other (in the order of decreasing pair natural occupation numbers), until a certain fraction of the PC-PAO (or PC-OSV if OSVs are used) pair energy is reached. The default energy threshold is

Table 1.3 Domain sizes and correlation energies relative to canonical MP2 (in %) for various LMP2 calculations.

	Average domain sizes				Relative LMP2 correlation energies			
Molecule	**PAO**	**OSV**	**PNO[a]**	**PNO[b]**	**PAO**	**OSV**	**PNO[a]**	**PNO[b]**
Testosterone	513.5	204.9	73.3	64.3	99.868	99.815	99.794	99.746
$C_{52}O_6H_{48}$	493.8	219.8	66.7	58.9	99.884	99.839	99.813	99.770
Nonactin	469.2	191.8	57.4	51.3	99.873	99.830	99.809	99.768
Elaiophylin	500.0	193.5	57.1	50.2	99.859	99.808	99.793	99.745

Basis set: cc-pVTZ, IEXT = 2, $T_{\mathrm{OSV}} = 10^{-9}$, $T^{\mathrm{occ}}_{\mathrm{PNO}} = 10^{-8}$, $T^{\mathrm{en}}_{\mathrm{PNO}} = 0.997$.
[a]Using PC-PAO amplitudes for PNO generation.
[b]Using PC-OSV amplitudes for PNO generation.

$T^{\mathrm{en}}_{\mathrm{PNO}} = 0.997$. This ensures that, for all pairs, at least 99.7% of the OSV pair correlation energies are obtained, even if the pair energies and natural occupation numbers are very small. One could of course set this threshold to 0.999 to reduce the error of the PNO domains, but this would increase the domains of distant pairs more than necessary.

Table 1.3 shows for some molecules the average PAO, OSV, and PNO pair domain sizes, the correlation energies (in percent relative to the canonical calculations) for PAO-LMP2, OSV-LMP2, and PNO-LMP2 calculations. In all cases, the same PAO orbital domains (`IEXT=2`) have been used. The PNOs are either generated from PC-PAO amplitudes or PC-OSV amplitudes. The corresponding timings are shown in Table 1.4. The average PAO pair domain sizes are for this basis (cc-pVTZ) around 500, and with these domains about 99.87% of the canonical correlation energy is recovered. The intermediate OSV step ($T_{\mathrm{OSV}} = 10^{-9}$) reduces the domain sizes to about 200, and this causes a 0.05% loss of correlation energy. The transformation from PAOs to PNOs ($T_{\mathrm{PNO}} = 10^{-8}$) reduces the correlation energy by about 0.07% and the domain sizes by one order of magnitude. Comparison of the last two columns of the table shows that the intermediate use of OSVs leads overall to a loss of 0.04–0.05% of the correlation energy, yielding in total

Table 1.4 Elapsed times for the various LMP2 calculations in Table 1.3 (excluding HF), using 10 cores of Xeon E5-2660 v3 @ 2.60GHz processors in one compute node.

			Elapsed times			
Molecule	**AOs**	**val**	**PAO**	**OSV**	**PNO[a]**	**PNO[b]**
Testosterone	1022	58	1711.2	164.9	122.7	63.7
$C_{52}O_6H_{48}$	2412	146	8557.5	1039.7	434.1	264.1
Nonactin	2456	148	6311.6	627.1	339.7	207.7
Elaiophylin	3392	206	13688.6	1279.7	668.5	392.2

[a]Using PC-PAO amplitudes for PNO generation.
[b]Using PC-OSV amplitudes for PNO generation.

about 99.75% of the canonical correlation energy. The elapsed time is reduced by roughly a factor of 2 and the memory usage by about 30%.

The 99.75% percent of correlation energy obtained with our default parameters is still below our goal to reproduce the canonical energy within 0.1%. Lowering the PNO threshold to 10^{-9} increases the PNO domains by about 20 orbitals, but improves the PNO-LMP2 correlation energy only by typically 0.05%. The remaining error is mainly due to the PAO domain approximation. Increasing the PAO domains further is possible but would significantly increase the computational effort for the integral transformation. However, as will be demonstrated in Section 1.11, a much better solution is to include explicitly correlated terms. This will reduce both the basis set incompleteness error as well as the domain error to negligible values and makes it possible to use even smaller PAO domains.

Finally, it is worthwhile to mention that for all three local MP2 methods, that is, PAO-LMP2, OSV-LMP2 and PNO-LMP2, Abelian point group symmetry can be used, provided the occupied orbitals localize in a symmetry-equivalent fashion [33]. This not only speeds up the calculation, but more importantly also reduces the memory requirements significantly.

1.8 Approximations for Distant Pairs

In this section, we will consider approximations for the integrals $(ai|bj)$ for distant pairs ij. We will assume that the four orbitals i, j, a, b are all local, and that $a \in [i]$, $b \in [j]$ are close to $|i\rangle$ and $|j\rangle$, respectively.

The integral $(ai|bj)$ describes the Coulomb interaction of the two local effective charge distributions $\rho_{ai}(\mathbf{r}_1) = \phi_a(\mathbf{r}_1)\phi_i(\mathbf{r}_1)$ and $\rho_{bj}(\mathbf{r}_2) = \phi_b(\mathbf{r}_2)\phi_j(\mathbf{r}_2)$, cf. equation (1.11). In general, Coulomb interactions of two non-overlapping and distant charge distributions can be described by a bipolar multipole expansion. Due to the orthogonality of the virtual and occupied orbitals, the charge distributions carry no charge, that is, $\int \rho_{ai}(\mathbf{r})d\mathbf{r} = 0$, $\int \rho_{bj}(\mathbf{r})d\mathbf{r} = 0$, and therefore the lowest-order contribution is the dipole–dipole interaction.

Charge-transfer type integrals $(ai|bj)$ with $a, b \in [i]$ or $a, b \in [j]$ as well as exchange type integrals with $a \in [j], b \in [i]$ decay exponentially with the distance of the local orbitals $|i\rangle$ and $|j\rangle$ and will be neglected. The effect of this approximation will be demonstrated later on.

1.8.1 Bipolar Multipole Approximations of Electron Repulsion Integrals

In the following, we will briefly summarize the theory for the bipolar multipole approximations as implemented in our program [12, 13]. We use cartesian multipole moment operators rather than traceless multipole operators as in other formulations [85, 202]. At least for low multipole orders this is simple and efficient.

Let $\mathbf{R}_i$ and $\mathbf{R}_j$ be the charge centers of two charge distributions, and $\mathbf{R} = \mathbf{R}_j - \mathbf{R}_i$ the vector between them. Since charge centers of the distributions $\rho_{ai}(\mathbf{r})$ and $\rho_{bj}(\mathbf{r})$ cannot be determined directly, we will use the charge centers of the two LMOs, $\mathbf{R}_i = \langle i|\mathbf{r}|i\rangle$ and $\mathbf{R}_j = \langle j|\mathbf{r}|j\rangle$. The distance between the charge centers is $R = |\mathbf{R}|$, and the unit vector in direction of $\mathbf{R}$ is denoted $\hat{\mathbf{R}} = \mathbf{R}/R$. Let $\mathbf{r}_1$ and $\mathbf{r}_2$ be the coordinates of the two electrons in the global coordinate system, and $\bar{\mathbf{r}}_1$ and $\bar{\mathbf{r}}_2$ the coordinates of the two electrons relative to the corresponding charge centers, that is, $\bar{\mathbf{r}}_1 = \mathbf{r}_1 - \mathbf{R}_i$ and $\bar{\mathbf{r}}_2 = \mathbf{r}_2 - \mathbf{R}_j$. We have to keep in mind that $\bar{\mathbf{r}}_1$ depends on the charge center $\mathbf{R}_i$ and $\bar{\mathbf{r}}_2$ on the charge center $\mathbf{R}_j$. Let further $\hat{\mathbf{r}} = (\bar{\mathbf{r}}_2 - \bar{\mathbf{r}}_1)/R$ and $\hat{r} = |\hat{\mathbf{r}}|$.

The vector between the two electrons is

$$\mathbf{r}_{12} = \mathbf{r}_2 - \mathbf{r}_1 = R(\hat{\mathbf{R}} + \hat{\mathbf{r}}). \tag{1.71}$$

The Coulomb operator can then be expressed as

$$\frac{1}{r_{12}} = \left[(\mathbf{r}_{12})^2\right]^{-\frac{1}{2}} = \frac{1}{R}\,[1+x]^{-\frac{1}{2}}, \tag{1.72}$$

where

$$x = 2\hat{\mathbf{R}}\hat{\mathbf{r}} + \hat{r}^2. \tag{1.73}$$

Expanding $(1+x)^{-\frac{1}{2}}$ in a Taylor series at $x = 0$ yields

$$\frac{1}{r_{12}} = \frac{1}{R}\sum_k^{\infty} c_k x^k \tag{1.74}$$

with coefficients

$$c_k = \left[\frac{1}{k!}\frac{d^k}{dx^k}(1+x)^{-\frac{1}{2}}\right]_{x=0} = (-1)^k\frac{1\cdot 3\cdots 2k-1}{2\cdot 4\cdots 2k}. \tag{1.75}$$

This Taylor series converges for $x < 1$, which should be fulfilled if R is sufficiently large, cf. equation (1.73).

Inserting the definition of x into equation (1.74) we can write the Coulomb operator in the general form

$$\frac{1}{r_{12}} = \sum_{n=0}^{\infty}\frac{1}{R^{n+1}}\sum_{m=0}^{n} V_{m,n-m}(\bar{\mathbf{r}}_1, \bar{\mathbf{r}}_2, \hat{\mathbf{R}}). \tag{1.76}$$

Truncating the expansion at $n = p + 1$ and integrating with the charge distributions, we obtain a series of integral approximations

$$(ai|bj)^{(p)} = \sum_{n=2}^{p+1}\frac{1}{R^{n+1}}\sum_{m=1}^{n-1}\int d\mathbf{r}_1\, \rho_{ai}(\mathbf{r}_1) \times \int d\mathbf{r}_2\, \rho_{bj}(\mathbf{r}_2) V_{m,n-m}(\bar{\mathbf{r}}_1, \bar{\mathbf{r}}_2, \hat{\mathbf{R}}). \tag{1.77}$$

Since our effective charge densities carry no charge, there are no contributions for $n = 0$ and $n = 1$, and the first non-vanishing term is $n = 2$. p is the highest order of multipole operators involved. With $p = 1$ only dipole operators (first moments) are needed, $p = 2$ also includes second moments, and so on.

It is straightforward to write down general expressions for $V_{m,n-m}$, which contain the contributions of powers of $\bar{\mathbf{r}}_1$ and $\bar{\mathbf{r}}_2$ with a fixed total power n (see Appendix B). The simplest approximation is obtained for $p = 1$ $(n = 2)$:

$$\begin{aligned} V_{1,1}(\bar{\mathbf{r}}_1, \bar{\mathbf{r}}_2, \hat{\mathbf{R}}) &= -2c_1 \bar{\mathbf{r}}_1 \cdot \bar{\mathbf{r}}_2 - 8c_2(\hat{\mathbf{R}} \cdot \bar{\mathbf{r}}_1)(\hat{\mathbf{R}} \cdot \bar{\mathbf{r}}_2) \\ &= \bar{\mathbf{r}}_1 \cdot \bar{\mathbf{r}}_2 - 3(\hat{\mathbf{R}} \cdot \bar{\mathbf{r}}_1)(\hat{\mathbf{R}} \cdot \bar{\mathbf{r}}_2). \end{aligned} \tag{1.78}$$

Integration over the charge densities yields

$$\begin{aligned} (ai|bj)^{(1)} = \frac{1}{R^3} \big[&\langle a|\mathbf{r}_1|i\rangle \cdot \langle b|\mathbf{r}_2|j\rangle \\ &-3(\hat{\mathbf{R}} \cdot \langle a|\mathbf{r}_1|i\rangle)(\hat{\mathbf{R}} \cdot \langle b|\mathbf{r}_2|j\rangle) \big]. \end{aligned} \tag{1.79}$$

Here we have used that due to the orthogonality of the virtual and occupied orbitals, $\langle a|i\rangle = 0$; the dipole moment integrals are origin-independent, since $\langle a|\bar{\mathbf{r}}_1|i\rangle = \langle a|\mathbf{r}_1 - \mathbf{R}_i|i\rangle = \langle a|\mathbf{r}_1|i\rangle$. $\langle a|\mathbf{r}_1|i\rangle$ denotes the dipole vector $(\langle a|x_1|i\rangle, \langle a|y_1|i\rangle, \langle a|z_1|i\rangle)$, and $\langle a|\mathbf{r}_1|i\rangle \cdot \langle b|\mathbf{r}_2|j\rangle$ the scalar product of the two vectors. This is called the dipole–dipole approximation. Note that we only need to compute and store the integrals for $a \in [i]$ and $b \in [j]$, and therefore the memory demand is low, even if higher multipole orders are included.

In some earlier works [54, 147] also a simplified form of the dipole–dipole approximation

$$(ai|bj) \approx \frac{\sqrt{2}}{R^3} \langle a|\mathbf{r}|i\rangle \cdot \langle b|\mathbf{r}|j\rangle \tag{1.80}$$

has been used. The differences between the results obtained with this or equation (1.79) are rather small. In this work, we will only use equation (1.79) which is formally more accurate.

Equation (1.79) can also be written as

$$(ai|bj)^{(1)} = \sum_{\alpha,\beta} M_{ai}^{(1,\alpha)} A_{\alpha,\beta}^{(1,1)} M_{bj}^{(1,\beta)}, \tag{1.81}$$

where $\alpha, \beta \in \{x, y, z\}$, $M_{ai}^{(1,\alpha)} = \langle a|\hat{\alpha}|i\rangle$ are the dipole integrals, and $\mathbf{A}^{(1,1)}$ is the dipole–dipole *interaction matrix*

$$A_{\alpha,\beta}^{(1,1)}(\hat{\mathbf{R}}) = \delta_{\alpha\beta} - 3\hat{R}_\alpha \hat{R}_\beta. \tag{1.82}$$

This formalism can be extended to any higher order, and the total approximation of the integrals can be written as

$$\begin{aligned} (ai|bj)^{(p)} \approx &\sum_{n=2}^{p+1} \frac{1}{R^{n+1}} \sum_{m=1}^{n-1} \\ &\times \sum_{\mu\nu} M_{ai}^{(m,\mu)}(\mathbf{R}_i) A_{\mu\nu}^{(m,n-m)}(\hat{\mathbf{R}}) M_{bj}^{(n-m,\nu)}(\mathbf{R}_j), \end{aligned} \tag{1.83}$$

where now μ runs over the $(m+1)(m+2)/2$ unique cartesian multipole moment components of order m and $M_{ai}^{(m,\mu)}(\mathbf{R}_i)$ are the integrals over these operator components evaluated at the origin $\mathbf{R}_i$. For example, for $m = 2$ the sum over μ runs over the six components of the second moment operators: xx, xy, xz, yy, yz, zz. The sum over ν is restricted accordingly. Note that for $p > 1$, this approximation does not include interactions of the highest multipoles with each other. For example, the $p = 3$ approximation includes the dipole–dipole, dipole–quadrupole, quadrupole–quadrupole, and dipole–octupole interactions, but the octupole–octupole interaction contributes only for $p = 4$.

The higher-order multipole moments depend on the origin $\mathbf{R}_i$ at which they are evaluated. Using $\mathbf{R}_i = (X_i, Y_i, Z_i)$ and $\mathbf{r} = (x, y, z)$, the integral $M_{ai}^{(m,\alpha)}(\mathbf{R}_i)$ can be written as

$$M_{ai}^{\{l_x,l_y,l_z\}}(\mathbf{R}_i) = \int (x - X_i)^{l_x}(y - Y_i)^{l_y}(z - Z_i)^{l_z}\rho_{ai}(\mathbf{r})d\mathbf{r}, \tag{1.84}$$

where $\{l_x, l_y, l_z\} \equiv (m, \alpha)$ with $m = l_x + l_y + l_z$ defines the operator component. For example, for the operator xxy, we have $m = 3$ and $l_x = 2$, $l_y = 1$, $l_z = 0$. In order to compute all integrals $(ai|bj)$, we need the multipole operators at all charge centers $\mathbf{R}_i$. Fortunately, if one has computed all multipole operators up to a given order at some origin $\mathbf{R}_0$, they can be shifted to any other position. Multiplying out equation (1.84) yields

$$\begin{aligned} M_{ai}^{\{l_x,l_y,l_z\}}(\mathbf{R}_i) = &\sum_{m_x=0}^{l_x} f_{m_x} \sum_{m_y=0}^{l_y} f_{m_y} \sum_{m_z=0}^{l_z} f_{m_z} \\ &\times M_{ai}^{\{l_x-m_x,l_y-m_y,l_z-m_z\}}(\mathbf{R}_0), \end{aligned} \tag{1.85}$$

where

$$f_{m_x} = \binom{l_x}{m_x} [X_0 - X_i]^{m_x}. \tag{1.86}$$

Similar definitions hold for f_{m_y} and f_{m_z}.

Closed-form expressions for the above formalism up to any order p have been derived and implemented by Hetzer *et al.* [12] and applied in PAO-LMP2 calculations. Increasing the order p improves the accuracy at long range, but leads to faster divergence at short range. Hetzer *et al.* therefore recommended to truncate the expansion at $p = 3$, which gives the best compromise of stability, accuracy, and efficiency. For technical reasons, the general program cannot be used in our current PNO-LMP2 program. We have therefore re-implemented the approximation up to $p = 3$; the required explicit expressions can be found in Appendix B.

A further improvement is possible by splitting the Coulomb operator into a short-range and a long-range part [13]. The integrals over the short range part, which decays quickly, are evaluated exactly. Effective screening techniques can be used to speed up this part of the calculation. The long-range part is treated by multipole approximations, which are then always convergent, and also the charge-transfer and exchange-type integrals can be included. The description of this method is beyond the scope of the current chapter and we refer to the original publication [13] for further details.

1.8.2 Approximations of Distant Pair Energies

In the following, we will assume that $|a\rangle$ and $|b\rangle$ are pseudo-canonical, with $|a\rangle \equiv |a^i\rangle$ and $|b\rangle \equiv |b^j\rangle$. It is possible to use either PC-PAO or PC-OSV domains, but except

Table 1.5 PAO-LMP2 intermolecular interaction energy of the ethylene dimer (in mE_h) computed at various multipole orders. The two molecules are displaced by R_{displ} (in Å) perpendicular to the molecular planes. The last line shows the PAO-LMP2 result without multipole approximation.

	R_{displ}		
p	4	5	6
1	−1.1991	−0.3500	−0.1232
2	−1.6055	−0.4335	−0.1447
3	−1.9825	−0.4913	−0.1569
4	−2.3366	−0.5121	−0.1593
5	−3.2181	−0.5363	−0.1608
Exact	−2.0689	−0.5257	−0.1613

Basis set: aug-cc-pVTZ.

for Table 1.5 we will only present results obtained with PC-OSVs using a threshold $T_{OSV} = 10^{-9}$.

Using the multipole approximations described in the previous section, the semi-canonical LMP2 pair energies (cf. equation 1.20) can be evaluated on the fly without storing the integrals. This is very fast, even for very large molecules. It is therefore convenient to compute the pair energies for all pairs for which $R_{ij} > R_{min}$ (using e.g., $R_{min} = 3\ a_0$) and then select those for which $E_{ij}^{sc} < T_{dist}$ as distant pairs. All remaining pairs are fully included in the LMP2 calculation. The approximated distant pair energies are summed up to the total distant pair energy E_{dist}^{sc}, which is added to the final LMP2 energy. The default threshold for selecting distant pairs is $T_{dist} = 10^{-6}\ E_h$.

Alternatively, a better approximation is obtained by storing the integrals $(ai|bj)$, and optimizing the corresponding amplitudes and pair energies iteratively along with all other pairs. For the multipole approximation of distant pairs, we use asymmetric OSV domains $a \in [i]_{OSV}$ and $b \in [j]_{OSV}$, as explained above. Each of these OSV domains typically comprise 100–200 orbitals, and in this large basis a full optimization of the OSV amplitudes would be very expensive. However, one can use the semi-canonical OSV distant pair amplitudes to determine for each distant pair PNO domains $[i]_{PNO}$ and $[j]_{PNO}$, and in most cases, only three to six orbitals in these domains are sufficient to recover more than 99.9% of the OSV distant pair energy. The additional cost for including the distant pairs in the iterative LMP2 optimization then becomes quite small. More details have been given elsewhere [203].

The corresponding distant pair energy will be denoted E_{dist}^{it}. Table 1.5 shows the convergence of the PAO-LMP2 interaction energy with the multipole order p for the ethylene dimer. The two molecules are parallel to each other and displaced by R_{displ} (in Å) perpendicular to the molecular planes. The domains in each C_2H_4 molecule include all PAOs at the corresponding molecule (`IEXT=2`), and all intermolecular pairs are treated as distant. The amplitudes are optimized iteratively. The table demonstrates that the multipole expansion diverges at high orders at $R_{displ} = 4$ Å, but converges nicely for $R_{displ} = 6$ Å. As already pointed out in Ref. [12], the dipole–dipole approximation

Table 1.6 PNO-LMP2 distant pair energies (in mE_h) with different distant pair approximations. The numbers over the columns are the multipole orders p used. NOEXCH means that all integrals are computed exactly, but charge-transfer and exchange-type integrals are neglected for distant pairs. FULL means that no distant pair approximations have been applied.

		Semi-canonical			Iterative				
Molecule	**Distant pairs**	**1**	**2**	**3**	**1**	**2**	**3**	**NOEXCH**	**FULL**
$T_{dist} = 10^{-5}\ E_h$									
Ester	1157	−2.39	−3.08	−3.47	−2.80	−3.62	−4.08	−4.27	−5.10
Precursor	1661	−2.96	−3.69	−4.01	−3.42	−4.29	−4.67	−4.80	−5.61
Auamin	4812	−10.08	−12.79	−13.58	−11.45	−14.31	−15.54	−15.56	−18.30
$T_{dist} = 10^{-6}\ E_h$									
Ester	598	−0.18	−0.21	−0.22	−0.21	−0.25	−0.26	0.27	−0.31
Precursor	951	−0.31	−0.36	−0.36	−0.36	−0.42	−0.42	0.43	−0.47
Auamin	2355	−0.88	−1.00	−1.02	−1.01	−1.15	−1.18	−1.19	−1.28

Basis set: VTZ-F12, IEXT=2.

significantly underestimates the interaction energy; even for $R_{displ} = 6$ Å, the error still amounts to −24%. This error is reduced to −10% and −3% for $p = 2$ and $p = 3$, respectively. Even for $R_{displ} = 4$ Å, the error for $p = 3$ is only −4%.

Distant pair energies computed with the semi-canonical (non-iterative) and iterative multipole approximation for larger molecules are presented in Table 1.6 for $p = 1, 2, 3$ and distant pair selection thresholds of $T_{dist} = 10\ \mu E_h$ and $T_{dist} = 1\ \mu E_h$. In all cases, the distant pairs have been selected using the semi-canonical $p = 1$ pair energies, that is, the number of distant pairs is independent of the approximation applied (with the more accurate $p = 3$ approximation less distant pairs would be selected and the results would become more accurate, but since we intended to compare the pure effect of the integral approximations this has not been used).

For comparison, also distant pair energies computed without multipole approximations are shown. In the first case, denoted NOEXCH, the charge-transfer and exchange type integrals are neglected, exactly as in the multipole approximations, but the remaining dispersion-type integrals are computed exactly (using density fitting). In the second case (FULL), all pairs are fully included in the LMP2 calculation without further approximations. Table 1.6 shows that with a large threshold of 10 μE_h, the approximate distant pair energies are in all cases significantly underestimated. There are three sources of errors: (i) the neglect of charge-transfer and exchange-type excitations for distant pairs (NOEXCH); (ii) the non-iterative semi-canonical approximation (SC) ; and (iii) the multipole approximation for the integrals (MLTP). The largest error (about −15%) arises from the NOEXCH approximation; the error caused by the SC approximation is also significant (−10% to −15%), while the error of the multipole approximation itself decreases quickly with p and is in all cases below 5% for $p = 3$. For example, for the Auamin molecule (cf. Section 1.1) the MLTP errors (relative to NOEXCH) amount to −26.4%, −8.0%, and −0.1% for $p = 1, 2, 3$, respectively. In total, the semi-canonical multipole approximation for $p = 1$ recovers only 55% of the exact LMP2 distant pair energy for Auamin. For the

Table 1.7 Total PNO-LMP2 reaction energies ΔE and distant pair contributions ΔE_{dist} without multipole approximations, and errors of the total reaction energies caused by various multipole approximations. All values in kJ mol^{-1}. The numbers over the columns correspond to the multipole order applied.

	PNO-LMP2		Errors of reaction energies						
	Reaction energies		Semi-canonical			Iterative			
Reaction	ΔE	ΔE_{dist}	1	2	3	1	2	3	NOEXCH
$T_{dist} = 10^{-5}\ E_h$									
I	−21.0	−3.2	1.8	1.4	1.2	1.5	0.9	0.7	0.5
II	18.2	4.5	−1.1	−0.7	−0.7	−0.6	−0.1	−0.1	−0.3
III	239.7	22.0	−10.2	−7.1	−6.2	−7.9	−4.1	−2.9	−2.9
$T_{dist} = 3 \times 10^{-6}\ E_h$									
III	239.7	6.9	−2.7	−1.9	−1.8	−1.8	−0.8	−0.6	−0.6
$T_{dist} = 10^{-6}\ E_h$									
III	239.7	2.0	−0.6	−0.5	−0.4	−0.4	−0.1	−0.1	−0.1

Basis set: VTZ-F12, IEXT=2.

other two molecules, the error is even larger (47% is recovered for the Ester and 53% for the Precursor).

For the smaller threshold (1 μE_h), the errors caused by the NOEXCH approximation are reduced, but still amount to about −7% for Auamin and −13% for the Ester. The relative errors of the other approximations are quite similar as for the larger threshold. Of course, the absolute errors are now reduced by one order of magnitude.

The errors in reaction energies caused by the distant pair approximations relative to the full LMP2 values are listed in Table 1.7 for three reactions involving the same molecules and thresholds. For the largest system (reaction **III**), the total contribution of the distant pairs to the dissociation energy amounts to 22 kJ mol^{-1} and 2.0 kJ mol^{-1} for the thresholds of 10 and 1 μE_h, respectively. This demonstrates that the numerous but very small energy contributions of distant pairs in large systems are by no means negligible when summed up. The errors in the reaction energies caused by the simple semi-canonical dipole–dipole approximations are −10.2 and −0.6 kJ mol^{-1}, respectively. With the iterative $p = 3$ approximations, these errors are reduced to −2.9 and −0.1 kJ mol^{-1}. Using the small threshold of 1 μE_h, the errors of the reaction energies for **I** and **II** are negligibly small (below 0.2 kJ mol^{-1}) and therefore not shown.

We conclude that the simple dipole–dipole approximation should only be applied with a small threshold T_{dist}; our default value of 1 μE_h is a safe choice. Larger thresholds should only be used with higher multipole orders (preferably $p = 3$). Including the distant pairs iteratively reduces the errors typically be a factor of 2. However, if *all* distant pairs are included in the iterations, linear scaling is no longer achieved in the LMP2. The linear scaling behavior could be kept by neglecting very distant pairs entirely, or by treating them non-iteratively.

1.9 Local Coupled-Cluster Methods (LCCSD)

The LCCSD wave function is defined as

$$\Psi_{\mathrm{CCSD}} = e^{\hat{T}}|0\rangle \tag{1.87}$$

with the cluster operator

$$\hat{T} = \sum_i \sum_{a\in[i]} t^i_a \hat{E}^a_i + \frac{1}{2} \sum_{ij\in P} \sum_{a,b\in[ij]} T^{ij}_{ab} \hat{E}^a_i \hat{E}^b_j. \tag{1.88}$$

Here a, b are virtual orbitals, which, depending on the method, can be PAOs, OSVs, or PNOs, and $[ij]$ denotes the corresponding pair domains. The singles domains $[i]$ are usually taken to be equal to the domains $[ii]$ of the "diagonal" pairs, but this restriction can also be lifted. The sum over pairs ij can be restricted to a pair list P. It is also possible to subdivide this list into different classes, which are then treated at different computational levels, for example, CCSD, CEPA (coupled electron pair approximation), or LMP2. Such pair approximations will be discussed in Section 1.9.1.

The amplitudes are determined by solving the usual LCCSD amplitude equations. Inserting the ansatz (1.87) into the Schrödinger equation, multiplying from the left with $e^{-\hat{T}}$, and projecting from the left with the reference function and the contravariant configurations yields

$$E_{\mathrm{CCSD}} = \langle 0|e^{-\hat{T}}\hat{H}e^{\hat{T}}|0\rangle, \tag{1.89}$$

$$r^i_a = \left\langle \tilde{\Phi}^a_i |e^{-\hat{T}}\hat{H}e^{\hat{T}}|0\right\rangle = 0, \qquad a \in [i], \tag{1.90}$$

$$R^{ij}_{ab} = \left\langle \tilde{\Phi}^{ab}_{ij} |e^{-\hat{T}}\hat{H}e^{\hat{T}}|0\right\rangle = 0, \qquad a, b \in [ij], \tag{1.91}$$

where the contravariant configurations for the single excitations are $|\tilde{\Phi}^a_i\rangle = \frac{1}{2}\hat{E}^a_i|0\rangle$; for the double excitations, they are defined as in LMP2, cf. equation (1.5). The equations can be solved iteratively exactly as described earlier for LMP2. In each iteration the residuals r^i_a and R^{ij}_{ab} have to be computed, and updates of the amplitudes can be obtained using equation (1.19) (and similarly for the singles amplitudes t^i_a).

The residuals r^i_a and R^{ij}_{ab} can be formulated compactly using amplitude and integral matrices. A proper factorization of these equations is important for efficiency and proper scaling, and various formulations have been described in the literature [18, 54, 198, 199]. For easy reference, the equations as implemented in our program are given in Appendix A.

The local coupled cluster correlation energy can generally (for any CC level) be written as

$$E^{\mathrm{CC}}_{\mathrm{corr}} = \sum_{i\geq j\in P} E_{ij} + E_{\mathrm{sing}} + E_{\mathrm{dist}}. \tag{1.92}$$

The pair energies E_{ij} are defined exactly as in LMP2:

$$E_{ij} = (2 - \delta_{ij}) \sum_{a,b\in[ij]} K^{ij}_{ab} \tilde{T}^{ij}_{ab}, \tag{1.93}$$

and E_{dist} is the energy contribution of the distant pairs, which can be approximated by the multipole treatment as described in Section 1.8. The singles energy contribution E_{sing} can be computed using "dressed" Fock matrix elements $\bar{f}_a^i$ (see Appendix A for their definition):

$$E_{\text{sing}} = 2 \sum_i \sum_{a \in [i]} t_a^i \bar{f}_a^i. \tag{1.94}$$

This can be seen as an orbital relaxation contribution to the Hartree–Fock energy.

The number of integrals needed in LCCSD is significantly larger than in LMP2. In LMP2, one only needs the integrals $(ai|bj)$ for $a, b \in [ij]$. In LCCSD, all other classes of transformed integrals are also required: 0-external $(ik|jl)$, 1-external $(ai|kl)$, 2-external $(ab|ij)$, and $(ai|bj)$, 3-external $(ab|ck)$, and 4-external $(ac|bd)$. For the 4-external integrals, the domains of all four virtual orbitals are the same, that is, $a, b, c, d \in [ij]$. However, for the other integral classes, the domain restrictions are much more complicated. Consider, for example, the 2-external exchange integrals $K_{cd}^{kl} = (ck|dl)$, which occur in several terms of the LCCSD equations. A typical term is

$$[\mathbf{R}^{ij}]_{ab} = \ldots \sum_{kl} \sum_{c \in [ik]} \sum_{d \in [lj]} [\mathbf{S}\mathbf{T}^{ik}]_{ac} [\mathbf{K}^{kl}]_{cd} [\mathbf{T}^{lj}\mathbf{S}]_{db}, \qquad a, b \in [ij]. \tag{1.95}$$

Let us assume that distant pairs are not included in the LCCSD *ansatz*. This means that ij, ik, and jl must be non-distant pairs, and for each ij, the index k must be "close" to i, and l be "close" to j (here "close" means not distant). This makes the list of operators $\mathbf{K}^{kl}$ that contribute to a pair ij asymptotically independent of the molecular size. However, the list of kl can be larger than the list of pairs ij. For a given kl, the indices c, d of K_{cd}^{kl} are restricted to $c \in [ik]$, $d \in [jl]$ for all i, j that are "close" to k and l, respectively. This also limits these indices to a number which is asymptotically independent of the molecular size. Thus, the total number of integrals as well as the computational effort will scale linearly with molecular size, but (i) the prefactors may be huge and (ii) the linear scaling regime be reached very late.

If the PAO basis is used (PAO-LCCSD), the dimension of the matrices $\mathbf{K}^{kl}$ can never become larger than the number of AOs (basis functions). However, in the PNO case (PNO-LCCSD) the total number of PNOs can be much larger than the number of AOs. Since the PNOs are different for each pair, one has to compute a huge number of blocks $\mathbf{K}_{[ik,jl]}^{kl}$, where $[ik, jl]$ denotes the domains of the indices c, d, respectively. However, the PNO domains are typically an order of magnitude smaller than in the PAO case, and therefore the total number of integrals in the PNO basis is usually comparable to that in the PAO basis. The most essential advantage of using PNOs is the small size of the matrix dimensions. Since the matrix multiplications scale cubically with the domain sizes, the evaluation of the above terms is in the PNO basis 2–3 orders of magnitude faster than in the PAO basis.

Similar considerations hold for the overlap matrices. Indicating again the domains of the indices of all matrices in square brackets, our example can be written as

$$\mathbf{R}_{[ij,ij]}^{ij} = \ldots \sum_{kl} \mathbf{S}_{[ij,ik]} \mathbf{T}_{[ik,ik]}^{ik} \mathbf{K}_{[ik,lj]}^{kl} \mathbf{T}_{[lj,lj]}^{lj} \mathbf{S}_{[lj,ij]}. \tag{1.96}$$

This means that one has to compute and store large lists of overlap and integral blocks. The blocks of the overlap matrices are addressed using two pair labels (e.g., ij, ik), the ones for the integrals by three pair labels *kl*, *ik*, *lj*. Efficient index packing and binary search algorithms are used to retrieve individual blocks. Similar considerations are possible for each individual term in the LCCSD equations.

Despite the enormous savings that are possible by using PNOs, for large molecules and good (at least triple-ζ) basis sets, the total amount of integrals and the computational cost can become exceedingly large if all non-distant pairs are fully included in the LCCSD residual equations. Fortunately, this problem can be alleviated by four further approximations, which take into account the fast decay of various long range contributions:

(i) Treatment of weak pairs at lower computational levels;
(ii) Neglect of terms in the LCCSD equations for weak pairs that cancel at long range;
(iii) Neglect of further integral contributions that decay exponentially with distance;
(iv) Projection approximations.

We refer to (i)–(iii) as "weak pair approximations", and we will briefly discuss these in the following sections.

1.9.1 Weak Pair Approximations

Apart from the distant pair approximations that have been discussed in Section 1.8, a distinction of "strong" and "weak" pairs has been introduced for LCCSD [7, 18, 31]. In these early models, "weak" pair amplitudes and corresponding pair energies were optimized at the LMP2 level and only "strong" pair amplitudes and pair energies were evaluated at the full coupled cluster level, that is, equation (1.92) is replaced by

$$\begin{aligned} E_{\text{corr}}^{\text{LCCSD}} &= E_{\text{sing}} + \sum_{i\geq j\in P_s} E_{ij}^{\text{CC}} + \sum_{i>j\in P_w} E_{ij}^{\text{LMP2}} + E_{\text{dist}} \\ &= E_{\text{sing}} + E_{\text{strong}} + E_{\text{weak}} + E_{\text{dist}} \end{aligned} \tag{1.97}$$

where P_s and P_w denote the lists of strong and weak pairs, respectively. The selection of the pairs in each class can be based either on their LMP2 energy contributions or on distance criteria. The reasoning behind this approximation is that the pair energies decay with R_{ij}^{-6}, and it is assumed that the LMP2 approximation for small weak pair energies does not much affect overall accuracy. Similar approximations were also adopted by Neese *et al.* in their DLPNO-CCSD method [54, 57], but they made the even cruder approximation that weak pair energies are computed using the non-iterative semi-canonical LMP2 approximation.

More recent work has shown, however, that this simple weak pair approximation can lead to unexpectedly large errors [31, 32, 145, 204]. Usually, the total correlation energy is overestimated. This is due to the neglect of the weak pair amplitudes in the residuals for strong pairs, which leads to an overestimation of the strong pair energy. Furthermore, MP2 is well known to overestimate long-range dispersion energies, and therefore the weak pair contribution is often also too large. The first effect can be reduced significantly by including the LMP2 weak pair amplitudes in the LCCSD residuals for strong pairs [26, 31]. But even this approximation may not be sufficient, in particular for systems with many aromatic rings and strong long-range dispersion interactions. An extreme case is the Auamin reaction, as will be demonstrated further below. Since

Table 1.8 Summary of pair classes and pair approximations.

Pair class	Energy criterion	Approximation of residual	Amplitudes included	
			Linear terms	Non-linear terms
Strong	$E_{ij}^{\mathrm{LMP2}} \geq T_{\mathrm{close}}$	Full LCCSD	Up to weak	Up to weak
Close	$T_{\mathrm{close}} > E_{ij}^{\mathrm{LMP2}} \geq T_{\mathrm{weak}}$	Approx. LCCSD[a]	Up to weak	Strong+close
Weak	$T_{\mathrm{weak}} > E_{ij}^{\mathrm{LMP2}} \geq T_{\mathrm{dist}}$	Approx. CEPA-2[a]	Up to weak	–
Distant	$T_{\mathrm{dist}} > E_{ij}^{\mathrm{SC-LMP2d}}$	LMP2(multipole)[b]	–	–

Basis set: VDZ-F12, $T_{\mathrm{LMO}} = 0.2$, $T_{\mathrm{PNO}}^{\mathrm{occ}} = 10^{-8}$.
[a] Approximations described in Section 1.9.2.
[b] LMP2 with non-iterative multipole approximation.

this system is not untypical for "real-life chemistry," it is important to apply higher-order approximations for weak pairs.

In order to keep the computational effort tractable we have therefore split the weak pair class further into "close" and "weak" pairs, leading to the final hierarchy of strong, close, weak, and distant pairs. The classification of the pairs is based on their LMP2 pair energies. A summary of the thresholds and approximations for each class is given in Table 1.8. The default values of the thresholds are (in E_{h}): $T_{\mathrm{dist}} = 10^{-6}$, $T_{\mathrm{weak}} = 10^{-5}$, $T_{\mathrm{close}} = 10^{-4}$. In addition, the non-linear terms in the LCCSD equations can be neglected for weak pairs.

The residuals for close and weak pairs employ approximations in which terms that cancel at long-range are neglected. These approximations will be outlined in the next section.

1.9.2 Long-Range Cancellations of Terms in the LCCSD Equations

In order to devise accurate weak pair approximations, it is important to consider the spatial decay properties of individual terms in the coupled-cluster equations [32, 204]. The contribution of individual terms in the residual to a pair energy can be obtained by using

$$\mathbf{R}^{ij} = \mathbf{K}^{ij} + \mathbf{O}^{ij}, \tag{1.98}$$

where $\mathbf{O}^{ij}$ is the part of the residual which depends on the amplitudes. For the converged LCCSD wave function the residuals vanish, and $\mathbf{K}^{ij} = -\mathbf{O}^{ij}$. Thus, we can rewrite the pair energies in equation (1.93) as

$$E_{ij} = -(2 - \delta_{ij}) \sum_{ab \in [ij]} O_{ab}^{ij} \tilde{T}_{ab}^{ij}. \tag{1.99}$$

The amplitudes $\tilde{T}_{ab}^{ij}$ decay asymptotically with R^{-3}, where R is the distance between the charge centers of orbitals i and j as defined in Section 1.8. There are also faster decaying contributions which contribute at shorter distances, but these contributions will be neglected in the following discussion.

Consider, for example, the contribution of 4-external integrals

$$R^{ij}_{ab} += \sum_{cd} (ac|bd) T^{ij}_{cd}. \tag{1.100}$$

As we have seen in Section 1.8, for large R only amplitudes where c is close to i and d is close to j will be significant. Due to the exponential decay of the integrals $(ac|bd)$ with the distances R_{ac} and R_{bd}, it follows that a must also be close to i and b close to j. The largest contributions will arise from $(aa|bb)T^{ij}_{ab}$. Since the charge distributions ρ_{aa} and ρ_{bb} carry charges of 1 (assuming normalized orbitals), the integral $(aa|bb)$ decays as R^{-1}, the residual contribution as R^{-4}, and the energy contribution with R^{-7}. It turns out, however, that these rather slowly decaying terms in the residual equations cancel at long range with other terms. In the following, this will be exemplified for the coupled-cluster doubles (CCD) residual [32].

The linear CCD terms containing slowly decaying integrals are

$$\begin{aligned} R^{ij}_{ab} &+= \sum_{cd} (ac|bd) T^{ij}_{cd} + \sum_{kl} (ik|lj) T^{kl}_{ab} \\ &- \sum_{kc} \left[T^{kj}_{ac}(cb|ik) + (ac|kj) T^{ik}_{cb} \right]. \end{aligned} \tag{1.101}$$

Equation (1.101) can be split into differently decaying parts

$$R^{ij}_{ab} += \mathcal{O}(R^{-4}) + \mathcal{O}(R^{-5}) + \mathcal{O}(R^{-6}), \tag{1.102}$$

where

$$\mathcal{O}(R^{-4}) = T^{ij}_{ab} \left[(aa|bb) + (ii|jj) - (aa|jj) - (ii|bb) \right], \tag{1.103}$$

$$\begin{aligned} \mathcal{O}(R^{-5}) = &\sum_{c \neq b} T^{ij}_{ac}[(aa|cb) - (ii|cb)] + \sum_{c \neq a} [(ac|bb) - (ac|jj)] T^{ij}_{cb} \\ &+ \sum_{l \neq j} T^{il}_{ab}[(ii|lj) - (aa|lj)] + \sum_{k \neq i} [(ik|jj) - (ik|bb)] T^{kj}_{ab}. \end{aligned} \tag{1.104}$$

At long-range, the positive and negative contributions in the $\mathcal{O}(R^{-4})$ and $\mathcal{O}(R^{-5})$ parts will cancel, leading to an overall R^{-6} decay for a given pair residual R^{ij}_{ab}, and therefore an overall R^{-9} decay of the contribution to the pair energy E_{ij}. Very similar considerations can be made for the additional terms in the CCSD residual equations. It can be shown that at long range all terms which contain integrals that decay with R^{-1} or R^{-2} cancel in any coupled cluster theory that respects the particle-hole symmetry.

The canceling contributions can be neglected in the close and weak pair residuals, leading to our approximate coupled-cluster treatment for those pairs [32]. The greatest advantage of this approximation is that the 4-external integrals $(ab|cd)$ are not needed for close and weak pairs. Furthermore, the summation $\sum_{kl}(ik|lj)T^{kl}_{ab}$ can be restricted to contributions where ik and jl are strong pairs. This leads to a balanced treatment of terms that are related by the particle-hole symmetry, and, in fact, we found that it improves the accuracy. More generally, approximations in further terms (e.g., those involving single excitations) should always treat contributions that are related by particle-hole symmetries on equal footing.

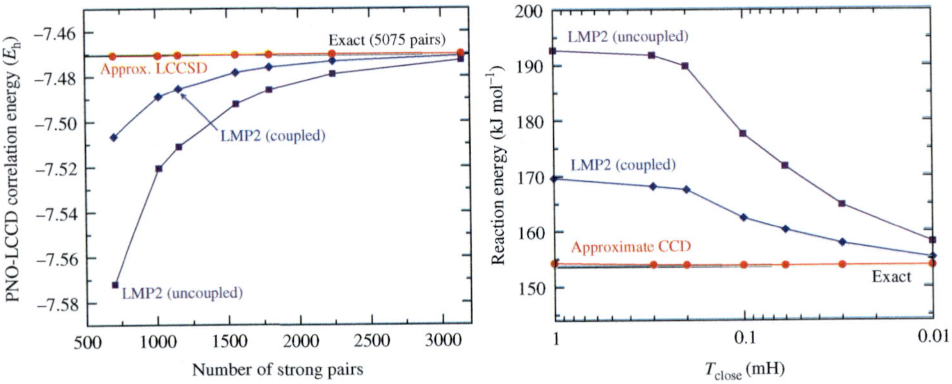

Figure 1.6 Effect of close pair approximations on the correlation and dissociation energies of Auamin. In both cases, distant pairs were excluded from the LCCD ($T_{dist} = 10^{-6}$), and no weak pair approximations were used ($T_{weak} = T_{dist}$). *Left panel*: Correlation energy as a function of the number of strong pairs. The remaining pairs are treated as close pairs. The data points correspond to the thresholds in the other right panel. *Right panel*: Dissociation energy (in kJ mol^{-1}) as a function of T_{close}. Basis set: VDZ-F12, `IEXT=1`. *Source*: Schwilk *et al.* 2015 [32]. Reproduced with permission of AIP Publishing.

In addition, the non-linear terms can be neglected for weak pairs. This corresponds to a linearized coupled-cluster or CEPA-0 method. We use CEPA-2, which differs from CEPA-0 by terms $-E_{ij} \cdot T^{ij}_{ab}$ in the residual R^{ij}_{ab}. Since the weak pair energies are small, the differences of CEPA-2 and CEPA-0 are tiny. Table 1.8 summarizes these approximations.

Figure 1.6 demonstrates for PNO-LCCD how these approximations converge with the thresholds that determine the pair classes. The left panel shows the convergence of the correlation energy of Auamin as a function of the number of strong pairs, and the right panel shows the corresponding convergence of the reaction (dissociation) energy. The data points in the left panel are located at the same thresholds as shown in the right panel, and so this also shows how the number of strong pairs depends on the threshold T_{close}. All remaining pairs (except distant pairs) have been treated as close pairs. For the cases denoted LMP2 (uncoupled) and LMP2 (coupled), the close pair amplitudes and pair energies were optimized in the preceding PNO-LMP2 calculation. In the "uncoupled" case, these are entirely neglected in the residuals for strong pairs, while in the "coupled case," they are included in the residuals (but not reoptimized). These figures demonstrate that in this case, large errors are caused by the use of LMP2 amplitudes and pair energies, and these errors decay very slowly with the threshold T_{close} (similar results are obtained if a distance criterion is used). For the largest threshold ($T_{close} = 1\,\mathrm{m}E_h$), the reaction energy is overestimated by nearly 40 kJ mol^{-1}. This is about half of the difference between the pure PNO-LMP2 and PNO-LCCD (without close or weak pair approximations) reaction energies (LMP2 overestimates the reaction energy by about 65 kJ mol^{-1} relative to LCCD). The error is more than halved if the LMP2 amplitudes are included in LCCD residuals for strong pairs. Note that this improvement is solely due to the strong pair energy, and this demonstrates that long-range couplings are very important. Even for the smallest threshold of $T_{close} = 0.01\,\mathrm{m}E_h$, the error of LMP2 (uncoupled) still amounts to 5 kJ mol^{-1}. These errors are dramatically

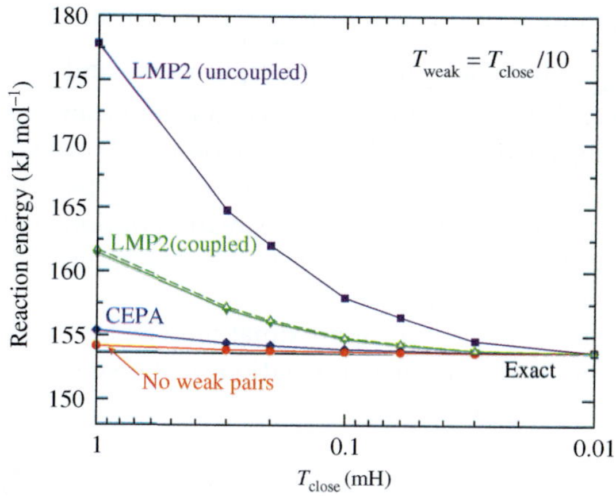

Figure 1.7 Effect of weak pair approximations on the dissociation energy of Auamin (in kJ mol^{-1}) as a function of T_{close}, $T_{weak} = T_{close}/10$. LMP2 (coupled), LMP2 (uncoupled), and approximate LCCD determine how the close pair amplitudes are optimized (see text). Basis set: VDZ-F12, `IEXT=1`. *Source*: Schwilk *et al.* 2015 [32]. Reproduced with permission of AIP Publishing.

reduced if the approximate coupled-cluster method is used for the close pairs. In this case, the error for $T_{close} = 1\,\mathrm{m}E_\mathrm{h}$ is reduced to about 2 kJ mol^{-1} and becomes entirely negligible for $T_{close} \leq 0.3\,\mathrm{m}E_\mathrm{h}$.

Figure 1.7 demonstrates the effect of the additional weak pair approximation. For simplicity, we have chosen fixed ratios T_{close}/T_{weak} equal to 10, and varied T_{close}. In this case, the close pairs are treated by approximate LCCD, and the weak pairs either by LMP2 (uncoupled), LMP2 (coupled), or CEPA. In the CEPA case, the weak pair amplitudes are included in the strong and close pair residuals, corresponding to LMP2 (coupled). The red curve corresponds to the approximate LCCD curve in the right panel of figure 1.6. The finding is that treating weak pairs by CEPA strongly reduces the error relative to LMP2 (coupled), and for $T_{close} \leq 0.3\,\mathrm{m}E_\mathrm{h}$ the error is below 1 kJ mol^{-1}.

Similar results have also been obtained for reactions **I** and **II**. The effects are qualitatively similar, but smaller in magnitude. Details can be found in the supplementary material of Ref. [32]. That the approximate coupled cluster treatment of close and weak pairs also works well if singles are included will be demonstrated in Section 1.11.2 for LCCSD-F12.

1.9.3 Projection Approximations

A further problem in PNO-LCCSD is the large number of integral blocks that have to be computed and stored, see for example equation (1.96) discussed above. Neese *et al.* [48, 49, 54] proposed projection approximations, in which an orbital in a particular domain, for example, $|a^{ik}\rangle$, is projected onto another domain, for example, $[kl]_{\mathrm{PNO}}$. This leads to the expansion

$$|a^{ik}\rangle \approx |\tilde{a}^{ik}\rangle = \sum_{c\in[kl]_{\mathrm{PNO}}} |c^{kl}\rangle\langle c^{kl}|a^{ik}\rangle = \sum_{c\in[kl]_{\mathrm{PNO}}} |c^{kl}\rangle S^{kl,ik}_{ca}. \tag{1.105}$$

It is easily shown that this also minimizes the square deviation $\langle a^{ik} - \tilde{a}^{ik}|a^{ik} - \tilde{a}^{ik}\rangle$. The term in equation (1.96) then becomes

$$R^{ij}_{[ij,ij]} = \dots \sum_{kl} \mathbf{S}_{[ij,ik]}\mathbf{T}^{ik}_{[ik,ik]}\mathbf{S}_{[ik,kl]}\mathbf{K}^{kl}_{[kl,kl]}\mathbf{S}_{[kl,lj]}\mathbf{T}^{lj}_{[lj,lj]}\mathbf{S}_{[lj,ij]}. \tag{1.106}$$

The number of integrals $\mathbf{K}^{kl}_{[kl,kl]}$ is drastically smaller than the number of all integrals $\mathbf{K}^{kl}_{[ik,lj]}$, but more matrix multiplications and additional S-matrix blocks are needed. Overall, the reduction of data is very significant, and the CPU time is comparable (due to savings in the integral transformation). The projection approximation might look rather crude since the domain $[kl]$ at best partly spans the domains $[ik]$ and $[jl]$. However, we have implemented this approximation optionally for all terms that are quadratic in the amplitudes, and found that the errors are surprisingly small. Detailed tests will be presented elsewhere [205].

Similar projection approximations can be applied to 3-external integrals [54, 57]. For example, the linear contribution of the singles amplitudes to the doubles residual is approximated as

$$R^{ij}_{ab} += \sum_{c\in[j]} (a^{ij}i|b^{ij}c^{j})t^{j}_{c} \approx \sum_{c\in[ij]} (a^{ij}i|b^{ij}c^{ij}) \left[\sum_{d\in[j]} S^{ij,j}_{cd} t^{j}_{d} \right] . \tag{1.107}$$

This should be a good approximation, since in the PAO basis the singles domain $[i]$ is contained in the pair domain $[ij]$ (which is the union of $[i]$ and $[j]$). It is not guaranteed that $[i]_{\text{PNO}}$ is contained in $[ij]_{\text{PNO}}$, but the error will become smaller with decreasing PNO threshold and zero if $T^{\text{occ}}_{\text{PNO}} = 0$. A similar approximation is also possible for the linear 3-external doubles contribution in the singles residual. In fact, we found that the error caused by these approximations is negligible. The big advantage is that only 3-index integrals need to be computed and stored in which all three virtual orbitals belong to the same PNO domain, which leads to significant savings in the memory usage and the CPU time.

In the contributions $\mathbf{T}^{ik}\mathbf{K}(\mathbf{E}^{kj})$ to the doubles residuals (see Appendix A), the 3-external integral contractions

$$\sum_{c\in[j]} (a^{ik}k|b^{ij}c^{j})t^{j}_{c} \approx \sum_{c\in[ij]} (a^{ik}k|b^{ij}c^{ij}) \left[\sum_{d\in[j]} S^{ij,j}_{cd} t^{j}_{d} \right] \tag{1.108}$$

occur. The 3-external integrals $(a^{ik}k|b^{ij}c^{ij})$ still contain two different domains (apart from the third index k). Neese *et al.* have also projected $|a^{ik}\rangle$ to domain $[ij]$. However, this changes the decay behavior of the integrals, since $|a^{ik}\rangle$ contains contributions close to k, but this may not be the case for $|a^{ij}\rangle$, and therefore this approximation might introduce larger errors.

In our current program, the latter approximation and the evaluation and storage of the large sets of 3-external integrals in equation (1.108) is avoided by computing the term as $K^{k\tilde{j}}_{a^{ik}b^{ij}} = (a^{ik}k|b^{ij}\tilde{j})$ where

$$|\tilde{j}\rangle = \sum_{c\in[j]_{\text{PNO}}} |c^{j}\rangle t^{j}_{c} = \sum_{\mu}^{\text{AO}} |\mu\rangle \tilde{t}^{j}_{\mu}, \tag{1.109}$$

$$\tilde{t}^{i}_{\mu} = \sum_{r\in[j]_{\text{PAO}}} P_{\mu r} \sum_{c\in[j]_{\text{PNO}}} \tilde{V}^{j}_{rc} t^{j}_{c}. \tag{1.110}$$

The integrals can then be computed by a partial integral transformation, just as the integrals K^{kj}_{ab}. However, the transformation must be done in each CCSD iteration. Similarly, the matrix $\tilde{\mathbf{G}}$, which contains slowly decaying terms (see Appendix A), is computed in each iteration in the AO basis and then transformed to the LMO/PNO basis. This proceeds in close analogy to the computation of a Fock matrix, and local density-fitting approximations can be used to achieve linear scaling of the exchange contribution [206]. Overall, this introduces contributions that scale cubically with the molecular size, but with a very small prefactor.

1.10 Triple Excitations

Local methods to compute the perturbative (T) correction for LCCSD have been described for PAO [16, 17, 19] OSV [167], and PNO [55] methods. Using PAO or OSV virtual orbitals, it is straightforward to define triples domains $[ijk]$, which are the union of the three orbital domains. It has been found that the list of triples ijk can be restricted so that at least one of the pairs ij, ik, or jk is a strong pair, and the others are strong or close pairs. Furthermore, only strong and close pair amplitudes need to be included in the triples calculation [16, 17]. This leads to linear scaling, and often the calculation of the triples correction takes much less time than the preceding LCCSD calculation. Since in a local orbital basis the Fock matrix is not diagonal, an iterative solution of the triples amplitude equations is in principle required, just as for LMP2. However, it was found that, in most cases, a semi-canonical non-iterative energy correction is accurate enough [16, 17]. Full iterative solutions are also available [17, 19], but much more expensive. The treatment of triples in PNO methods is more involved, since the union of the three pair domains is non-orthogonal, large, and not well suitable. Riplinger *et al.* have proposed to use so-called "triples-natural orbitals" [55]. A detailed description of these methods is beyond the scope of the current article, and the reader is referred to the original publications.

1.11 Local Explicitly Correlated Methods

Electron correlation methods generally suffer from the problem of slow basis set convergence, and very large basis sets are often required to obtain converged results. The slow convergence is due to the fact that the expansion in products of limited 1-electron functions (orbitals) do not describe well the shape of the wave function at small to intermediate values of interelectron distance r_{12}, which should satisfy the electronic wave function cusp condition [207, 208]:

$$\left.\frac{\partial \Psi}{\partial r_{12}}\right|_{r_{12}=0} = \frac{1}{2}\Psi(r_{12}=0). \tag{1.111}$$

This problem can be avoided by including terms in the wave function that depend explicitly on r_{12} and thus can describe the cusp properly [209]. Early implementations used a linear R_{12} correlation factor, and a number of so-called R12-methods were developed [97–108]. It was found later that a Slater-type function $F_{12} \sim \exp(-\gamma r_{12})$ yields much better basis set convergence and numerical stability [109]. This family of methods is called the F12 methods [109–153].

In this section, we will summarize the essential theory in local MP2-F12 and CCSD-F12 methods [142–148] as implemented in MOLPRO. We will focus on the approximations most suitable for efficient, parallel implementations for PNO based local correlation methods, namely the 3*A and the CCSD-F12x (x = a, b) approximations.

1.11.1 PNO-LMP2-F12

The first-order MP2-F12 wave function can be written as

$$\Psi^{(1)} = \frac{1}{2}\sum_{i,j}^{\text{val}}\left(\sum_{a,b}^{\text{virt}}|\Phi_{ij}^{ab}\rangle T_{ab}^{ij} + \sum_{\alpha,\beta}^{\text{complete}}|\Phi_{ij}^{\alpha\beta}\rangle T_{\alpha\beta}^{ij}\right) \tag{1.112}$$

with

$$T_{\alpha\beta}^{ij} = \sum_{kl}^{\text{val}}\langle\Phi_{ij}^{\alpha\beta}|\hat{Q}_{12}F_{12}|kl\rangle\, T_{kl}^{ij}, \tag{1.113}$$

and α, β denote a formally complete orthonormal orbital basis.

$$F_{12} = -\frac{1}{\gamma}e^{-\gamma r_{12}} \tag{1.114}$$

is the correlation factor, which depends on a length-scale parameter γ. For valence-shell correlation $\gamma = 1\, a_0^{-1}$ is usually a good choice. $\hat{Q}_{12}$ is a projector that keeps the explicitly correlated terms orthogonal to the reference Hartree–Fock functions as well as to the conventional configurations $|\Phi_{ij}^{ab}\rangle$. The projector can be chosen from several different forms [118]:

$$\textit{Ansatz}\ \mathbf{1}:\quad \hat{Q}_{12} = (1-\hat{r}_1)(1-\hat{r}_2), \tag{1.115}$$

$$\textit{Ansatz}\ \mathbf{2}:\quad \hat{Q}_{12} = (1-\hat{o}_1)(1-\hat{o}_2), \tag{1.116}$$

$$\textit{Ansatz}\ \mathbf{3}:\quad \hat{Q}_{12} = (1-\hat{o}_1)(1-\hat{o}_2) - \hat{v}_1\hat{v}_2, \tag{1.117}$$

where $\hat{o} = \sum_m |m\rangle\langle m|$, $\hat{v} = \sum_a |a\rangle\langle a|$, and $\hat{r} = \hat{o} + \hat{v}$ are 1-electron projection operators onto the occupied, virtual, and full orbital spaces, respectively, and the subscripts refer to the electron coordinates on which they act. *Ansatz* **1** is unsatisfactory since it neglects the configurations in which one electron is excited to a virtual orbital and the other to an orbital outside of the MO basis. *Ansätze* **2** and **3** are equivalent before further approximations are introduced. *Ansatz* **3** leads to simpler working equations and is used in our work as well as in most other implementations.

The Hylleraas functional for MP2-F12 can be obtained by inserting equation (1.112) into equation (1.3):

$$E_2 = \sum_{i,j}^{\text{val}}\sum_{a,b}^{\text{virt}}\left(K_{ab}^{ij} + R_{ab}^{ij}\right)\tilde{T}_{ab}^{ij} + \sum_{i,j}^{\text{val}}\sum_{k,l}^{\text{val}}\left(V_{kl}^{ij} + R_{kl}^{ij}\right)\tilde{T}_{kl}^{ij}, \tag{1.118}$$

where

$$V_{kl}^{ij} = \left\langle kl|F_{12}\hat{Q}_{12}r_{12}^{-1}|ij\right\rangle, \tag{1.119}$$

$$R_{kl}^{ij} = V_{kl}^{ij} + \left\langle\tilde{\Phi}_{ij}^{kl}|\hat{H}^{(0)} - E^{(0)}|\Psi^{(1)}\right\rangle. \tag{1.120}$$

The contravariant configurations $|\tilde{\Phi}_{ij}^{kl}\rangle$ and amplitudes $\tilde{T}_{kl}^{ij}$ are defined in analogy of those described in Section 1.3. For optimized amplitudes the residuals R_{ab}^{ij} and R_{kl}^{ij} must vanish. The explicit forms of the residuals are [118]

$$R_{ab}^{ij} = K_{ab}^{ij} + f_{ac}T_{cb}^{ij} + T_{ac}^{ij}f_{cb} - f_{io}T_{ab}^{oj} - T_{ab}^{io}f_{oj} + T_{mn}^{ij}C_{ab}^{mn} \tag{1.121}$$

and

$$R_{kl}^{ij} = V_{kl}^{ij} + B_{kl,mn}T_{mn}^{ij} - S_{kl,mn}(f_{io}T_{mn}^{oj} + T_{mn}^{io}f_{oj}) + C_{ab}^{kl}T_{ab}^{ij}, \tag{1.122}$$

where i, j, k, l, m, n, o denote valence orbitals and the summations over repeated dummy indices are implied. The matrix elements are defined as

$$B_{kl,mn} = \langle kl|F_{12}\hat{Q}_{12}(\hat{f}_1 + \hat{f}_2)\hat{Q}_{12}F_{12}|mn\rangle, \tag{1.123}$$

$$S_{kl,mn} = \langle kl|F_{12}\hat{Q}_{12}F_{12}|mn\rangle, \tag{1.124}$$

$$C_{ab}^{kl} = \langle kl|F_{12}\hat{Q}_{12}(\hat{f}_1 + \hat{f}_2)|ab\rangle. \tag{1.125}$$

Up to this point, no approximations are introduced within the second-order Møller-Plesset treatment. However, expanding the projector in equations (1.123) through (1.125) leads to 3- and 4-electron integrals such as

$$\left\langle ij|r_{12}^{-1}\hat{o}_1F_{12}|kl\right\rangle = \sum_{m}^{\text{occ}} \left\langle ijm|r_{12}^{-1}F_{32}|mkl\right\rangle. \tag{1.126}$$

The many-electron integrals can be avoided by applying resolutions of the identity (RI) on electron 2 in the operator $\hat{o}_1$:

$$\hat{o}_1 \rightarrow \hat{o}_1\hat{u}_2, \tag{1.127}$$

where $\hat{u} = |\alpha\rangle\langle\alpha|$. Equation (1.126) can then be factorized as

$$\left\langle ij|r_{12}^{-1}\hat{o}_1F_{12}|kl\right\rangle = \sum_{m}^{\text{occ}}\sum_{\alpha}^{\text{complete}} \left\langle ij|r_{12}^{-1}|m\alpha\right\rangle\langle m\alpha|\hat{F}_{12}|kl\rangle. \tag{1.128}$$

Note that it is common to use the so-called "physicist's notation" for the 2-electron integrals in F12 theory, that is, $\langle ij|r_{12}^{-1}|m\alpha\rangle = (im|j\alpha)$. The RI is exact when an infinite basis $|\alpha\rangle$ is used. In practice, the RI is approximated by an auxiliary basis set. With this approximation, the projector (*Ansatz* **3**) can be written as

$$\hat{Q}_{12} = 1 - \sum_{a,b}^{\text{virt}} |ab\rangle\langle ab| + \sum_{m,n}^{\text{occ}} |mn\rangle\langle mn| - \sum_{m}^{\text{occ}}\sum_{\alpha}^{\text{RI}} (|m\alpha\rangle\langle m\alpha| + |\alpha m\rangle\langle\alpha m|). \tag{1.129}$$

In most canonical F12 methods, a complementary auxiliary basis set (CABS) approach [112] is used, in which the operator $\hat{u}$ in equation (1.127) is written as a sum of the contributions from the orbital basis and the auxiliary basis. The CABS approach leads to simpler equations in canonical F12 methods but is less well suited in local methods, since due to the necessary orthogonalization steps, it is more difficult to apply local RI approximations. However, when using the union of the AO basis and the auxiliary basis as the RI basis, equation (1.129) is equivalent to the projector in the CABS formulation. For example, this can be achieved by using the union of the correlation consistent orbital basis sets and the corresponding `OPTRI` auxiliary bases [154].

In the PNO-LMP2-F12 method, we introduce domain approximations on the projector:

$$\hat{Q}_{12}^{ij} = 1 - \sum_{a,b\in[ij]_{\mathrm{PNO}}} |ab\rangle\langle ab| + \sum_{m,n\in[ij]_{\mathrm{LMO}}} |mn\rangle\langle mn| - \sum_{m\in[ij]_{\mathrm{LMO}},\bar{\alpha}\in[ij]_{\mathrm{RI}}} (|m\bar{\alpha}\rangle\langle m\bar{\alpha}| + |\bar{\alpha}m\rangle\langle\bar{\alpha}m|), \tag{1.130}$$

where $[ij]_{\mathrm{PNO}}$, $[ij]_{\mathrm{LMO}}$, and $[ij]_{\mathrm{RI}}$ denote domains of orthonormal PNOs, LMOs, and RI functions for an LMO pair ij, respectively. The RI orbitals $\bar{\alpha}$ are orthogonalized for each domain $[ij]_{\mathrm{RI}}$ separately, that is,

$$|\bar{\alpha}\rangle = \sum_{\beta\in[ij]_{\mathrm{RI}}} |\beta\rangle \left[\mathbf{S}_{\mathrm{RI}}^{-1/2}\right]_{\beta\alpha}. \tag{1.131}$$

The projector is pair-specific. Limiting $a, b \in [ij]_{\mathrm{PNO}}$ in equation (1.130) ensures that only doubly excited configurations included in the PNO-LMP2 first-order wave function are projected out from the explicitly correlated part. The contributions of the remaining parts of the virtual orbital space (outside the domains) are implicitly approximated by the F12 terms, and this leads to a reduction of the domain errors by typically one order of magnitude [142, 143, 145–148]. Thus, as will be demonstrated later, smaller domains can be used in explicitly correlated methods.

The domain approximations in the summations over the LMOs and the RI basis are applied to reduce the scaling of the method. They do cause slight errors in projecting out the contributions of the occupied-occupied block. However, the integrals such as $\langle ij|\hat{F}_{12}|m\alpha\rangle$, which occur in equations (1.134) and (1.135) below, decay exponentially with the distance between the LMOs i and m as well as with the distances between the LMO j and the RI basis functions α (and similarly with i and j exchanged). As a result, LMOs m and RI functions α will contribute significantly only if they are spatially close to either i or j, and the domain approximations lead to only small errors [145, 148].

As proposed by Ten-no [109, 110], the F12 amplitudes T_{kl}^{ij} can be determined from the first-order wave function cusp conditions and then take the fixed values

$$T_{kl}^{ij} = \frac{3}{8}\delta_{ik}\delta_{jl} + \frac{1}{8}\delta_{il}\delta_{jk}. \tag{1.132}$$

The fixed amplitudes have the advantage that no amplitude equations have to be solved for the F12 correction. In combination with $\gamma = 1\,a_0^{-1}$, this approximation leads to negligible errors in treating valence–valence correlations, although in treating core and core–valence correlations in heavy atoms, different geminal exponents are required [138].

Straightforward insertion of the projector $\hat{Q}^{ij}$ into the expressions for $B_{kl,mn}$ (equation 1.123) and C_{ab}^{kl} (equation 1.125) leads to RI approximations for products of the Fock operator and the F12 correlation factor. These may be slowly convergent, in particular for the contribution of the kinetic energy operators $\hat{t}_1$ and $\hat{t}_2$ in the Fock operators $\hat{f}_1$ and $\hat{f}_2$. To reduce this problem, various commutator approximations have been proposed [107, 114], which involve matrix elements over commutators such as $[\hat{t}_{12}, \hat{F}_{12}]$, where $\hat{t}_{12} = \hat{t}_1 + \hat{t}_2$. Historically, these approximations are denoted A, B, C. Approximations B and C are formally most accurate and equivalent for complete RI basis sets. Approximation A is simplified and does not lead to exactly the same result. Optionally, further simplifications can be achieved by assuming that the extended Brillouin condition (EBC) $f_{ax} = 0$ is valid, were x denotes orbitals in the complementary auxiliary space (i.e., outside the orbital basis). This approximation is indicated by a star (*). For a more complete discussion, see for example Refs. [107, 118].

We use the 3*A approximation [118] in our LMP2-F12 and LCCSD-F12 methods. This is the simplest of all approximations, but extensive benchmarks have shown that it yields excellent results. The EBC approximation leads to $C^{kl}_{ab} = 0$, and this decouples the F12 contribution from the standard LMP2 amplitude equations. Furthermore, the exchange commutator terms $[\hat{k}_1 + \hat{k}_2, F_{12}]$ are neglected. The most important advantage of approximation A* is that the entire theory depends only linearly on the size of the RI basis. Furthermore, no RI approximations involving Fock matrices occur, and therefore local RI approximations can be introduced most easily.

Combined with the fixed amplitudes, the computation of F12 corrections are non-iterative and completely decoupled from the PNO-LMP2 calculations. The F12 residual equation (1.122) becomes

$$R^{ij}_{kl} = V^{ij}_{kl} + B_{kl,mn} T^{ij}_{mn}. \tag{1.133}$$

When using fixed amplitudes, these F12 residuals do not vanish and should be included explicitly in the energy expression, that is, a Hylleraas functional should be used (since this depends only quadratically on the differences between the fixed and optimized amplitudes). Due to the δ-factors in equation (1.132), the summations over kl and mn in equations (1.118) and (1.133) are removed, and the only quantities required for the F12 energy corrections are (for details see Refs. [118, 144, 148]):

$$\begin{aligned} V^{ij}_{ij} = \langle ij | r^{-1}_{12} F_{12} | ij \rangle + \sum_{m,n \in [ij]_{\text{LMO}}} K^{ij}_{mn} F^{ij}_{mn} - \sum_{a,b \in [ij]_{\text{PNO}}} K^{ij}_{ab} F^{ij}_{ab} \\ - \sum_{m \in [ij]_{\text{LMO}}} \sum_{\alpha \in [ij]_{\text{RI}}} (K^{ij}_{\alpha m} \bar{F}^{ij}_{\alpha m} + K^{ji}_{\alpha m} \bar{F}^{ji}_{\alpha m}), \end{aligned} \tag{1.134}$$

$$\begin{aligned} B_{ij,ij} = \langle ij | [\hat{F}_{12}, \hat{t}_{12}] \hat{F}_{12} | ij \rangle + \sum_{m,n \in [ij]_{\text{LMO}}} U^{ij}_{mn} F^{ij}_{mn} - \sum_{a,b \in [ij]_{\text{PNO}}} U^{ij}_{ab} F^{ij}_{ab} \\ - \sum_{m \in [ij]_{\text{LMO}}} \sum_{\alpha \in [ij]_{\text{RI}}} (U^{ij}_{\alpha m} \bar{F}^{ij}_{\alpha m} + U^{ji}_{\alpha m} \bar{F}^{ji}_{\alpha m}), \end{aligned} \tag{1.135}$$

and similar quantities V^{ij}_{ji} and $B_{ij,ji}$. In the above equations, we used the notations $K^{ij}_{mn} = \langle ij | r^{-1}_{12} | mn \rangle$, $F^{ij}_{mn} = \langle ij | \hat{F}_{12} | mn \rangle$, and $U^{ij}_{mn} = \langle ij | [\hat{F}_{12}, \hat{t}_{12}] | mn \rangle$ (and similarly with mn replaced by αm or ab). $\bar{F}^{ij}_{\alpha m}$ is introduced to account for the nonorthogonality of the RI basis functions with the overlap matrix $S_{\alpha\beta} = \langle \alpha | \beta \rangle$. It is obtained by solving the linear equations

$$F^{ijp}_{\alpha m} = \sum_{\beta \in [ij]_{\text{RI}}} S_{\alpha\beta} \bar{F}^{ijp}_{\beta m}. \tag{1.136}$$

Since all domain sizes are asymptotically independent of the molecular size, and the number of pairs *ij* scales linearly with the molecular size (with pair approximations applied, cf. Section 1.8), the number of integrals required for computing the F12 correction also scales linearly.

The approximation 3*A leads to a slight overestimation of the correlation energies, but the impact on energy differences such as reaction energies is minor [118, 133, 210, 211]. The 3*A approximation does not require double RIs, and efficient domain approximations can be easily applied without introducing significant errors. Recently, more rigorous treatments of MP2-F12 in local or fragmentation methods have been used by others [152, 153], but there is no evidence yet that these methods are more efficient or more accurate than the 3*A approximation we used.

Figures 1.8 and 1.9 demonstrate the convergence of the PNO-LMP2 and PNO-LMP2-F12 correlation energies of adrenaline as a function of the PAO extension parameter `IEXT` and the PNO domain thresholds, respectively. It can be seen that the F12 correction improves the convergence with the domain sizes dramatically. Most importantly, with F12 much smaller PAO domains are necessary to reach convergence, and `IEXT=1` is sufficient. Later, it will be demonstrated that this also holds for LCCSD-F12 and much larger molecules. The reduction of the PAO domain size is very important for the efficiency, since all integrals are first computed in the PAO basis and then transformed to the PNO basis. Similarly, the convergence with the PNO domain thresholds is also strongly improved, cf. Figure 1.9. The left panel shows the convergence as a function of $T_{\text{PNO}}^{\text{occ}}$ without the additional completeness criterion $T_{\text{PNO}}^{\text{en}}$, while in the right panel, $T_{\text{PNO}}^{\text{en}} = 0.997$ has been used. Accordingly, about 99.7% of the correlation energy is recovered even with the largest threshold $T_{\text{PNO}}^{\text{occ}} = 10^{-6}$. With the same thresholds, PNO-LMP2-F12 yields already 99.96% of the canonical MP2-F12 value. Note the

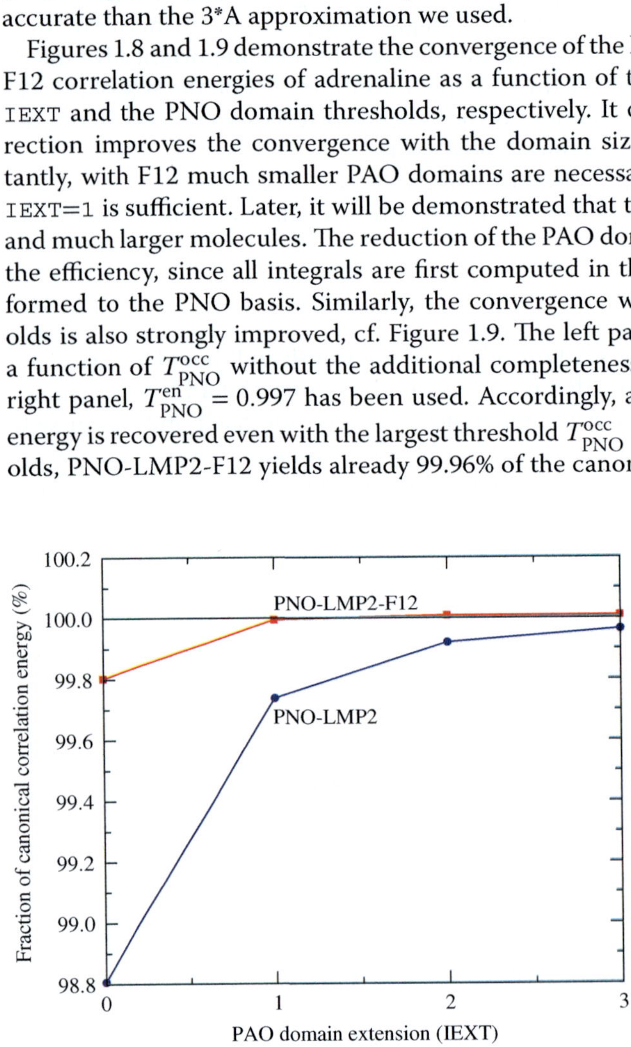

Figure 1.8 PNO-LMP2 and PNO-LMP2-F12 correlation energies of adrenaline relative to the canonical values (in percent) as a function of the PAO domain extenion parameter `IEXT` (see text). The PNO occupation number threshold $T_{\text{PNO}} = 10^{-10}$ and the completeness criterion $T_{\text{PNO}}^{\text{en}} = 0.997$ have been used. Basis set: VTZ-F12.

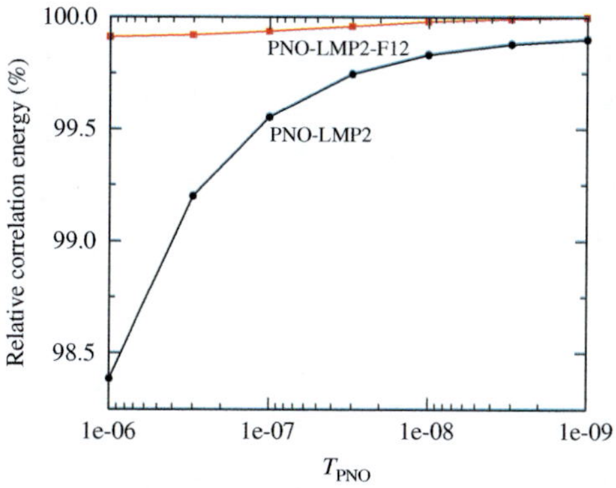

Figure 1.9 PNO-LMP2 and PNO-LMP2-F12 correlation energies of adrenaline relative to the canonical values (in percent) as a function of the PNO threshold T_{PNO}. *Left panel*: PNO domain selection using the natural occupation number threshold T_{PNO} only. *Right panel*: In addition, the completeness criterion $T^{en}_{PNO} = 0.997$ has been used (see text). `IEXT=2`, basis set: VTZ-F12.

different scales of the left and right panels of figure 1.9. It should be noted that the additional PNO domain error can also be corrected quite efficiently by adding the difference of the semi-canonical PAO-LMP2 (or OSV-LMP2) and PNO-LMP2 correlation energies to the LMP2 or LCCSD correlation energies, as done in the PNO methods of Neese *et al.* [48–57]. However, in contrast to the F12 treatment, this does not correct for errors caused by the incomplete PAO domains nor for the basis set incompleteness error.

Absolute PNO-LMP2 and PNO-LMP2-F12 correlation energies for various PAO domain sizes are presented in Table 1.9 for adrenaline and some other molecules. These calculations confirm that with PNO-LMP2-F12, `IEXT=1` is sufficient to reach $\geq 99.96\%$ of the canonical MP2-F12 correlation energy, while with the same domains, PNO-LMP2 only yields 99.0–99.5% of the canonical MP2 correlation energy. Note also the large differences between the absolute values of the PNO-LMP2 and PNO-LMP2-F12 correlation energies. This is due to the basis set incompleteness error, which is greatly reduced by the F12 correction. The PNO-LMP2-F12 correlation energies for the VDZ-F12 basis are slightly larger than those obtained with the cc-pVTZ-F12 (VTZ-F12) basis. This is due to the well-known overshooting of the 3*A approximation.

A comparison of the MP2 and MP2-F12 reaction energies of our three benchmark reactions with their local counterparts is presented in Table 1.10. While the MP2 and PNO-LMP2 differ significantly (in particular for reaction **III**), the MP2-F12 and PNO-LMP2-F12 results are in very close agreement. This table also demonstrates the effect of the additional completeness energy criterion T^{en}_{PNO} for the PNO domain selection. For reaction **III**, this increases the PNO-LMP2 reaction energy by 2.1 kJ mol^{-1}, which is mainly attributed to long-range effects. This effect is reduced to 0.5 kJ mol^{-1} by the F12 treatment. Note that for reaction **I**, the canonical MP2 overestimates the correlation effect relative to the MP2-F12 result (which is in close agreement with the extrapolated MP2 complete basis set (CBS) limit, cf. Ref. [147]). This is attributed to intramolecular basis set superposition errors (BSSE). These are reduced by LMP2 [14, 212], and indeed the PNO-LMP2/VTZ-F12 result is much closer to the CBS limit than the MP2 one.

Table 1.9 PNO-LMP2 and PNO-LMP2-F12 correlation energies using various PAO domain sizes (basis set: VDZ-F12, $T_{LMO} = 0.2$, $T_{PNO}^{occ} = 10^{-8}$). The fraction of correlation energy relative to the canonical values is given in parenthesis. `IEXT=Full` corresponds to the canonical MP2-F12 calculation.

	VDZ-F12				VTZ-F12			
IEXT	PAO	PNO	LMP2	LMP2-F12	PAO	PNO	LMP2	LMP2-F12
Naphtalene								
0	61	53.1	−1.434336 (97.799)	−1.754813 (99.787)	108	84.6	−1.607515 (98.820)	−1.751784 (99.793)
1	144	67.1	−1.459509 (99.516)	−1.758781 (100.013)	257	101.9	−1.622407 (99.736)	−1.755287 (99.993)
2	246	69.8	−1.464503 (99.856)	−1.759206 (100.037)	441	105.9	−1.624709 (99.877)	−1.755356 (99.997)
Full	338		−1.466615 (100.000)	−1.758552 (100.000)	674	0.0	−1.626705 (100.000)	−1.755415 (100.000)
Adrenaline								
0	49	44.1	−2.173019 (97.721)	−2.708837 (99.780)	88	70.4	−2.457964 (98.759)	−2.701751 (99.785)
1	121	56.4	−2.211627 (99.458)	−2.714507 (99.989)	216	86.7	−2.480927 (99.682)	−2.706874 (99.974)
2	223	60.2	−2.219818 (99.826)	−2.715291 (100.017)	402	91.7	−2.485388 (99.861)	−2.707183 (99.986)
Full	458		−2.223690 (100.000)	−2.714817 (100.000)	923		−2.488848 (100.000)	−2.707570 (100.000)
Testosterone								
0	48	38.9	−3.301752 (96.725)	−4.110938 (99.764)	85	61.3	−3.734627 (98.395)	−4.103197 (99.793)
1	133	51.4	−3.380683 (99.037)	−4.119281 (99.967)	239	77.5	−3.776146 (99.489)	−4.110686 (99.975)
2	286	55.4	−3.401254 (99.640)	−4.120945 (100.007)	517	82.5	−3.786582 (99.764)	−4.110783 (99.978)
Full	882		−3.413553 (100.000)	−4.120653 (100.000)	1617		−3.795542 (100.000)	−4.111706 (100.000)
Progesterone								
0	48	38.7	−3.591953 (96.754)	−4.466493 (99.761)	86	60.9	−4.059920 (98.412)	−4.458090 (99.792)
1	132	51.0	−3.676732 (99.038)	−4.475547 (99.963)	238	76.8	−4.104467 (99.491)	−4.466212 (99.974)
2	282	54.8	−3.698765 (99.631)	−4.477428 (100.005)	510	81.6	−4.115578 (99.761)	−4.466408 (99.978)
Full	960	0.0	−3.712464 (100.000)	−4.477214 (100.000)	1759		−4.125452 (100.000)	−4.467379 (100.000)

Table 1.10 Reaction energies (in kJ mol^{-1}) for the androstendione (**I**), testosterone (**II**), and gold (**III**) reactions.

Method	$T_{\text{PNO}}^{\text{en}}$	Reaction I	Reaction II	Reaction III
HF		−4.9	−26.8	91.9
MP2		22.4	−23.0	255.9
PNO-LMP2	0	17.5	−20.4	237.4
PNO-LMP2	0.997	18.2	−20.9	239.5
HF+CABS		−5.8	−26.2	92.2
MP2-F12		18.8	−19.7	253.0
PNO-LMP2-F12	0	18.5	−19.9	251.6
PNO-LMP2-F12	0.997	18.6	−19.9	252.1

Basis set: VTZ-F12, $T_{\text{PNO}}^{\text{occ}} = 10^{-8}$, `IEXT=2`, $T_{\text{dist}} = 10^{-6}$. The iterative multipole approximation with $p = 3$ has been used. HF+CABS includes the CABS singles correction [120,123] to the Hartree-Fock energy.

1.11.2 PNO-LCCSD-F12

The CCSD-F12 wave function is defined as

$$\Psi_{\text{CCSD-F12}} = \exp\left(\hat{T}_1 + \hat{T}_2\right)|0\rangle, \tag{1.137}$$

where the excitation operators are defined as

$$\hat{T}_1 = \sum_i^{\text{val}} \sum_a^{\text{virt}} t_a^i \hat{E}_i^a, \tag{1.138}$$

$$\hat{T}_2 = \frac{1}{2} \sum_{i,j}^{\text{val}} \sum_{a,b}^{\text{virt}} T_{ab}^{ij} \hat{E}_{ij}^{ab} + \frac{1}{2} \sum_{i,j}^{\text{val}} \sum_{\alpha,\beta}^{\text{complete}} \sum_{k,l}^{\text{val}} T_{kl}^{ij} \langle kl|F_{12}\hat{Q}_{12}|\alpha\beta\rangle \hat{E}_{ij}^{\alpha\beta}. \tag{1.139}$$

These operators differ from the CCSD ones (cf. Section 1.9) by the additional F12 term. The F12 operator, projector, and amplitudes are the same as defined in Section 1.11.1.

A formally complete treatment of the CCSD-F12 method introduces terms in which the unity operator in the projector must be approximated with a double RI, practically limiting its application to very small molecules. In the PNO-LCCSD-F12 method we use the so-called CCSD-F12*x* (x = a, b) approximations [120,133], in which only contributions that are linear in the F12 amplitudes are considered (beyond LMP2-F12). Furthermore, there are no contributions to the residuals that require double RI approximations. The CCSD-F12*x* method has been extensively benchmarked, and is found to give similar or even better results than more complete treatments [115, 129]. This is due to certain systematic error cancellations and the avoidance of multiple RI approximations [133]. In addition, we also use the approximation 3*A and fixed amplitudes in PNO-LCCSD-F12. The PNO-LCCSD-F12*x* and the PNO-LCCSD residual equations then differ only slightly:

$$\Delta r_a^i = \sum_j^{ij\in P_s} \sum_{b\in[ij]_{\text{PNO}}} S_{ab}^{i,ij}\left(2\tilde{V}_{bj}^{ij} - \tilde{V}_{bj}^{ji}\right) \quad \forall a \in [ii]_{\text{PNO}}, \tag{1.140}$$

and

$$\Delta R^{ij}_{ab} = \tilde{V}^{ij}_{ab} - \sum_{k\in[ij]_{\mathrm{LMO}}} \left[\tilde{V}^{ij}_{ak}(\mathbf{S}^{ij,k}\mathbf{t}^{k})_b + (\mathbf{S}^{ij,k}\mathbf{t}^{k})_a \tilde{V}^{ij}_{bk} \right]$$
$$\forall i,j \in P_s, \quad \forall a,b \in [ij]_{\mathrm{PNO}}, \tag{1.141}$$

where

$$S^{ij,k}_{ab} = \langle a^{ij} | b^{kk} \rangle, \tag{1.142}$$
$$\tilde{V}^{ij}_{ab} = \left\langle ab | r_{12}^{-1} Q^{ij}_{12} F_{12} | \tilde{ij} \right\rangle. \tag{1.143}$$

The quantity $\tilde{V}^{ij}_{ak}$ is defined in a similar way. The modified kets

$$|\tilde{ij}\rangle = \frac{3}{8}\,|ij\rangle + \frac{1}{8}\,|ji\rangle \tag{1.144}$$

account for the fixed amplitudes defined in equation (1.132). We treat the F12 correction at the CCSD-F12*x* level only for strong pairs *ij*. For close and weak pairs, the F12 corrections at the MP2-F12 level are included, and previous benchmark calculations [133] have shown that this simple treatment is a surprisingly good approximation. We note that in equation (1.140), a projection approximation has been used to project the contributions of $\tilde{V}^{ij}_{bj}$ from PNO domain $b \in [ij]$ to the singles domain $a \in [i]$.

The PNO-LCCSD-F12a energy expression then differs from the PNO-LCCSD one (cf. equation 1.97) only by adding the MP2-F12 energy correction $\Delta E^{\mathrm{MP2}}_{\mathrm{F12}}$:

$$E^{\mathrm{LCCSD-F12a}}_{\mathrm{corr}} = E^{\mathrm{LCCSD}}_{\mathrm{corr}} + \Delta E^{\mathrm{MP2}}_{\mathrm{F12}}. \tag{1.145}$$

But note that due to the additional terms in the residuals, the amplitudes are different, and therefore the $E^{\mathrm{LCCSD}}_{\mathrm{corr}}$ contribution is different than for LCCSD. The F12b approximation adds further terms which introduce an additional coupling with the conventional amplitudes:

$$E^{\mathrm{LCCSD-F12b}}_{\mathrm{corr}} = E^{\mathrm{LCCSD-F12a}}_{\mathrm{corr}} + \sum_{i\geq j\in P_s} (2-\delta_{ij}) \sum_{a,b\in[ij]_{\mathrm{PNO}}} \left[\left(2D^{ij}_{ab} - D^{ji}_{ab}\right) \tilde{V}^{ij}_{ab} \right.$$
$$\left. + \left(2\tilde{V}^{ij}_{aj} - \tilde{V}^{ji}_{aj}\right)(\mathbf{S}^{ij,i}\mathbf{t}^{i})_a + \left(2\tilde{V}^{ji}_{bi} - \tilde{V}^{ij}_{bi}\right)(\mathbf{S}^{ij,j}\mathbf{t}^{j})_b \right], \tag{1.146}$$

where

$$[\mathbf{D}^{ij}]_{ab} = [\mathbf{T}^{ij}]_{ab} + [\mathbf{S}^{ij,i}\mathbf{t}^{i}]_a[\mathbf{S}^{ij,j}\mathbf{t}^{j}]_b, \quad a,b \in [ij]. \tag{1.147}$$

The F12b approximation is more rigorous than F12a. Nevertheless, it has been found that the F12a approximation sometimes gives more accurate energy differences if small AO basis sets are used. This is due to a systematic overestimation of the F12a energy correction, which favorably compensates for basis set incompleteness errors. For accurate calculations with larger basis sets (triple-ζ and larger), the F12b approximation should be used [133].

The quantity $\tilde{V}^{ij}_{ab}$ is computed as [143, 145]

$$\tilde{V}^{ij}_{ab} = \langle ab|r_{12}^{-1}F_{12}|\tilde{ij}\rangle - \sum_{c,d\in[ij]_{\mathrm{PNO}}} K^{cd}_{ab}\tilde{F}^{ij}_{cd} - \sum_{m,n\in[ij]_{\mathrm{LMO}}} K^{mn}_{ab}\tilde{F}^{ij}_{mn}$$
$$- \sum_{m\in[ij]_{\mathrm{LMO}},c\in[ij]_{\mathrm{PNO}}} \left(K^{mc}_{ab}\tilde{F}^{ij}_{mc} + K^{cm}_{ab}\tilde{F}^{ij}_{cm}\right), \tag{1.148}$$

where

$$\tilde{F}^{ij}_{xy} = \langle xy|F_{12}|\tilde{ij}\rangle, \quad (x, y \text{ representing PNOs or LMOs}) \tag{1.149}$$

and the other 2-electron integrals are defined as in section 1.11.1. The $\tilde{V}^{ij}_{ak}$ operators can be computed in a similar fashion.

In equations (1.140) through (1.148), we have applied similar domain approximations as in the PNO-LMP2-F12 method. In addition, the last summation in equation (1.148), which should strictly run over the complete virtual space (including the CABS part), is restricted to the PNOs in the domain $[ij]_{\mathrm{PNO}}$ [143, 145]. This avoids very expensive contributions of 3-external integrals involving the CABS space. Furthermore, the F12 terms are only added to the CCSD residual equations for strong pairs, and the F12 corrections for the close and weak pairs are thus limited to the MP2-F12 level. These approximations significantly reduce the computational cost of the PNO-LCCSD-F12 corrections, since all 3- and 4-external integrals required for the F12 terms are also needed for the PNO-LCCSD calculations. The effect of the approximations on the accuracy has been shown to be small [143, 145].

In the PNO-LCCSD-F12 method, all summations are restricted to domains that asymptotically have a constant size. Using the local density-fitting technique for the integral evaluations and transformations to be discussed in the following section, the method scales linearly with the molecule size. With our present implementation, the inclusion of the F12 terms increases the overall computational time of a PNO-LCCSD calculation only by 10–20%. More than 50% of this is due to the MP2-F12 correction. The cost of LCCSD-F12a and LCCSD-F12b is virtually the same. As will be demonstrated in the following, the F12 corrections very much improve the results and should therefore always be included.

Table 1.11 shows the computed reaction energies for our three benchmark reactions for PNO-LMP2, PNO-LMP2-F12, PNO-LCCSD, and PNO-LCCSD-F12 using the VDZ-F12 and VTZ-F12 basis sets. For simplicity, the prefix PNO- is omitted. In order to minimize the effect of DF and RI approximations, the aug-cc-pVTZ auxiliary basis sets (cf. Section 1.2) have been used for both orbital basis sets. The F12 corrections have a significant effect on the reaction energies and lead to much better convergence with the basis set. Furthermore, using the F12 correction, the PAO domain size (`IEXT=1` or `IEXT=2`) has a much smaller effect on the explicitly correlated results than without the F12 correction. Thus, it may often be sufficient to use `IEXT=1` in LCCSD-F12 calculations, which saves considerable time and memory. In Table 1.11 also different LCCSD-F12 approximations are compared: LCCSD+ΔF12 means that the F12 correction computed at the PNO-LMP2-F12 level is simply added to the PNO-LCCSD result. F12a and F12b additionally include couplings between the LCCSD amplitudes and the F12 terms (see above). As expected, the (absolute) F12 corrections systematically decrease

Table 1.11 Reaction energies (in kJ mol^{-1}) for the androstendione (**I**), testosterone (**II**), and gold (**III**) reactions for different values of `IEXT` and two basis sets.

Reaction	Basis	LMP2	LMP2-F12	LCCSD	LCCSD+ΔF12	LCCSD-F12a	LCCSD-F12b
`IEXT=1`							
I	VDZ-F12	12.3	18.9	13.9	20.4	20.1	18.9
	VTZ-F12	15.4	18.2	16.4	19.2	18.9	18.4
II	VDZ-F12	−18.3	−19.3	−22.7	−23.8	−23.9	−23.9
	VTZ-F12	−20.4	−19.7	−26.4	−25.7	−25.7	−25.7
III	VDZ-F12	214.4	249.8	153.3	188.6	188.1	185.1
	VTZ-F12	234.1	250.6	171.6	188.1	187.5	185.6
`IEXT=2`							
I	VDZ-F12	17.8	19.2	19.2	20.5	20.2	19.1
	VTZ-F12	18.2	18.6	18.8	19.3	19.1	18.7
II	VDZ-F12	−20.2	−19.8	−24.5	−24.1	−24.3	−24.3
	VTZ-F12	−20.9	−19.9	−26.8	−25.7	−25.8	−25.8
III	VDZ-F12	225.2	253.0	162.4	190.2	189.7	186.6
	VTZ-F12	239.5	252.1	176.2	188.7	188.3	186.5

$T_{close} = 10^{-4}$, $T_{weak} = 10^{-5}$, $T_{dist} = 10^{-6}$. The CABS singles corrections are included in all F12 results. SCS-LMP2 distant pair energies obtained using the iterative multipole approximation ($p = 3$) were used in the LCCSD and LCCSD-F12 calculations.

in the order ΔF12>F12a>F12b. It is known from conventional coupled cluster calculations that the basis set effect is usually somewhat smaller at the CCSD than at the MP2 level, and this effect is reflected in the current results. Overall, the final LCCSD-F12b reaction energies are least sensitive to the basis set, and we therefore believe that the LCCSD-F12b/VTZ-F12 results are most accurate. Except for reaction **III**, the differences between the F12a and F12b results are negligible. For reaction **III** the difference still amounts to about 2 kJ mol^{-1}. This can be seen as a measure of the uncertainty. In this very difficult case, the LMP2-F12 and LCCSD-F12b results differ by 65 kJ mol^{-1}, and the F12b correction amounts to 10.3 kJ mol^{-1} (`IEXT=2`, VTZ-F12).

Figure 1.10 shows the convergence of the LCCSD-F12b correlation energy of Auamin (left panel) and of the reaction energy (right panel) as a function of the close and weak pair thresholds T_{close} and T_{weak}. We have chosen ratios of T_{close}/T_{weak} to be 10, 30, or $T_{weak} = T_{dist}$ (no weak pairs). The variations of the reaction energies caused by the close pair approximation below a threshold of $T_{close} = 3 \times 10^{-4}\ E_h$ are smaller than 0.5 kJ mol^{-1}, while for this threshold the correlation energy of the Auamin molecule differs by only 0.15 mE_h from the converged value. The weak pair CEPA approximation somewhat increases the errors, but for a threshold of $T_{weak} = 3 \times 10^{-5}\ E_h$ the additional error is still below 1 kJ mol^{-1}.

In summary, the benchmarks presented in this and previous sections have shown that it is possible to control all local approximations, and to converge the results toward the canonical values. Of course a direct comparison with the canonical results is only

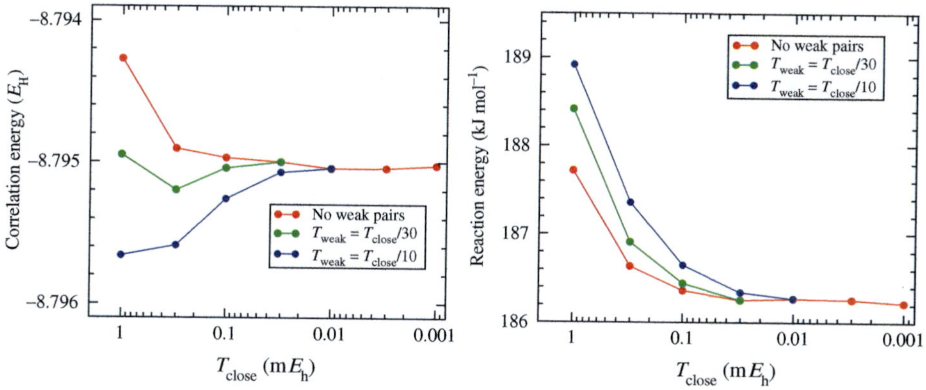

Figure 1.10 Convergence of the PNO-LCCSD-F12a absolute energies for the Auamin molecule (left) and reaction energies for reaction **III** (right) depending on T_{close} and T_{weak} pair selection thresholds. Basis set: VTZ-F12 (for Au cc-pVTZ-PP), $T_{\text{dist}}3\,10^{-6}E_{\text{h}}$, distant pair energies computed with the iterative multipole approximation p=3, `IEXT=1`.

possible for rather small molecules. For large molecules, the only way to check the convergence is to lower the thresholds, as has been shown in figure 1.10 for close and weak pair approximations. If tight close and weak pair thresholds are applied, it appears likely that the canonical CCSD-F12 results (which are impossible to compute) would not differ more than 1–2 kJ mol^{-1} from the PNO-LCCSD-F12 values. Since the implementation of our PNO-LCCSD-F12 method has been completed very recently, further benchmarks and improvements are still under way. The results will be presented in a future publication.

1.12 Technical Aspects

1.12.1 Local Density Fitting

Density fitting (DF) is a widely used approximation for the evaluation of 2-electron integrals in electron correlation methods. The principle of DF is to use an auxiliary basis to approximate the 1-electron charge densities in equation (1.11):

$$\rho_{ai}(\mathbf{r}) = \phi_a(\mathbf{r})\phi_i(\mathbf{r}) \approx \tilde{\rho}_{ai} = \sum_{A}^{\text{fit}} d_{ai}^{A}\phi_A(\mathbf{r}), \tag{1.150}$$

and similar for ρ_{bj}. Equation (1.11) then becomes

$$(ai|bj) \approx \sum_{A,B}^{\text{fit}} d_{ai}^{A} d_{bj}^{B} \int_{\mathbb{R}^3} d\mathbf{r}_1 \int_{\mathbb{R}^3} d\mathbf{r}_2 \phi_A(\mathbf{r}_1)\phi_B(\mathbf{r}_2) = \sum_{A,B}^{\text{fit}} d_{ai}^{A} d_{bj}^{B} (A|B). \tag{1.151}$$

As shown by Dunlap *et al.* [213], the density-fitting coefficient d_{ai}^{A} can be obtained by minimizing

$$\Delta_{ai} = \int_{\mathbb{R}^3} d\mathbf{r}_1 \int_{\mathbb{R}^3} d\mathbf{r}_2 [\rho_{ai}(\mathbf{r}_1) - \tilde{\rho}_{ai}(\mathbf{r}_1)] \frac{1}{r_{12}} [\rho_{ai}(\mathbf{r}_2) - \tilde{\rho}_{ai}(\mathbf{r}_2)], \tag{1.152}$$

which is equivalent to solving a linear equation system

$$(A|ai) = \sum_{B}^{\text{fit}} (A|B) d_{ai}^{A}, \tag{1.153}$$

where $(A|ai)$ is a 3-index, 2-electron integral

$$(A|ai) = \int_{\mathbb{R}^3} \mathrm{d}\mathbf{r}_1 \int_{\mathbb{R}^3} \mathrm{d}\mathbf{r}_2 \phi_A(\mathbf{r}_1) \frac{1}{r_{12}} \rho_{ai}(\mathbf{r}_2) = (ai|A). \tag{1.154}$$

Similarly, we have

$$(A|bj) = \sum_{B}^{\text{fit}} (A|B) d_{bj}^{A}. \tag{1.155}$$

Inserting equation (1.155) into equation (1.151) gives a simplified equation for the 2-electron integrals

$$(ai|bj) \approx \sum_{A}^{\text{fit}} d_{ai}^{A} (A|bj) = \sum_{A,B} (A|ai)[\mathbf{J}^{-1}]_{AB} (B|bj), \qquad [\mathbf{J}]_{AB} = (A|B). \tag{1.156}$$

The density-fitting technique significantly reduces the computational cost of canonical MP2 calculations without introducing significant errors. In local methods, an additional advantage is that the summations over the fitting functions can be limited to those close to the orbitals i and j, which reduces the scaling of the method. In the PNO-LMP2 and PNO-LCCSD methods, we restrict the summations over the fitting functions to domains $[ij]_{\text{fit}}$, determined in a similar fashion as for the domains $[ij]_{\text{PAO}}$ (see Section 1.6.2). The parameters for extending the fitting domains are denoted `IDFDOM` and `RDFDOM` (corresponding to `IEXT` and `REXT`, respectively). Our benchmark calculations show that using the default values `IDFDOM=3` and `RDFDOM=7` (all distances in a_0 and with `IEXT=2` and `REXT=5`) usually lead to negligible errors in PNO-LMP2 calculations [147, 148]. Even smaller domains `IDFDOM=2` and `RDFDOM=5` give satisfactory results when explicit correlation is included, and these values were used for PNO-LCCSD and PNO-LCCSD-F12 calculations.

The 3-index integrals can only be evaluated in the AO basis and must be transformed to the LMO or PNO bases in our local correlation methods. In the PNO-LMP2 method, only 2-external K_{ab}^{ij} integrals in pseudo-canonical PAO and OSV bases are needed. We first transform the 3-index integrals to the LMO and PAO bases

$$(A|\mu i) = \sum_{\nu}^{\text{AO}} (A|\mu\nu) L_{\nu i}, \tag{1.157}$$

$$(A|\tilde{r}i) = \sum_{\mu}^{\text{AO}} (A|\mu i) P_{\mu\tilde{r}}. \tag{1.158}$$

The $K_{\tilde{r}\tilde{s}}^{ij}$ integrals are then assembled pairwise

$$K_{\tilde{r}\tilde{s}}^{ij} = \sum_{A,B\in[ij]_{\text{fit}}} (A|\tilde{r}i)[\mathbf{J}^{-1}]_{AB}(B|\tilde{s}j) \tag{1.159}$$

before they are transformed to the PC-PSV basis as

$$K^{ij}_{a^{ij}b^{ij}} = \sum_{r,s\in[ij]_{\mathrm{PAO}}} \tilde{V}^{ij}_{ra^{ij}} K^{ij}_{\tilde{r}\tilde{s}} \tilde{V}^{ij}_{sb^{ij}}. \tag{1.160}$$

Due to the large number of fitting functions, this strategy is found to be more efficient than transforming the 3-index integrals to the PC-PSV basis [147]. To achieve linear scaling, the sparsity of the LMO and PAO expansion coefficients must be exploited to limit the summations in equations (1.157) and (1.158).

The PNO-LCCSD method requires a number of different types of integrals and the strategy described above for transforming the $K^{ij}_{a^{ij}b^{ij}}$ integrals may not be optimal. For example, the 4-external integrals $(a^{ij}b^{ij}|c^{ij}d^{ij})$ are assembled after transforming 3-index integrals to the PNO basis

$$(a^{ij}b^{ij}|c^{ij}d^{ij}) = \sum_{A,B\in[ij]_{\mathrm{fit}}} (A|a^{ij}b^{ij})(\mathbf{J}^{-1})_{AB}(B|c^{ij}d^{ij}). \tag{1.161}$$

The reason for transforming to the PNO basis before the assembly step is performed is that the number of the 4-external integrals scales with the fourth power in the number of the virtual orbitals per pair, and the memory and CPU requirements for assembling them in the PAO basis would be much larger. Furthermore, a transformation of the 4-index integrals from the PAO to the PNO basis is avoided.

For integrals other than the Coulomb integrals, it is essential to use the robust fitting formula [108, 214] to ensure sufficient accuracy. For example, in our PNO-LMP2-F12 method, local density fitting of an F12 integral takes the form

$$\begin{aligned} F^{ij}_{\tilde{r}\tilde{s}} \approx & \sum_{A\in[i]_{\mathrm{fit}}} d^{A}_{\tilde{r}i}(A|F_{12}|\tilde{s}j) + \sum_{B\in[j]_{\mathrm{fit}}} (B|F_{12}|\tilde{r}i)d^{B}_{\tilde{s}j} \\ & - \sum_{A\in[i]_{\mathrm{fit}}} \sum_{B\in[j]_{\mathrm{fit}}} d^{A}_{\tilde{r}i}(A|F_{12}|B)d^{B}_{\tilde{s}j}. \end{aligned} \tag{1.162}$$

The robust fitting formula ensures that the errors of the integrals are quadratic in the fitting errors. Equation (1.162) can also be used in assembling the Coulomb integrals. However, if we use the union of fitting domains $A, B \in [ij]_{\mathrm{fit}}$ and replace the kernel F_{12} with r_{12}^{-1}, the last two terms in equation (1.162) cancel out, and the equation reduces to equation (1.159). In other words, the simple local fitting formula (1.159) is robust for Coulomb integrals. We note that the F12-specific commutator integrals take a more complicated form [108].

The local DF method ensures that the integral evaluation and transformation in our PNO methods asymptotically scale linearly with the size of the molecule. There is a non-negligible overhead due to the pairwise linear equation system solving, and therefore there is a certain molecular size below which local DF will be more expensive than the standard DF procedure. The overhead is partially compensated in a parallel environment, since in a standard DF method, high parallel efficiency is more difficult to achieve due to increased I/O. More details about the parallelization of the program will be given in the following section.

1.12.2 Parallelization

Our PNO methods are parallelized with the help of MPI and the Global Arrays (GA) toolkit [215]. The GA toolkit provides a portable one-sided remote memory access interface so that a process can access remote data without intervening the corresponding remote process. Currently, we do not use shared memory parallelization.

We use static as well as dynamic parallelization techniques. Static parallelization means that predetermined tasks are allocated to certain processors (processing cores). This is used, for example, in the iterative solution of the LMP2 or LCCSD equations, where groups of pair residuals are assigned to each processor. After computing the residuals, each processor updates the amplitudes for the corresponding pairs and copies them to a GA. Subsequently, other processors read from this GA all amplitudes which are needed in their residuals and which are not local to them. Thus, communication and synchronization is necessary only once per iteration after the amplitudes have been copied to the GA.

This static parallelization strategy is especially useful when calculated quantities (e.g., integrals or amplitudes) can remain local to the processors during a procedure, and only little communication of data is necessary. In order to minimize the memory demands, the group of pairs allocated to one processing core should share as many data as possible. Furthermore, the total amount of memory (and/or CPU time) on each compute core should be similar, that is, a good load balancing should be achieved. The optimal distribution of pairs to the processors is a highly non-trivial task. We use the METIS graph partitioning program [216] for this purpose, but even then the load balancing is often not optimal, and therefore dynamic parallelization may be advantageous, even if this leads to more communication.

In the dynamic parallelization model, the tasks are allocated to the processors on a first-come-first-serve basis. This is achieved using a task list and a shared counter, which is incremented whenever a task is allocated to a processor. In order to obtain optimal load balancing, larger tasks should be processed first and small ones at the end. This model yields a very good load balancing, provided that the number of tasks is significantly larger than the number of processing cores. An example is the evaluation and the transformation of the 2-electron integrals. We first compute all required 3-index PAO/LMO integrals sets such as $(A|ij)$, $(A|\tilde{r}i)$, and $(A|\tilde{r}\tilde{s})$ and store them in GAs. This step is parallelized over blocks of fitting functions A to avoid redundancies in the evaluation of the AO integrals $(A|\mu\nu)$. In the following step, the different types of 4-index integrals are assembled and transformed to the PC-PNO basis. This step is parallelized over LMO pairs, and the 3-index integrals are retrieved from the GAs as required. The resulting 4-index PC-PNO integrals are also stored in GAs and retrieved when needed in the calculations of the residuals. An exception is the calculation of the F12 correction in the PNO-LMP2-F12 method, in which the individual contributions are computed on the fly from the assembled 4-index integrals, and the integrals are then discarded immediately.

Our PNO methods are designed to minimize computational redundancy by computing (nearly) all types of 3-index and 4-index integrals only once for the whole molecule in a PNO-LCCSD-F12 calculation. This leads to minimum CPU-time but requires a lot of communication of 3-index and 4-integrals between the processors (via the GAs). The communication overhead is reasonably small if a fast network (e.g., InfiniBand) is available. It is also possible to minimize the communication at the prize of redundant integral evaluations. For example, in LMP2 one can compute groups of operators $\mathbf{K}^{ij}$ on different processors, without any communication. This requires redundant calculations of the 3-index integrals. Using our compute cluster with a few hundred CPU cores and QDR-InfiniBand connections between the nodes, this is much slower than the dynamical calculation. However, this may not be true when massively parallel computers or a slow network connection is used.

1.13 Comparison of Local Correlation and Fragment Methods

In fragmentation methods [60–87, 151], the correlation of electrons in subsets of LMOs or LMO pairs is treated independently in different fragments or increments. Various ways to choose these fragments (or increments) and to assemble the total correlation energy have been proposed, and this is the subject of other chapters in this book. One possibility is to assemble the total correlation energy using pair energies E_{ij} obtained in different fragments. This is for example the case in the "Divide-Expand-Consolidate" (DEC) approach [66–74]. Alternatively, the correlation energy can also be rewritten as a sum of orbital contributions E_i, which are computed in independent fragments. This is for example used in the Cluster in Molecules (CIM) approach [75–85, 151]. Incremental methods [88–96] are somewhat different and use a many-body expansion. It is also possible to split the preceding HF calculation into fragments [60–65], but this is more difficult and has to take the electrostatic interactions with the environment into account. Furthermore, it may be necessary to introduce artificial "link atoms" in order to saturate each fragment. Here we will only discuss the case that a single HF calculation is carried out for the whole system, just as in local correlation methods.

In order to obtain an accurate value for a particular pair energy E_{ij} in a fragment, the interaction with neighboring pairs must be taken into account. We can distinguish between "primary" pairs, which are those for which the pair energies are computed and used in the final energy assembly, and "secondary" pairs, whose amplitudes are needed to get accurate residuals for the primary pairs, but which do not directly contribute to the energy of the fragment. Similarly, if the correlation energy is expressed as a sum of orbital contributions, one can distinguish between "central MOs" and "buffer" or "environmental" MOs [79, 81]. In any case, each group of primary pairs needs an interaction region of secondary pairs around it. The number of primary and secondary pairs together (or the number of LMOs in a fragment) determines the total size and cost of each fragment calculation. The interaction regions of different fragments overlap strongly, and this leads to a high redundancy in fragmentation treatments. The advantage, however, is that the fragments can be treated independently, and thousands of fragments can be computed in parallel on a supercomputer. Furthermore, conventional (non-local) methods can be used. For this, one can determine a single PAO domain for each fragment and then use the pseudo-canonical orbitals of this domain as virtual space for the fragment. No further domain or pair approximations are applied, and this makes such methods technically much simpler than the local correlation treatments as described in this chapter.

The most important question is how big the interaction regions have to be in order to obtain accurate correlation energies, since the cost of a CCSD calculation for a single fragment scales with $\mathcal{O}(\mathcal{N}^6)$, where $\mathcal{N}$ is a measure of the fragment size (e.g., the number of correlated orbitals in it, including the buffer region). If triples are included, the scaling is even $\mathcal{O}(\mathcal{N}^7)$. If all fragments would be calculated in parallel on a supercomputer, the total elapsed time would depend on the time needed for the largest fragment.

The smallest fragments one can think of are obtained if each one contains only a single primary pair ij. Our local correlation treatment allows us to estimate the size of the interaction regions that are needed to describe the pair energies E_{ij} accurately, at least on the average. The size of the fragments depends on the number of amplitude matrices $\mathbf{T}^{ik}$ and $\mathbf{T}^{jk}$ that are taken into account in the computation of the residual $\mathbf{R}^{ij}$. This number depends on the distant pair threshold T_{dist} in our PNO-LCCSD method.

In Figure 1.11, the dependence of the PNO-LCCSD reaction energy of the three benchmark reactions is shown as a function of T_{dist}. Either distant pairs are neglected entirely or an LMP2 correction for distant pairs is added. When the distant pairs are entirely neglected, the convergence with decreasing threshold T_{dist} is very slow. The LMP2 correction somewhat improves the accuracy, but overshoots significantly, as has been shown earlier for PNO-LCCD. Significantly better results are obtained by using spin-component scaled LMP2 (SCS-LMP2) [217] for the distant pair correction. This is consistent with the well-known fact that SCS-LMP2 avoids the overestimation of dispersion interactions by LMP2. But for the Auamin reaction, the error still amounts to about 2 kJ mol^{-1} if a threshold of $T_{\text{dist}} = 10^{-5}\,E_\text{h}$ is used. It should be noted that for distant pairs, when the charge-transfer and exchange contributions are neglected (NOEX approximation), the SCS-LMP2 pair correlation energies are obtained simply by scaling the LMP2 ones with a factor of $23/30 = 0.766$.

Table 1.12 shows the average and maximum numbers of LMOs k that occur in $\mathbf{T}^{ik}$ and $\mathbf{T}^{jk}$ contributing to a residual $\mathbf{R}^{ij}$ as a function of two thresholds: T_{primary} and $T_{\text{secondary}}$. T_{primary} determines the selection of primary pairs, while $T_{\text{secondary}}$ controls the selection of the secondary pairs for each primary pair (Figure 1.11 corresponds to $T_{\text{dist}} = T_{\text{secondary}} = T_{\text{primary}}$). We restrict here $T_{\text{secondary}} \leq T_{\text{primary}}$. These thresholds determine the minimum fragment size that would be necessary for a fragment calculation with comparable accuracy. Naturally, the number of LMOs k increases with decreasing threshold T_{primary}, since for small thresholds more pairs with distant orbitals i and j are included. For the molecules shown in Table 1.12, with the smallest considered threshold of $T_{\text{primary}} = 3 \times 10^{-6}$, the largest fragment comprises the whole molecule. Even with a threshold of $T_{\text{primary}} = 3 \times 10^{-5}$, the largest fragments still comprise 60–100 LMOs. Unfortunately, conventional CCSD or even CCSD(T) calculations of this size with reasonable basis sets are extremely challenging at present, and this limits the accuracy or applicability of fragmentation approaches.

On the other hand, in local correlation methods, only a non-redundant set of integrals and residuals is computed. Furthermore, the CPU time per residual is greatly reduced through the use of domain and pair approximations. Therefore, such calculations should be much faster than fragmentation calculations. However, the size of molecules that can be treated is mainly limited by the required memory and/or disk space. Our methods are very well parallelized, and in order to avoid I/O bottlenecks, currently all data is kept in distributed memory (global arrays or distributed local data). The number of computer nodes that are necessary for a particular calculation depends on the memory per node, the size of the molecule, and the basis set. The molecular size that can be handled is in principle scalable with the number of nodes, but in practice, some quantities such as overlap or Fock matrices must be kept locally on each node. This currently limits the

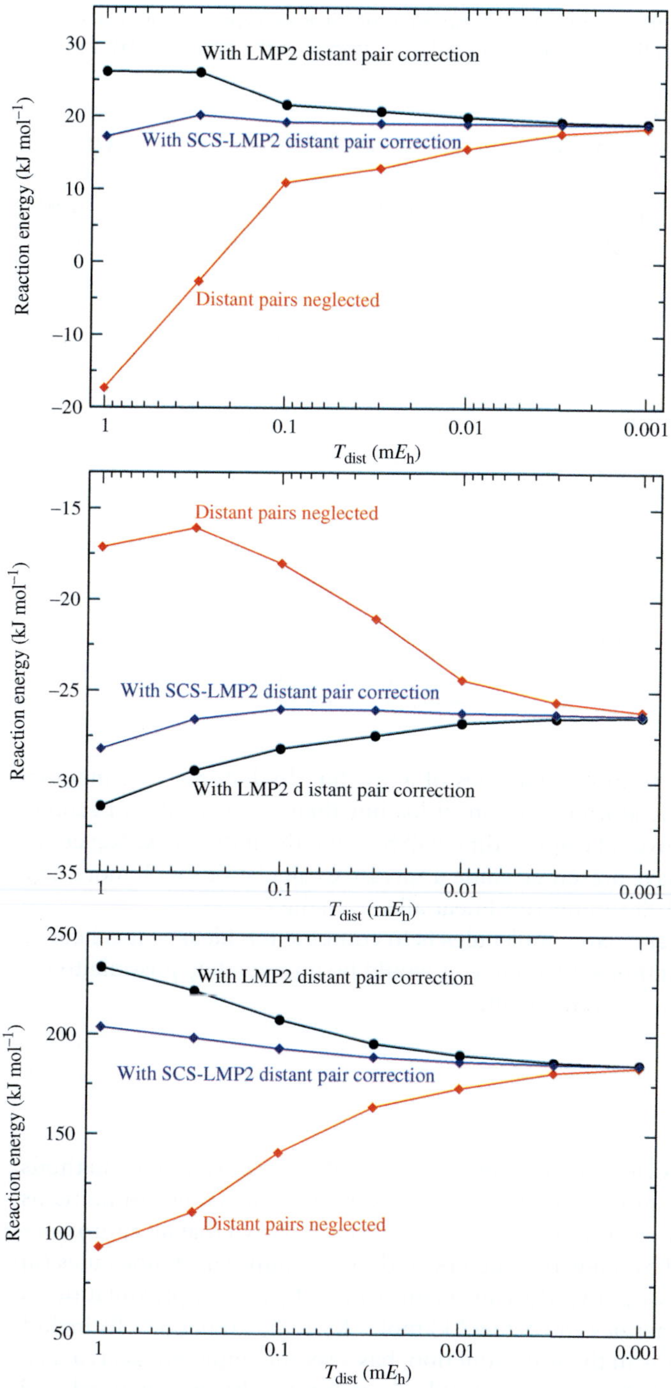

Figure 1.11 LCCSD-F12 reaction energies for the reactions **I** (top), **II** (middle), and **III** (bottom) as a function of the distant pair threshold T_{dist}. All pairs which have LMP2 pair energies $\geq T_{dist}$ are included. The remaining pairs are either neglected or treated by LMP2. Basis set: VTZ-F12, `IEXT=1`.

Table 1.12 The average (n_k^{avg}) and the maximum (n_k^{max}) numbers of LMOs that are associated with an LMO pair ij with the restriction that the pair energies $E_{ij} \geq T_{primary}$ (shown in the table header), and $E_{ik} \geq T_{secondary}$ or $E_{kj} \geq T_{secondary}$.

		3×10^{-4}		3×10^{-5}		3×10^{-6}	
Molecule	$T_{secondary}$	n_k^{avg}	n_k^{max}	n_k^{avg}	n_k^{max}	n_k^{avg}	n_k^{max}
Auamin	3×10^{-4}	24.5	55				
	3×10^{-5}	49.4	100	56.3	102		
	3×10^{-6}	81.5	122	87.9	122	91.7	122
Precursor	3×10^{-4}	22.7	38				
	3×10^{-5}	37.7	64	41.1	71		
	3×10^{-6}	51.7	78	54.8	78	58.0	78
Auamin2s	3×10^{-4}	23.8	56				
	3×10^{-5}	41.5	68	45.9	71		
	3×10^{-6}	58.7	76	61.9	76	63.8	76
Ester	3×10^{-4}	22.6	41				
	3×10^{-5}	39.0	56	42.2	64		
	3×10^{-6}	50.5	66	53.5	69	55.7	69

applicability of our method to molecular sizes of a few hundred atoms. The memory problem can be alleviated by storing data on disks, but then parallelization becomes much less efficient and the overall elapsed time will become the bottleneck. Nevertheless, Neese and coworkers have presented some impressive calculations with more than 20,000 basis functions and 1000 atoms (for linear alkane chains).

A way forward to treat "very" large molecules is to combine the ideas of local correlation and fragmentation methods. Of course, it would be immediately possible to use local correlation methods for each fragment.

1.14 Summary

In this chapter, we have reviewed important aspects of local correlation methods, namely the construction of most suitable local orbital spaces, the choice of domains, and pair approximations for weak and distant pairs. We have shown that all approximations can be well controlled by a few parameters, and for medium-sized molecules (up to about 100 atoms), it is possible to reduce the errors caused by local approximations to the mE_h scale for absolute energies and below 1 kJ mol^{-1} for relative energies. By including explicitly correlated terms in the wave function, basis set incompleteness errors are also strongly reduced, and better than quintuple-ζ quality results are obtained with triple-ζ basis sets. This makes it possible to compute highly accurate coupled-cluster energies for much larger molecules than can be treated by conventional coupled-cluster methods.

Without an F12 treatment, basis set incompleteness errors are in many cases much larger than those caused by local approximations, and therefore it is strongly recommended to always use the explicitly correlated variants. With our new PNO-LCCSD-F12 method, the additional cost is small, usually less than 20% of the total computation time, but the improvements of the results are drastic. Besides the reduction of basis set errors, also the errors caused by the local domain approximation are strongly reduced by the F12 terms. The reason is twofold: first, intramolecular basis superposition errors (BSSE) can be significant in canonical calculations unless very large basis sets are used. The local domain approximation strongly reduces these errors [14, 147, 212], which leads to smaller absolute values of the correlation energy, and so apparently to domain errors. These decay slowly with increasing domain sizes, since then the BSSE is recovered. This view is corroborated by the finding that the convergence of relative energies and other properties toward the CBS limit is often significantly better for local than for canonical methods [25, 147, 218, 219]. The BSSE effects are strongly reduced by the explicit correlation treatment, and this contributes to the reduction of the (apparent) domain errors. The second reason for the reduction of these errors is that the F12 terms in the wave function can implicitly approximate the energy contributions of the orbitals that are omitted in the domain approximation.

In many early papers on local correlation methods, it was assumed that the largest errors arise from the domain approximation, and that the effect of pair approximations is small. In our current methods, it is likely that the opposite is true. The F12 corrections reduce the domain errors typically by one order of magnitude (see above). Furthermore, much larger PAO domains can now be used than in early PAO-LCCSD programs. With triple-ζ basis sets, our current default PAO domains comprise 300–600 PAOs per pair. Solving the PAO-LCCSD amplitude equations with such large domains would be extremely demanding. However, the domain sizes are reduced by a factor of 5–10 (even more for weak pairs) when PNOs are used, and this dramatically reduces the CPU and memory demands. The integrals are still computed in the PAO basis using large domains, but transformed on the fly to the much smaller pair-specific PNO domains. Thus, 4-index quantities in the PAO basis are never stored. For example, the number of 4-external integrals depends on the fourth power of the domain sizes, and is therefore in the PNO basis 3–4 orders of magnitude smaller than in the PAO basis. This makes it possible to keep all quantities such as integrals, amplitudes, and intermediates in distributed high-speed memory, which is a prerequisite for efficient parallelization.

Despite this improvement, PNO-LCCSD calculations without further approximations still become prohibitively expensive for very large molecules. This is due to the fact that the number of pairs increases quadratically with the number of correlated electrons. Even if distant pairs with energy contributions $\leq 10^{-6}\ E_\mathrm{h}$ are treated by the multipole approximation, there may still be a very large number of "weak" pairs with individually small energy contributions of $10^{-6} \leq E_{ij} \leq 10^{-4}\ E_\mathrm{h}$. The sum of these pair energies is significant and also important in order to obtain accurate relative energies. Approximating the amplitudes and energies of weak pairs by LMP2, as was done in the older PAO-LCCSD methods developed in our group [7, 18, 31], and also adopted in the DLPNO-LCCSD methods of Riplinger *et al.* [54, 57], may unfortunately lead to unexpectedly large errors. There are two sources for these errors. First, long-range correlation effects are significantly overestimated by MP2. This effect can be significantly reduced by using SCS-MP2 pair energies, but tests have shown that the accuracy is still unsatisfactory for weak pairs. Second, it is very important to include the amplitudes of the weak pairs into the strong-pair residuals, that is, to account for the coupling of strong and weak pairs accurately. Without this, the strong pair energies are overestimated. Using LMP2 amplitudes for the weak pairs is insufficient to remove this error.

It is therefore important to apply higher-order approximations for simplifying the treatment of close and weak pairs. We have shown that an approximate CCSD treatment [32, 204, 205], in which terms that cancel at long-range are neglected, leads to excellent results, and nevertheless to a significant reduction of the computational effort and storage demands. Most importantly, the expensive class of 4-external integrals is then needed only for the strong pairs. Furthermore, a linearized coupled-cluster (CEPA) treatment is sufficient for weak pairs.

The finding that the neglect of weak pairs in LCCSD treatments causes large errors is also highly relevant for fragmentation methods. It means that the fragments must be very large in order to obtain converged results. Even in the extreme case that each fragment is used to compute only a single pair energy, the necessary fragment sizes may comprise 100 or more LMOs. Conventional calculations of this size may be extremely expensive, if possible at all.

The computational cost and storage demands of local correlation methods increase at least linearly with the molecular size. It is difficult to avoid the storage of some quantities that scale quadratically on each computing core, as for example the Fock and overlap matrices in the full PAO basis, and this becomes a limiting factor for calculations with more than $\approx 10,000$ basis functions. Furthermore, due to the relatively slow decay of some singles terms it is advantageous to compute in each iteration a "dressed" Fock matrix that depends on the singles amplitudes. This part requires some operations that scale cubically with the molecular size, though with a very low prefactor. The applicability of our current program is mainly limited by the memory demands. Since the data is distributed over many processors, the method is well suited for small compute clusters as available in many groups. Currently, it allows accurate LCCSD-F12 calculations for molecules with 100–200 atoms in a few hours of elapsed time using 100–200 compute cores. A perturbative correction for triple excitations is still under development.

Calculations for larger molecules would be possible by combining local correlation and fragmentation methods. The overall memory and storage demands then become independent of the molecular size, at the prize of introducing a lot of redundancy. We believe, however, that for the simulation of most real chemistry problems, calculations for molecules with more that 100–200 atoms are hardly necessary. One should also keep in mind that in large systems, dynamical and entropic effects become increasingly important, and therefore single-point calculations loose their justification. The environment in large systems can be taken into account by QM/MM methods or other embedding techniques. Local correlation methods in combination with QM/MM approaches have for example been successfully used to study some enzyme reactions [29, 220–222]. Local correlation methods also allow treating the active region of a reaction at a high level [e.g., LCCSD(T)], and a larger environment at a lower level, for example, LMP2 [30]. This can save a huge amount of computer time without losing much accuracy.

Finally, it should be mentioned that the implementation of analytical energy gradients for high-level local correlation methods is very challenging. Efficient analytical energy gradients are available for PAO-LMP2 [23, 223] and Laplace-transformed LMP2 methods [42], but these still cannot compete with the speed of DFT gradient calculations.

Although rather straightforward in principle [224], a manual implementation of LCCSD gradients would be cumbersome. This is particularly true for PNO-based methods, since various auxiliary conditions, which are related to the PNO definition, must be taken into account. The implementation of such complex methods can be facilitated using tools like the integrated tensor framework (ITF) of Knizia [182, 225], which has recently been extended to local PAO, OSV, and PNO methods by Kats [34, 176]. This is a program which can directly execute a long sequence of tensor expressions, which are provided in a readable formula file. The expressions can be generated and factorized with the help of computer algebra programs. However, such gradient calculations would at least be twice as expensive as an energy calculation. They could be used for accurate geometry optimizations of medium-sized molecules, but it is unlikely that in the near future high-level wave function methods can be applied for direct dynamics calculations for large systems.

Appendix A: The LCCSD Equations

In the following, summations over repeated indices are implied. r, s, t, u will denote non-orthogonal virtual orbitals (e.g., PAOs or PNOs), and $\mathbf{S}$ their overlap matrix $S_{rs} = \langle r|s\rangle$. We define amplitude and integral matrices (upper case bold face letters)

$$[\mathbf{T}^{kl}]_{rs} = T^{kl}_{rs}, \qquad \tilde{\mathbf{T}}^{kl} = 2\mathbf{T}^{kl} - \mathbf{T}^{lk} \tag{A.1}$$

$$[\mathbf{D}^{ij}]_{rs} = T^{ij}_{rs} + t^i_r t^j_s, \qquad \tilde{\mathbf{D}}^{kl} = 2\mathbf{D}^{kl} - \mathbf{D}^{lk} \tag{A.2}$$

$$[\mathbf{K}^{kl}]_{rs} = (rk|ls), \qquad \mathbf{L}^{kl} = 2\mathbf{K}^{kl} - \mathbf{K}^{lk} \tag{A.3}$$

$$[\mathbf{J}^{kl}]_{rs} = (rs|kl), \qquad [\mathbf{F}]_{rs} = F_{rs} \tag{A.4}$$

as well as vectors (lower case bold face letters)

$$[\mathbf{t}^i]_r = t^i_r, \tag{A.5}$$

$$[\mathbf{k}^{kli}]_r = (rk|li), \qquad \mathbf{l}^{kli} = 2\mathbf{k}^{kli} - \mathbf{k}^{lki} \tag{A.6}$$

$$[\bar{\mathbf{k}}^{kli}]_r = [\mathbf{k}^{kli} + \mathbf{K}^{kl}\mathbf{t}^i]_r, \qquad \bar{\mathbf{l}}^{kli} = \mathbf{l}^{kli} + \mathbf{L}^{kl}\mathbf{t}^i, \tag{A.7}$$

$$[\mathbf{f}^k]_r = F_{rk}. \tag{A.8}$$

The dimension of all these matrices and vectors corresponds to the virtual space (or, in practice, domains in the virtual space). The domains are implied by the superscripts of the residual and amplitude matrices or vectors, and will not be shown explicitly. The following contractions of amplitudes and integrals will be used and have to be computed in each iteration:

$$[\mathbf{K}(\mathbf{D}^{ij})]_{rs} = (rt|su)D^{ij}_{tu} \tag{A.9}$$

$$[\mathbf{K}(\mathbf{D}^{ij})^k]_r = [\mathbf{K}(\mathbf{D}^{ij})]_{rk} = (rt|ku)D^{ij}_{tu} \tag{A.10}$$

$$[\mathbf{J}(\mathbf{E}^{ik})]_{rs} = (rs|it)t^k_t, \tag{A.11}$$

$$[\mathbf{K}(\mathbf{E}^{ik})]_{rs} = (ri|st)t^k_t, \tag{A.12}$$

$$[\tilde{\mathbf{G}}]_{rs} = [2(rs|ku) - (rk|su)]t^k_u. \tag{A.13}$$

Note that in previous papers [18, 31], we denoted the latter matrix as $[\mathbf{G}(\mathbf{E})]_{rs}$. This notation stems from the formal definition $[\mathbf{E}^{ik}]_{tu} = \delta_{ti} t_u^k$ and $[\mathbf{G}(\mathbf{E})]_{rs} = \sum_k [2\mathbf{J}(\mathbf{E}^{kk}) - \mathbf{K}(\mathbf{E}^{kk})]_{rs}$. Here we will use the simplified notation.

In the calculation of $\mathbf{D}^{ij}$ in the PNO basis, the singles must be projected

$$D^{ij}_{a^{ij}b^{ij}} = T^{ij}_{a^{ij}b^{ij}} + \langle a^{ij}|c^{ii}\rangle t^i_{c^{ii}} t^j_{d^{jj}} \langle d^{jj}|b^{ij}\rangle. \tag{A.14}$$

In terms of these quantities, the singles and the doubles residuals are

$$\mathbf{r}^i = \tilde{\mathbf{f}}^i + \bar{\mathbf{F}}^\dagger \mathbf{t}^i - \mathbf{S}\mathbf{T}^{kl}\mathbf{l}^{lki} + \mathbf{S}\tilde{\mathbf{T}}^{ik}\bar{\mathbf{f}}^k + \mathbf{K}(\tilde{\mathbf{T}}^{ik})^k - \beta_{ik}\mathbf{S}\mathbf{t}^k, \tag{A.15}$$

$$\mathbf{R}^{ij} = \mathbf{K}^{ij} + \mathbf{K}(\mathbf{D}^{ij}) + \alpha_{ij,kl}\mathbf{S}\mathbf{D}^{kl}\mathbf{S} + \mathbf{G}^{ij} + \mathbf{G}^{ji^\dagger}, \tag{A.16}$$

$$\begin{aligned}\mathbf{G}^{ij} = {} & \mathbf{K}(\mathbf{E}^{ij}) + \mathbf{S}\mathbf{T}^{ij}\mathbf{X} + \mathbf{S}\left[\tilde{\mathbf{T}}^{ik}\mathbf{Y}^{kj} - \frac{1}{2}\mathbf{T}^{ki}\mathbf{Z}^{kj}\right] - (\mathbf{S}\mathbf{T}^{ki}\mathbf{Z}^{kj})^\dagger \\ & - \left[\mathbf{K}(\mathbf{D}^{ij})^k + \bar{\mathbf{k}}^{ikj} + \mathbf{J}^{jk}\mathbf{t}^i\right](\mathbf{S}\mathbf{t}^k)^\dagger - \beta_{ik}[\mathbf{S}\mathbf{T}^{kj}\mathbf{S}]\end{aligned} \tag{A.17}$$

with the intermediates

$$\alpha_{ij,kl} = (ik|jl) + \mathbf{t}^{i^\dagger}\mathbf{k}^{klj} + \mathbf{t}^{j^\dagger}\mathbf{k}^{lki} + tr(\mathbf{D}^{ij}\mathbf{K}^{lk}), \tag{A.18}$$

$$\beta_{ik} = f_{ik} + \mathbf{t}^{i^\dagger}\mathbf{f}^k + \mathbf{t}^{l^\dagger}\bar{\mathbf{l}}^{lki} + tr(\mathbf{L}^{kl}\mathbf{T}^{li}), \tag{A.19}$$

$$\tilde{\mathbf{f}}^i = \mathbf{f}^i + (2\mathbf{K}^{ik} - \mathbf{J}^{ik})\mathbf{t}^k, \tag{A.20}$$

$$\bar{\mathbf{f}}^k = \mathbf{f}^k + \mathbf{L}^{kl}\mathbf{t}^l, \tag{A.21}$$

$$\bar{\mathbf{F}} = \mathbf{F} + \tilde{\mathbf{G}} - \mathbf{L}^{lk}\mathbf{T}^{kl}\mathbf{S}, \tag{A.22}$$

$$\mathbf{X} = \bar{\mathbf{F}} - \bar{\mathbf{f}}^k(\mathbf{S}\mathbf{t}^k)^\dagger, \tag{A.23}$$

$$\mathbf{Y}^{kj} = \mathbf{K}^{kj} + \mathbf{K}(\mathbf{E}^{kj}) - \frac{1}{2}\left[\mathbf{J}^{kj} + \mathbf{J}(\mathbf{E}^{kj}) + \bar{\mathbf{l}}^{klj}(\mathbf{S}\mathbf{t}^l)^\dagger\right] + \frac{1}{4}\mathbf{L}^{kl}\tilde{\mathbf{T}}^{lj}\mathbf{S}, \tag{A.24}$$

$$\mathbf{Z}^{kj} = \mathbf{J}^{kj} + \mathbf{J}(\mathbf{E}^{kj}) - \frac{1}{2}\mathbf{K}^{lk}\mathbf{T}^{jl}\mathbf{S} - \bar{\mathbf{k}}^{lkj}(\mathbf{S}\mathbf{t}^l)^\dagger. \tag{A.25}$$

In a canonical CCSD or in a PAO-LCCSD program, the intermediates $\mathbf{X}$, $\mathbf{Y}^{kj}$, and $\mathbf{Z}^{kj}$ can be precomputed and then multiplied from the left with the appropriate amplitude matrices (in the PAO case the intermediates $\mathbf{Y}^{kj}$ and $\mathbf{Z}^{kj}$ are stored using united domains, see Ref. [147] for details). This leads to the smallest number of matrix multiplications. In a PNO-LCCSD method, this is not possible, since the left indices of the intermediates depend on the pair with which they are multiplied, and therefore each term in the intermediates must be recomputed for each i.

In our program, $\tilde{\mathbf{G}}$ is computed in each iteration in the full AO basis similar to a Fock matrix and subsequently transformed to the LMO/PAO|PNO basis. The intermediates $\bar{\mathbf{f}}^k$, $\tilde{\mathbf{f}}^i$, and β_{ik} can then be simplified to

$$\bar{f}^k_r = f^k_r + \tilde{G}_{rk}, \tag{A.26}$$

$$\tilde{f}^k_r = f^k_r + \tilde{G}_{kr}, \tag{A.27}$$

$$\beta_{ik} = f_{ik} + \tilde{G}_{ik} + \mathbf{t}^{i^\dagger}\bar{\mathbf{f}}^k + tr(\mathbf{L}^{kl}\mathbf{T}^{li}). \tag{A.28}$$

This also avoids summations over slowly decaying terms. Furthermore, the singles energy contribution can be compactly written as

$$E_{\text{sing}} = \mathbf{t}^{k^\dagger}(\mathbf{f}^k + \mathbf{L}^{kl}\mathbf{t}^l) = \mathbf{t}^{k^\dagger}\bar{\mathbf{f}}^k. \tag{A.29}$$

Note that all terms are now expressed by a "dressed" Fock matrix $\mathbf{F}_{\mathrm{AO}} + \tilde{\mathbf{G}}_{\mathrm{AO}}$. We compute the two contributions separately. $\mathbf{F}_{\mathrm{AO}}$ is taken from the preceding HF calculation, which uses the JKFIT auxiliary basis, while $\tilde{\mathbf{G}}_{\mathrm{AO}}$ is computed in each iteration using the MP2FIT auxiliary basis (for consistency with other terms). The sum of both contributions is subsequently transformed to the appropriate LMO/PAO spaces.

Appendix B: Derivation of the Interaction Matrices

The powers of x defined in equation (1.73) are

$$x = (\hat{\mathbf{r}} \cdot \hat{\mathbf{r}}) + 2(\hat{\mathbf{r}} \cdot \hat{\mathbf{R}}), \tag{B.1}$$

$$x^2 = (\hat{\mathbf{r}} \cdot \hat{\mathbf{r}})^2 + 4(\hat{\mathbf{r}} \cdot \hat{\mathbf{R}})^2 + 4(\hat{\mathbf{r}} \cdot \hat{\mathbf{r}})(\hat{\mathbf{r}} \cdot \hat{\mathbf{R}}), \tag{B.2}$$

$$x^3 = (\hat{\mathbf{r}} \cdot \hat{\mathbf{r}})^3 + 8(\hat{\mathbf{r}} \cdot \hat{\mathbf{R}})^3 + 6(\hat{\mathbf{r}} \cdot \hat{\mathbf{r}})^2(\hat{\mathbf{r}} \cdot \hat{\mathbf{R}}) + 12(\hat{\mathbf{r}} \cdot \hat{\mathbf{r}})(\hat{\mathbf{r}} \cdot \hat{\mathbf{R}})^2, \tag{B.3}$$

$$\begin{aligned} x^4 = {} & (\hat{\mathbf{r}} \cdot \hat{\mathbf{r}})^4 + 16(\hat{\mathbf{r}} \cdot \hat{\mathbf{R}})^4 + 24(\hat{\mathbf{r}} \cdot \hat{\mathbf{r}})^2(\hat{\mathbf{r}} \cdot \hat{\mathbf{R}})^2 \\ & + 8(\hat{\mathbf{r}} \cdot \hat{\mathbf{r}})^4(\hat{\mathbf{r}} \cdot \hat{\mathbf{R}}) + 32(\hat{\mathbf{r}} \cdot \hat{\mathbf{r}})(\hat{\mathbf{r}} \cdot \hat{\mathbf{R}})^3. \end{aligned} \tag{B.4}$$

The different interactions are obtained by replacing $\hat{\mathbf{r}}$ by the appropriate $-\bar{\mathbf{r}}_1/R$ or $\bar{\mathbf{r}}_2/R$. Whenever $(\bar{\mathbf{r}}_1 \cdot \bar{\mathbf{r}}_2)$ occurs, a factor of 2 applies, since the contributions of $(\bar{\mathbf{r}}_1 \cdot \bar{\mathbf{r}}_2)$ and $(\bar{\mathbf{r}}_2 \cdot \bar{\mathbf{r}}_1)$ are the same. Similarly, appropriate factors apply for powers of $(\hat{\mathbf{r}} \cdot \hat{\mathbf{R}})$. This yields for the dipole—dipole interaction:

$$\begin{aligned} V_{1,1} &= -2c_1(\bar{\mathbf{r}}_1 \cdot \bar{\mathbf{r}}_2) - 8c_2(\bar{\mathbf{r}}_1 \cdot \hat{\mathbf{R}})(\bar{\mathbf{r}}_2 \cdot \hat{\mathbf{R}}) \\ &= \bar{\mathbf{r}}_1 \cdot \bar{\mathbf{r}}_2 - 3(\bar{\mathbf{r}}_1 \cdot \hat{\mathbf{R}})(\bar{\mathbf{r}}_2 \cdot \hat{\mathbf{R}}). \end{aligned} \tag{B.5}$$

Using the factors $c_1 = -\frac{1}{2}$, $c_2 = \frac{3}{8}$, $c_3 = -\frac{5}{16}$, $c_4 = \frac{35}{128}$, the terms describing the interaction of higher moments are derived similarly. The results are summarized in the following. The interaction potentials in equation (1.77) are

Dipole–Quadrupole:

$$\begin{aligned} V_{1,2} &= -8c_2(\bar{\mathbf{r}}_1 \cdot \bar{\mathbf{r}}_2)(\bar{\mathbf{r}}_2 \cdot \hat{\mathbf{R}}) - 4c_2(\bar{\mathbf{r}}_2 \cdot \bar{\mathbf{r}}_2)(\bar{\mathbf{r}}_1 \cdot \hat{\mathbf{R}}) - 24c_3(\bar{\mathbf{r}}_1 \cdot \hat{\mathbf{R}})(\bar{\mathbf{r}}_2 \cdot \hat{\mathbf{R}})^2 \\ &= -3(\bar{\mathbf{r}}_1 \cdot \bar{\mathbf{r}}_2)(\bar{\mathbf{r}}_2 \cdot \hat{\mathbf{R}}) - \frac{3}{2}(\bar{\mathbf{r}}_2 \cdot \bar{\mathbf{r}}_2)(\bar{\mathbf{r}}_1 \cdot \hat{\mathbf{R}}) + \frac{15}{2}(\bar{\mathbf{r}}_1 \cdot \hat{\mathbf{R}})(\bar{\mathbf{r}}_2 \cdot \hat{\mathbf{R}})^2; \end{aligned} \tag{B.6}$$

Quadrupole–Quadrupole:

$$\begin{aligned} V_{2,2} = {} & 2c_2\left[2(\bar{\mathbf{r}}_1 \cdot \bar{\mathbf{r}}_2)(\bar{\mathbf{r}}_1 \cdot \bar{\mathbf{r}}_2) + (\bar{\mathbf{r}}_1 \cdot \bar{\mathbf{r}}_1)(\bar{\mathbf{r}}_2 \cdot \bar{\mathbf{r}}_2)\right] \\ & + 12c_3\left[(\bar{\mathbf{r}}_1 \cdot \bar{\mathbf{r}}_1)(\bar{\mathbf{r}}_2 \cdot \bar{\mathbf{r}})^2 + (\bar{\mathbf{r}}_2 \cdot \bar{\mathbf{r}}_2)(\bar{\mathbf{r}}_1 \cdot \bar{\mathbf{r}})^2\right] \\ & + 48c_3(\bar{\mathbf{r}}_1 \cdot \hat{\mathbf{R}})(\bar{\mathbf{r}}_1 \cdot \bar{\mathbf{r}}_2)(\bar{\mathbf{r}}_2 \cdot \hat{\mathbf{R}}) + 96c_4(\bar{\mathbf{r}}_1 \cdot \hat{\mathbf{R}})^2(\bar{\mathbf{r}}_2 \cdot \hat{\mathbf{R}})^2 \\ = {} & \frac{3}{2}(\bar{\mathbf{r}}_1 \cdot \bar{\mathbf{r}}_2)(\bar{\mathbf{r}}_1 \cdot \bar{\mathbf{r}}_2) + \frac{3}{4}(\bar{\mathbf{r}}_1 \cdot \bar{\mathbf{r}}_1)(\bar{\mathbf{r}}_2 \cdot \bar{\mathbf{r}}_2) \\ & - \frac{15}{4}\left[(\bar{\mathbf{r}}_1 \cdot \bar{\mathbf{r}}_1)(\bar{\mathbf{r}}_2 \cdot \hat{\mathbf{R}})^2 + (\bar{\mathbf{r}}_2 \cdot \bar{\mathbf{r}}_2)(\bar{\mathbf{r}}_1 \cdot \hat{\mathbf{R}})^2\right] \\ & - 15(\bar{\mathbf{r}}_1 \cdot \hat{\mathbf{R}})(\bar{\mathbf{r}}_1 \cdot \bar{\mathbf{r}}_2)(\bar{\mathbf{r}}_2 \cdot \hat{\mathbf{R}}) + \frac{105}{4}(\bar{\mathbf{r}}_1 \cdot \hat{\mathbf{R}})^2(\bar{\mathbf{r}}_2 \cdot \hat{\mathbf{R}})^2; \end{aligned} \tag{B.7}$$

Octupole–Dipole:

$$\begin{aligned} V_{1,3} &= -4c_2\,(\bar{\mathbf{r}}_1\cdot\bar{\mathbf{r}}_2)(\bar{\mathbf{r}}_2\cdot\bar{\mathbf{r}}_2) \\ &\quad -24c_3\big[(\bar{\mathbf{r}}_1\cdot\bar{\mathbf{r}}_2)(\bar{\mathbf{r}}_2\cdot\hat{\mathbf{R}})^2-(\bar{\mathbf{r}}_1\cdot\hat{\mathbf{R}})(\bar{\mathbf{r}}_2\cdot\bar{\mathbf{r}}_2)(\bar{\mathbf{r}}_2\cdot\hat{\mathbf{R}})\big] \\ &\quad -64c_4(\bar{\mathbf{r}}_2\cdot\hat{\mathbf{R}})^3(\bar{\mathbf{r}}_1\cdot\hat{\mathbf{R}}) \\ &= -\frac{3}{2}(\bar{\mathbf{r}}_1\cdot\bar{\mathbf{r}}_2)(\bar{\mathbf{r}}_2\cdot\bar{\mathbf{r}}_2)+\frac{15}{2}\big[(\bar{\mathbf{r}}_1\cdot\bar{\mathbf{r}}_2)(\bar{\mathbf{r}}_2\cdot\hat{\mathbf{R}})^2+(\bar{\mathbf{r}}_1\cdot\hat{\mathbf{R}})(\bar{\mathbf{r}}_2\cdot\bar{\mathbf{r}}_2)(\bar{\mathbf{r}}_2\cdot\hat{\mathbf{R}})\big] \\ &\quad -\frac{35}{2}(\bar{\mathbf{r}}_2\cdot\hat{\mathbf{R}})^3(\bar{\mathbf{r}}_1\cdot\hat{\mathbf{R}})\big]. \end{aligned} \tag{B.8}$$

The remaining two interactions are obtained using

$$V_{2,1}(\bar{\mathbf{r}}_1,\bar{\mathbf{r}}_2,\hat{\mathbf{R}}) = -V_{1,2}(\bar{\mathbf{r}}_2,\bar{\mathbf{r}}_1,\hat{\mathbf{R}}), \tag{B.9}$$

$$V_{3,1}(\bar{\mathbf{r}}_1,\bar{\mathbf{r}}_2,\hat{\mathbf{R}}) = V_{1,3}(\bar{\mathbf{r}}_2,\bar{\mathbf{r}}_1,\hat{\mathbf{R}}). \tag{B.10}$$

In order to obtain the interaction matrices, one has to expand the scalar products and collect the contributions of $\bar{\mathbf{r}}_1$ and $\bar{\mathbf{r}}_2$. In the following, $\alpha,\beta,\gamma,\mu,\nu\in\{x,y,z\}$. For example,

$$(\bar{\mathbf{r}}_1\cdot\bar{\mathbf{r}}_2)^2=\sum_{\alpha\beta}(\alpha_1\beta_1)(\alpha_2\beta_2)=\sum_{\alpha\le\beta}\frac{2}{l_\alpha!l_\beta!}(\alpha_1\beta_1)(\alpha_2\beta_2).$$

The factor $2/(l_\alpha!l_\beta!)=2-\delta_{\alpha\beta}$ occurs since we restrict the summations in equation (1.83) to the unique second-moment operators, that is, $\alpha\le\beta$ (the l_α, l_β are defined as explained below equation (1.84)). Similarly, for third moments, there may be up to $3!=6$ equal contributions, and factors of $6/(l_\alpha!l_\beta!l_\gamma!)$ arise.

The interaction matrices up to $p=3$ read explicitly:

$$A^{(1,1)}_{\alpha,\beta}=\delta_{\alpha\beta}-3\hat{R}_\alpha\hat{R}_\beta, \tag{B.11}$$

$$\begin{aligned} A^{(2,1)}_{\alpha\beta,\gamma} &= \frac{1}{l_\alpha l_\beta}\Big[15\hat{R}_\alpha\hat{R}_\beta\hat{R}_\gamma-3(\hat{R}_\alpha\delta_{\beta\gamma}+\hat{R}_\beta\delta_{\alpha\gamma}+\hat{R}_\gamma\delta_{\alpha\beta})\Big], \\ A^{(1,2)}_{\gamma,\alpha\beta} &= -A^{(2,1)}_{\alpha\beta,\gamma}, \qquad \alpha\le\beta, \end{aligned} \tag{B.12}$$

$$\begin{aligned} A^{(2,2)}_{\alpha\beta,\mu\nu} &= \frac{1}{l_\alpha!l_\beta!l_\mu!l_\nu!}\Big[3(\delta_{\alpha\mu}\delta_{\beta\nu}+\delta_{\alpha\nu}\delta_{\beta\mu}+\delta_{\alpha\beta}\delta_{\mu\nu}) \\ &\quad -15\,(\delta_{\alpha\beta}\hat{R}_\mu R_\nu+\delta_{\alpha\mu}\hat{R}_\beta R_\nu+\delta_{\alpha\nu}\hat{R}_\beta R_\mu \\ &\quad +\delta_{\beta\mu}\hat{R}_\alpha\hat{R}_\nu+\delta_{\beta\nu}\hat{R}_\alpha\hat{R}_\mu+\delta_{\mu\nu}\hat{R}_\alpha\hat{R}_\beta) \\ &\quad +105\,\hat{R}_\alpha\hat{R}_\beta\hat{R}_\mu\hat{R}_\nu\Big] \qquad \alpha\le\beta,\ \mu\le\nu, \end{aligned} \tag{B.13}$$

$$A^{(3,1)}_{\alpha\beta\gamma,\mu}=A^{(2,2)}_{\alpha\beta,\gamma\mu}, \tag{B.14}$$

$$A^{(1,3)}_{\mu,\alpha\beta\gamma}=A^{(3,1)}_{\alpha\beta\gamma,\mu}, \qquad \alpha\le\beta\le\gamma. \tag{B.15}$$

The interaction matrices for the quadrupole–quadrupole and octupole–dipole interactions are formally identical but differ by the prefactors and index restrictions.

Acknowledgments

This work has been funded by the ERC Advanced Grant 320723 (ASES).

References

1 Pulay, P. (1983) Localizability of dynamic electron correlation. *Chem. Phys. Lett.*, **100**, 151–154.

2 Saebø, S. and Pulay, P. (1985) Local configuration-interaction: An efficient approach for larger molecules. *Chem. Phys. Lett.*, **113**, 13–18.

3 Pulay, P. and Saebø, S. (1986) Orbital-invariant formulation and second-order gradient evaluation in Møller-Plesset perturbation theory. *Theor. Chim. Acta*, **69**, 357–368.

4 Saebø, S. and Pulay, P. (1987) Fourth-order Møller-Plesset perturbation theory in the local correlation treatment. I. Method. *J. Chem. Phys.*, **86**, 914–922.

5 Saebø, S. and Pulay, P. (1988) The local correlation treatment. II. Implementation and tests. *J. Chem. Phys.*, **88**, 1884–1890.

6 Saebø, S. and Pulay, P. (1993) Local treatment of electron correlation. *Annu. Rev. Phys. Chem.*, **44**, 213–236.

7 Hampel, C. and Werner, H.J. (1996) Local treatment of electron correlation in coupled cluster theory. *J. Chem. Phys.*, **104**, 6286–6297.

8 Maslen, P.E. and Head-Gordon, M. (1998) Non-iterative local second order Moller-Plesset theory. *Chem. Phys. Lett.*, **283**, 102–180.

9 Maslen, P.E. and Head-Gordon, M. (1998) Noniterative local second order Moller-Plesset theory: Convergence with local correlation space. *J. Chem. Phys.*, **109**, 7093–7099.

10 Ayala, P.Y. and Scuseria, G.E. (1999) Linear scaling second-order Møller-Plesset theory in the atomic orbital basis for large molecular systems. *J. Chem. Phys.*, **110**, 3660–3671.

11 Scuseria, G.E. and Ayala, P.Y. (1999) Linear scaling coupled cluster and perturbation theories in the atomic orbital basis. *J. Chem. Phys.*, **111**, 8330–8343.

12 Hetzer, G., Pulay, P., and Werner, H.J. (1998) Multipole approximation of distant pair energies in local MP2 calculations. *Chem. Phys. Lett.*, **290**, 143–149.

13 Hetzer, G., Schütz, M., Stoll, H., and Werner, H.J. (2000) Low-order scaling local correlation methods II: Splitting the Coulomb operator in linear scaling local second-order Møller-Plesset perturbation theory. *J. Chem. Phys.*, **113**, 9443–9455.

14 Schütz, M., Rauhut, G., and Werner, H.J. (1998) Local treatment of electron correlation in molecular clusters: Structures and stabilities of $(H_2O)_n$, $n = 2$–4. *J. Phys. Chem. A*, **102**, 5997–6003.

15 Schütz, M., Hetzer, G., and Werner, H.J. (1999) Low-order scaling local electron correlation methods. I. Linear scaling local MP2. *J. Chem. Phys.*, **111**, 5691–5705.

16 Schütz, M. and Werner, H.J. (2000) Local perturbative triples correction (T) with linear cost scaling. *Chem. Phys. Lett.*, **318**, 370–378.

17 Schütz, M. (2000) Low-order scaling local electron correlation methods. III. Linear scaling local perturbative triples correction (T). *J. Chem. Phys.*, **113**, 9986–10001.

18 Schütz, M. and Werner, H.J. (2001) Low-order scaling local electron correlation methods. IV. Linear scaling local coupled-cluster (LCCSD). *J. Chem. Phys.*, **114**, 661–681.
19 Schütz, M. (2002) Linear scaling local connected triples beyond local (T): Local CCSDT-1b with O(N) scaling. *J. Chem. Phys.*, **116**, 8772–8785.
20 Schütz, M. (2002) A new, fast, semi-direct implementation of linear scaling local coupled cluster theory. *Phys. Chem. Chem. Phys.*, **4**, 3941–3947.
21 Schütz, M. and Manby, F.R. (2003) Linear scaling local coupled cluster theory with density fitting. I : 4-external integrals. *Phys. Chem. Chem. Phys.*, **5**, 3349–3358.
22 Werner, H.J., Manby, F.R., and Knowles, P. (2003) Fast linear scaling second-order Møller-Plesset perturbation theory (MP2) using local and density fitting approximations. *J. Chem. Phys.*, **118**, 8149–8160.
23 Schütz, M., Werner, H.J., Lindh, R., and Manby, F.R. (2004) Analytical energy gradients for local second-order Møller-Plesset perturbation theory using density fitting approximations. *J. Chem. Phys.*, **121**, 737–750.
24 Auer, A. and Nooijen, M. (2006) Dynamically screened local correlation method using enveloping localized orbitals. *J. Chem. Phys.*, **125**, 024 104.
25 Hrenar, T., Rauhut, G., and Werner, H.J. (2006) Impact of local and density fitting approximations on harmonic vibrational frequencies. *J. Phys. Chem. A*, **110**, 2060–2064.
26 Werner, H.J. and Pflüger, K. (2006) On the selection of domains and orbital pairs in local correlation treatments. *Ann. Reports Comput. Chem.*, **2**, 53–80.
27 Mata, R. and Werner, H.J. (2006) Calculation of smooth potential energy surfaces using local electron correlation methods. *J. Chem. Phys.*, **125**, 184110.
28 Mata, R. and Werner, H.J. (2007) Local correlation methods with a natural localized molecular orbital basis. *Mol. Phys.*, **105**, 2753–2761.
29 Mata, R.A., Werner, H.J., Thiel, S., and Thiel, W. (2008) Towards accurate barriers for enzymatic reactions: QM/MM case study on *p*-hydroxybenzoate hydroxylase. *J. Chem. Phys.*, **128**, 025104.
30 Mata, R., Werner, H.J., and Schütz, M. (2008) Correlation regions within a localized molecular orbital approach. *J. Chem. Phys.*, **128**, 144106.
31 Werner, H.J. and Schütz, M. (2011) An efficient local coupled cluster method for accurate thermochemistry of large systems. *J. Chem. Phys.*, **135**, 144116.
32 Schwilk, M., Usvyat, D., and Werner, H.J. (2015) Communication: Improved pair approximations in local coupled-cluster methods. *J. Chem. Phys.*, **142**, 121102.
33 Köppl, C. and Werner, H.J. (2015) On the use of Abelian point group symmetry in density-fitted local MP2 using various types of virtual orbitals. *J. Chem. Phys.*, **142**, 164108.
34 Kats, D. and Manby, F.R. (2013) Sparse tensor framework for implementation of general local correlation methods. *J. Chem. Phys.*, **138**, 144101.
35 Kats, D. (2014) Speeding up local correlation methods. *J. Chem. Phys.*, **141**, 244101.
36 Russ, N.J. and Crawford, T.D. (2004) Local correlation in coupled cluster calculations of molecular response properties. *Chem. Phys. Lett.*, **400**, 104–111.
37 Maslen, P.E., Dutoi, A.D., Lee, M.S., Shao, Y., and Head-Gordon, M. (2005) Accurate local approximations to the triples correlation energy: Formulation, implementation and tests of 5th-order scaling models. *Mol. Phys.*, **103**, 425–437.

38 DiStasio, R.A., Jung, Y.S., and Head-Gordon, M. (2005) A resolution-of-the-identity implementation of the local triatomics-in-molecules model for second-order Møller-Plesset perturbation theory with application to alanine tetrapeptide conformational energies. *J. Chem. Theory Comput.*, **1**, 862–876.

39 Subotnik, J.E. and Head-Gordon, M. (2005) A local correlation model that yields intrinsically smnooth potential-energy surfaces. *J. Chem. Phys.*, **123**, 064108.

40 Lawler, K.V., Parkhill, J.A., and Head-Gordon, M. (2008) Penalty functions for combining coupled-cluster and perturbation amplitudes in local correlation methods with optimized orbitals. *Mol. Phys.*, **106**, 2309–2324.

41 Schweizer, S., Kussmann, J., Doser, B., and Ochsenfeld, C. (2008) Linear-scaling Cholesky decomposition. *J. Comp. Chem.*, **29**, 1004–1010.

42 Schweizer, S., Doser, B., and Ochsenfeld, C. (2008) An atomic orbital-based reformulation of energy gradients in second-order Møller-Plesset perturbation theory. *J. Chem. Phys.*, **128**, 154101.

43 Doser, B., Lambrecht, D.S., Kussmann, J., and Ochsenfeld, C. (2009) Linear-scaling atomic orbital-based second-order Møller–Plesset perturbation theory by rigorous integral screening criteria. *J. Chem. Phys.*, **130**, 064107.

44 Doser, B., Zienau, J., Clin, L., Lambrecht, D.S., and Ochsenfeld, C. (2010) A linear-scaling MP2 method for large molecules by rigorous integral-screening criteria. *Z. Phys. Chem.*, **224**, 397–412.

45 Kats, D. and Schütz, M. (2009) A multistate local coupled cluster CC2 response method based on the Laplace transform. *J. Chem. Phys.*, **131**, 124117.

46 Freundorfer, K., Kats, D., Korona, T., and Schütz, M. (2010) Local CC2 response method for triplet states based on Laplace transform: Excitation energies and first-order properties. *J. Chem. Phys.*, **133**, 244110.

47 Maschio, L. (2011) Local MP2 with density fitting for periodic systems: A parallel implementation. *J. Chem. Theory Comput.*, **7**, 2818–2830.

48 Neese, F., Wennmohs, F., and Hansen, A. (2009) Efficient and accurate local approximations to coupled-electron pair approaches: An attempt to revive the pair natural orbital method. *J. Chem. Phys.*, **130**, 114108.

49 Neese, F., Hansen, A., and Liakos, D.G. (2009) Efficient and accurate approximations to the local coupled cluster singles doubles method using a truncated pair natural orbital basis. *J. Chem. Phys.*, **131**, 064103.

50 Hansen, A., Liakos, D.G., and Neese, F. (2011) Efficient and accurate local single reference correlation methods for high-spin open-shell molecules using pair natural orbitals. *J. Chem. Phys.*, **135**, 214102.

51 Liakos, D.G., Hansen, A., and Neese, F. (2011) Weak molecular interactions studied with parallel implementations of the local pair natural orbital coupled pair and coupled cluster methods. *J. Chem. Theory Comput.*, **7**, 76–87.

52 Izsak, R., Hansen, A., and Neese, F. (2012) The resolution of identity and chain of spheres approximations for the LPNO-CCSD singles Fock term. *Mol. Phys.*, **110**, 2413–2417.

53 Schmitz, G., Helmich, B., and Hättig, C. (2013) A O(N-3) scaling PNO-MP2 method using a hybrid OSV-PNO approach with an iterative direct generation of OSVs. *Mol. Phys.*, **111**, 2463–2476.

54 Riplinger, C. and Neese, F. (2013) An efficient and near linear scaling pair natural orbital based local coupled cluster method. *J. Chem. Phys.*, **138**, 034106.
55 Riplinger, C., Sandhoefer, B., Hansen, A., and Neese, F. (2013) Natural triple excitations in local coupled cluster calculations with pair natural orbitals. *J. Chem. Phys.*, **139**, 134101.
56 Pinski, P., Riplinger, C., Valeev, E.F., and Neese, F. (2015) Sparse maps—A systematic infrastructure for reduced-scaling electronic structure methods. I. An efficient and simple linear scaling local MP2 method that uses an intermediate basis of pair natural orbitals. *J. Chem. Phys.*, **143**, 034108.
57 Riplinger, C., Pinski, P., Becker, U., Valeev, E.F., and Neese, F. (2016) Sparse maps—A systematic infrastructure for reduced-scaling electronic structure methods. II. Linear scaling domain based pair natural orbital coupled cluster theory. *J. Chem. Phys.*, **144**, 024109.
58 Maurer, S.A., Clin, L., and Ochsenfeld, C. (2014) Cholesky-decomposed density MP2 with density fitting: Accurate MP2 and double-hybrid DFT energies for large systems. *J. Chem. Phys.*, **140**, 224112.
59 Schurkus, H.F. and Ochsenfeld, C. (2016) Communication: An effective linear-scaling atomic-orbital reformulation of the random-phase approximation using a contracted double-Laplace transformation. *J. Chem. Phys.*, **144**, 031101.
60 Fedorov, D.G. and Kitaura, K. (2005) Coupled-cluster theory based upon the fragment molecular-orbital method. *J. Chem. Phys.*, **123**, 134103.
61 Fedorov, A., Couzijn, E.P.A., Nagornova, N.S., Boyarkin, O.V., Rizzo, T.R., and Chen, P. (2010) Structure and bonding of isoleptic coinage metal (Cu, Ag, Au) dimethylaminonitrenes in the gas phase. *J. Am. Chem. Soc.*, **132**, 13789–13798.
62 Hughes, T.F., Flocke, N., and Bartlett, R.J. (2008) Natural linear-scaled coupled-cluster theory with local transferable triple excitations: Applications to peptides. *J. Phys. Chem. A*, **112**, 5994–6003.
63 Gordon, M.S., Mullin, J.M., Pruitt, S.R., Roskop, L.B., Slipchenko, L.V., and Boatz, J.A. (2009) Accurate methods for large molecular systems. *J. Phys. Chem. B*, **113**, 9646–9663.
64 Gordon, M.S., Fedorov, D.G., Pruitt, S.R., and Slipchenko, L.V. (2012) Fragmentation methods: A route to accurate calculations on large systems. *Chem. Rev.*, **112**, 632–672.
65 Findlater, A.D., Zahariev, F., and Gordon, M.S. (2015) Combined fragment molecular orbital cluster in molecule approach to massively parallel electron correlation calculations for large systems. *J. Phys. Chem. A*, **119**, 3587–3593.
66 Ziółkowski, M., Jansìk, B., Kjærgaard, T., and Jørgensen, P. (2010) Linear scaling coupled cluster method with correlation energy based error control. *J. Chem. Phys.*, **133**, 014107.
67 Kristensen, K., Ziółkowski, M., Jansìk, B., Kjærgaard, T., and Jørgensen, P. (2011) A locality analysis of the divide-expand-consolidate coupled cluster amplitude equations. *J. Chem. Theory Comput.*, **7**, 1677–1694.
68 Høyvik, I.M., Kristensen, K., Jansìk, B., and Jørgensen, P. (2012) The divide-expand-consolidate family of coupled cluster methods: Numerical illustrations using second order Møller-Plesset perturbation theory. *J. Chem. Phys.*, **136**, 014105.
69 Kristensen, K., Jørgensen, P., Jansìk, B., Kjærgaard, T., and Reine, S. (2012) Molecular gradient for second-order Møller-Plesset perturbation theory using the divide-expand-consolidate (DEC) scheme. *J. Chem. Phys.*, **137**, 114102.

70 Kristensen, K., Kjærgaard, T., Høyvik, I.M., Ettenhuber, P., Jørgensen, P., Jansik, B., Reine, S., and Jakowski, J. (2013) The divide-expand-consolidate MP2 scheme goes massively parallel. *Mol. Phys.*, **111**, 1196–1210.

71 Eriksen, J.J., Baudin, P., Ettenhuber, P., Kristensen, K., Kjærgaard, T., and Jørgensen, P. (2015) Linear-scaling coupled cluster with perturbative triple excitations: The divide–expand–consolidate CCSD(T) model. *J. Chem. Theory Comput.*, **11**, 2984–2993.

72 Baudin, P., Ettenhuber, P., Reine, S., Kristensen, K., and Kjaergaard, T. (2016) Efficient linear-scaling second-order Møller-Plesset perturbation theory: The divide-expand-consolidate RI-MP2 model. *J. Chem. Phys.*, **144**, 054102.

73 Bykov, D., Kristensen, K., and Kjaergaard, T. (2016) The molecular gradient using the divide-expand-consolidate resolution of the identity second-order Møller-Plesset perturbation theory: The DEC-RI-MP2 gradient. *J. Chem. Phys.*, **145**, 024106.

74 Ettenhuber, P., Baudin, P., Kjaergaard, T., Jorgensen, P., and Kristensen, K. (2016) Orbital spaces in the divide-expand-consolidate coupled cluster method. *J. Chem. Phys.*, **144**, 164116.

75 Li, W., Piecuch, P., Gour, J.R., and Li, S. (2009) Local correlation calculations using standard and renormalized coupled-cluster approaches. *J. Chem. Phys.*, **131**, 114109.

76 Li, W. and Piecuch, P. (2010) Improved design of orbital domains within the cluster-in-molecule local correlation framework: Single-environment cluster-in-molecule ansatz and its application to local coupled-cluster approach with singles and doubles. *J. Phys. Chem. A*, **114**, 8644–8657.

77 Li, W. and Piecuch, P. (2010) Multilevel extension of the cluster-in-molecule local correlation methodology: Merging coupled-cluster and Møller-Plesset perturbation theories. *J. Phys. Chem. A*, **114**, 6721–6727.

78 Li, W., Guo, Y., and Li, S. (2012) A refined cluster-in-molecule local correlation approach for predicting the relative energies of large systems. *Phys. Chem. Chem. Phys.*, **14**, 7854–7862.

79 Guo, Y., Li, W., and Li, S. (2014) Improved cluster-in-molecule local correlation approach for electron correlation calculation of large systems. *J. Phys. Chem. A*, **118**, 8996–9004.

80 Li, W., Chen, C., Zhao, D., and Li, S. (2015) LSQC: Low scaling quantum chemistry program. *Int. J. Quantum Chem.*, **115**, 641–646.

81 Li, W., Ni, Z., and Li, S. (2016) Cluster-in-molecule local correlation method for post-Hartree-Fock calculations of large systems. *Mol. Phys.*, **114**, 1447–1460.

82 Rolik, Z. and Kallay, M. (2011) A general-order local coupled-cluster method based on the cluster-in-molecule approach. *J. Chem. Phys.*, **135**, 104111.

83 Rolik, Z., Szegedy, L., Ladjánszki, I., Ladóczki, B., and Kallay, M. (2013) An efficient linear-scaling CCSD(T) method based on local natural orbitals. *J. Chem. Phys.*, **139**, 094105.

84 Kallay, M. (2015) Linear-scaling implementation of the direct random-phase approximation. *J. Chem. Phys.*, **142**, 204105.

85 Nagy, P.R., Samu, G., and Kallay, M. (2016) An integral-direct linear-scaling second-order Møller-Plesset approach. *J. Chem. Theory Comput.*, **12**, 4897–4914.

86 Saha, A. and Raghavachari, K. (2014) Dimers of dimers (DOD): A new fragment-based method applied to large water clusters. *J. Chem. Theory Comput.*, **10**, 58–67.

87 Yoshikawa, T. and Nakai, H. (2015) Linear-scaling self-consistent field calculations based on divide-and-conquer method using resolution-of-identity approximation on graphical processing units. *J. Comput. Chem.*, **36**, 164–170.

88 Stoll, H. (1992) On the correlation energy of graphite. *J. Chem. Phys.*, **97**, 8449–8454.

89 Friedrich, J., Hanrath, M., and Dolg, M. (2007) Fully automated implementation of the incremental scheme: Application to CCSD energies for hydrocarbons and transition metal compounds. *J. Chem. Phys.*, **126**, 154110.

90 Friedrich, J., Hanrath, M., and Dolg, M. (2008) Evaluation of incremental correlation energies for open-shell systems: Application to the intermediates of the 4-exo cyclization, Arduengo carbenes and an anionic water cluster. *J. Phys. Chem. A*, **112**, 8762–8766.

91 Friedrich, J. and Dolg, M. (2008) Implementation and performance of a domain-specific basis set incremental approach for correlation energies: Applications to hydrocarbons and a glycine oligomer. *J. Chem. Phys.*, **129**, 244105.

92 Friedrich, J., Coriani, S., Helgaker, T., and Dolg, M. (2009) Implementation of the incremental scheme for one-electron first-order properties in coupled-cluster theory. *J. Chem. Phys.*, **131**, 154102.

93 Friedrich, J. and Dolg, M. (2009) Fully automated incremental evaluation of MP2 and CCSD(T) energies: Application to water clusters. *J. Chem. Theory Comput.*, **5**, 287–294.

94 Friedrich, J. and Hänchen, J. (2013) Incremental CCSD(T)(F12*)|MP2: A black box method to obtain highly accurate reaction energies. *J. Chem. Theory Comput.*, **9**, 5381–5394.

95 Meitei, O.R. and Hesselmann, A. (2016) Molecular energies from an incremental fragmentation method. *J. Chem. Phys.*, **144**, 084109.

96 Hoyau, S., Maynau, D., and Malrieu, J.P. (2011) A regionally contracted multireference configuration interaction method: General theory and results of an incremental version. *J. Chem. Phys.*, **134**, 054125.

97 Kutzelnigg, W. (1985) r_{12}-Dependent terms in the wave function as closed sums of partial wave amplitudes for large *l*. *Theor. Chim. Acta*, **68**, 445–469.

98 Klopper, W. and Kutzelnigg, W. (1987) Møller-Plesset calculations taking care of the correlation CUSP. *Chem. Phys. Lett.*, **134**, 17–22.

99 Klopper, W. and Kutzelnigg, W. (1990) MP2-R12 calculations on the relative stability of carbocations. *J. Phys. Chem.*, **94**, 5625.

100 Klopper, W., Röhse, R., and Kutzelnigg, W. (1991) CID and CEPA calculations with linear r_{12} terms. *Chem. Phys. Lett.*, **178**, 455–461.

101 Kutzelnigg, W. and Klopper, W. (1991) Wave functions with terms linear in the interelectronic coordinates to take care of the correlation cusp. I. General theory. *J. Chem. Phys.*, **94**, 1985–2001.

102 Termath, V., Klopper, W., and Kutzelnigg, W. (1991) Wave functions with terms linear in the interelectronic coordinates to take care of the correlation cusp. II. Second-order Møller-Plesset (MP2-R12) calculations on closed-shell atoms. *J. Chem. Phys.*, **94**, 2002–2019.

103 Klopper, W. and Kutzelnigg, W. (1991) Wave functions with terms linear in the interelectronic coordinates to take care of the correlation cusp. III. Second-order Møller-Plesset (MP2-R12) calculations on molecules of first row atoms. *J. Chem. Phys.*, **94**, 2020–2030.

104 Klopper, W. (1991) Orbital-invariant formulation of the MP2-R12 method. *Chem. Phys. Lett.*, **186**, 583–585.

105 Noga, J., Kutzelnigg, W., and Klopper, W. (1992) CC-R12, a correlation cusp corrected coupled-cluster method with a pilot application to the Be_2 potential curve. *Chem. Phys. Lett.*, **199**, 497–504.

106 Noga, J. and Kutzelnigg, W. (1994) Coupled cluster theory that takes care of the correlation cusp by inclusion of linear terms in the interelectronic coordinates. *J. Chem. Phys.*, **101**, 7738–7762.

107 Klopper, W. and Samson, C.C.M. (2002) Explicitly correlated second-order Møller-Plesset methods with auxiliary basis sets. *J. Chem. Phys.*, **116**, 6397–6410.

108 Manby, F.R. (2003) Density fitting in second-order linear-r_{12} Møller-Plesset perturbation theory. *J. Chem. Phys.*, **119**, 4607–4613.

109 Ten-no, S. (2004) Initiation of explicitly correlated Slater-type germinal theory. *Chem. Phys. Lett.*, **398**, 56–61.

110 Ten-no, S. (2004) Explicitly correlated second order perturbation theory: Introduction of a rational generator and numerical quadratures. *J. Chem. Phys.*, **121**, 117–129.

111 May, A.J. and Manby, F.R. (2004) An explicitly correlated second order Møller-Plesset theory using a frozen Gaussian geminal. *J. Chem. Phys.*, **121**, 4479–4485.

112 Valeev, E.F. (2004) Improving on the resolution of the identity in linear R12 ab initio theories. *Chem. Phys. Lett.*, **395**, 190–195.

113 Tew, D.P. and Klopper, W. (2005) New correlation factors for explicitly correlated electronic wave functions. *J. Chem. Phys.*, **123**, 074101.

114 Kedžuch, S., Milko, M., and Noga, J. (2005) Alternative formulation of the matrix elements in MP2-R12 theory. *Int. J. Quantum Chem.*, **105**, 929–936.

115 Fliegl, H., Klopper, W., and Hättig, C. (2005) Coupled-cluster theory with simplified linear-r_{12} corrections: The CCSD(R12) model. *J. Chem. Phys.*, **122**, 084107.

116 Fliegl, H., Hättig, C., and Klopper, W. (2006) Inclusion of the (T) triples correction into the linear-r12 corrected coupled-cluster model CCSD(R12). *Int. J. Quantum Chem.*, **106**, 2306–2317.

117 Manby, F.R., Werner, H.J., Adler, T.B., and May, A.J. (2006) Explicitly correlated local second-order perturbation theory with a frozen geminal correlation factor. *J. Chem. Phys.*, **124**, 094103.

118 Werner, H.J., Adler, T.B., and Manby, F.R. (2007) General orbital invariant MP2-F12 theory. *J. Chem. Phys.*, **126**, 164102.

119 Noga, J., Kedžuch, S., and Šimunek, J. (2007) Second order explicitly correlated R12 theory revisited: A second quantization framework for treatment of the operators' partitionings. *J. Chem. Phys.*, **127**, 034106.

120 Adler, T.B., Knizia, G., and Werner, H.J. (2007) A simple and efficient CCSD(T)-F12 approximation. *J. Chem. Phys.*, **127**, 221106.

121 Ten-no, S. (2007) A simple F12 geminal correction in multi-reference perturbation theory. *Chem. Phys. Lett.*, **447**, 175–179.

122 Tew, D.P., Klopper, W., Neiss, C., and Hättig, C. (2007) Quintuple-zeta quality coupled-cluster correlation energies with triple-zeta basis sets. *Phys. Chem. Chem. Phys.*, **9**, 1921–1930.

123 Knizia, G. and Werner, H.J. (2008) Explicitly correlated RMP2 for high-spin open-shell reference states. *J. Chem. Phys.*, **128**, 154103.

124 Shiozaki, T., Kamiya, M., Hirata, S., and Valeev, E.F. (2008) Explicitly correlated coupled-cluster singles and doubles method based on complete diagrammatic equations. *J. Chem. Phys.*, **129**, 071101.

125 Shiozaki, T., Kamiya, M., Hirata, S., and Valeev, E.F. (2008) Equations of explicitly-correlated coupled-cluster methods. *Phys. Chem. Chem. Phys.*, **10**, 3358–3370.

126 Noga, J., Kedžuch, S., Šimunek, J., and Ten-no, S. (2008) Explicitly correlated coupled cluster F12 theory with single and double excitations. *J. Chem. Phys.*, **128**, 174103.

127 Tew, D.P., Klopper, W., and Hättig, C. (2008) A diagonal orbital-invariant explicitly correlated coupled-cluster method. *Chem. Phys. Lett.*, **452**, 326–332.

128 Valeev, E.F. (2008) Coupled-cluster methods with perturbative inclusion of explicitly correlated terms: A preliminary investigation. *Phys. Chem. Chem. Phys.*, **10**, 106–113.

129 Valeev, E.F. and Crawford, T.D. (2008) Simple coupled-cluster singles and doubles method with perturbative inclusion of triples and explicitly correlated geminals: The $CCSD(T)_{\overline{R12}}$ model. *J. Chem. Phys.*, **128**, 244113.

130 Torheyden, M. and Valeev, E.F. (2008) Variational formulation of perturbative explicitly-correlated coupled-cluster methods. *Phys. Chem. Chem. Phys.*, **10**, 3410.

131 Bokhan, D., Ten-no, S., and Noga, J. (2008) Implementation of the CCSD(T)-F12 method using cusp conditions. *Phys. Chem. Chem. Phys.*, **10**, 3320–3326.

132 Köhn, A., Richings, G.W., and Tew, D.P. (2008) Implementation of the full explicitly correlated coupled-cluster singles and doubles model CCSD-F12 with optimally reduced auxiliary basis dependence. *J. Chem. Phys.*, **129**, 201103.

133 Knizia, G., Adler, T.B., and Werner, H.J. (2009) Simplified CCSD(T)-F12 methods: Theory and benchmarks. *J. Chem. Phys.*, **130**, 054104.

134 Shiozaki, T., Kamiya, M., Hirata, S., and Valeev, E.F. (2009) Higher-order explicitly correlated coupled-cluster methods. *J. Chem. Phys.*, **130**, 054101.

135 Bokhan, D., Bernadotte, S., and Ten-no, S. (2009) Implementation of the CCSD(T)(F12) method using numerical quadratures. *Chem. Phys. Lett.*, **469**, 214–218.

136 Bischoff, F.A., Wolfsegger, S., Tew, D.P., and Klopper, W. (2009) Assessment of basis sets for F12 explicitly-correlated molecular electronic-structure methods. *Mol. Phys.*, **107**, 963–975.

137 Hättig, C., Tew, D.P., and Köhn, A. (2010) Accurate and efficient approximations to explicitly correlated coupled-cluster singles and doubles, CCSD-F12. *J. Chem. Phys.*, **132**, 231102.

138 Werner, H.J., Knizia, G., and Manby, F.R. (2011) Explicitly correlated coupled cluster methods with pair-specific geminals. *Mol. Phys.*, **109**, 407–417.

139 Shiozaki, T., Knizia, G., and Werner, H.J. (2011) Explicitly correlated multireference configuration interaction: MRCI-F12. *J. Chem. Phys.*, **134**, 034113.

140 Werner, H.J. and Manby, F.R. (2006) Explicitly correlated second-order perturbation theory using density fitting and local approximations. *J. Chem. Phys.*, **124**, 054114.

141 Manby, F.R., Werner, H.J., Adler, T.B., and May, A.J. (2006) Explicitly correlated local second-order perturbation theory with a frozen geminal correlation factor. *J. Chem. Phys.*, **124**, 094103.

142 Werner, H.J. (2008) Eliminating the domain error in local explicitly correlated second-order Møller-Plesset perturbation theory. *J. Chem. Phys.*, **129**, 101103.

143 Adler, T.B. and Werner, H.J. (2009) Local explicitly correlated coupled-cluster methods: Efficient removal of the basis set incompleteness and domain errors. *J. Chem. Phys.*, **130**, 241101.

144 Adler, T.B., Werner, H.J., and Manby, F.R. (2009) Local explicitly correlated second-order perturbation theory for the accurate treatment of large molecules. *J. Chem. Phys.*, **130**, 054106.

145 Adler, T.B. and Werner, H.J. (2011) An explicitly correlated local coupled cluster method for calculations of large molecules close to the basis set limit. *J. Chem. Phys.*, **135**, 144117.

146 Krause, C. and Werner, H.J. (2012) Comparison of explicitly correlated local coupled-cluster methods with various choices of virtual orbitals. *Phys. Chem. Chem. Phys.*, **14**, 7591–7604.

147 Werner, H.J., Knizia, G., Krause, C., Schwilk, M., and Dornbach, M. (2015) Scalable electron correlation methods. I. PNO-LMP2 with linear scaling in the molecular size and near-inverse-linear scaling in the number of processors. *J. Chem. Theory Comput.*, **11**, 484–507.

148 Ma, Q. and Werner, H.J. (2015) Scalable electron correlation methods. II. Parallel PNO-LMP2-F12 with near linear scaling in the molecular size. *J. Chem. Theory Comput.*, **11**, 5291–5304.

149 Tew, D.P., Helmich, B., and Hättig, C. (2011) Local explicitly correlated second-order Moller-Plesset perturbation theory with pair natural orbitals. *J. Chem. Phys.*, **135**, 074107.

150 Hättig, C., Tew, D.P., and Helmich, B. (2012) Local explicitly correlated second- and third-order Møller-Plesset perturbation theory with pair natural orbitals. *J. Chem. Phys.*, **136**, 204105.

151 Li, W. (2013) Linear scaling explicitly correlated MP2-F12 and ONIOM methods for the long-range interactions of the nanoscale clusters in methanol aqueous solutions. *J. Chem. Phys.*, **138**, 014106.

152 Wang, Y.M., Hättig, C., Reine, S., Valeev, E., Kjaergaard, T., and Kristensen, K. (2016) Explicitly correlated second-order Møller-Plesset perturbation theory in a divide-expand-consolidate (DEC) context. *J. Chem. Phys.*, **144**, 204112.

153 Pavošević, F., Pinski, P., Riplinger, C., Neese, F., and Valeev, E.F. (2016) Sparse maps—A systematic infrastructure for reduced-scaling electronic structure methods. IV. Linear-scaling second-order explicitly correlated energy with pair natural orbitals. *J. Chem. Phys.*, **144**, 144109.

154 Peterson, K.A., Adler, T.B., and Werner, H.J. (2008) Systematically convergent basis sets for explicitly correlated wavefunctions: The atoms H, He, B-Ne, and Al-Ar. *J. Chem. Phys.*, **128**, 084102.

155 Hill, J.G., Peterson, K.A., Knizia, G., and Werner, H.J. (2009) Extrapolating MP2 and CCSD explicitly correlated correlation energies to the complete basis set limit with first and second row correlation consistent basis sets. *J. Chem. Phys.*, **131**, 194105.

156 Edmiston, C. and Krauss, M. (1965) Configuration-interaction calculation of H_3 and H_2. *J. Chem. Phys.*, **42**, 1119–1120.

157 Meyer, W. (1971) Ionization energies of water from PNO-CI calculations. *Int. J. Quantum Chem.*, **5**, 341–348.

158 Meyer, W. (1973) PNO-CI Studies of electron correlation effects. I. Configuration expansion by means of nonorthogonal orbitals, and application to the ground state and ionized states of methane. *J. Chem. Phys.*, **58**, 1017–1035.

159 Ahlrichs, R., Driessler, F., Lischka, H., Staemmler, V., and Kutzelnigg, W. (1975) PNO-CI (pair natural orbital configuration interaction) and CEPA-PNO (coupled electron pair approximation with pair natural orbitals) calculations of molecular systems. II. The molecules BeH_2, BH, BH_3, CH_4, CH_3^-, NH_3 (planar and pyramidal), H_2O, OH_3^+, HF and the Ne atom. *J. Chem. Phys.*, **62**, 1235–1247.

160 Taylor, P.R., Bacskay, G., Hush, N., and Hurley, A. (1976) The coupled-pair approximation in a basis of independent-pair natural orbitals. *Chem. Phys. Lett.*, **41**, 444–449.

161 Taylor, P.R. (1981) A rapidly convergent CI expansion based on several reference configurations, using optimized correlating orbitals. *J. Chem. Phys.*, **74**, 1256–1270.

162 Staemmler, V. and Jaquet, R. (1981) CEPA calculations on open-shell molecules. I. Outline of the method. *Theor. Chim. Acta.*, **59**, 487–500.

163 Fink, R. and Staemmler, V. (1993) A multi-configuration reference CEPA method based on pair natural orbitals. *Theor. Chem. Acc.*, **87**, 129–145.

164 Yang, J., Kurashige, Y., and Manby, F.R. (2011) Tensor factorizations of local second-order Møller–Plesset theory. *J. Chem. Phys.*, **134**, 044123.

165 Kurashige, Y., Yang, J., Chan, G.K.L., and Manby, F.R. (2012) Optimization of orbital-specific virtuals in local Møller-Plesset perturbation theory. *J. Chem. Phys.*, **136**, 124106.

166 Yang, J., Chan, G.K.L., Manby, F.R., Schütz, M., and Werner, H.J. (2012) The orbital-specific-virtual local coupled cluster singles and doubles method. *J. Chem. Phys.*, **136**, 144105.

167 Schütz, M., Yang, J., Chan, G.K.L., Manby, F.R., and Werner, H.J. (2013) The orbital-specific virtual local triples correction: OSV-L(T). *J. Chem. Phys.*, **138**, 054109.

168 Walter, D., Venkatnathan, A., and Carter, E. (2003) Local correlation in the virtual space in multireference singles and doubles configuration interaction. *J. Chem. Phys.*, **118**, 8127–8139.

169 Chwee, T.S., Szilva, A.B., Lindh, R., and Carter, E.A. (2008) Linear scaling multireference singles and doubles configuration interaction. *J. Chem. Phys.*, **128**, 224106.

170 Chwee, T.S. and Carter, E.A. (2010) Density fitting of two-electron integrals in local multireference single and double excitation configuration interaction calculations. *Mol. Phys.*, **108**, 2519–2526.

171 Chwee, T.S. and Carter, E.A. (2010) Cholesky decomposition within local multireference singles and doubles configuration interaction. *J. Chem. Phys.*, **132**, 074104.

172 Krisiloff, D.B. and Carter, E.A. (2012) Approximately size extensive local multireference singles and doubles configuration interaction. *Phys. Chem. Chem. Phys.*, **14**, 7710–7717.

173 Krisiloff, D.B., Krauter, C.M., Ricci, F.J., and Carter, E.A. (2015) Density fitting and cholesky decomposition of the two-electron integrals in local multireference configuration interaction theory. *J. Chem. Theory Comput.*, **11**, 5242–5251.

174 Demel, O., Pittner, J., and Neese, F. (2015) A local pair natural orbital-based multireference Mukherjee's coupled cluster method. *J. Chem. Theory Comput.*, **11**, 3104–3114.

175 Guo, Y., Sivalingam, K., Valeev, E.F., and Neese, F. (2016) Sparse maps—A systematic infrastructure for reduced-scaling electronic structure methods. III. Linear-scaling multireference domain-based pair natural orbital N-electron valence perturbation theory. *J. Chem. Phys.*, **144**, 094111.

176 Menezes, F., Kats, D., and Werner, H.J. (2016) Local complete active space second-order perturbation theory using pair natural orbitals (PNO-CASPT2). *J. Chem. Phys.*, **145**, 124115.

177 Fedorov, A., Batiste, L., Couzijn, E.P.A., and Chen, P. (2010) Experimental and theoretical study of a gold(I) aminonitrene complex in the gas phase. *ChemPhysChem*, **11**, 1002–1005.

178 Figgen, D., Rauhut, G., Dolg, M., and Stoll, H. (2005) Energy-consistent pseudopotentials for group 11 and 12 atoms: adjustment to multi-configuration Dirac–Hartree–Fock data. *Chem. Phys.*, **311**, 227–244.

179 Peterson, K. and Puzzarini, C. (2005) Systematically convergent basis sets for transition metals. II. Pseudopotential-based correlation consistent basis sets for the group 11 (Cu, Ag, Au) and 12 (Zn, Cd, Hg) elements. *Theor. Chem. Acc.*, **114**, 283–296.

180 Weigend, F., Köhn, A., and Hättig, C. (2002) Efficient use of the correlation consistent basis sets in resolution of the identity MP2 calculations. *J. Chem. Phys.*, **116**, 3175–3183.

181 Weigend, F. (2002) A fully direct RI-HF algorithm: Implementation, optimised auxiliary basis sets, demonstration of accuracy and efficiency. *Phys. Chem. Chem. Phys.*, **4**, 4285–4291.

182 Werner, H.J., Knowles, P.J., Knizia, G., Manby, F.R., and Schütz, M. (2011) Molpro: A general-purpose quantum chemistry program package. *WIREs Comput. Mol. Sci.*, **2**, 242–253.

183 Werner, H.J., Knowles, P.J., Knizia, G., Manby, F.R., Schütz, M., and and others (2016) Molpro, development version 2015.2, a package of ab initio programs. See http://www.molpro.net.

184 Hoyvik, I.M. and Jorgensen, P. (2016) Characterization and generation of local occupied and virtual Hartree–Fock orbitals. *Chem. Rev.*, **116**, 3306–3327.

185 Pipek, J. and Mezey, P.G. (1989) A fast intrinsic localization procedure applicable for ab initio and semiempirical linear combination of atomic orbital wave functions. *J. Chem. Phys.*, **90**, 4916–4926.

186 Edmiston, C. and Ruedenberg, K. (1963) Localized atomic and molecular orbitals. *Rev. Mod. Phys.*, **35**, 457–465.

187 Edmiston, C. and Ruedenberg, K. (1965) Localized atomic and molecular orbitals. II. *J. Chem. Phys.*, **43**, S97–S116.

188 Knizia, G. (2013) Intrinsic atomic orbitals: An unbiased bridge between quantum theory and chemical concepts. *J. Chem. Theory Comput.*, **9**, 4834–4843.

189 Janowski, T. (2014) Near equivalence of intrinsic atomic orbitals and quasiatomic orbitals. *J. Chem. Theory Comput.*, **10**, 3085–3091.

190 Lu, W.C., Wang, C.Z., Schmidt, M.W., Bytautas, L., Ho, K.M., and Ruedenberg, K. (2004) Molecule intrinsic minimal basis sets. I. Exact resolution of ab initio optimized molecular orbitals in terms of deformed atomic minimal-basis orbitals. *J. Chen. Phys.*, **120**, 2629–2637.

191 West, A.C., Schmidt, M.W., Gordon, M.S., and Ruedenberg, K. (2013) A comprehensive analysis of molecule-intrinsic quasi-atomic, bonding, and correlating orbitals. I. Hartree-Fock wave functions. *J. Chem. Phys.*, **139**, 234107.

192 Lehtola, S. and Jónsson, H. (2013) Unitary optimization of localized molecular orbitals. *J. Chem. Theory Comput.*, **9**, 5365–5372.

193 Reed, A.E., Weinstock, R.B., and Weinhold, F. (1985) Natural population analysis. *J. Chem. Phys.*, **83**, 735–746.

194 Reed, A.E. and Weinhold, F. (1985) Natural localized molecular orbitals. *J. Chem. Phys.*, **83**, 1736–1740.

195 Reed, A.E. and Weinhold, F. (1983) Natural bond orbital analysis of near-Hartree-Fock water dimer. *J. Chem. Phys.*, **78**, 4066–4073.

196 Foster, J.P. and Weinhold, F. (1980) Natural hybrid orbitals. *J. Am. Chem. Soc.*, **102**, 7211–7218.

197 Pulay, P., Saebø, S., and Meyer, W. (1984) An efficient reformulation of the closed-shell self-consistent electron pair theory. *J. Chem. Phys.*, **81**, 1901–1905.

198 Scuseria, G.E., Janssen, C.L., and Schaefer III, H.F. (1988) An efficient reformulation of the closed-shell coupled cluster single and double excitation (CCSD) equations. *J. Chem. Phys*, **89**, 7382–7387.

199 Hampel, C., Peterson, K.A., and Werner, H.J. (2011) A comparison of the efficiency and accuracy of the quadratic configuration interaction (QCISD), coupled cluster (CCSD), and Brueckner coupled cluster (BCCD) methods. *J. Chem. Phys.*, **135**, 144116.

200 Boughton, J.W. and Pulay, P. (1993) Comparison of the Boys and Pipek-Mezey localizations in the local correlation approach and automatic virtual basis selection. *J. Comput. Chem.*, **14**, 736–740.

201 Kats, D. (2016) Speeding up local correlation methods: System-inherent domains. *J. Chem. Phys.*, **145**, 014103.

202 Stone, A. (1975) *Theoretical Models of Chemical Bonding, Part 4* (ed. Z.B. Maksić), Springer, Heidelberg, 102.

203 Werner, H.J. (2016) Communication: Multipole approximations of distant pair energies in local correlation methods with pair natural orbitals. *J. Chem. Phys.*, **145**, 201101.

204 Schütz, M., Masur, O., and Usvyat, D. (2014) Efficient and accurate treatment of weak pairs in local CCSD (T) calculations. II. Beyond the ring approximation. *J. Chem. Phys.*, **140**, 244107.

205 Schwilk, M., Ma, Q., Köppl, C., and Werner, H.J. (2017) A parallel PNO-LCCSD method. *To be published.*

206 Köppl, C. and Werner, H.J. (2015) Parallel and low-order scaling implementation of Hartree-Fock exchange using local density fitting. *J. Chem. Theory Comput.*, **12**, 3122–3134.

207 Kato, T. (1957) On the eigenfunctions of many-particle systems in quantum mechanics. *Comm. Pure Appl. Math.*, **10**, 151–177.

208 Pack, R.T. and Byers Brown, W. (1966) Cusp conditions for molecular wavefunctions. *J. Chem. Phys.*, **45**, 556–559.

209 Hylleraas, E.A. (1929) Neue Berechnung der Energie des Heliums im Grundzustande, sowie des tiefsten Terms von Ortho-Helium. *Z. Phys.*, **54**, 347–366.

210 Werner, H.J., Adler, T.B., Knizia, G., and Manby, F.R. (2010) Efficient explicitly correlated coupled-cluster approximations, in *Recent Progress in Coupled Cluster Methods* (eds. P. Čársky, J. Paldus, and J. Pittner), Springer, Heidelberg, pp. 573–619.

211 Werner, H.J., Knizia, G., Adler, T.B., and Marchetti, O. (2010) Benchmark studies for explicitly correlated perturbation- and coupled cluster theories. *Z. Phys. Chem.*, **224**, 493–511.

212 Saebø, S., Tong, W., and Pulay, P. (1993) Efficient elimination of basis set superposition errors by the local correlation method: Accurate ab initio studies of the water dimer. *J. Chem. Phys.*, **98**, 2170–2175.

213 Dunlap, B.I., Connolly, J.W.D., and Sabin, J.R. (1979) On some approximations in applications of $X\alpha$ theory. *J. Chem. Phys.*, **71** (8), 3396–3402.

214 Dunlap, B.I. (2000) Robust and variational fitting. *Phys. Chem. Chem. Phys.*, **2**, 2113–2116.

215 Nieplocha, J., Palmer, B., Tipparaju, V., Krishnan, M., Trease, H., and Aprà, E. (2006) Advances, applications and performance of the global arrays shared memory programming toolkit. *International Journal of High Performance Computing Applications*, **20**, 203–231.

216 Karypis, G. and Kumar, V. (1998) A fast and high quality multilevel scheme for partitioning irregular graphs. *SIAM J. Sci. Comput.*, **20**, 359–392.

217 Grimme, S. (2003) Improved second-order Møller-Plesset perturbation theory by separate scaling of parallel- and antiparallel-spin pair correlation energies. *J. Chem. Phys.*, **118**, 9095–9102.

218 Rauhut, G., Azhary, A.E., Eckert, F., Schumann, U., and Werner, H.J. (1999) Impact of local approximations on MP2 vibrational frequencies. *Spectrochim. Acta*, **A 55**, 647–658.

219 Rauhut, G. and Werner, H.J. (2003) The vibrational spectra of furoxan and dichlorofuroxan: a comparative theoretical study using density functional theory and Local Electron Correlation Methods. *Phys. Chem. Chem. Phys.*, **5**, 2001–2008.

220 Claeyssens, F., Harvey, J.N., Manby, F.R., Mata, R.A., Mulholland, A.J., Ranaghan, K.E., Schütz, M., Thiel, S., Thiel, W., and Werner, H.J. (2006) High-Accuracy Computation of Reaction Barriers in Enzymes. *Angew. Chem. Int. Ed.*, **45**, 6856–6859.

221 Dieterich, J.M., Werner, H.J., Mata, R.A., Metz, S., and Thiel, W. (2010) Reductive half-eeaction of aldehyde oxidoreductase toward acetaldehyde: Ab initio and free energy QM/MM calculations. *J. Chem. Phys.*, **132**, 035101.

222 Rommel, J.B., Liu, Y., Werner, H.J., and Kästner, J. (2012) Role of Tunneling in the Enzyme Glutamate Mutase. *J. Phys. Chem. B*, **116**, 13682–13689.

223 ElAzhary, A., Rauhut, G., Pulay, P., and Werner, H.J. (1998) Analytical energy gradients for local second-order Møller-Plesset perturbation theory. *J. Chem. Phys.*, **108**, 5185–5193.

224 Rauhut, G. and Werner, H.J. (2001) Analytical energy gradients for local coupled-cluster methods. *Phys. Chem. Chem. Phys.*, **3**, 4853–4862.

225 Shamasundar, K.R., Knizia, G., and Werner, H.J. (2011) A new internally contracted multi-reference configuration interaction method. *J. Chem. Phys.*, **135**, 054101.

2

Density and Potential Functional Embedding: Theory and Practice

Kuang Yu,[1] Caroline M. Krauter,[1] Johannes M. Dieterich,[1] and Emily A. Carter[2]

[1] Department of Mechanical and Aerospace Engineering, Princeton University, USA
[2] School of Engineering and Applied Science, Princeton University, USA

2.1 Introduction

The class of methodologies known as embedding theories refers to methods that partition a total physical system into two or more subsystems, simulating them individually while accounting for inter-subsystem interactions. Various strategies to describe these subsystem interactions give rise to different families of embedding theories, that is, embedding theories are grouped first based on how they model the subsystem interaction and then based on the level of theory on which the subsystems are modeled. Some members of these families [1–5] are discussed in other chapters of this book. Here, we introduce orbital-free embedding theories (OFETs) [6, 7] and, in particular, our advanced formulations, implementations, and applications thereof [8–19]. Our aim is to provide an overview of these algorithms, thereby allowing the reader to confidently carry out state-of-the-art embedding simulations. We also aim to help readers improve upon these simulations by developing novel algorithms through understanding both successful approaches and current limitations. All of the algorithms developed by our group discussed within this chapter are freely available from github [20], and we encourage external use, improvement, and extensions.

We give a general introduction to OFETs in Section 2.2 and introduce our density functional embedding theory (DFET) in Section 2.3. This theory is formally exact and the embedding potential replacing the interaction between the subsystems is determined at the density functional theory (DFT) level. Once obtained, the embedding potential is used for highly accurate embedded correlated wavefunction (CW) calculations on the subsystems of most interest, to either improve upon the DFT description or access electronically excited states. We discuss theoretical and algorithmic details of DFET in Sections 2.3.1 and 2.3.2 and explain how to use the resulting embedding potential in embedded CW calculations in Section 2.3.3. Finally, we discuss our potential functional embedding theory (PFET) in Section 2.4, allowing a self-consistent mutual polarization between subsystems described at different levels of theories (i.e., CW or DFT). PFET improves upon DFET, as the latter keeps the embedding potential fixed at

Fragmentation: Toward Accurate Calculations on Complex Molecular Systems, First Edition.
Edited by Mark S. Gordon.

the DFT/DFT level during the embedded CW calculation, neglecting mutual polarization effects at the CW/DFT level.

2.2 Theoretical Background

OFETs are based on partitioning the density of the total system ρ_{tot} into a sum of subsystem densities ρ_i:

$$\rho_{tot} = \sum_{i=0}^{N} \rho_i \tag{2.1}$$

and representing subsystem interactions through an embedding potential, $V_{emb}(\vec{r})$, instead of calculating their explicit couplings on the wavefunction level [6, 7]. Logically, these methods do not require knowledge of orbitals and/or density matrices of either subsystems or the entire system, and hence are compatible with subsystem descriptions on a wide range of levels of theory. This intrinsic flexibility gives rise to two major advantages: first, it allows for modular and elegant implementations, as will be discussed in Sections 2.3.1 and 2.3.2; and second, OFETs have the potential to reduce both prefactor and scaling compared to a simulation of the full system, while essentially still retaining the same accuracy depending on how subsystem interactions are modeled [8, 21]. Additionally, embedding methods allow us to set up simulations that strike different balances in the space spanning computational efficiency versus accuracy with ease, as will be shown in Section 2.3.3. Indeed, starting with the very first account by Cortona in 1991 of what we now call an OFET* [6, 7], discussions initially focused on making DFT studies feasible for large systems, where DFT's $O(N^3)$ scaling proved prohibitive at the time. To this end, Cortona proposed to partition a crystal into its atoms and to express the interaction of these subsystems by a so-called crystalline potential V_{cryst} [6, 7]. We shall, for the sake of simplicity, from now on only assume two subsystems denoted by A and B, where subsystem A has the character of an "embedded region" or "cluster," that is, a region of increased interest, and subsystem B represents an "environment." Extensions to arbitrary numbers of regions are straightforward.

Assuming this separation, the Schrödinger equation for the cluster region of all OFETs becomes:

$$(H_A + V_{emb}[\rho_A, \rho_{tot}])\Psi_A = E_A \Psi_A \tag{2.2}$$

where H_A is the electronic Hamiltonian of the cluster region, E_A and Ψ_A are respectively the energy and wavefunction of the cluster region, and $V_{emb}[\rho_A, \rho_{tot}]$ is the potential designed to represent the environment's influence on the cluster. We typically further

*A counter-argument can be made that the arguments brought forth for the first, non-OFET derived by Whitten and Pakkanen in 1980 [22, 23] did focus less on performance and more on bringing together mean-field and CW theories to improve surface-adsorbate descriptions. Their work also should be considered seminal in this aspect.

partition this interaction potential into its kinetic T_s, exchange-correlation (XC) V_{xc}, Hartree V_J, and nuclear-electron V_{NE} potential contributions [8, 13]:

$$V_{emb}[\rho_A, \rho_{tot}] = \left(\frac{\delta T_s[\rho_{tot}]}{\delta \rho_{tot}} - \frac{\delta T_s[\rho_A]}{\delta \rho_A} \right) + V_{xc}[\rho_{tot}] - V_{xc}[\rho_A] \\ + V_J[\rho_{tot}] - V_J[\rho_A] + V_{NE}[\rho_{tot}] - V_{NE}[\rho_A]. \tag{2.3}$$

Different OFETs vary in one or more of three key features readily identifiable from the above: (a) how we partition the total system density into subsystems, (b) how we construct the interaction potential, and (c) on which level of theory we solve the Schrödinger equation (2.2) for the subsystem of interest including the interaction potential. Choices made on these questions affect resource consumption and accuracy of the resulting embedding calculation. Another important feature of OFET is whether the resulting embedding potential is unique [16]. Embedding methods guaranteeing a unique converged potential hold a very practical advantage, as their result will be independent of the initial guess for the embedding potential. All of these features can also serve as metrics to rank embedding methods, which we will use in this capacity as follows.

The initial partitioning of the entire system into subsystems at the atomic level is typically dictated by physical needs to properly capture the physical feature or phenomenon under study, and hence is not part of the embedding theory per se. What is important, however, is the way in which the electron density is partitioned and whether responses to density changes are allowed to propagate from one subsystem to another. Our discussions therefore will be on the electron density level.

The first flexible embedding theory framework for OFETs is frozen density embedding (FDE) as introduced by Wesolowski and Warshel [1, 24, 25]. The original FDE algorithm freezes the environment density ρ_B and solves equation (2.2) self-consistently at the DFT level. The embedding potential is adjusted in each iteration using the updated cluster and frozen environment densities. The resulting kinetic energy term is calculated approximately with a kinetic energy density functional (KEDF) [6, 7]. Two major sources of error arise additionally to the limitation of pure DFT-in-DFT embedding and the accuracy associated with DFT approximations. First, using an approximate KEDF to evaluate the kinetic energy term is fast, but typically insufficient for systems with anything other than simple electronic structures [26–36]. Second, the environment density is kept frozen and thus unresponsive to changes in the cluster density. If the environment density could be determined exactly prior to the embedding calculation, FDE would be a formally exact theory. However, one realistically needs to use an approximate density for the environment, differing significantly from the actual density. Last but not least, FDE embedding methods have not yet been rigorously proven to feature unique converged embedding potentials [16].

Algorithms have proposed freezing and thawing the cluster and environment densities until convergence to overcome the lack of response of the environment density for cluster density changes [25, 37, 38]. This is a pragmatic and, as proven by applications, successful solution to achieve polarization of the environment. However, the converged solution for the subsystem densities is not unique and therefore relies on the initial guess [16].

The Carter group pioneered periodic embedded CW (ECW) methods [8–19], to improve on the accuracy of periodic DFT through CW methods. Although they do not originally hail from FDE, ECW methods can be discussed within the FDE framework as solving equation (2.2) at a CW level of theory. Depending on the CW method used, ECW theory can thereby extend the validity of embedding methods beyond the applicability of DFT to the treatment of, for example, excited states [10, 12, 13, 21, 39–42]. The embedding potential is still constructed at a DFT level, the environment density is kept frozen, and a KEDF is used for the interaction kinetic energy term. Density inconsistencies may arise between the total density evaluated at the DFT level and the cluster density evaluated on the CW level. A simple remedy is to construct the total density from both the DFT environment and CW cluster densities. One successful choice is to iteratively compute the total density from the initial, non-embedded frozen environment density $\rho_B^{(0)}$ and the CW cluster density [13]:

$$\rho_{\text{tot}} = \rho_B^{(0)} + \rho_A^{CW}. \tag{2.4}$$

Since this approach does not keep the total system density fixed, we can retain either the environment or total density within this framework.

ECW-style methods have been developed also strictly within the FDE framework [43, 44]. Similar to ECW, these WFT/DFT methods have been successfully used to study electronic excitations in molecules. They nevertheless suffer from the same source of inaccuracy present in both FDE and this first ECW theory: the approximate KEDF to obtain the kinetic energy.

In recent years, use of optimized effective potential (OEP) theory within embedding methods has shown promise [45, 46]. Within FDE, the OEP replaces both KEDFs and iterative subsystem density cycles, achieving accurate reproduction of the total density of the system. It therefore must be considered a crucial ingredient in advanced embedding methods, and we will highlight its features and implementation in Section 2.3.1.2.

We next introduce OFETs designed to address all of the challenges discussed above. These allow for density changes in the cluster partition to propagate to the environment, use an OEP for an accurately reproduced total density, and have a unique embedding potential while still retaining maximum flexibility in order to solve the resulting Schrödinger equation at a variety of levels of theory.

2.3 Density Functional Embedding Theory

2.3.1 Basic Theory

As stated above, consider two interacting subsystems A (cluster) and B (environment). To describe the interactions between these two subsystems, DFET contains two fundamental steps: (i) constructing V_{emb} using relatively cheap DFT and (ii) performing embedded cluster calculations at a more expensive CW level. In the first step, we have to explicitly compute the electronic structures of the total system and the environment using DFT (*vide infra*). Both the total system and environment are in practice large, extended periodic systems (e.g., a slab representing metal surfaces, a bulk crystal lattice, etc.), so the first step is typically performed using planewave (PW) basis sets. The cluster is, however, smaller and non-periodic, thus the embedded CW calculations in the second step are typically carried out using Gaussian type orbital (GTO) basis sets.

Due to the different basis set setups, these two steps thus adopt different algorithms and are performed using different programs. The challenges in self-consistent field (SCF) convergence behavior and numerical integrations are also very different. Therefore, the calculations in the two steps will be discussed separately as two different topics below. In this section, we focus on the first step (V_{emb} construction at the DFT level) and leave the algorithmic details for the second step in Section 2.3.3.

2.3.1.1 Definition of the Embedding Potential

DFET (as well as PFET, *vide infra*) assumes that the effect of subsystem B on A and the effect of A on B can be represented by a common embedding (interaction) potential:

$$V_{emb}^{A\to B}(\vec{r}) = V_{emb}^{B\to A}(\vec{r}) = V_{emb}(\vec{r}). \tag{2.5}$$

V_{emb} is determined by imposing the constraint that the DFT density of the total system is reproduced by the sum of the two subsystem DFT densities optimized in the presence of V_{emb}:

$$\rho_{tot}(\vec{r}) = \rho_A(\vec{r}, V_{emb}) + \rho_B(\vec{r}, V_{emb}). \tag{2.6}$$

Here, $\rho_{K=A,B}(\vec{r}, V_{emb})$ means the subsystem densities are computed using DFT calculations embedded in the potential V_{emb}.

Equation (2.5) is somewhat contradictory to the conventional understanding that $V_{emb}^{A\to B}(\vec{r})$ and $V_{emb}^{B\to A}(\vec{r})$ should be separate potentials describing different physics. However, the partitioning of $\rho_{tot}(\vec{r})$ is arbitrary without equation (2.5), even with a fixed number of electrons for both subsystems ($\int \rho_A(\vec{r})d\vec{r} = N_A$ and $\int \rho_B(\vec{r})d\vec{r} = N_B$). In principle, multiple sets of $\rho_A(\vec{r})$ and $\rho_B(\vec{r})$ exist to satisfy equation (2.6). Due to the one-to-one correspondence between ρ and V(the first Hohenberg–Kohn theorem) [47], non-unique density partitions lead to non-unique embedding potentials ($V_{emb}^{A\to B}(\vec{r})$ and $V_{emb}^{B\to A}(\vec{r})$). Equation (2.5) removes this arbitrariness in DFET. Both the density partition and the embedding potentials therefore are unique under this constraint, as proven in Ref. [16]. As discussed in Section 2.2, this is a very beneficial characteristic for embedding theories.

2.3.1.2 Optimization Procedure

Solving for the unique embedding potential defined by equations (2.5) and (2.6) is similar to an OEP problem. In OEP theory, one needs to find the potential reproducing a particular reference density for a single quantum system. In DFET, one needs to consider the summation of the two subsystem densities. We therefore can use the idea of the Wu–Yang functional (W–Y functional) [48] that has been used to solve for OEPs, extending it to a two-body form to solve for an embedding potential in DFET. The extended W–Y functional is defined as [16, 48]

$$W[V_{emb}] = \tilde{E}_A[\rho_A, V_{emb}] + \tilde{E}_B[\rho_B, V_{emb}] - \int \rho_{tot}(\vec{r}) \cdot V_{emb}(\vec{r})d\vec{r}. \tag{2.7}$$

Here, $\tilde{E}_K[\rho, V], (K = A, B)$ is the subsystem energy, including its interaction with the embedding potential ($\tilde{E}_K[\rho, V] = E_K[\rho] + \int \rho(\vec{r}) \cdot V(\vec{r})d\vec{r}$).

Since the subsystem densities are obtained using a fully variational DFT method in the presence of V_{emb}, we have $\left.\frac{\delta \tilde{E}_K}{\delta \rho_K(\vec{r})}\right|_{V_{\mathrm{emb}}} = 0$, then

$$\begin{aligned}\frac{\delta \tilde{E}_K}{\delta V_{\mathrm{emb}}(\vec{r})} &= \left.\frac{\delta \tilde{E}_K}{\delta V_{\mathrm{emb}}(\vec{r})}\right|_{\rho_K} + \int d\vec{r}' \cdot \left.\frac{\delta \tilde{E}_K}{\delta \rho_K(\vec{r}')}\right|_{V_{\mathrm{emb}}} \frac{\delta \rho_K(\vec{r}')}{\delta V_{\mathrm{emb}}(\vec{r})} \\ &= \left.\frac{\delta \tilde{E}_K}{\delta V_{\mathrm{emb}}(\vec{r})}\right|_{\rho_K} = \rho_K(\vec{r}).\end{aligned} \tag{2.8}$$

The total functional derivative of $W[V_{\mathrm{emb}}]$ therefore is directly related to the density difference between the total system and subsystems:

$$\frac{\delta W[V_{\mathrm{emb}}]}{\delta V_{\mathrm{emb}}(\vec{r})} = \rho_A(\vec{r}) + \rho_B(\vec{r}) - \rho_{\mathrm{tot}}(\vec{r}). \tag{2.9}$$

The functional derivative given by equation (2.9) vanishes for any V_{emb} satisfying equation (2.6), thus the unique, optimal V_{emb} corresponds to the only stationary point for W. Additionally, the second derivative of W is rigorously negative [16], so W defined by equation (2.7) is universally concave. The optimized V_{emb} therefore must correspond to the only maximum for the extended W–Y functional.

We first perform a PW-DFT calculation on the entire system to obtain the total reference density (ρ_{tot}) used in equation (2.7). Then, with a given V_{emb}, we perform separate embedded PW-DFT calculations on the cluster and environment to obtain $\tilde{E}_A$, ρ_A and $\tilde{E}_B$, ρ_B, respectively. The functional derivative of W is computed and the value of W maximized with respect to V_{emb}. The optimal embedding potential V_{emb} can be constructed at the PW-DFT level via this procedure, and then used for the embedded CW calculations in the second step.

Compared to previous FDE methods, we see two significant advantages of DFET: (i) as proven, the density partition is not arbitrary but unique, and (ii) it is unnecessary to compute the kinetic interaction energy in this scheme, thus an approximate KEDF is not needed. The disadvantage, however, is that the cluster-environment interactions are computed at a pure DFT level, while the cluster is described using CW theory, somewhat inconsistent from the theoretical perspective.

2.3.2 Embedding Potential Construction—Implementations in Planewave Codes

All calculations in the embedding potential construction step are performed using a PW basis set and include periodic boundary conditions. In our implementation, V_{emb} is always discretized and stored on a three-dimensional uniform grid in real space, conforming to the representation of electron densities in PW-DFT codes. The wavefunction forms cusps around the nucleus region due to the singularity in the Coulomb potential. These are typically too sharp to represent on any uniform grid with reasonable grid size. Two widely used techniques in PW-DFT codes to treat the sharp wavefunctions are the pseudopotential (PP) [49, 50] and the projector-augmented-wave (PAW) [51, 52] methods. The implementation of the embedding potential solver is different depending on the core region treatment and thus will be discussed separately in Sections 2.3.2.1 and 2.3.2.2, respectively.

2.3.2.1 Implementation with Pseudopotentials in ABINIT

The norm-conserving PP method is an important approach widely used in solid-state electronic structure calculations due to its simplicity and accuracy. The implementation of DFET within the norm-conserving PP framework is relatively straightforward and will be introduced in this section. Here, we use the ABINIT code [53–55] as an example to discuss the implementation details.

When employing the norm-conserving PP method, the sharp core electron wavefunction is completely removed from the system and its effects are represented by a nonlocal PP (NLPP). Furthermore, the core region of the valence electron wavefunction is also pseudized to a smoother nodeless form. In this formalism, the density and wavefunction therefore are smooth enough to calculate the density-potential integral using a direct summation over the real-space uniform grid

$$\Delta\tilde{E} = \int \rho(\vec{r})V_{\text{emb}}(\vec{r})d\vec{r} \approx \sum_i \rho(\vec{r}_i)\cdot V_{\text{emb}}(\vec{r}_i)\cdot dV. \tag{2.10}$$

Here, i runs over all real-space grid points and dV is the volume per grid point. This equation implies that the uniform grid used for the embedding potential must conform to the uniform grid used for the density in the code. Since the grid used for the Kohn–Sham (KS) potential (V_{KS}) in ABINIT is identical to the density grid, a modification of the potential term can also be done in a straightforward point-by-point way:

$$V(\vec{r}_i) = V_{\text{KS}}(\vec{r}_i) + V_{\text{emb}}(\vec{r}_i). \tag{2.11}$$

Correspondingly, based on equation (2.10), the discretized functional derivative of $\Delta\tilde{E}$ with respect to V_{emb} is exactly the discretized density grid ($\rho(\vec{r}_i)$) and therefore, the derivative of the extended W–Y functional can be easily evaluated according to equation (2.9):

$$\frac{\partial W}{\partial V_{\text{emb}}(\vec{r}_i)} = \rho_A(\vec{r}_i) + \rho_B(\vec{r}_i) - \rho_{\text{tot}}(\vec{r}_i). \tag{2.12}$$

The W–Y functional value and the derivative are then fed to a gradient-based local optimization engine (we use and recommend the L-BFGS optimizer) [56, 57] to obtain the optimal embedding potential. To run the optimization, one has to first run a PW-DFT calculation on the entire system to obtain a total density file. One then prepares standard ABINIT input files for both the cluster and environment in separate folders, as well as an input file containing all optimization parameters (i.e., maximum iterations, convergence thresholds, initial guesses for embedding potentials, etc.). The optimizer is implemented in a modified version of ABINIT-6.0.3, which performs DFT calculations on both the cluster and environment as well as reads the necessary information (i.e., energies and densities) to conduct optimization.

2.3.2.2 Implementation with PAW in VASP

Compared to norm-conserving PPs, PAW theory formulates a more *ab initio* methodology while also leading to softer wavefunctions, which require a lower basis set energy cutoff. PAW has become one of the mainstream methods used in PW-DFT programs due to these advantages. Unlike the PP method, PAW theory adopts a frozen core approximation, in which the full core electron wavefunction is explicitly included and kept fixed during the SCF iterations. To handle the sharp cusps of both the core and valence electrons, the PAW method [51, 52] establishes a one-to-one mapping between the realistic all-electron (AE) wavefunctions and soft pseudized (PS) wavefunctions representable on the uniform grid

$$|\Psi_n\rangle = |\tilde{\Psi}_n\rangle + \sum_i (|\phi_i\rangle - |\tilde{\phi}_i\rangle)\langle\tilde{p}_i \mid \tilde{\Psi}_n\rangle. \tag{2.13}$$

Here, n is the index for the n'th band, while $|\phi_i\rangle$ and $|\tilde{\phi}_i\rangle$ are the onsite AE and PS basis sets, respectively. In PAW theory, we define the core region using an augmentation sphere with a radius of R_{aug}. The AE and PS basis sets are forcibly identical in the interstitial region outside and different within the augmentation sphere. $|\tilde{p}_i\rangle$ are the projector functions orthonormal to the PS basis: $\langle\tilde{p}_i \mid \tilde{\phi}_j\rangle = \delta_{ij}$. Within this theoretical framework, the total electron density can be decomposed into three terms:

$$\rho = \tilde{\rho} + \rho^1 - \tilde{\rho}^1. \tag{2.14}$$

The first term $\tilde{\rho}$ is the PS density calculated directly from PS wavefunctions on a uniform grid. The second (ρ^1) and third ($\tilde{\rho}^1$) terms are the one-center densities (denoted by the superscript 1) defined within the augmentation sphere and associated with the AE and PS basis functions, respectively. Unlike the first term, the latter two terms are expanded into different angular channels using real spherical harmonics, and the radial parts are expressed on an ultrafine radial grid. Similarly, the KS Hamiltonian for the PAW ansatz is also composed of soft global and sharp one-center parts, respectively:

$$H = -\frac{1}{2}\nabla^2 + \tilde{v}_{\text{eff}} + \sum_{ij} |\tilde{p}_i\rangle \left(\hat{D}_{ij} + D^1_{ij} - \tilde{D}^1_{ij}\right) \langle\tilde{p}_j| \tag{2.15}$$

$$\hat{D}_{ij} = \sum_L \int \tilde{v}_{\text{eff}}(\vec{r})\hat{Q}^L_{ij}(\vec{r})d\vec{r} \tag{2.16}$$

$$D^1_{ij} = \left\langle \phi_i \left| -\frac{1}{2}\nabla^2 + v^1_{\text{eff}} \right| \phi_j \right\rangle \tag{2.17}$$

$$\tilde{D}^1_{ij} = \left\langle \tilde{\phi}_i \left| -\frac{1}{2}\nabla^2 + \tilde{v}^1_{\text{eff}} \right| \tilde{\phi}_j \right\rangle + \sum_L \int_{\Omega_r} \tilde{v}^1_{\text{eff}}(\vec{r})\hat{Q}^L_{ij}(\vec{r})d\vec{r}. \tag{2.18}$$

Here, a compensation charge $\hat{\rho}$ and its angular-momentum-decomposed components $\hat{Q}^L_{ij}(\vec{r})$ were introduced in Refs. [51, 52] to compensate for the multipole moments difference between the AE and PS densities within the core region. The PAW Hamiltonian is also expressed on two sets of grids: equations (2.15) and (2.16) involve the Hartree and XC potentials of the soft PS nuclear/electron density ($\tilde{v}_{\text{eff}}(\vec{r})$), computed and stored on the uniform grid; the one-center AE and PS potentials (v^1_{eff} and $\tilde{v}^1_{\text{eff}}$) in equations (2.17) and (2.18) are projected onto angular momentum channels and stored on a fine radial grid.

The implementation of the embedding potential becomes more complicated due to this dual grid strategy adopted in PAW theory. Here, we only give a general description of the existing problems and solutions, while the reader is referred to Ref. [19] for more details. Adding the embedding potential to $\tilde{v}_{\text{eff}}(\vec{r})$ is relatively straightforward following equation (2.11), since both potentials are stored in the same uniform grid. However, for the modification of the one-center potentials, we need to first project $V_{\text{emb}}(\vec{r})$ from the uniform grid to the radial grids using real spherical harmonics ($S_{\text{LM}}(\hat{r})$):

$$V_{\text{emb}}(\vec{r}) = \sum_{\text{LM}} V_{\text{emb}}^{\text{LM}}(r) S_{\text{LM}}(\hat{r}). \tag{2.19}$$

We then add the projected $V_{\text{emb}}^{\text{LM}}(r)$ to the corresponding LM-component of v_{eff}^1 and $\tilde{v}_{\text{eff}}^1$:

$$v_{\text{eff,emb}}^{1,\text{LM}}(r) = v_{\text{eff}}^{1,\text{LM}}[r] + V_{\text{emb}}^{\text{LM}} \tag{2.20}$$

$$\tilde{v}_{\text{eff,emb}}^{1,\text{LM}}(r) = \tilde{v}_{\text{eff}}^{1,\text{LM}}(r) + V_{\text{emb}}^{\text{LM}}. \tag{2.21}$$

Note that the radial grid is much finer than the uniform grid, so the projection is associated with an interpolation scheme. A profound consequence arises: the change of a potential value $V_{\text{emb}}(\vec{r}_i)$ on one uniform grid point ($\vec{r}_i$) effectively leads to changes of $V_{\text{emb}}^{\text{LM}}(r)$ values on multiple radial grid points (r). The resulting energy change therefore is no longer determined by the density value on that particular uniform grid point ($\rho(\vec{r}_i)$), but rather a smeared average of the densities in the neighboring region. In other words, equations (2.8) and (2.12) break down ($\frac{\delta \tilde{E}_K}{\delta V_{\text{emb}}(\vec{r}_i)} \neq \rho_K(\vec{r}_i)$) for dual-grid setups with extra effort involved in computing the exact energy derivatives ($\frac{\delta \tilde{E}_K}{\delta V_{\text{emb}}(\vec{r}_i)}$) to obtain the W–Y functional gradient needed for the optimization algorithm:

$$\frac{\delta W}{\delta V_{\text{emb}}(\vec{r}_i)} = \frac{\delta \tilde{E}_A}{\delta V(\vec{r}_i)} + \frac{\delta \tilde{E}_B}{\delta V(\vec{r}_i)} - \frac{\delta \tilde{E}_{\text{tot}}}{\delta V(\vec{r}_i)}. \tag{2.22}$$

We have shown in Ref. [19] that the exact derivative can be evaluated using:

$$\frac{\delta \tilde{E}}{\delta V_{\text{emb}}(\vec{r})} = \tilde{\rho} + \hat{\rho} + \sum_{i,j} \rho_{ij} \frac{\delta (D_{ij}^1 - \tilde{D}_{ij}^1)}{\delta V_{\text{emb}}(\vec{r})}, \tag{2.23}$$

within which i, j run over all the radial basis sets; ρ_{ij} is the local density matrix; and

$$\frac{\delta (D_{ij}^1 - \tilde{D}_{ij}^1)}{\delta V_{\text{emb}}(\vec{r})} = \sum_{\text{LM}} C_{lml'm'}^{\text{LM}} \int dr \left(Q_{lml'm'}(r) - Q_{lml'm'}^L(r) g_L(r) \right) \frac{\partial V_{\text{emb}}^{\text{LM}}(r)}{\partial V_{\text{emb}}(\vec{r})}. \tag{2.24}$$

Here, $C_{lml'm'}^{\text{LM}} = \int S_{\text{LM}}(\hat{r}) S_{lm}(\hat{r}) S_{l'm'}(\hat{r}) d\hat{r}$, while $Q_{lml'm'}(r)$, $Q_{lml'm'}^L(r)$, and $g_L(r)$ are the respective depletion and compensation charge quantities defined in the standard PAW algorithm (see Refs. [19, 51, 52] for definitions). It therefore is critical to evaluate the quantity $\frac{\partial V_{\text{emb}}^{\text{LM}}(r)}{\partial V_{\text{emb}}(\vec{r})}$, which is related to the projection algorithm used in equation (2.19) (*vide supra*) and is in general expensive to compute. Fortunately, $\frac{\delta \tilde{E}}{\delta V_{\text{emb}}(\vec{r}_i)}$ differs from $\rho(\vec{r}_i)$ only in the region where $|\Psi\rangle$ is extremely sharp, so we do not have to compute equation (2.23) at every uniform grid point. We instead introduce a factor F_{corr}, and the exact derivatives are evaluated only inside the sphere with a radius of $F_{\text{corr}} R_{\text{aug}}$; we directly use $\rho(\vec{r})$ outside this region. Empirically, the result is good enough when the value of F_{corr} is within the range of 0.67–0.90.

With the modified Hamiltonian and energy derivative equations, we now discuss details of the projection algorithm used in equation (2.19). We adopt a real-space projection algorithm for the LM-decomposition of V_{emb} to increase the computational efficiency. The general idea is to project V_{emb} using both angular and radial basis functions:

$$V_{\text{emb}}(\vec{r}) = \sum_{\text{LM}} S_{\text{LM}}(\hat{r}) \sum_{K} R_K(r) V_{\text{LMK}}. \tag{2.25}$$

We then reassemble the radial expansion to obtain $V^{\text{LM}}(r)$:

$$V^{\text{LM}}_{\text{emb}}(r) = \sum_{K} R_K(r) V_{\text{LMK}}. \tag{2.26}$$

Here, we use sinc functions for the radial basis set R_K, which, in combination with the real spherical harmonics $S_{\text{LM}}(\hat{r})$, forms a complete orthonormal basis set within a spherical region defined by radius R_{cut}:

$$R_K(r) = \sqrt{\frac{2}{R_{\text{cut}}}} \sin\left(\frac{K\pi}{R_{\text{cut}}} r\right) / r, \quad K = 1, 2, \ldots, K_{\max}. \tag{2.27}$$

$K_{\max} = 20$ generates reliable results for most test calculations. To find the expansion coefficient V_{LMK}, we perform the real-space integration:

$$\begin{aligned} V_{\text{LMK}} &= \int_{\Omega} S_{\text{LM}}(\hat{r}) R_K(r) V_{\text{emb}}(\vec{r}) d\vec{r} \\ &\approx \sum_{\vec{r}_m \in \Omega} S_{\text{LM}}(\hat{r}_m) R_K(r_m) V_{\text{emb}}(\vec{r}_m) dv. \end{aligned} \tag{2.28}$$

We use a secondary uniform grid $\{\vec{r}_m\}$ only defined inside the core region to compute equation (2.28). We typically use a 50 × 50 × 50 size for the secondary uniform grid, which is much finer than the primary uniform grid $\{\vec{r}_i\}$ defined over the entire box, since it is distributed only in the small core region. The embedding potential value on the secondary uniform grid ($V_{\text{emb}}(\vec{r}_m)$) is interpolated from the primary grid ($V_{\text{emb}}(\vec{r}_i)$) using a three-dimensional cubic B-spline method [19]. The term $\frac{\partial V^{\text{LM}}_{\text{emb}}(r)}{\partial V_{\text{emb}}(\vec{r})}$ (and the energy derivative $\frac{\delta \tilde{E}}{\delta V_{\text{emb}}(\vec{r}_i)}$) can be evaluated using very localized integrations with this real-space projection algorithm. The entire PAW-DFET calculation becomes efficient and affordable.

In our implementation of the PAW-DFET in VASP [58–60], a new tag "LEXTPOT = .true." in the INCAR file enables VASP to read the embedding potential from an external file named "EXTPOT" and to perform embedded DFT calculations. With this tag set, the code also computes the exact energy derivatives ($\frac{\delta \tilde{E}}{\delta V_{\text{emb}}(\vec{r}_i)}$) and stores it in CHGCAR format (file "DERIV" in Figure 2.1). F_{corr} can be set in INCAR, which controls the spatial range in which the exact derivative calculations are performed. In practice, the value of F_{corr} should be large enough such that the integration of DERIV equals the total number of valence electrons ($\int \frac{\delta \tilde{E}}{\delta V_{\text{emb}}(\vec{r})} d\vec{r} \approx \sum_i \frac{\delta \tilde{E}}{\delta V_{\text{emb}}(\vec{r}_i)} dV \approx N_e$). As in the ABINIT implementation, a reference calculation is done at zero embedding potential to obtain the

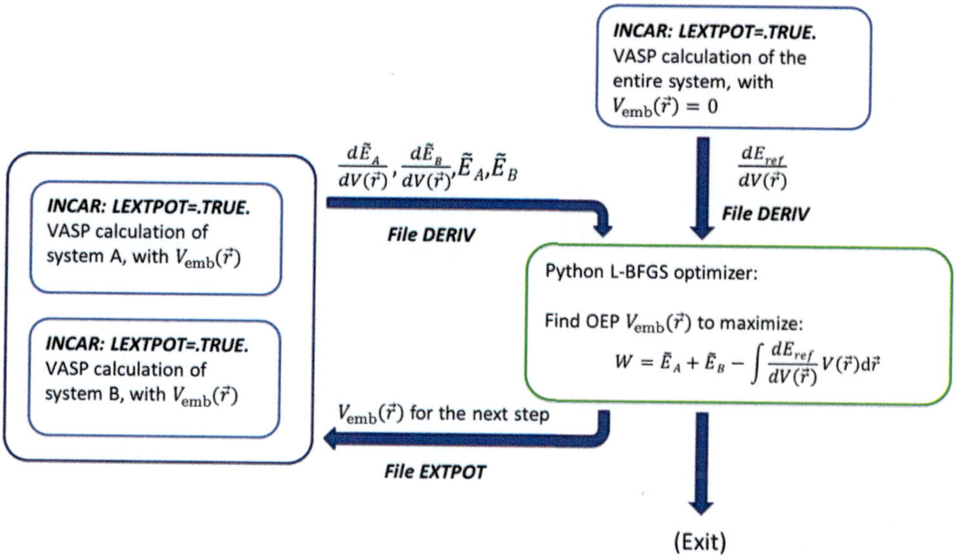

Figure 2.1 The runtime flowchart of the PAW-DFET code implemented in VASP.

reference DERIV for the entire system. The energies and derivatives of the embedded cluster and environment are then computed and fed into a python L-BFGS optimizer [56, 57, 61] to obtain the optimal embedding potential. The entire process is described in Figure 2.1.

2.3.2.3 Penalty Functions—Damping the Unphysical Oscillations

Similar to the conventional OEP calculation with GTO basis sets [62], the extended W–Y OEP in the DFET scheme often leads to unphysical oscillations in the resulting embedding potential. We take the Cl_2 molecule as an example demonstration of this oscillation. We partition the Cl_2 molecule into two subsystems, each containing a single Cl atom, and then perform the PAW-DFET calculations according to the procedure described above. The resulting V_{emb} along the bond axis of Cl_2 is plotted as the red dashed line in Figure 2.2. Comparing the density difference ($\rho_A + \rho_B - \rho_{tot}$) before and after optimization (black and blue curves in Figure 2.3), we see that this embedding potential reproduces the reference density with high fidelity, although it demonstrates large oscillations especially in the core region.

The oscillation observed in the shape of the embedding potential is extremely sensitive to the numerical details of the calculation, and thus is unphysical. This is a well-known problem [62] initiated by the unbalanced description of the wavefunction and potential (more specifically, the wavefunction is unable to respond to arbitrary unphysical potential changes, since the wavefunction is described using a finite basis set. This causes the insensitivity of the density with respect to the unphysical oscillations in the potential, leading to the numerical problem stated above). It causes transferability problems when one tries to apply the embedding potential derived from a PW basis to an embedded CW

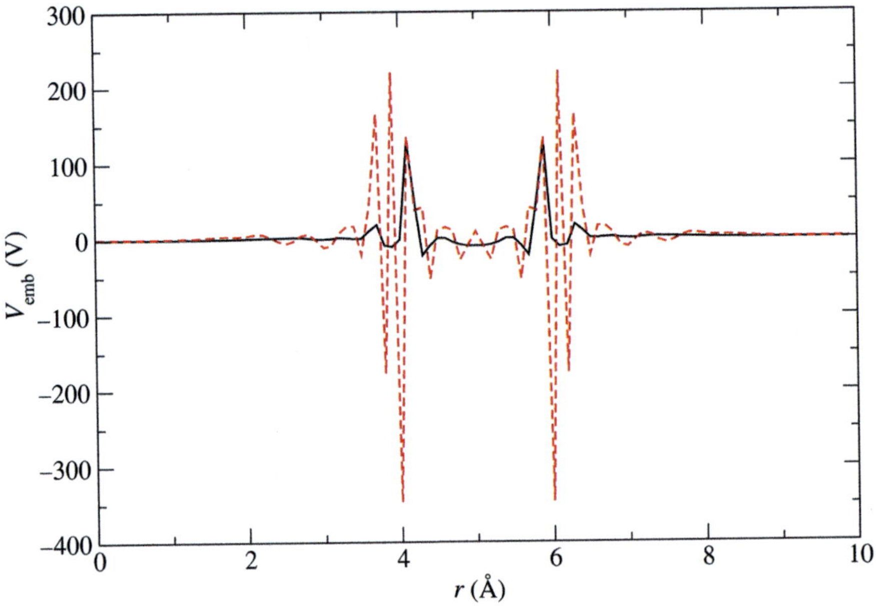

Figure 2.2 Embedding potentials along the bond axis of Cl_2. Black solid curve is the potential obtained with the penalty function and red dashed curve is the potential obtained without the penalty function. The two Cl atoms are located at 4 Å and 6 Å. *Source*: Yu *et al.* 2015 [19]. Reprinted with permission of AIP Publishing.

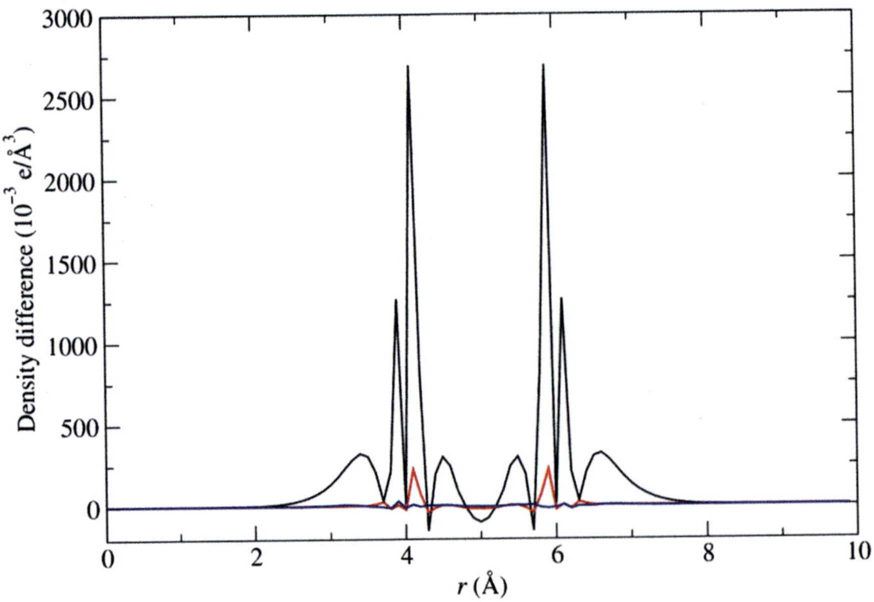

Figure 2.3 Differences between summations of the subsystem and reference densities. Black curves are the differences before optimization (without embedding potential), red curves are the differences after optimization with penalty functions, and blue curves are the differences after optimization without penalty functions. *Source*: Yu *et al.* 2015 [19]. Adapted with permission of AIP Publishing.

calculation in a GTO basis. In order to damp the unphysical oscillation, we introduce an extra penalty function in the extended W–Y functional:

$$W[V_{\text{emb}}, \lambda] = W - \lambda \int V_{\text{emb}} \nabla^2 V_{\text{emb}} d\vec{r}, \tag{2.29}$$

with an associated prefactor λ to control the strength of the penalty. The optimal value of λ can be system dependent, and careful tests need to be done to choose a proper value. Here, we take λ equal to 10^{-5} eV/(V^2/Å^2) and obtain an embedding potential, plotted in black in Figure 2.2. The unphysical oscillations are largely damped, leaving all physical features intact, including negative (attractive) features in the bonding region and positive (repulsive) features in the core region, reproducing expected chemical bonding effects. We also plot the density differences resulting from this embedding potential in Figure 2.3 (red curve). Except for small differences near the nuclei, the reference densities of the total system have been well reproduced, proving our damping function does not significantly damp the embedding potential's physical structure. We found that the penalty function is very important in both PP- and PAW-DFET calculations; we thus suggest users include it by default.

2.3.2.4 Illustrative Example

With all of the algorithm's ingredients presented, we show an illustrative example of the DFET embedding potential in this section. We consider an Al_{12} cluster embedded in a $5 \times 5 \times 4$ Al_{100} slab, representing the (111) surface of the face-centered cubic (fcc) Al (Figures 2.4a and 2.4b). We used a $126 \times 126 \times 150$ uniform grid for densities and embedding potentials. The value of F_{corr} was set to be 0.67 using a 700 eV PW kinetic

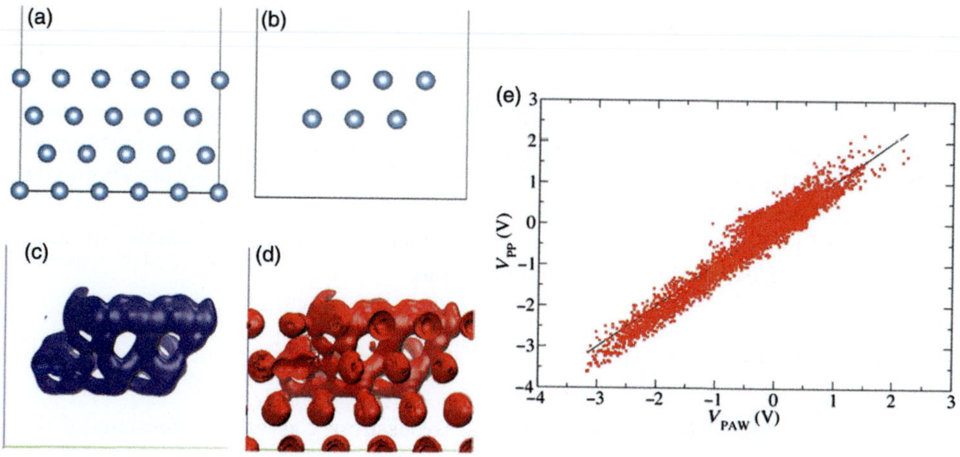

Figure 2.4 Embedding potentials for Al slab: (a) and (b), side views of the Al slab and the Al_{12} cluster; (c) isosurface of the embedding potential from ABINIT-PP calculations, isovalue = −1.206 V; (d) same as (c) for VASP-PAW; (e) scatter plot comparing $V_{\text{PP}}^{\text{ABINIT}}$ vs. $V_{\text{PAW}}^{\text{VASP}}$. Points within 1.2 Å (≈$F_{\text{corr}} \cdot R_{\text{aug}}$) from the atomic centers are omitted due to the large difference between the PAW and PP wavefunction at the atomic centers. The RMSD between the $V_{\text{PP}}^{\text{ABINIT}}$ and the $V_{\text{PAW}}^{\text{VASP}}$ showed in (e) is 0.2 V. *Source*: Yu *et al.* 2015 [19]. Reprinted with permission of AIP Publishing.

energy cutoff. The embedding potentials are constructed using both PP-ABINIT and PAW-VASP setups (Figures 2.4c and 2.4d) for comparison.

The embedding potential is negative in the boundary region of both ABINIT-PP and VASP-PAW calculations, representing attractive interactions of the metallic bond. The VASP-PAW potential shows extra features in the core region because the PAW algorithm explicitly includes core wavefunctions. The difference of embedding potentials in the core regions is a direct reflection of how the core wavefunctions are treated. If we disregard the core and only compare the interstitial region on a point-by-point basis (Figure 2.4e), the values of the two embedding potentials at all grid points are highly correlated, meaning these two potentials are quantitatively very similar. This is expected because these two potentials should reflect the same physics, regardless of the core electron treatment used. This comparison clearly indicates the numerical robustness of the embedding potential derived using the DFET scheme.

2.3.3 Embedded Cluster Calculations

Once an embedding potential is optimized, it can be used for embedded cluster calculations to obtain a correction for the DFT energy from CW methods. These calculations are typically carried out using GTOs, and the programs employed are unlike those typically used to obtain the embedding potential. We will show how to include the embedding potential in a cluster calculation and to evaluate the CW corrected total energy, while discussing some illustrative applications.

2.3.3.1 Calculation of Embedding Integrals

As discussed in Section 2.2, the Hamiltonian H in embedded cluster calculations has to be replaced by:

$$\mathrm{H}_{\mathrm{emb}} = \mathrm{H} + V_{\mathrm{emb}}, \tag{2.30}$$

where V_{emb} is the embedding potential optimized as described in Section 2.3.2. Consequently, V_{emb} enters the calculation during the formation of the one-electron integrals and the setup of the calculation. The necessary modifications in the quantum chemistry codes can be done with minimal changes.

Some attention has to be paid to the specific algorithm used to evaluate the one-electron integrals I_{ab} over the embedding potential:

$$I_{ab} = \langle a|V_{\mathrm{emb}}|b\rangle = \int \phi_a(\vec{r})V_{\mathrm{emb}}(\vec{r})\phi_b(\vec{r})dr^3. \tag{2.31}$$

The $\phi_{k=a,b,\ldots}$ denotes GTOs (real-valued), and we assume Cartesian Gaussians of the form

$$\phi_k(\vec{r}) = N_k x^{l_k} y^{m_k} z^{n_k} e^{-\alpha_k(\vec{r}-\vec{R}_k)^2}, \tag{2.32}$$

with N_k being a normalization constant; l_k, m_k, and n_k the Cartesian angular components; α_k the exponent; and $\vec{R}_k$ the center of ϕ_k. With V_{emb} defined on an equally spaced

Table 2.1 Energy shift ΔE (at the HF level) induced in a Cl_2 molecule in the center of a $10 \times 10 \times 10$ Å box (bond length 1.99 Å) by a constant embedding potential $V_{\text{emb}} = 1$ represented on a $150 \times 150 \times 150$ equally spaced grid.

Integration scheme	Cl basis set	ΔE [E_h]	$\Delta E_{\text{expected}}$ [E_h]
Real space	ECP	13.999967	14
	cc-pVDZ	40.735507	34
Reciprocal space	ECP	13.999999	14
	cc-pVDZ	33.999971	34

Results are shown for two different basis sets: Dunning's cc-pVDZ [63] all-electron basis set and the LANL2DZ [64] ECP (10 core electrons) + basis set including (d,p) polarization functions [65] for the seven valence electrons.

three-dimensional grid, the most straightforward way to evaluate the necessary integrals is given by:

$$I_{ab} \approx \sum_{n=1}^{N} \phi_a(\vec{r}_n) V_{\text{emb}}(\vec{r}_n) \phi_b(\vec{r}_n) dv. \tag{2.33}$$

Here, the index n runs over all N grid points, and the volume element is $dv = V_{\text{cell}}/N$. This simple approach works when a PP or effective core potential (ECP) approach is used and the atomic orbitals do not feature any sharp peaks. On the other hand, for an AE calculation with the products $\phi_a(\vec{r})\phi_b(\vec{r})$ containing cusps, the typical grid sizes do not allow for an accurate calculation of the one-electron embedding integrals.

Let us consider again the Cl_2 molecule in the center of a cubic $10 \times 10 \times 10$ Å box as an example. When applying a constant embedding potential with $V_{\text{emb}} = 1$ at all points, an energy shift of $\Delta E = 34 E_h$ (equal to the number of electrons in the system) would be expected. Table 2.1 shows the results obtained with the integration scheme defined in equation (2.33) on a grid of typical size and with different types of basis sets. While the expected result is well reproduced when using an ECP for Cl, it deviates strongly when an AE basis set is employed.

The situation can be remedied by switching to a reciprocal space integration scheme for cases where the product $\phi_a(\vec{r})\phi_b(\vec{r})$ is sharp. The product of two Gaussians is again a Gaussian at the combined center $\vec{R}_I = \frac{\alpha_a \vec{r}_a + \alpha_b \vec{r}_b}{\alpha_a + \alpha_b}$ and with the combined exponent $\alpha = \alpha_a + \alpha_b$. The combined exponent α thus is a measure for the sharpness of the product $\phi_a(\vec{r})\phi_b(\vec{r})$. The values of the integrals can be greatly improved using the following integration scheme in reciprocal space for cases in which $\alpha > 20.0$ and $\vec{R}_I = \vec{r}_a = \vec{r}_b$ [19].

After Fourier-transformation of the embedding potential into reciprocal space (yielding $\tilde{V}_{\text{emb}}(\vec{k})$), the integral is given by:

$$\begin{aligned} I_{ab} \approx \sum_{k_x,k_y,k_z} & \tilde{V}_{\text{emb}}(\vec{k}) e^{ikR_I} f(k_x, l_a + l_b, \alpha_a + \alpha_b) \\ & \cdot f(k_y, m_a + m_b, \alpha_a + \alpha_b) f(k_z, n_a + n_b, \alpha_a + \alpha_b) \end{aligned} \tag{2.34}$$

and the functions $f(k,l,\alpha) = \int e^{ikx}x^l e^{-\alpha x^2} dx$ can be evaluated with the following recursion relations:

$$
\begin{aligned}
f(k,0,\alpha) &= \sqrt{\frac{\pi}{\alpha}}\, e^{-\dfrac{k^2}{4\alpha}} \\
f(k,1,\alpha) &= \frac{ik}{2\alpha} f(k,0,\alpha) \\
f(k,l,\alpha) &= \frac{l-1}{2\alpha} f(k,l-2,\alpha) + \frac{ik}{2\alpha} f(k,l-1,\alpha).
\end{aligned}
\tag{2.35}
$$

The integrals over Gaussians at different centers or with combined exponents $\alpha < 20.0$ are still evaluated according to equation (2.33). If necessary, the integrals can then be transformed to a representation based on spherical and/or contracted Gaussian basis functions.

The results for the example of a Cl_2 molecule in a constant potential with this modified integration scheme are also shown in Table 2.1, the values of which reproduce the expected ones very well. Such a reciprocal space integration scheme thus should be applied especially in the case of embedded AE calculations.

We have implemented the integration scheme on a purely real-space grid via equation (2.33) as well as on the reciprocal-space scheme as given by equation (2.34) in developer's versions of the quantum chemistry software packages NWChem-6.5 [66], GAMESS (May 1, 2013 version) [67, 68], and MOLCAS 7.8 [69, 70]. We have also implemented a standalone version that can produce output files containing the final integrals adapted for usage in combination with MOLCAS, GAMESS, MOLPRO [71, 72], and TigerCI [73]. Interfacing with the program package MOLPRO is particularly convenient as it has a matrix manipulation module that reads an arbitrary matrix from an input file and adds it to the Hamiltonian so that no changes in the code are necessary in this case. We are working on increasing the number of quantum chemistry codes to be included in our standalone version. We alternatively provide a version that can be interfaced as a library. Our software is released under an open-source license, and we encourage contributions from the community to extend its functionality and available interfaces.

With the embedding integrals at hand, we can now determine a correction to the DFT energy from CW methods.

2.3.3.2 Evaluation of the Total Energy

An energy expression can be derived by decomposing the total energy [19]

$$
\begin{aligned}
E_{\text{tot}} = &\left(E_B^{\text{DFT}}\left[\rho_B^{\text{DFT}}\right] + E_{\text{int}}^{\text{DFT}}\left[\rho_A^{\text{DFT}}, \rho_B^{\text{DFT}}\right] + E_A^{\text{DFT}}\left[\rho_A^{\text{DFT}}\right]\right) \\
&+ \left(E_A^{\text{CW}}\left[\rho_A^{\text{CW}}\right] - E_A^{\text{DFT}}\left[\rho_A^{\text{DFT}}\right]\right).
\end{aligned}
\tag{2.36}
$$

The first term in parentheses is the DFT total energy, expressed as the DFT energy of each subsystem and $E_{\text{int}}^{\text{DFT}}$, the interaction energy between subsystems A and B evaluated at the DFT level. $E_X^Y(\rho_X^Y)$ is the energy of subsystem X evaluated at either the CW or DFT level ($Y = CW, DFT$). The second term in parentheses contains the correction to the DFT total energy that should be calculated by performing embedded CW and DFT calculations on the cluster (A) and then subtracting the term $\int V_{\text{emb}}(\vec{r})\rho(\vec{r})dr^3$. This term is already contained in E_{int}^{DFT}, so we exclude it in the cluster energies to avoid double counting.

A simpler expression can, however, be obtained by applying a first-order correction to E_{int}^{DFT} accounting for the CW treatment of subsystem A [19]:

$$\begin{aligned} E_{int}^{DFT}\left[\rho_B^{DFT},\rho_A^{CW}\right] &= E_{int}^{DFT}\left[\rho_A^{DFT},\rho_B^{DFT}\right] + \int \left.\frac{\partial E_{int}^{DFT}}{\partial \rho}\right|_{\rho=\rho_A^{DFT}} \left(\rho_A^{CW} - \rho_A^{DFT}\right) dr^3 \\ &= E_{int}^{DFT}\left[\rho_A^{DFT},\rho_B^{DFT}\right] + \int V_{emb}(\vec{r})\left(\rho_A^{CW} - \rho_A^{DFT}\right) dr^3. \end{aligned} \tag{2.37}$$

Inserting this into equation (2.36) yields:

$$\begin{aligned} E_{tot} = &\left(E_B^{DFT}\left[\rho_B^{DFT}\right] + E_{int}^{DFT}\left[\rho_A^{DFT},\rho_B^{DFT}\right] + E_A^{DFT}\left[\rho_A^{DFT}\right]\right) \\ &+ \left(E_A^{CW}\left[\rho_A^{CW}\right] + \int V_{emb}\rho_A^{CW} dr^3\right) \\ &- \left(E_A^{DFT}\left[\rho_A^{DFT}\right] + \int V_{emb}\rho_A^{DFT} dr^3\right). \end{aligned} \tag{2.38}$$

Consequently, an energy including the local CW correction for subsystem A can be obtained by performing self-consistent embedded CW and DFT calculations on the subsystem of interest, here subsystem A, and then evaluating the energy according to [19]

$$E_{tot}^{DFET} = E_{tot}^{DFT} + \left(\tilde{E}_{emb,A}^{CW} - \tilde{E}_{emb,A}^{DFT}\right). \tag{2.39}$$

This ansatz formally resembles the ONIOM [74] approach ($E_{tot}^{ONIOM} = E_{tot}^{DFT} + (E_A^{CW} - E_A^{DFT})$), where the correction for the total DFT energy is obtained by performing isolated CW and DFT cluster calculations ($V_{emb} = 0$) on the subsystem of interest. In contrast to the ONIOM approach, equation (2.39) includes the environment's polarization effect on the cluster under consideration. This is especially important when considering metallic systems with highly delocalized electron distributions.

2.3.3.3 Examples

We now show some applications of the DFET embedding scheme. We first discuss a typical field of application for our embedding methods, the interaction of small molecules with metallic surfaces. We then demonstrate the application to the treatment of defect states in a semiconductor system.

Application to Reactions on Metallic Surfaces

When investigating the reaction of molecules on metal surfaces using DFET, the first step is to carve out a cluster from the metal slab. Typical clusters contain approximately 10–12 atoms, which is a compromise of the physics that must be represented in the cluster and the size limit for the CW method employed. Figure 2.5 shows some examples for cluster sizes and shapes successfully employed to treat the interaction of small molecules (e.g., O_2 and H_2) at different adsorption sites on the (111) surface of fcc metals (e.g., Al or Au) [18, 75–80].

The embedding potential is usually determined without the adsorbate present. This approach is well suited for the case of metallic systems due to their short screening

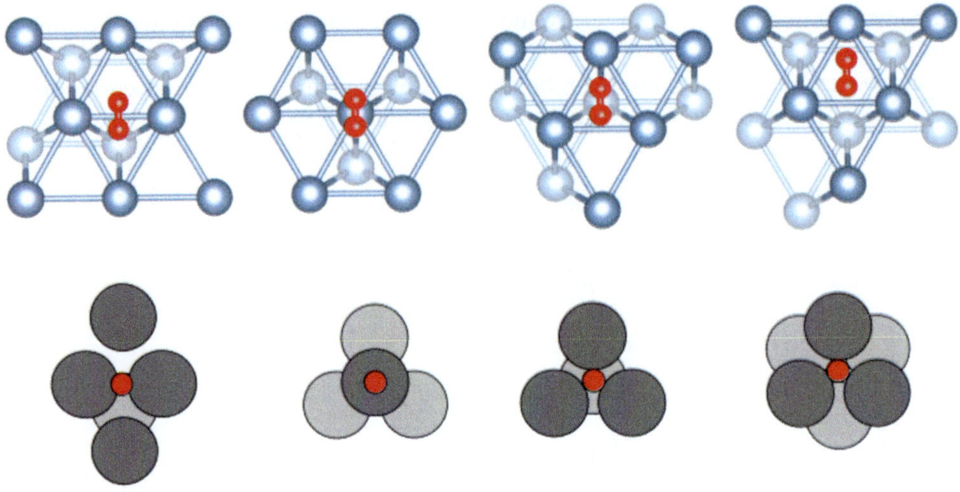

Figure 2.5 Typical cluster sizes and shapes used to model reactions of small molecules (here O_2) at different adsorption sites on the (111) surface of cubic close-packed metals such as Au or Al. *Source:* Adapted with permission from Ref. 75. Copyrighted by the American Physical Society.

length. Charge transfer induced by the adsorbate is usually screened on the scale of typical cluster sizes; nevertheless, convergence with respect to cluster size should always be checked, especially in the case of strong interactions [75]. Moreover, the interaction with the molecule may involve charge-transfer processes that are poorly described by DFT because most conventional XC functionals exhibit a lack of a derivative discontinuity, resulting in an artificial, long-range charge transfer [81]. If the adsorbate was included during the generation of V_{emb}, then this failure of DFT might be carried over into the CW calculation and influence the boundary regions. Lastly, the sheer number of necessary single-point calculations prohibits the re-evaluation of V_{emb} for each point when investigating a potential energy surface (PES) for a surface reaction.

Embedded CW and DFT calculations can be performed for molecules approaching the surface once V_{emb} has been obtained for a given cluster with the implementation in either VASP or ABINIT. Here, we will show results obtained for the dissociative adsorption of O_2 on Al as an example. Experimentally, a significantly reduced sticking probability is observed for oxygen impinging on Al at low incident energies, suggesting the presence of an energy barrier [82–84]. In contrast, a pure KS-DFT description of the system does not yield a barrier for oxygen dissociation on Al [85–87]. CW calculations on isolated clusters were able to reproduce a barrier for dissociative O_2 adsorption [88–90], but it is unclear whether this is due to the CW description or an artifact caused by replacing the periodic metal with an isolated cluster. The situation is even further complicated by experiments suggesting the height of the reaction barrier is strongly dependent on the oxygen molecule orientation relative to the surface [91]; seemingly contradictory experiments either finding pairs [92] (indicative of dissociative adsorption) or isolated [83, 93] adsorbed oxygen atoms (indicative of abstraction of one oxygen atom while the other is ejected into the gas phase) at low kinetic energies.

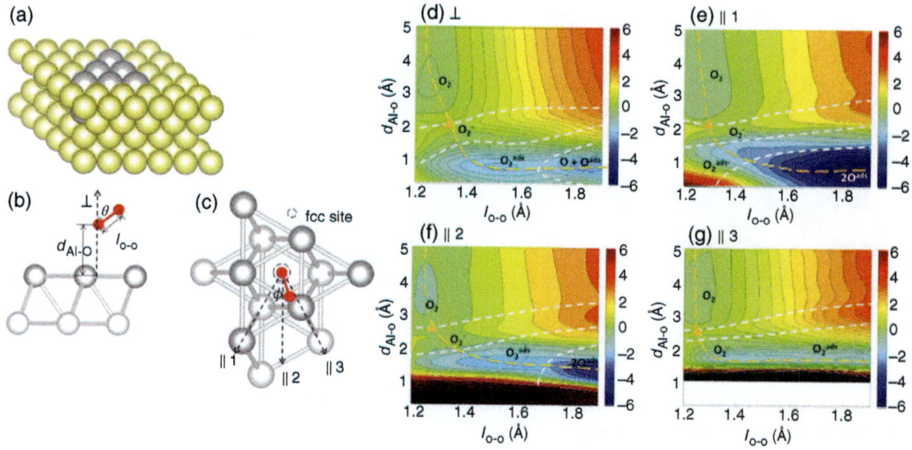

Figure 2.6 (*Left*) Structure of the (a) slab and the Al_{12} cluster (in (b) side and (c) top view) used to describe the dissociative adsorption of O_2 at the fcc site on Al(111). (*Right*) Examples for PESs for O_2 approaching the Al(111) surface in a (d) perpendicular or (e–g) different parallel configurations. *Source:* Cheng *et al.* 2015 [78]. Adapted with permission of American Chemical Society.

The Al_{12} cluster selected to model O_2 adsorption on the fcc site of Al(111) is shown in Figure 2.6. The embedding potential was obtained from the implementation in ABINIT as described in Section 2.3.2.1. The general structure of the optimized V_{emb} (not shown here) is the same as for the Al_{12} cluster depicted in Figure 2.4 and features an attractive shell at the boundary of the cluster toward the environment, representing the metallic bonds between the cluster and the remainder of the slab.

The PES for an O_2 molecule approaching the surface in different orientations can be obtained with the embedding potential in hand (Figure 2.6). The energies are evaluated according to equation (2.39) and multireference perturbation theory of second order (CASPT2) [94], based on complete active space SCF (CASSCF) [95–97] calculations, is used for the CW method (as implemented in MOLCAS). Reaction barriers of 200–600 meV [75, 78] are predicted for different pathways, corresponding well with experiments [82]. Analysis of charge evolution on the O_2 molecule along the reaction pathway, as predicted by embedded CW theory, shows that the origin of the barrier is due to charge transfer from the surface to the molecule.

By including the effect of different tilt and azimuthal angles of the O_2 molecule relative to the surface, embedded CW calculations can also explain the surprising result of finding both isolated and pairs of oxygen atoms due to two reaction pathways with similarly low activation energy following the initial charge transfer [78]. In summary, the embedded CW method as presented above is able to reliably correct the spurious PESs obtained from DFT due to the lack of derivative discontinuity [81] and self-interaction errors [98] on the surface of Al. More examples of applying the embedded CW method to surface reactions (including electronically excited states) can be found in Refs. [18, 76, 77, 79, 80].

The Case of ZnS Defects

As a second type of application, we demonstrate the use of DFET to treat defect states in semiconductors. The bonding situation here is different from the metallic system

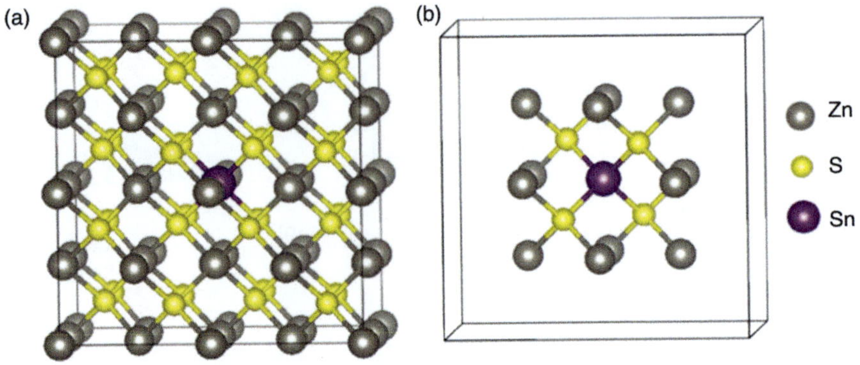

Figure 2.7 Structure of the (a) bulk and (b) cluster model for Sn-doped ZnS. *Source*: Yu *et al.* 2015 [19]. Adapted with permission of AIP Publishing.

discussed above, so consideration of the charge distribution between cluster and environment in addition to selecting the cluster size will be important.

Let us consider a $2 \times 2 \times 2$ supercell of zincblende ZnS (64 atoms) with a Sn_{Zn} antisite (Sn replacing Zn) defect in the center to introduce an n-type defect (compare Figure 2.7a) [19]. The cluster for the embedding calculations is set up to include the Sn defect with the first and second shell of neighbors, resulting in $SnZn_{12}S_4$ (compare Figure 2.7b).

A natural choice is to consider both the cluster and environment as neutral for metallic systems. The DFET optimization process works for any atom or charge partitioning, but they need to be chosen *a priori*. For bonds with both ionic and covalent character, such as Zn–S bonds, it is difficult to decide if the bonds should be cleaved homolytically or heterolytically. One possibility is to use Bader charges [99–103] from the bulk crystal to determine a physically meaningful electron partitioning. Optimizing V_{emb} for different electron distributions (with PAW-DFET in VASP), performing PW-embedded DFT-PBE [104] calculations with the optimized potential (also in VASP), and comparing the Bader charge on the central atom to the bulk value show that the optimal cluster configuration is $[SnZn_{12}S_4]^{18-}$, corresponding to heterolytic bond cleavage. As another example, a $[CuZn_{12}S_4]^0$cluster reproduces the bulk Bader charge on the central Cu atom when a Cu_{Zn} antisite is introduced to represent a p-type defect [19].

We compare the projected density of states (PDOS) at the central Sn defect atom obtained from an embedded cluster calculation to the bulk PDOS in order to assess the quality of DFET for such defect states in semiconductors and to analyze the ability to correctly reproduce properties beyond relative energies [19]. PW DFT-PBE [104] was used in both cases for consistency. In the bulk calculation, the PDOS features a characteristic peak at −0.22 eV relative to the Fermi level (Figure 2.8a). This peak corresponds to the Sn 4s orbital and constitutes the occupied n-type defect state. The PDOS obtained from DFET is displayed in Figure 2.8b and shows that the position of the defect state is well reproduced at −0.23 eV with an appropriate charge on the embedded cluster (Table 2.2). By contrast, bare cluster calculations ($V_{emb} = 0$) are not able to predict the correct Sn 4s band energy (Table 2.2) [19].

While the position of the defect state is well reproduced by DFET in these examples, other states often shift or additional states are introduced resulting from the dangling

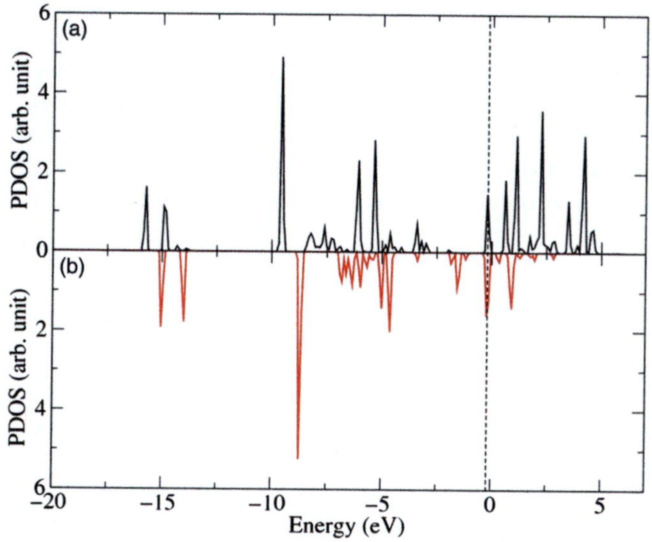

Figure 2.8 PDOS of a Sn defect atom in (a) bulk ZnS and (b) in an embedded $[SnZn_{12}S_4]^{18-}$ cluster. All energies are relative to the Fermi level (at 0 eV). The position of the Sn 4s peak is labeled with a dashed line. *Source*: Yu *et al.* 2015 [19]. Adapted with permission of AIP Publishing.

Table 2.2 Energy of the Sn defect state relative to the Fermi level (at 0 eV). Values in parentheses denote the Bader charges (in *e*) for the central Sn atom.

	Defect band position [eV]	
System	**Bare cluster**	**Embedded cluster**
Supercell with Sn_{Zn}	−0.22 (1.25)	
$[SnZn_{12}S_4]^0$	0.45	0.12 (1.35)
$[SnZn_{12}S_4]^{-18}$	−0.55	−0.23 (1.23)

Source: Yu *et al.* 2015 [19]. Adapted with permission of AIP Publishing.

bonds; for example, the embedded cluster core levels are all shifted with respect to the bulk values. Nevertheless, DFET is able to reliably produce the desired quantity of the defect state as the state of interest in this example.

2.4 Potential Functional Embedding Theory

In DFET, we first construct the embedding potential at the PW-DFT level of theory, and then apply the DFT-derived potential to the CW calculation on the isolated cluster. The environment-cluster interaction therefore is computed at the pure DFT level even though the cluster is described using a CW method. More specifically, approximations in the description of the environment–cluster interactions are twofold: (i) we use the

DFT, rather than CW, cluster density to compute the interaction and (ii) the interaction itself is treated at the DFT level. If the CW cluster density significantly differs from the DFT cluster density, then DFET cannot correctly describe the response of the environment to the cluster. To solve this problem, we introduce PFET [15] in this section, which effectively fixes the first approximation while keeping the second approximation and formulates a truly self-consistent hybrid CW/DFT embedding theory. Currently, PFET is only implemented with PPs, so the discussions below will be based on the PP formulae.

2.4.1 Basic Theories and Technical Details

2.4.1.1 Definition of Energies

Similar to DFET, PFET assumes that both the cluster (subsystem A) and environment (subsystem B) share a common embedding potential V_{emb}. Instead of trying to reproduce a reference density, PFET focuses on the energy of the total system, which can be written as a functional of V_{emb}:

$$E_{\text{tot}}[V_{\text{emb}}] = \tilde{E}_A[V_{\text{emb}}] + \tilde{E}_B[V_{\text{emb}}] + E_{\text{int}}[\rho_A, \rho_B]. \tag{2.40}$$

Similar to previous definitions, $\tilde{E}_{K=A,B}[V_{\text{emb}}]$ are the energies of each subsystem containing its interaction with V_{emb}, and $E_{\text{int}}[\rho_A, \rho_B]$ is the interaction energy between the two subsystems. Differing from the two-step procedure utilized in DFET, we simply minimize $E_{\text{tot}}[V_{\text{emb}}]$ to obtain the optimal V_{emb} in PFET. The subsystem energies and densities ($\tilde{E}_K$ and ρ_K) can be computed using different methods (typically, CW for cluster and DFT for environment).

The accuracy of the PFET method depends on the accuracy of the interaction energy, which is defined as

$$\begin{aligned} E_{\text{int}}[\{\rho_K\}] &= J_{\text{int}}[\{\rho_K\}] + E^{\text{xc}}_{\text{int}}[\{\rho_K\}] + T_{S,\text{int}}[\{\rho_K\}] \\ &\quad + \sum_{K_1 \neq K_2}^{A,B} \int V^{K_1}_{\text{ion}}(\vec{r})\rho_{K_2}(\vec{r})d\vec{r} - \int V_{\text{emb}}(\vec{r}) \sum_{K=A,B} \rho_K(\vec{r})d\vec{r}. \end{aligned} \tag{2.41}$$

On the right-hand side, $J_{\text{int}}[\{\rho_K\}]$ is the classical Coulomb interaction between the two subsystem densities:

$$J_{\text{int}}[\{\rho_K\}] = \int \frac{\rho_A(\vec{r}) \cdot \rho_B(\vec{r}')}{|\vec{r} - \vec{r}'|} d\vec{r}d\vec{r}' \tag{2.42}$$

and $E^{\text{xc}}_{\text{int}}\left[\{\rho_K\}\right]$ is the XC interaction energy, which is computed using an approximate XC density functional $E^{\text{xc}}[\rho]$ at the DFT level:

$$E^{\text{xc}}_{\text{int}}[\{\rho_K\}] = E^{\text{xc}}[\rho_{\text{tot}}] - E^{\text{xc}}[\rho_A] - E^{\text{xc}}[\rho_B]. \tag{2.43}$$

The third term $T_{S,\text{int}}[\{\rho_K\}]$ is the kinetic interaction energy

$$T_{S,\text{int}}[\{\rho_K\}] = T_S[\rho_{\text{tot}}] - T_S[\rho_A] - T_S[\rho_B]. \tag{2.44}$$

As introduced in Section 2.2, the kinetic energy term ($T_S[\rho]$) can be computed using either approximate KEDFs or the exact OEP procedure, as we will do here. The exact OEP procedure is computationally more expensive but features a higher accuracy compared to current KEDFs. We will discuss the OEP procedure in detail in the next section.

The fourth term $\left(\sum_{K_1 \neq K_2}^{A,B} \int V_{\text{ion}}^{K_1}(\vec{r}) \rho_{K_2}(\vec{r}) d\vec{r} \right)$ represents the interaction between the nuclei in subsystem K_1 and the electron density of subsystem K_2. For the nuclear Coulomb potential and the local part of the PP, this interaction can be directly computed using the integral shown in equation (2.41). However, the interaction between the NLPP and the electrons is more difficult to calculate. We need explicit orbitals rather than just densities to evaluate the NLPP–electron interaction. This is similar to the kinetic interaction energy, which also requires computing explicit orbitals. The NLPP–electron interaction therefore is combined with the kinetic energy interaction term and computed using the OEP procedure (*vide infra*).

The last term $(\int V_{\text{emb}}(\vec{r}) \sum_{K=A,B} \rho_K(\vec{r}) d\vec{r})$ is subtracted to remove the subsystem density-V_{emb} interaction energies contained in $\tilde{E}_K$. The subsystem density-V_{emb} integral effectively represents the subsystem interactions, which is explicitly included as the E_{int} term in equation (2.41), and has to be removed to avoid double counting.

When both subsystems are described at the DFT level of theory, and when E_{int} is computed using exact methods such as OEP (*vide infra*), the PFET energy $E_{\text{tot}}[V_{\text{emb}}]$ reproduces the exact DFT energy of the total system. Minimizing the PFET energy at the DFT/DFT level therefore must lead to the ground-state DFT density of the total system$\rho_{\text{tot}}^{\text{DFT}}$, due to the one-to-one correspondence between the energy and density. We thus obtain the following equation: $\rho_A(\vec{r}, V_{\text{emb}}) + \rho_B(\vec{r}, V_{\text{emb}}) = \rho_{\text{tot}}^{\text{DFT}}(\vec{r})$, showing that the resulting PFET potential naturally satisfies the constraint enforced in DFET. Since the DFET potential is strictly unique, the two methods are in principle equivalent at the DFT/DFT level, while PFET provides a new theoretical framework that is extendable to the hybrid CW/DFT level of theory.

2.4.1.2 Optimized Effective Potential (OEP) Scheme for Exact Kinetic Energy

In this section, we introduce the OEP procedure used in PFET to compute the exact kinetic energy and kinetic potential.

For a given arbitrary density ρ_{in}, which may be the result of either a GTO-CW calculation (e.g., ρ_A), PW-DFT (e.g., ρ_B) calculation, or summation of DFT and CW densities (e.g., $\rho_{\text{tot}} = \rho_A + \rho_B$), we aim to compute its kinetic energy $T_S[\rho_{\text{in}}]$ and the corresponding kinetic potential

$$V_{T_S}(\vec{r}) = \frac{\delta T_S[\rho_{\text{in}}]}{\delta \rho_{\text{in}}(\vec{r})}. \tag{2.45}$$

In PFET, we follow the W–Y OEP procedure introduced in Ref. [48] and define the W–Y functional of V_{OEP}:

$$W[V_{\text{OEP}}] = 2\sum_{i=1}^{n} \left\langle \varphi_i \right| -\frac{1}{2}\nabla^2 \left| \varphi_i \right\rangle + \int V_{\text{OEP}}(\vec{r}) \left(\rho(\vec{r}) - \rho_{\text{in}}(\vec{r})\right) d\vec{r}. \tag{2.46}$$

Within equation (2.46), we compute the orbitals $\varphi_i[V_{\text{OEP}}]$ by solving the SCF equations with V_{OEP}:

$$\left[-\frac{1}{2}\nabla^2 + V_{\text{OEP}}(\vec{r})\right] \varphi_i = \varepsilon_i \varphi_i. \tag{2.47}$$

In the OEP procedure, we first maximize $W\left[V_{\text{OEP}}\right]$ to obtain the optimal V_{OEP} and the associated orbitals $\{\varphi_i\}$, which allows us to compute the kinetic energy and potential via

$$T_s[\rho_{\text{in}}] = 2\sum_{i=1}^{n}\left\langle \varphi_i \left| -\frac{1}{2}\nabla^2 \right| \varphi_i \right\rangle \tag{2.48}$$

$$V_{T_S}(\vec{r}) = \left.\frac{\delta T[\rho]}{\delta \rho}\right|_{\rho=\rho_{\text{in}}} = \mu[\rho_{\text{in}}] - V_{\text{OEP}}[\rho_{\text{in}}]. \tag{2.49}$$

Here, the chemical potential $\mu[\rho_{\text{in}}]$ is the Fermi energy obtained by solving equation (2.47).

With the orbitals in hand, evaluation of the NLPP-density interactions ($E_{\text{NLPP}}[\rho_{\text{in}}]$) is straightforward using the NLPP energy formula. This term is then combined with the kinetic energy ($T_S[\rho_{\text{in}}] + E_{\text{NLPP}}[\rho_{\text{in}}]$) to calculate the total orbital-dependent contributions to the interaction energy E_{int}.

We emphasize here that, unlike the original W–Y OEP scheme, we express V_{OEP} using a three-dimensional uniform grid instead of expanding it using Gaussian basis functions. Correspondingly, we perform all OEP calculations with a PW basis set, using a modified version of ABINIT. As with the original W–Y OEP scheme, the PW-OEP utilized here also suffers from the unbalanced description of the potential and wavefunction (a similar problem also appears in DFET, which was discussed in Section 2.3.2.3). We therefore must introduce a penalty function in the W–Y functional (equation 2.46) to damp the unphysical oscillation in the resulting effective potential. The penalty function utilized here takes the same form as the one used in DFET (equation 2.29). However, due to the inclusion of the penalty function, the input density ρ_{in} cannot be reproduced exactly by the resulting $\rho\left[V_{\text{OEP}}\right]$, leading to small errors in the kinetic interaction energy. A linear correction to the kinetic/NLPP energy is introduced to reduce this error:

$$T_S[\rho_{\text{in}}] = T_S[\rho[V_{\text{OEP}}]] + E_{\text{NLPP}}[\rho[V_{\text{OEP}}]] + \int V_{\text{OEP}}[\rho_{\text{in}}](\rho[V_{\text{OEP}}] - \rho_{\text{in}})d\vec{r}. \tag{2.50}$$

With the PW-OEP method, we can compute the energy of the total system exactly on the hybrid CW/DFT level of theory (equations 2.40 and 2.41). To perform energy minimization, we need the corresponding energy gradient with respect to V_{emb}, the computation of which will be discussed in Section 2.4.1.3.

2.4.1.3 Energy Gradient

The gradients of the first two terms in equation (2.40) ($\tilde{E}_A$ and $\tilde{E}_B$) are simply the densities of the two subsystems (ρ_A and ρ_B) (equation 2.8). An analytical formula for the interaction energy gradient ($\frac{\delta E_{\text{int}}}{\delta V_{\text{emb}}}$) is not available because there is no simple formula to compute the gradient of the kinetic energies obtained via OEP. We therefore adopt a different numerical approach to compute the energy gradient. Utilizing the chain rule and considering $\frac{\rho_K(\vec{r}')}{\delta V(\vec{r})} = \frac{\delta E_K}{\delta V(\vec{r}')\delta V(\vec{r})} = \frac{\rho_K(\vec{r})}{\delta V(\vec{r}')}$, we obtain the following relationship [15, 105]:

$$\frac{\delta E_{\text{int}}}{\delta V_{\text{emb}}} = \sum_{K=A,B}\int \left.\frac{\delta E_{\text{int}}}{\delta \rho_K(\vec{r}')}\right|_{V_{\text{emb}}} \cdot \frac{\delta \rho_K(\vec{r})}{\delta V_{\text{emb}}(\vec{r}')} d\vec{r}'. \tag{2.51}$$

Here, the term $\frac{\delta E_{\text{int}}}{\delta \rho_K(\vec{r}')}$ is directly related to the Coulombic, XC, ionic, and kinetic potentials of both the total system and subsystems:

$$\frac{\delta E_{\text{int}}}{\delta \rho_K} = (V_J[\rho_{\text{tot}}] - V_J[\rho_K]) + (V_{\text{xc}}[\rho_{\text{tot}}] - V_{\text{xc}}[\rho_K]) + (V_{T_S}[\rho_{\text{tot}}] - V_{T_S}[\rho_K]) + \sum_{K' \neq K} V_{\text{ion}}^{K'} - V_{\text{emb}}. \tag{2.52}$$

We follow the standard procedures of PW-DFT programs to evaluate the Coulombic potential V_J, the XC potential V_{xc}, and the ionic potential $V_{\text{ion}}^{K'}$. The kinetic potential V_{T_S} can be derived via the OEP procedure as discussed in Section 2.4.1.2. We then use the term $\frac{\delta E_{\text{int}}}{\delta \rho_K(\vec{r}')}$ to evaluate the interaction energy gradient according to equation (2.51). Considering a small potential variation $\delta V_{\text{emb}}(\vec{r}') = \frac{\delta E_{\text{int}}}{\delta \rho_K(\vec{r}')} \cdot \lambda$, where $\lambda \to 0$, the response of the density $\rho_K(\vec{r})$ with respect to this potential change is

$$\delta \rho_K(\vec{r}) = \int \delta V_{\text{emb}}(\vec{r}') \cdot \frac{\delta \rho_K(\vec{r})}{\delta V_{\text{emb}}(\vec{r}')} d\vec{r}' = \lambda \int \frac{\delta E_{\text{int}}}{\delta \rho_K(\vec{r}')} \cdot \frac{\delta \rho_K(\vec{r})}{\delta V_{\text{emb}}(\vec{r}')} d\vec{r}'. \tag{2.53}$$

Comparing equation (2.53) with (2.51), we see the interaction energy gradient can be written as [15, 105]

$$\frac{\delta E_{\text{int}}}{\delta V_{\text{emb}}} = \frac{1}{\lambda} \sum_K \delta \rho_K(\vec{r}) \approx \frac{1}{2\lambda} \sum_K \left\{ \rho_K \left[V_{\text{emb}}^1 \right] - \rho_K \left[V_{\text{emb}}^2 \right] \right\} \tag{2.54}$$

$$V_{\text{emb}}^1 = V_{\text{emb}} + \lambda \frac{\delta E_{\text{int}}}{\delta \rho_K(\vec{r})} \tag{2.55}$$

$$V_{\text{emb}}^2 = V_{\text{emb}} - \lambda \frac{\delta E_{\text{int}}}{\delta \rho_K(\vec{r})}. \tag{2.56}$$

We therefore only need to alter $V_{\text{emb}}(\vec{r})$ by $\lambda \frac{\delta E_{\text{int}}}{\delta \rho_K(\vec{r})}$ (equations 2.55 and 2.56) to obtain the responses of $\rho_K(\vec{r})$, so that the interaction energy gradient $\frac{\delta E_{\text{int}}}{\delta V_{\text{emb}}}$ can be evaluated numerically via equation (2.54). The total energy gradient $\frac{\delta E_{\text{tot}}}{\delta V_{\text{emb}}}$ can be subsequently calculated by adding ρ_A, ρ_B, and $\frac{\delta E_{\text{int}}}{\delta V_{\text{emb}}}$. We use the L-BFGS optimizer to minimize the energy with the total energy gradient and obtain the optimal embedding potential at the PFET level of theory.

2.4.1.4 Summary of the Code Structure

A summary of the PFET code structure is given in Figure 2.9.

The core part of the PFET code is an L-BFGS optimizer, which takes the energy and gradient information as input and optimizes the embedding potential. In each iteration, the optimizer generates the next V_{emb} and feeds it to electronic structure programs to perform embedding calculations. As in DFET, the MOLCAS and ABINIT programs were modified to enable embedding calculations using either GTO or PW basis sets, respectively. In Figure 2.9, we assume a GTO-CW/PW-DFT setup, with MOLCAS used for the cluster (subsystem A) and ABINIT for the environment (subsystem B). (The code also allows use of ABINIT for both the cluster and environment to perform

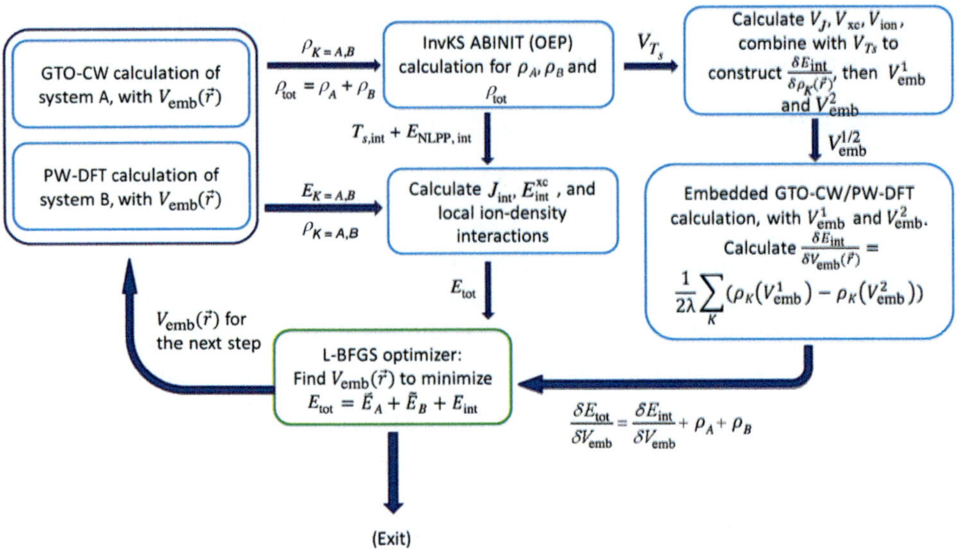

Figure 2.9 Runtime flowchart of the PFET scheme, implemented in the MOLCAS/ABINIT setup.

PFET calculations at the DFT/DFT level, which is essentially equivalent to DFET. One could also run the code using MOLCAS for both subsystems to enable a GTO/GTO embedding calculation at up to the CW/CW level of theory, although the PW ABINIT code is still needed in this situation to evaluate interaction energies at the PW-DFT level.) The resulting subsystem and total densities ($\rho_{K=A,B}$ and $\rho_A + \rho_B$) are fed to a modified "invKS" (inverse Kohn–Sham) [15] ABINIT-7.0.5 code to perform the OEP calculations. The resulting kinetic and NLPP interaction energies are combined with all other energy components in equations (2.40) and (2.41) to compute the PFET total energy. Meanwhile, the resulting kinetic potentials V_{T_S} are combined with all other potential terms in equation (2.52) to obtain $\frac{\delta E_{int}}{\delta \rho_K}$. The shifted embedding potentials (V^1_{emb} and V^2_{emb}) are constructed according to equations (2.55) and (2.56). These two potentials are again fed to the embedded MOLCAS or ABINIT program to compute the density responses ($\rho_K[V^{1/2}_{emb}]$) plugged into equation (2.54), leading to the total energy gradient needed for the L-BFGS optimizer.

The PFET code only supports the use of ABINIT to optimize the embedding potential up to now, as all real-space integrals are performed using direct summations and the dual-grid setup in PAW is not yet implemented for PFET.

2.4.2 Illustrative Examples

In this section, we provide some preliminary examples of PFET. Although it provides a theoretical framework extendable to the hybrid CW/DFT level of theory, here we only show results at the PW-DFT/PW-DFT level as a proof of principle because (i) numerical problems remain in hybrid PFET calculations that will be discussed in the Outlook section and (ii) the PFET result is equivalent to the DFET and total KS-DFT results at the PW-DFT/PW-DFT level, so the latter two methods provide rigorous benchmarks and facilitate assessment of PFET.

2.4.2.1 AlP Diatomic

We first show an example of a PFET calculation from Ref. [15] at the PW-DFT/PW-DFT level for a diatomic consisting of an Al atom and a P atom separated by 2.2 Å. The PW91 XC functional [106], along with bulk-derived local PPs [33, 107], was used in combination with a 800 eV PW kinetic energy cutoff for all DFT and OEP calculations. The diatomic is naturally partitioned into two single-atom subsystems on which the PFET optimization is performed.

As previously mentioned in Sections 2.2 and 2.4.1.1, the kinetic interaction energy can be computed using either KEDFs or the OEP, though the OEP procedure should yield a higher accuracy. To verify this point, we compare results obtained by the OEP (labeled as emb-OEP in the figures) and two KEDFs including Thomas–Fermi [108, 109] (labeled as emb-TF) and HC10[110] (labeled as emb-HC10). Because PFET is formally equivalent to DFET at the PW-DFT/PW-DFT level, we can use agreement of the densities ($\rho_{\text{tot}}^{\text{DFT}}$ compared with $\rho_A + \rho_B$) as a criterion to evaluate the quality of the resulting embedding potential. The densities along the Al–P bond axis resulting from all three methods are plotted together with the benchmark DFT density in Figure 2.10. It is clear that both the TF and HC10 KEDFs cause significant errors compared to the benchmark, while the OEP procedure reproduces the KS-DFT density with high fidelity. This shows the significant improvement of the OEP compared to KEDFs, as well as the equivalence of both embedding methods at the DFT/DFT level.

In Figure 2.11, we show the contour plot of V_{emb} in a plane containing the two atoms. Similar to the DFET case, we observe repulsion in the nuclear region and attraction in the bonding region, capturing the effects of chemical bond formation.

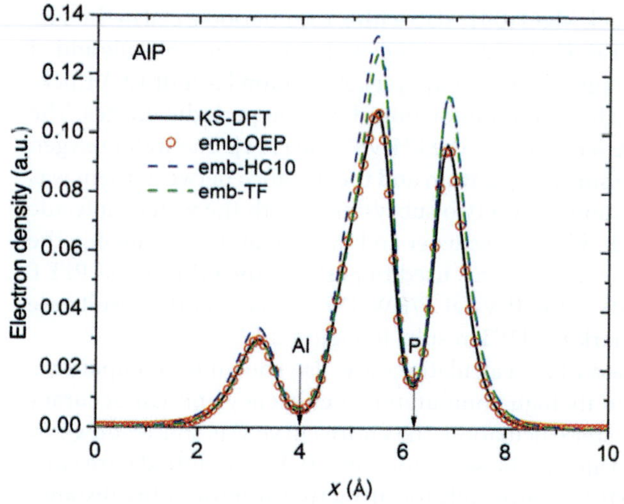

Figure 2.10 Comparison of electron densities along the AlP bond axis: KS-DFT density (benchmark) is the black curve. The emb-OEP, emb-HC10, and emb-TF results are shown as red circles, blue dashes, and green dash-dots, respectively. *Source*: Huang and Carter 2011 [15]. Reprinted with permission of AIP Publishing.

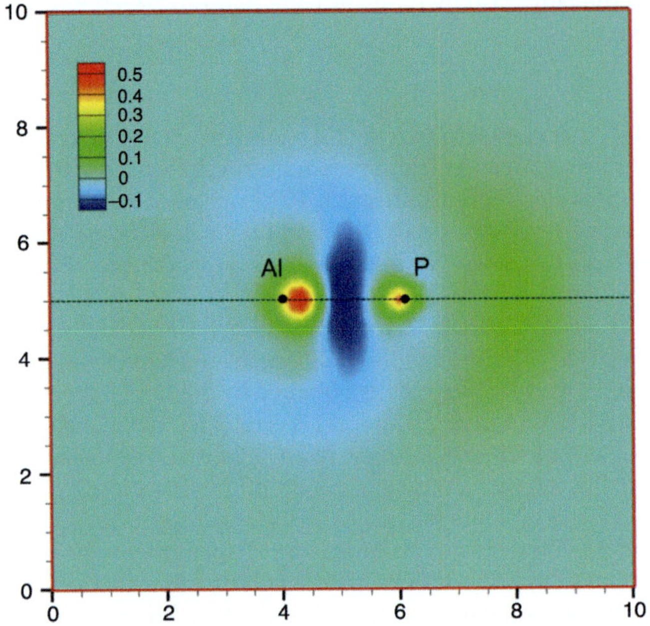

Figure 2.11 Contour plot of the embedding potential (in hartree·bohr [3]) in a plane containing the AlP molecule, with the coordinates in Å. *Source*: Huang and Carter 2011 [15]. Reprinted with permission of AIP Publishing.

2.4.2.2 H_2O on MgO (001)

In this example, we consider the interaction between a water molecule and the MgO (001) surface [15]. A three-layer thick MgO (001) slab with a 1×1 periodic cell in the surface plane containing two Mg and two O atoms was used, with one water molecule added to one side of the slab (water coverage of 0.25 monolayer). A vacuum layer of 12 Å thickness was added above the MgO slab to limit interactions between periodic images. The NLPP used in this test case was generated with the FHI98 code [111]. The water oxygen atom was fixed directly above one surface Mg atom and the distance between them was varied. The total system was partitioned into two subsystems, with the water molecule and first layer of the MgO slab included in the cluster and the remainder included in the environment. The PBE XC functional [104] was used to perform the OEP-based PFET calculations (labeled as emb-OEP) at the PW-DFT/PW-DFT level, and the results are plotted together with the benchmark KS-DFT results in Figure 2.12.

Bare cluster and TF-KEDF-based PFET calculations are also shown for comparison. The TF KEDF was selected due to its high computational efficiency and fair accuracy for ionic systems. For the bare cluster calculation, the entire curve was shifted down by 1819.879 eV to match the energy value at the separation of 4.0 Å. The shifted bare cluster energy (labeled as shift-bare-ML) is universally too attractive over the entire distance range. Similarly, the TF KEDF-based PFET calculation yields different energies than the KS-DFT benchmark. After shifting down by 2.539 eV to match the energy value at 4.0 Å (shifted curve is labeled as shift-emb-TF in Figure 2.12), TF-KEDF-based PFET agrees quite well with KS-DFT. The OEP-based PFET generates energies directly comparable with the KS-DFT benchmark without a shift required. The OEP-based PFET energy

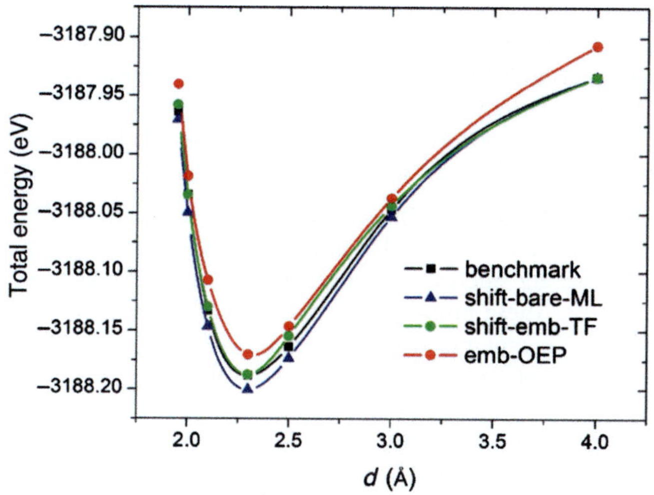

Figure 2.12 Total energy versus the distance d between the water molecule and the MgO(001) surface. Comparison between the benchmark (black squares), shifted non-embedded results (labeled "shift-bare-ML," blue triangles), shifted emb-TF results (labeled "shift-emb-TF," green circles), and emb-OEP results (red circles). Both shift-bare-ML (shifted downward by 1819.879 eV) and shift-emb-TF (shifted downward by 2.539 eV) data are shifted to match the benchmark at $d = 4.0$ Å. The emb-OEP results are not shifted and are consistently higher than the benchmark, but with a small absolute error of less than 30 meV. *Source*: Huang and Carter 2011 [15]. Reprinted with permission of AIP Publishing.

agrees with the benchmark reasonably well, with a slight overestimation of approximately 30 meV across the entire distance range. These results demonstrate the consistency of the PFET method compared to the KS-DFT benchmark, verifying the effectiveness of the embedding theory at the DFT/DFT level.

2.5 Summary and Outlook

The major new idea in both DFET and PFET compared to previous embedding methodologies is to apply a common, numerically exact (within DFT) potential to all subsystems. As discussed, this assumption removes arbitrariness in the density partitioning of the total system and importantly generates a unique embedding potential. DFET and PFET were developed based on this assumption, focusing on reproducing total densities and energy minimization, respectively. Comparing these two methods, DFET builds the embedding potential at a pure DFT level, while PFET provides the opportunity for a truly self-consistent hybrid CW/DFT theory to describe the subsystem interactions. Although formally less consistent, DFET is more robust, computationally cheaper, and more convenient to perform, while PFET faces more numerical problems before it can become practical (*vide infra*). Currently, DFET has been successfully applied in a series of applications to describe surface reactions as shown in Section 2.3.3.3. A truly hybrid CW/DFT level of PFET is still under development.

Some major difficulties at the CW/DFT level of PFET are currently associated with the OEP procedure. First, the computational cost of the OEP method significantly decreases the efficiency of PFET compared to DFET. Second, the PW-based OEP procedure is not numerically robust enough for PFET optimization. More specifically, the OEP demonstrates extraordinary sensitivity to basis set choice. For example, the $V_{T_S} + V_J + V_{xc} = \mu - V_{OEP} + V_J + V_{xc}$ term (from equation 2.49) derived using the PW-OEP method for a density obtained from a GTO calculation often suffers from severe oscillations, even with the damping function. This unstable performance of the OEP eventually makes evaluation of the interaction energy gradient ($\frac{\delta E_{int}}{\delta \rho_K}$) problematic. Furthermore, convergence of the OEP is often insufficient, leading to non-smooth character for the PFET energy functional $E[V_{emb}]$. The situation is exacerbated by adopting a numerical, instead of an analytical, procedure to evaluate the energy gradient, which essentially relies on the smoothness of the energy surface. These numerical problems are still tolerable in PW-DFT/PW-DFT calculations, but become major challenges at GTO-CW/PW-DFT level. A more robust OEP method is therefore needed for further development of PFET.

Current OEP procedures fail for excited-state calculations, which thus prohibit the application of PFET to such problems. In general, the excited-state density has completely different features compared to the ground-state density. For example, for a one-dimensional single-particle system with any finite potential, one can easily prove that the ground- and excited-state wavefunctions have different numbers of nodes. It therefore is theoretically impossible to use a ground-state OEP procedure to invert the excited-state density in such a situation. Unfortunately, the current OEP can only be performed for the ground state, and further studies are needed to develop a robust procedure to evaluate the kinetic interaction energy of excited states.

The other problem residing in PFET is related to the evaluation of the energy derivative for subsystems. We typically use $\frac{\delta \tilde{E}_K}{\delta V_{emb}} = \rho_K$ in PFET, which is guaranteed to be correct in all variational methods such as KS-DFT, HF, CASSCF, and configuration interaction (CI). However, this relationship does not hold for non-variational methods such as second-order Møller–Plesset perturbation theory (MP2), CASPT2, and coupled cluster (CC). More work therefore needs to be done in order to overcome this theoretical difficulty and to combine these non-variational CW methods with the variational PFET procedure.

Besides refining the numerical techniques in PFET, another direction has led to extending the PFET to a time-dependent (TD) version. The TD-PFET provides a new divide-and-conquer approach to study the propagation of a quantum system interacting with a changing external field. The theoretical and algorithmic details for TD-PFET are described in Ref. [17].

Although DFET is more mature than PFET, problems exist that require further improvements. One particular issue is the unsystematic determination of formal charges of the subsystems. While the partition of electron numbers is intuitively straightforward in certain cases (such as metals), it is highly nontrivial in other cases (such as semiconductors; see Section 2.3.3.3). Unfortunately, many cluster properties depend heavily on the assigned formal charges, so a reasonable charge distribution is critical to ensure the quality of the DFET results.

Besides the formal charge assignment, the numerical performance of DFET also depends heavily on the electronic structure properties of all subsystems. DFET works on clusters without capping atoms at the boundary, so the electronic structure of the cluster can be ill-defined. This is especially serious for covalently bonded systems. In these situations, the partitioning of the total system inevitably introduces dangling bonds, which severely perturb the electronic structure of the cluster. These dangling bonds embedded in DFET potentials demonstrate very strong multireference/quasi-degenerate character, often causing convergence difficulties in GTO-based embedded CW calculations. Most available codes were not developed to solve such strongly correlated systems, so more robust numerical techniques are needed in this scenario. Besides the frequently observed convergence problems, the embedded GTO-CW calculations also feature higher basis-set sensitivity compared to normal CW calculations.

Despite these current limitations, DFET and PFET enable the combination of multiple quantum mechanical tools to conduct accurate *ab initio* simulations of chemistry that is otherwise untreatable. Both methods are currently under active development, and we expect to see more applications of both methods in a variety of fields (e.g., heterogeneous catalysis, local defects or exciton studies in semiconductors, etc.) in the future.

Acknowledgments

We are grateful to the U.S. Department of Energy, Office of Science, Basic Energy Sciences under Award No. DE-SC0002120 and the U.S. National Science Foundation under Grant No. 1265700 for support of this work. C.M.K. acknowledges support by a fellowship within the Postdoc-Program of the German Academic Exchange Service (DAAD).

References

1 Wesolowski, T. A.; Warshel, A. Frozen density functional approach for ab initio calculations of solvated molecules. *J. Phys. Chem.* **1993**, 97, 8050–8053.

2 Neugebauer, J.; Louwerse, M. J.; Baerends, E. J.; Wesolowski, T. A. The merits of the frozen-density embedding scheme to model solvatochromic shifts. *J. Chem. Phys.* **2005**, 122, 094115.

3 Elliott, P.; Burke, K.; Cohen, M. H.; Wasserman, A. Partition density-functional theory. *Phys. Rev. A* **2010**, 82, 024501.

4 Knizia, G.; Chan, G. K.-L. Density matrix embedding: A simple alternative to dynamical mean-field theory. *Phys. Rev. Lett.* **2012**, 109, 186404.

5 Manby, F. R.; Stella, M.; Goodpaster, J. D.; Miller, T. F. A simple, exact density-functional-theory embedding scheme. *J. Chem. Theory Comput.* **2012**, 8, 2564–2568.

6 Cortona, P. Self-consistently determined properties of solids without band-structure calculations. *Phys. Rev. B* **1991**, 44, 8454–8458.

7 Cortona, P. Direct determination of self-consistent total energies and charge densities of solids: A study of the cohesive properties of the alkali halides. *Phys. Rev. B* **1992**, 46, 2008–2014.

8 Govind, N.; Wang, Y. A.; da Silva, A. J. R.; Carter, E. A. Accurate ab initio energetics of extended systems via explicit correlation embedded in a density functional environment. *Chem. Phys. Lett.* **1998**, 295, 129–134.

9 Govind, N.; Wang, Y. A.; Carter, E. A. Electronic-structure calculations by first-principles density-based embedding of explicitly correlated systems. *J. Chem. Phys.* **1999**, 110, 7677–7688.

10 Klüner, T.; Govind, N.; Wang, Y. A.; Carter, E. Prediction of electronic excited states of adsorbates on metal surfaces from first principles. *Phys. Rev. Lett.* **2001**, 86, 5954–5957.

11 Klüner, T.; Govind, N.; Wang, Y. A.; Carter, E. A. Klüner *et al.* Reply. *Phys. Rev. Lett.* **2002**, 88, 209702. doi: 10.1103/PhysRevLett.88.209702

12 Klüner, T.; Govind, N.; Wang, Y. A.; Carter, E. A. Periodic density functional embedding theory for complete active space self-consistent field and configuration interaction calculations: Ground and excited states. *J. Chem. Phys.* **2002**, 116, 42–54.

13 Huang, P.; Carter, E. A. Self-consistent embedding theory for locally correlated configuration interaction wave functions in condensed matter. *J. Chem. Phys.* **2006**, 125, 084102.

14 Sharifzadeh, S.; Huang, P.; Carter, E. A. All-electron embedded correlated wavefunction theory for condensed matter electronic structure. *Chem. Phys. Lett.* **2009**, 470, 347–352.

15 Huang, C.; Carter, E. A. Potential-functional embedding theory for molecules and materials. *J. Chem. Phys.* **2011**, 135, 194104.

16 Huang, C.; Pavone, M.; Carter, E. A. Quantum mechanical embedding theory based on a unique embedding potential. *J. Chem. Phys.* **2011**, 134, 154110.

17 Huang, C.; Libisch, F.; Peng, Q.; Carter, E. A. Time-dependent potential-functional embedding theory. *J. Chem. Phys.* **2014**, 140, 124113.

18 Libisch, F.; Huang, C.; Carter, E. A. Embedded correlated wavefunction schemes: Theory and applications. *Acc. Chem. Res* **2014**, 47, 2768–2775.

19 Yu, K.; Libisch, F.; Carter, E. A. Implementation of density functional embedding theory within the projector-augmented-wave method and applications to semiconductor defect states. *J. Chem. Phys.* **2015**, 143, 102806.

20 Princeton University github. www.github.com/EACcodes

21 Huang, P.; Carter, E. A. Advances in correlated electronic structure methods for solids, surfaces, and nanostructures. *Annu. Rev. Phys. Chem.* **2008**, 59, 261–290.

22 Pakkanen, T. A.; Whitten, J. L. Electron density and density matrix approximations using potential energy error bounds. *J. Chem. Phys.* **1978**, 69, 2168–2174.

23 Whitten, J. L.; Pakkanen, T. A. Chemisorption theory for metallic surfaces: Electron localization and the description of surface interactions. *Phys. Rev. B* **1980**, 21, 4357–4367.

24 Wesolowski, T.; Muller, R. P.; Warshel, A. Ab initio frozen density functional calculations of proton transfer reactions in solution. *J. Phys. Chem.* **1996**, 100, 15444–15449.

25 Wesolowski, T. A. Density functional theory with approximate kinetic energy functionals applied to hydrogen bonds. *J. Chem. Phys.* **1997**, 106, 8516–8526.

26 Chacón, E.; Alvarellos, J. E.; Tarazona, P. Nonlocal kinetic energy functional for nonhomogeneous electron systems. *Phys. Rev. B* **1985**, 32, 7868–7877.

27 Wang, L.-W.; Teter, M. P. Kinetic-energy functional of the electron density. *Phys. Rev. B* **1992**, 45, 13196–13220.

28 Wang, Y. A.; Govind, N.; Carter, E. A. Orbital-free kinetic-energy density functionals with a density-dependent kernel. *Phys. Rev. B* **1999**, 60, 16350–16358.

29 Wang, Y. A.; Govind, N.; Carter, E. A. Erratum: Orbital-free kinetic-energy density functionals with a density-dependent kernel [*Phys. Rev. B* 1999, *60*, 16350–16358]. *Phys. Rev. B* **2001**, 64, 089903.

30 Wang, Y. A.; Carter, E. A. Orbital-free kinetic-energy density functional theory. In: *Theoretical Methods in Condensed Phase Chemistry*, Schwartz, S. D., Ed. Springer, Dordrecht, 2002; pp 117–184.

31 Zhou, B.; Ligneres, V. L.; Carter, E. A. Improving the orbital-free density functional theory description of covalent materials. *J. Chem. Phys.* **2005**, 122, 044103.

32 Götz, A. W.; Beyhan, S. M.; Visscher, L. Performance of kinetic energy functionals for interaction energies in a subsystem formulation of density functional theory. *J. Chem. Theory Comput.* **2009**, 5, 3161–3174.

33 Huang, C.; Carter, E. A. Nonlocal orbital-free kinetic energy density functional for semiconductors. *Phys. Rev. B* **2010**, 81, 045206.

34 Xia, J.; Carter, E. A. Density-decomposed orbital-free density functional theory for covalently bonded molecules and materials. *Phys. Rev. B* **2012**, 86, 235109.

35 Shin, I.; Carter, E. A. Enhanced von Weizsäcker Wang-Govind-Carter kinetic energy density functional for semiconductors. *J. Chem. Phys.* **2014**, 140, 18A531.

36 Xia, J.; Carter, E. A. Single-point kinetic energy density functionals: A pointwise kinetic energy density analysis and numerical convergence investigation. *Phys. Rev. B* **2015**, 91, 045124.

37 Wesolowski, T. A.; Weber, J. Kohn-Sham equations with constrained electron density: An iterative evaluation of the ground-state electron density of interacting molecules. *Chem. Phys. Lett.* **1996**, 248, 71–76.

38 Wesołowski, T. A.; Ellinger, Y.; Weber, J. Density functional theory with an approximate kinetic energy functional applied to study structure and stability of weak van der Waals complexes. *J. Chem. Phys.* **1998**, 108, 6078–6083.

39 Huang, P.; Carter, E. A. Local electronic structure around a single Kondo impurity. *Nano Lett.* **2006**, 6, 1146–1150.

40 Huang, P.; Carter, E. A. Ab initio explanation of tunneling line shapes for the Kondo impurity state. *Nano Lett.* **2008**, 8, 1265–1269.

41 Sharifzadeh, S.; Huang, P.; Carter, E. A. Origin of tunneling lineshape trends for Kondo states of Co adatoms on coinage metal surfaces. *J. Phys. Condens. Mat.* **2009**, 21, 355501.

42 Kanan, D. K.; Sharifzadeh, S.; Carter, E. A. Quantum mechanical modeling of electronic excitations in metal oxides: Magnesia as a prototype. *Chem. Phys. Lett.* **2012**, 519–520, 18–24.

43 Daday, C.; König, C.; Valsson, O.; Neugebauer, J.; Filippi, C. State-specific embedding potentials for excitation-energy calculations. *J. Chem. Theory Comput.* **2013**, 9, 2355–2367.

44 Daday, C.; König, C.; Neugebauer, J.; Filippi, C. Wavefunction in density functional theory embedding for excited states: Which wavefunctions, which densities? *Chem. Phys. Chem.* **2014**, 15, 3205–3217.

45 Wang, Y.; Parr, R. G. Construction of exact Kohn-Sham orbitals from a given electron density. *Phys. Rev. A* **1993**, 47, R1591–R1593.

46 Wu, Q.; Yang, W. A direct optimization method for calculating density functionals and exchange–correlation potentials from electron densities. *J. Chem. Phys.* **2003**, 118, 2498–2509.

47 Hohenberg, P.; Kohn, W. Inhomogeneous electron gas. *Phys. Rev.* **1964**, 136, B864–B871.

48 Wu, Q.; Yang, W. T. A direct optimization method for calculating density functionals and exchange-correlation potentials from electron densities. *J. Chem. Phys.* **2003**, 118, 2498–2509.

49 Hamann, D. R.; Schluter, M.; Chiang, C. Norm-conserving pseudopotentials. *Phys. Rev. Lett.* **1979**, 43, 1494–1497.

50 Troullier, N.; Martins, J. L. Efficient pseudopotentials for plane-wave calculations. *Phys. Rev. B* **1991**, 43, 1993–2006.

51 Blöchl, P. E. Projector augmented-wave method. *Phys. Rev. B* **1994**, 50, 17953–17979.

52 Kresse, G.; Joubert, D. From ultrasoft pseudopotentials to the projector augmented-wave method. *Phys. Rev. B* **1999**, 59, 1758–1775.

53 Gonze, X.; Beuken, J.-M.; Caracas, R.; Detraux, F.; Fuchs, M.; Rignanese, G.-M.; Sindic, L.; Verstraete, M.; Zerah, G.; Jollet, F.; Torrent, M.; Roy, A.; Mikami, M.; Ghosez, P.; Raty, J.-Y.; Allan, D. C. First-principles computation of material properties: The ABINIT software project. *Comput. Mat. Sci.* **2002**, 25, 478–492.

54 Gonze, X.; Rignanese, G.-M.; Verstraete, M.; Beuken, J.-M.; Pouillon, Y.; Caracas, R.; Jollet, F.; Torrent, M.; Zerah, G.; Mikami, M.; Ghosez, P.; Veithen, M.; Raty, J.-Y.; Olevano, V.; Bruneval, F.; Reining, L.; Godby, R.; Onida, G.; Hamann, D. R.; Allan, D. C. A brief introduction to the ABINIT software package. *Zeit. Kristallogr.* **2005**, 220, 558–562.

55 Gonze, X.; Amadon, B.; Anglade, P. M.; Beuken, J. M.; Bottin, F.; Boulanger, P.; Bruneval, F.; Caliste, D.; Caracas, R.; Cote, M.; Deutsch, T.; Genovese, L.; Ghosez, P.; Giantomassi, M.; Goedecker, S.; Hamann, D. R.; Hermet, P.; Jollet, F.; Jomard, G.; Leroux, S.; Mancini, M.; Mazevet, S.; Oliveira, M. J. T.; Onida, G.; Pouillon, Y.; Rangel, T.; Rignanese, G. M.; Sangalli, D.; Shaltaf, R.; Torrent, M.; Verstraete, M. J.; Zerah, G.; Zwanziger, J. W. ABINIT: First-principles approach to material and nanosystem properties. *Comput. Phys. Commun.* **2009**, 180, 2582–2615.

56 Zhu, C.; Byrd, R. H.; Lu, P.; Nocedal, J. Algorithm 778: L-BFGS-B: Fortran subroutines for large-scale bound-constrained optimization. *ACM Trans. Math. Softw.* **1997**, 23, 550–560.

57 Morales, J. L.; Nocedal, J. Remark on "Algorithm 778: L-BFGS-B: Fortran subroutines for large-scale bound constrained optimization". *ACM Trans. Math. Softw.* **2011**, 38, Article 7.

58 Kresse, G.; Hafner, J. Ab initio molecular dynamics for liquid metals. *Phys. Rev. B* **1993**, 47, 558–561.

59 Kresse, G.; Furthmüller, J. Efficiency of ab-initio total energy calculations for metals and semiconductors using a plane-wave basis set. *Comput. Mat. Sci.* **1996**, 6, 15–50.

60 Kresse, G.; Furthmüller, J. Efficient iterative schemes for ab initio total-energy calculations using a plane-wave basis set. *Phys. Rev. B* **1996**, 54, 11169–11186.

61 Byrd, R. H.; Lu, P. H.; Nocedal, J.; Zhu, C. Y. A limited memory algorithm for bound constrained optimization. *Siam J. Sci. Comput.* **1995**, 16, 1190–1208.

62 Heaton-Burgess, T.; Bulat, F. A.; Yang, W. T. Optimized effective potentials in finite basis sets. *Phys. Rev. Lett.* **2007**, 98, 256401.

63 Woon, D. E.; Dunning, T. H. Gaussian basis sets for use in correlated molecular calculations. III. The atoms aluminum through argon. *J. Chem. Phys.* **1993**, 98, 1358–1371.

64 Wadt, W. R.; Hay, P. J. Ab initio effective core potentials for molecular calculations. Potentials for main group elements Na to Bi. *J. Chem. Phys.* **1985**, 82, 284–298.

65 Check, C. E.; Faust, T. O.; Bailey, J. M.; Wright, B. J.; Thomas M. Gilbert; Sunderlin, L. S. Addition of polarization and diffuse functions to the LANL2DZ basis set for p-block elements. *J. Phys. Chem. A* **2001**, 105, 8111–8116.

66 Valiev, M.; Bylaska, E. J.; Govind, N.; Kowalski, K.; Straatsma, T. P.; Dam, H. J. J. V.; Wang, D.; Nieplocha, J.; Apra, E.; Windus, T. L.; de Jong, W. A. NWChem: A comprehensive and scalable open-source solution for large scale molecular simulations. *Comput. Phys. Commun.* **2010**, 181, 1477–1489.

67 Schmidt, M. W.; Baldridge, K. K.; Boatz, J. A.; Elbert, S. T.; Gordon, M. S.; Jensen, J. H.; Koseki, S.; Matsunaga, N.; Nguyen, K. A.; Su, S.; Windus, T. L.; Dupuis, M.; Montgomery, J. A. General atomic and molecular electronic structure system. *J. Comput. Chem.* **1993**, 14, 1347–1363.

68 Gordon, M. S.; Schmidt, M. W. Advances in electronic structure theory: GAMESS a decade later. In: *Theory and Applications of Computational Chemistry: The First Forty Years*, Dykstra, C. E.; Frenking, G.; Kim, K. S.; Scuseria, G. E., Eds. Elsevier, 2005; pp 1176–1189.

69 Karlström, G.; Lindh, R.; Malmqvist, P.-Å.; Roos, B. O.; Ryde, U.; Veryazov, V.; Widmark, P.-O.; Cossi, M.; Schimmelpfennig, B.; Neogrady, P.; Seijo, L. MOLCAS: A program package for computational chemistry. *Comput. Mater. Sci.* **2003**, 28, 222–239.

70 Aquilante, F.; De Vico, L.; Ferré, N.; Ghigo, G.; Malmqvist, P.-Å.; Neogrády, P.; Pedersen, T. B.; Pitoňák, M.; Reiher, M.; Roos, B. O.; Serrano-Andrés, L.; Urban, M.; Veryazov, V.; Lindh, R. MOLCAS 7: The next generation. *J. Comput. Chem.* **2010**, 31, 224–247.

71 Baksht, E. K.; Lomaev, M. I.; Rybka, D.; Tarasenko, V. F. Discharge current and ultrashort avalanche electron beam current in a volume nanosecond gas discharge in inhomogeneous electric field. *Tech. Phys. Lett.* **2007**, 33, 216–219.

72 Werner, H.-J.; Knowles, P. J.; Knizia, G.; Manby, F. R.; Schütz, M. Molpro: A general-purpose quantum chemistry program package. *WIREs Comput. Mol. Sci.* **2012**, 2, 242–253.

73 Krisiloff, D. B.; Dieterich, J. M.; Libisch, F.; Carter, E. A. Numerical challenges in a Cholesky-decomposed local correlation quantum chemistry framework. In: *Numerical Challenges in a Cholesky-Decomposed Local Correlation Quantum Chemistry Framework*, Melnick, R., Ed. John Wiley & Sons, 2015; pp 59–91.

74 Vreven, T.; Morokuma, K. Chapter 3—Hybrid methods: ONIOM(QM:MM) and QM/MM. In: *Annual Reports in Computational Chemistry*, Vol. 2. Elsevier, pp 35–51.

75 Libisch, F.; Huang, C.; Liao, P.; Pavone, M.; Carter, E. A. Origin of the energy barrier to chemical reactions of O2 on Al(111): Evidence for charge transfer, not spin selection. *Phys. Rev. Lett.* **2012**, 109, 198303.

76 Libisch, F.; Cheng, J.; Carter, E. A. Electron-transfer-induced dissociation of H2 on gold nanoparticles: Excited-state potential energy surfaces via embedded correlated wavefunction theory. *Z. Phys. Chem.* **2013**, 227, 1455–1466.

77 Mukherjee, S.; Libisch, F.; Large, N.; Neumann, O.; Brown, L. V.; Cheng, J.; Lassiter, J. B.; Carter, E. A.; Nordlander, P.; Halas, N. J. Hot electrons do the impossible: Plasmon-induced dissociation of H2 on Au. *Nano Lett.* **2013**, 13, 240–247.

78 Cheng, J.; Libisch, F.; Carter, E. A. Dissociative adsorption of O2 on Al(111): The role of orientational degrees of freedom. *J. Phys. Chem. Lett.* **2015**, 6, 1661–1665.

79 Zhou, L.; Zhang, C.; McClain, M. J.; Manjavacas, A.; Krauter, C. M.; Tian, S.; Berg, F.; Everitt, H. O.; Carter, E. A.; Nordlander, P.; Halas, N. J. Aluminum nanocrystals as a plasmonic photocatalyst for hydrogen dissociation. *Nano Lett.* **2016**, 16, 1478–1484.
80 Libisch, F.; Krauter, C. M.; Carter, E. A. Corrigendum to: Plasmon-driven dissociation of H2 on gold nanoclusters. *Z. Phys. Chem.* **2016**, 230, 131–132.
81 Perdew, J. P.; Parr, R. G.; Levy, M.; Balduz, J. L. Density-functional theory for fractional particle number: Derivative discontinuities of the energy. *Phys. Rev. Lett.* **1982**, 49, 1691–1694.
82 Österlund, L.; Zorić, I.; Kasemo, B. Dissociative sticking of O2 on Al(111). *Phys. Rev. B* **1997**, 55, 15452–15455.
83 Komrowski, A. J.; Sexton, J. Z.; Kummel, A. C.; Binetti, M.; Weiße, O.; Hasselbrink, E. Oxygen abstraction from dioxygen on the Al(111) surface. *Phys. Rev. Lett.* **2001**, 87, 246103.
84 Binetti, M.; Hasselbrink, E. Abstraction of oxygen from dioxygen on Al(111) revealed by resonant multiphoton ionization laser spectrometry. *J. Phys. Chem. B* **2004**, 108, 14677–14684.
85 Sasaki, T.; Ohno, T. Dissociation process of O2 on the Al(111) surface. *Surf. Sci.* **1999**, 433–435, 172–175.
86 Honkala, K.; Laasonen, K. Oxygen molecule dissociation on the Al(111) surface. *Phys. Rev. Lett.* **2000**, 84, 705–708.
87 Yourdshahyan, Y.; Razaznejad, B.; Lundqvist, B. I. Adiabatic potential-energy surfaces for oxygen on Al(111). *Phys. Rev. B* **2002**, 65, 075416.
88 C. Mosch; C. Koukounas; N. Bacalis; A. Metropoulos; A. Gross; A. Mavridis. Interaction of dioxygen with Al clusters and Al(111): A comparative theoretical study. *J. Phys. Chem. C* **2008**, 112, 6924–6932.
89 Livshits, E.; Baer, R.; Kosloff, R. Deleterious effects of long-range self-repulsion on the density functional description of O2 sticking on Aluminum. *J. Phys. Chem. A* **2009**, 113, 7521–7527.
90 Bacalis, N. C.; Metropoulos, A.; Gross, A. Theoretical study of the O2 interaction with a tetrahedral Al4 cluster. *J. Phys. Chem. A* **2010**, 114, 11746–11750.
91 Kurahashi, M.; Yamauchi, Y. Steric effect in O2 sticking on Al(111): Preference for parallel geometry. *Phys. Rev. Lett.* **2013**, 110, 246102.
92 Schmid, M.; Leonardelli, G.; Tschelieẞnig, R.; Biedermann, A.; Varga, P. Oxygen adsorption on Al(111): Low transient mobility. *Surf. Sci.* **2001**, 478, L355–L362.
93 Binetti, M.; Weiße, O.; Hasselbrink, E.; Komrowski, A. J.; Kummel, A. C. Abstractive chemisorption of O2 on Al(111). *Faraday Discuss.* **2000**, 117, 313–320.
94 Andersson, K.; Malmqvist, P.-Å.; Roos, B. O.; Sadlej, A. J.; Wolinski, K. Second-order perturbation theory with a CASSCF reference function. *J. Phys. Chem.* **1990**, 94, 5483–5488.
95 Roos, B. O. The complete active space SCF method in a fock-matrix-based super-CI formulation. *Int. J. Quant. Chem.* **1980**, 18, 175–189.
96 Roos, B. O.; Taylor, P. R.; Siegbahn, P. E. M. A complete active space SCF method (CASSCF) using a density matrix formulated super-CI approach. *Chem. Phys.* **1980**, 48, 157–173.
97 Siegbahn, P. E. M.; Almlöf, J.; Heiberg, A.; Roos, B. O. The complete active space SCF (CASSCF) method in a Newton–Raphson formulation with application to the HNO molecule. *J. Chem. Phys.* **1981**, 74, 2384–2396.

98 Perdew, J. P.; Zunger, A. Self-interaction correction to density-functional approximations for many-electron systems. *Phys. Rev. B* **1981**, 23, 5048–5079.

99 Bader, R. F. W. *Atoms in Molecules: A Quantum Theory*. Oxford University Press, New York, 1990.

100 Henkelman, G.; Arnaldsson, A.; Jónsson, H. A fast and robust algorithm for Bader decomposition of charge density. *Comput. Mat. Sci.* **2006**, 36, 354–360.

101 Sanville, E.; Kenny, S. D.; Smith, R.; Henkelman, G. Improved grid-based algorithm for Bader charge allocation. *J. Comput. Chem.* **2007**, 28, 899–908.

102 Tang, W.; Sanville, E.; Henkelman, G. A grid-based Bader analysis algorithm without lattice bias. *J. Phys. Condens. Matter* **2009**, 21, 084204.

103 Yu, M.; Trinkle, D. R. Accurate and efficient algorithm for Bader charge integration. *J. Chem. Phys.* **2011**, 134, 064111.

104 Perdew, J. P.; Burke, K.; Ernzerhof, M. Generalized gradient approximation made simple. *Phys. Rev. Lett.* **1996**, 77, 3865–3868.

105 Huang, C.; Carter, E. A. Direct minimization of the optimized effective problem based on efficient finite differences. *Phys. Rev. B* **2011**, 84, 165122.

106 Perdew, J. P.; Wang, Y. Accurate and simple analytic representation of the electron-gas correlation-energy. *Phys. Rev. B* **1992**, 45, 13244–13249.

107 Huang, C.; Carter, E. A. Transferable local pseudopotentials for magnesium, aluminum and silicon. *Phys. Chem. Chem. Phys.* **2008**, 10, 7109–7120.

108 Fermi, E. Un metodo statistico per la determinazione di alcune priopriretà dell'atomo. *Rend. Accad. Naz. Lincei* **1927**, 6, 602–607.

109 Thomas, L. H. The calculation of atomic fields. *Proc. Cambridge Phil. Soc.* **1927**, 23, 542–548.

110 Huang, C.; Carter, E. A. Nonlocal orbital-free kinetic energy density functional for semiconductors. *Phys. Rev. B* **2010**, 81, 045206.

111 Fuchs, M.; Scheffler, M. Ab initio pseudopotentials for electronic structure calculations of poly-atomic systems using density-functional theory. *Comput. Phys. Commun.* **1999**, 119, 67–98.

3

Modeling and Visualization for the Fragment Molecular Orbital Method with the Graphical User Interface FU, and Analyses of Protein–Ligand Binding

Dmitri G. Fedorov[1] *and Kazuo Kitaura*[2]

[1] *Research Center for Computational Design of Advanced Functional Materials (CD-FMat), National Institute of Advanced Industrial Science and Technology (AIST), Tsukuba, Japan*
[2] *Graduate School of System Informatics, Kobe University, Japan*

3.1 Introduction

Many biological processes involve large molecular systems containing thousands of atoms. Although force fields have been applied to many biological systems with considerable success, the application of quantum-mechanical methods is also becoming feasible with the development of fragment-based methods [1–3]. There are several difficulties in performing large-scale quantum-mechanical calculations: (a) molecular modeling, (b) result visualization, and (c) analysis of results in terms of well-defined physical concepts. All of these problems are addressed in this chapter.

Several graphical user interfaces (GUI) have been developed for FMO: Biostation [4], Facio [5], and PAICS View [6]. There are also some web-based interfaces, FragIt, and fragment molecular orbital (FMO) portal [7]. The main reason for our developing a new GUI is the need for an easy addition of new features. The new GUI is called FU, the origin of the name being in the FMO input file maker called FMOutil [8]. FU is an open-source software distributed free of charge [9]. Here, its current version 0.4.0 is described.

Several analyses of fragment pair interaction energies have been proposed [10–13] in the FMO [14, 15] framework, and FMO has been applied to protein–ligand binding [16–23]. However, drug discovery studies [24] may benefit from a new analysis with an explicit separation of the desolvation penalty [25].

When a protein binds a ligand in solution, a part of the protein and ligand surface is desolvated, so that protein atoms can come close to those of the ligand. This constitutes the desolvation penalty. It is typically endothermic and cancels a large part of the exothermic protein–ligand interactions, so that without considering the penalty, protein–ligand binding would be greatly overestimated. It is therefore useful to devise an analysis that explicitly defines the value of the desolvation penalty in the total binding.

Fragmentation: Toward Accurate Calculations on Complex Molecular Systems, First Edition.
Edited by Mark S. Gordon.

Although the protein–ligand binding is explicitly studied here, the developed analyses are general and can be applied to other binding processes in solution.

3.2 Overview of FMO

A detailed description of the FMO method can be found in reviews and other publications [26–32]. Here, a very concise overview is given for the convenience of the reader. FMO can be applied to proteins, DNA, lipids, and many other kinds of systems. In FMO, a molecular system is divided into fragments, and these fragments are calculated in the presence of the electrostatic potential (ESP) exerted by the whole system. The ESP is computed from the electronic state of fragments, and thus the fragment calculations have to be iterated until their self-consistent convergence. After that, pairs of fragments (and, sometimes, triples) are calculated in the electrostatic field, frozen at this point.

The total energy of the system in the two-body FMO (FMO2) method in the most simple representation is

$$E^{\text{FMO2}} = \sum_{I} E_I + \sum_{I>J} (E_{IJ} - E_I - E_J) \tag{3.1}$$

where E_I and E_{IJ} are the energies of fragments I and pairs IJ, respectively.

FMO is often used with the polarizable continuum model (PCM) [33] describing the solvent [34–36]. In PCM, the solute is placed into a cavity representing the solvent. The cavity surface is divided into pieces called tesserae, and a point charge is placed on each tessera. These solvent charges are polarized by the solute and they in turn polarize the solute, so that the determination of the solvent charges is done self-consistently with the electronic state. In addition to these electrostatic solvent effects, the following non-polar terms are also added to the total energy: the solute-solvent dispersion (disp) and repulsion (rep) energies, as well as the solvent cavitation (cav) energy.

3.3 Methodology

3.3.1 FMO/PCM Formulation in the Presence of Dummy Atoms

In order to define desolvation penalty in new kinds of analyses below, it is necessary to extend the FMO/PCM formulation, allowing dummy atoms in the cavity definition. Dummy atoms have no charge, but they contribute to the definition of the cavity shape. The cavity surface around dummy atoms is also divided into tesserae, and a point charge is placed on each tessera. In FMO, all dummy atoms are assigned to fragment 0, treated separately. All sums over fragments below run from fragment 1 to N (the total number of fragments), that is, they do not include this extra special fragment.

The FMO2/PCM energy expression is

$$E^{\text{FMO2/PCM}} = \sum_{I} E'_I + \sum_{I>J} \left[\left(E'_{IJ} - E'_I - E'_J \right) + Tr(\Delta \mathbf{D}^{IJ} \mathbf{V}^{IJ}) \right] + \Delta E_{\text{cav}} + \Delta E_{\text{disp}} + \Delta E_{\text{rep}} \tag{3.2}$$

where $E_X = E'_X - Tr(\Delta\mathbf{D}^X\mathbf{V}^X)$, X = I, J, or IJ and $\Delta\mathbf{D}^{IJ} = \mathbf{D}^{IJ} - (\mathbf{D}^I \oplus \mathbf{D}^J)$. $\mathbf{D}^X$ and $\mathbf{V}^X$ are the density matrix and ESP of X, respectively. By separating the solute-solvent electrostatic (es) energy, one obtains

$$\begin{aligned} E^{\text{FMO2/PCM}} &= \sum_I E''_I + \sum_{I>J}\left[\left(E''_{IJ} - E''_I - E''_J\right) + Tr(\Delta\mathbf{D}^{IJ}\mathbf{V}^{IJ})\right] \\ &\quad + \sum_I \Delta E^{\text{es}}_I + \sum_{I>J}\left(\Delta E^{\text{es}}_{IJ} - \Delta E^{\text{es}}_I - \Delta E^{\text{es}}_J\right) + \Delta E_{\text{cav}} + \Delta E_{\text{disp}} + \Delta E_{\text{rep}} \end{aligned} \tag{3.3}$$

where $E''_X = E'_X - \Delta E^{\text{es}}_X$.

The solute-solvent electrostatic (es) energy is

$$\Delta E^{\text{es}}_I = \Delta E^{\text{es}}_{I(I)} + \sum_{J\neq I} \Delta E^{\text{es}}_{I(J)} + \Delta E^{\text{es}}_{I(0)} = \Delta\tilde{E}^{\text{es}}_{I(I)} + \sum_{J\neq I} \Delta E^{\text{es}}_{I(J)} \tag{3.4}$$

$$\Delta\tilde{E}^{\text{es}}_{I(I)} = \Delta E^{\text{es}}_{I(I)} + \Delta E^{\text{es}}_{I(0)} \tag{3.5}$$

Here and below, $\Delta E^{\text{A}}_{I(J)}$ for A = es, disp, rep corresponds to the interaction of solute atoms in fragment I with the solvent surface of fragment J. $\Delta E^{\text{es}}_{I(0)}$ is the interaction of nuclei and electrons in I with ASCs on the dummy atom surface (0).

The cavitation energy is

$$\Delta E_{\text{cav}} = \sum_I \Delta E^{\text{cav}}_I + \Delta E^{\text{cav}}_0 \tag{3.6}$$

The solute-solvent dispersion (disp) and repulsion (rep) energies are

$$\Delta E_{\text{disp}} = \sum_I \Delta E^{\text{disp}}_I = \sum_I \left(\Delta\tilde{E}^{\text{disp}}_{I(I)} + \sum_{J\neq I} \Delta E^{\text{disp}}_{I(J)}\right) \tag{3.7}$$

$$\Delta E_{\text{rep}} = \sum_I \Delta E^{\text{rep}}_I = \sum_I \left(\Delta\tilde{E}^{\text{rep}}_{I(I)} + \sum_{J\neq I} \Delta E^{\text{rep}}_{I(J)}\right) \tag{3.8}$$

Here, similar to es, for A = disp and rep,

$$\Delta\tilde{E}^{\text{A}}_{I(I)} = \Delta E^{\text{A}}_{I(I)} + \Delta E^{\text{A}}_{I(0)} \tag{3.9}$$

where $\Delta E^{\text{A}}_{I(0)}$ describes the interaction of atoms in I with the dummy atom surface. Note that there is no $\Delta E^{\text{A}}_{0(I)}$ and no $\Delta E^{\text{A}}_{0(0)}$ for A = es, disp, and rep, because atoms in "0" are dummy. The FMO2/PCM energy can be rewritten using decompositions of the solvation energy components as [34]

$$\begin{aligned} E^{\text{FMO2/PCM}} &= \sum_I E''_I + \sum_{I>J}\left[\left(E''_{IJ} - E''_I - E''_J\right) + Tr(\Delta\mathbf{D}^{IJ}\mathbf{V}^{IJ})\right. \\ &\quad \left. + \Delta E^{\text{es2}}_{IJ} + \Delta E^{\text{es3}}_{IJ} + \Delta E^{\text{disp}}_{IJ} + \Delta E^{\text{rep}}_{IJ}\right] \\ &\quad + \sum_I \Delta\tilde{E}^{\text{solv}}_I + \Delta E^{\text{cav}}_0 \end{aligned} \tag{3.10}$$

The fragment solvation energy is

$$\Delta\tilde{E}^{\text{solv}}_I = \Delta E^{\text{cav}}_I + \Delta\tilde{E}^{\text{es}}_{I(I)} + \Delta\tilde{E}^{\text{disp}}_{I(I)} + \Delta\tilde{E}^{\text{rep}}_{I(I)} \tag{3.11}$$

The coupling of the solute density transfer to the solvent potential $\mathbf{W}_{\mathrm{e}}^{IJ}$ (this potential includes the contribution of ASCs for both dummy 0 and normal fragments) is

$$\Delta E_{IJ}^{\mathrm{es3}} = \Delta E_{IJ}^{\mathrm{es}} - \Delta E_{I}^{\mathrm{es}} - \Delta E_{J}^{\mathrm{es}} = \frac{1}{2} Tr\left(\Delta \mathbf{D}^{IJ} \mathbf{W}_{\mathrm{e}}^{IJ}\right) \tag{3.12}$$

For A = disp or rep, the solvent screenings to pair interaction energies (PIEs) are

$$\Delta E_{IJ}^{\mathrm{A}} = \Delta E_{I(J)}^{\mathrm{A}} + \Delta E_{J(I)}^{\mathrm{A}} \tag{3.13}$$

And the two-body contribution to the electrostatic solvent screening is

$$\Delta E_{IJ}^{\mathrm{es2}} = \Delta E_{I(J)}^{\mathrm{es}} + \Delta E_{J(I)}^{\mathrm{es}} \tag{3.14}$$

Then, the total energy is

$$E_{\mathrm{FMO2/PCM}} = \sum_{I} E_{I}'' + \sum_{I} \Delta \tilde{E}_{I}^{\mathrm{solv}} + \sum_{I>J} \Delta E_{IJ}^{\mathrm{int}} + \Delta E_{0}^{\mathrm{cav}} \tag{3.15}$$

where the PIE in solution and the total solvent screening are, respectively,

$$\Delta E_{IJ}^{\mathrm{int}} = \left(E_{IJ}'' - E_{I}'' - E_{J}''\right) + Tr(\Delta \mathbf{D}^{IJ} \mathbf{V}^{IJ}) + \Delta E_{IJ}^{\mathrm{solv}} \tag{3.16}$$

$$\Delta E_{IJ}^{\mathrm{solv}} = \Delta E_{IJ}^{\mathrm{es2}} + \Delta E_{IJ}^{\mathrm{es3}} + \Delta E_{IJ}^{\mathrm{disp}} + \Delta E_{IJ}^{\mathrm{rep}} \tag{3.17}$$

Note that in practice, one calculates $\Delta E_{I(J)}^{\mathrm{A}}$ and $\Delta E_{I}^{\mathrm{A}}$ for A = es, disp, and rep and then

$$\Delta \tilde{E}_{I(I)}^{\mathrm{A}} = \Delta E_{I}^{\mathrm{A}} - \sum_{J \neq I} \Delta E_{I(J)}^{\mathrm{A}} \tag{3.18}$$

3.3.2 New Analyses Defining the Desolvation Penalty in the Protein–Ligand Binding

The process of binding is thought of as consisting of two steps: a partial desolvation and then bonding between desolvated protein and ligand. Two kinds of decomposition of the binding energy are done: (a) asymmetric and (b) symmetric. In the latter, likewise contributions are defined for both protein and ligand, resulting in three terms, the desolvation energy of the protein, the desolvation energy of the ligand, and the bonding energy. In the asymmetric formulation, the desolvation energy of the ligand is not defined (instead, it is added to the effective bonding energy). For the symmetric analysis, one can choose to separate the cavitation energy from the rest of terms.

These analyses proposed here are very general—they can be combined with any method, such as a force field or a quantum-mechanical method such as FMO. The only requirement is the availability of the energies such as the energy of the complex and individual protein and ligand. After the general formulation, the analyses are specifically elaborated for FMO to define fragment contributions.

3.3.2.1 Asymmetric Binding Analysis (ABA)

The protein–ligand binding energy is computed as the difference of the complex energy $\mathrm{P} \cdot \mathrm{L}$ from the energies of the protein P and ligand L. The expression is modified by adding and subtracting the energy $\mathrm{P} \cdot \ell$. It represents the energy of the protein in the cavity of the complex, so that "ℓ" stands for dummy atoms of the ligand. Likewise, E^{ℓ} is the energy of the ligand when all atoms are dummy, $E^{\ell} = E^{\text{cav,L}}$.

$$\Delta E = E^{\text{P}\cdot\text{L}} - E^{\text{P}} - E^{\text{L}} = E^{\text{P}\cdot\text{L}} - E^{\text{P}} - E^{\text{L}} + E^{\text{P}\cdot\ell} - E^{\text{P}\cdot\ell} + E^{\ell} - E^{\ell} \tag{3.19}$$

$$\Delta E = \Delta E^{\text{des,P}} + \Delta E^{\text{bon,P}\cdot\text{L+L}\cdot\text{solv}} - \Delta E^{\text{L}\cdot\text{solv}} = \Delta E^{\text{des,P}} + \Delta\tilde{E}^{\text{bon,P}\cdot\text{L}} \tag{3.20}$$

where the protein desolvation $\Delta E^{\text{des,P}}$, protein–ligand plus ligand–solvent bonding $\Delta E^{\text{bon,P}\cdot\text{L+L}\cdot\text{solv}}$, and ligand-solvent interaction $\Delta E^{\text{L}\cdot\text{solv}}$ energies are

$$\Delta E^{\text{des,P}} = E^{\text{P}\cdot\ell} - E^{\text{P}} - E^{\ell} \tag{3.21}$$

$$\Delta E^{\text{bon,P}\cdot\text{L+L}\cdot\text{solv}} = E^{\text{P}\cdot\text{L}} - E^{\text{P}\cdot\ell} \tag{3.22}$$

$$\Delta E^{\text{L}\cdot\text{solv}} = E^{\text{L}} - E^{\ell} \tag{3.23}$$

$$\Delta\tilde{E}^{\text{bon,P}\cdot\text{L}} = \Delta E^{\text{bon,P}\cdot\text{L+L}\cdot\text{solv}} - \Delta E^{\text{L}\cdot\text{solv}} \tag{3.24}$$

Summarizing, in this approach the binding energy is decomposed as the protein desolvation energy and the effective protein–ligand bonding energy. The latter includes the ligand desolvation.

3.3.2.2 Symmetric Binding Analysis (SBA)

It can be thought of as a modification of the asymmetric analysis when one term $\Delta\tilde{E}^{\text{bon,P}\cdot\text{L}}$ is further elaborated, separating the ligand desolvation energy from it.

$$\Delta E = E^{\text{P}\cdot\text{L}} - E^{\text{P}} - E^{\text{L}} = E^{\text{P}\cdot\text{L}} - E^{\text{P}} - E^{\text{L}} + E^{\text{P}\cdot\ell} - E^{\text{P}\cdot\ell} + E^{\text{p}\cdot\text{L}} - E^{\text{p}\cdot\text{L}} + E^{\text{p}} - E^{\text{p}} + E^{\ell} - E^{\ell} \tag{3.25}$$

$$\Delta E = \Delta E^{\text{des,P}} + \Delta E^{\text{des,L}} + \Delta E^{\text{bon,P}\cdot\text{L}} \tag{3.26}$$

where

$$\Delta E^{\text{des,L}} = E^{\text{p}\cdot\text{L}} - E^{\text{p}} - E^{\text{L}} \tag{3.27}$$

$$\Delta E^{\text{bon,P}\cdot\text{L}} = E^{\text{P}\cdot\text{L}} - E^{\text{P}\cdot\ell} - E^{\text{p}\cdot\text{L}} + E^{\text{p}} + E^{\ell} \tag{3.28}$$

Superscript "p" indicates protein made of dummy atoms.

Summarizing, the binding energy is given as the bonding energy, and the desolvation energy of the protein and ligand.

3.3.2.3 Symmetric Binding Analysis with Separated Cavitation (SBAC)

This analysis can be derived from considering

$$\Delta E = E^{\text{P}\cdot\text{L}} - E^{\text{P}} - E^{\text{L}} = E^{\text{P}\cdot\text{L}} - E^{\text{P}} - E^{\text{L}} + E^{\text{P}\cdot\ell} - E^{\text{P}\cdot\ell} + E^{\text{p}\cdot\text{L}} - E^{\text{p}\cdot\text{L}} \tag{3.29}$$

and then separating the cavitation energy contributions.

$$\Delta E = \Delta\bar{E}^{\text{des,P}} + \Delta\bar{E}^{\text{des,L}} + \Delta\bar{E}^{\text{bon,P}\cdot\text{L}} + \Delta\Delta E^{\text{cav}} \tag{3.30}$$

where

$$\Delta\bar{E}^{\text{des,P}} = \Delta E^{\text{des,P}} - \Delta\Delta E^{\text{cav}} \tag{3.31}$$

$$\Delta\bar{E}^{\text{des,L}} = \Delta E^{\text{des,L}} - \Delta\Delta E^{\text{cav}} \tag{3.32}$$

$$\Delta\bar{E}^{\text{bon,P}\cdot\text{L}} = \Delta E^{\text{bon,P}\cdot\text{L}} + \Delta\Delta E^{\text{cav}} \tag{3.33}$$

The combination of all cavitation energy terms in $\Delta E^{\mathrm{des,P}}$, $\Delta E^{\mathrm{des,L}}$, and $\Delta E^{\mathrm{bon,P\cdot L}}$ is exactly the same, $\Delta\Delta E^{\mathrm{cav}}$, with either a positive or negative sign, so subtracting or adding it, one obtains $\Delta\tilde{E}^{\mathrm{des,P}}$, $\Delta\tilde{E}^{\mathrm{des,L}}$, and $\Delta\tilde{E}^{\mathrm{bon,P\cdot L}}$, which have no cavitation energy contributions.

$$\Delta\Delta E^{\mathrm{cav}} = \Delta E^{\mathrm{cav,P\cdot L}} - \Delta E^{\mathrm{cav,P}} - \Delta E^{\mathrm{cav,L}} \tag{3.34}$$

Summarizing, the binding energy in SBAC is similar to SBA, but the cavitation energy is explicitly separated, whereas in SBA, for the completeness of considering full solvated energies, it is added and subtracted from several terms in a somewhat redundant manner. On the other hand, it is done in a manner consistent with the definition of the total PCM energies.

3.3.2.4 Fragment-Wise Elaboration of SBA in FMO

$\Delta E^{\mathrm{des,P}}$ and $\Delta E^{\mathrm{des,L}}$ are left without further decomposition here, although one could elaborate it. Using the FMO/PCM formulation with dummy atoms,

$$\begin{aligned}
\Delta E^{\mathrm{bon,P\cdot L}} &= E^{\mathrm{P\cdot L}} + E^{\mathrm{P}} + E^{\ell} - E^{\mathrm{P\cdot\ell}} - E^{\mathrm{p\cdot L}} \\
&= \sum_I E_I''^{\mathrm{P\cdot L}} + \sum_I \Delta\tilde{E}_I^{\mathrm{solv,P\cdot L}} + \sum_{I>J} \Delta E_{IJ}^{\mathrm{int,P\cdot L}} + E^{\mathrm{P}} + E^{\ell} \\
&\quad - \sum_I E_I''^{\mathrm{P\cdot\ell}} - \sum_I \Delta\tilde{E}_I^{\mathrm{solv,P\cdot\ell}} - \sum_{I>J} \Delta E_{IJ}^{\mathrm{int,P\cdot\ell}} - \Delta E_0^{\mathrm{cav,P\cdot\ell}} \\
&\quad - \sum_I E_I''^{\mathrm{p\cdot L}} - \sum_I \Delta\tilde{E}_I^{\mathrm{solv,p\cdot L}} - \sum_{I>J} \Delta E_{IJ}^{\mathrm{int,p\cdot L}} - \Delta E_0^{\mathrm{cav,p\cdot L}} \\
&= \sum_{I\in \mathrm{P\cdot L}} \Delta E_I'' + \sum_{I\in \mathrm{P\cdot L}} \Delta\Delta\tilde{E}_I^{\mathrm{solv}} + \sum_{I>J\in P} \Delta\Delta E_{IJ}^{\mathrm{int}} + \sum_{I>J\in L} \Delta\Delta E_{IJ}^{\mathrm{int}} \\
&\quad + \sum_{I\in P, J\in L} \Delta E_{IJ}^{\mathrm{int}}
\end{aligned} \tag{3.35}$$

Here,

$$\Delta E_I'' = E_I''^{\mathrm{P\cdot L}} - E_I''^{\mathrm{P\cdot\ell}} \quad \text{for} \quad I \in P; \quad \Delta E_I'' = E_I''^{\mathrm{P\cdot L}} - E_I''^{\mathrm{p\cdot L}} \quad \text{for} \quad I \in L \tag{3.36}$$

$$\begin{aligned}
\Delta\Delta E_{IJ}^{\mathrm{int}} &= \Delta E_{IJ}^{\mathrm{int,P\cdot L}} - \Delta E_{IJ}^{\mathrm{int,P\cdot\ell}} \quad \text{for} \quad I,J \in P; \\
\Delta\Delta E_{IJ}^{\mathrm{int}} &= \Delta E_{IJ}^{\mathrm{int,P\cdot L}} - \Delta E_{IJ}^{\mathrm{int,p\cdot L}} \quad \text{for} \quad I,J \in L
\end{aligned} \tag{3.37}$$

The definition of the fragment solvation energy change is

$$\begin{aligned}
&\sum_I \Delta\tilde{E}_I^{\mathrm{solv,P\cdot L}} + E^{\mathrm{P}} + E^{\ell} - \sum_I \Delta\tilde{E}_I^{\mathrm{solv,P\cdot\ell}} - \Delta E_0^{\mathrm{cav,P\cdot\ell}} - \sum_I \Delta\tilde{E}_I^{\mathrm{solv,p\cdot L}} - \Delta E_0^{\mathrm{cav,p\cdot L}} \\
&= -\sum_I \Delta\Delta E_I^{\mathrm{cav}} + \sum_I \Delta\Delta\tilde{E}_I^{\mathrm{es}} + \sum_I \Delta\Delta\tilde{E}_I^{\mathrm{disp}} + \sum_I \Delta\Delta\tilde{E}_I^{\mathrm{rep}} = \sum_I \Delta\Delta\tilde{E}_I^{\mathrm{solv}}
\end{aligned} \tag{3.38}$$

For A = es, disp, and rep,

$$\text{if } I \in P,\ \Delta\Delta\tilde{E}_I^{\mathrm{A,P\cdot L}} = \Delta E_I^{\mathrm{A,P\cdot L}} - \Delta\tilde{E}_I^{\mathrm{A,P\cdot\ell}};\ \text{if } I \in L,\ \Delta\Delta\tilde{E}_I^{\mathrm{A,P\cdot L}} = \Delta E_I^{\mathrm{A,P\cdot L}} - \Delta\tilde{E}_I^{\mathrm{A,p\cdot L}} \tag{3.39}$$

The cavitation energy of fragments in the complex obeys the relations

$$\sum_I \Delta E_I^{\mathrm{cav,P\cdot L}} = \sum_{I\in P} \Delta E_I^{\mathrm{cav,P\cdot\ell}} + \Delta E_0^{\mathrm{cav,P\cdot\ell}} = \sum_{I\in L} \Delta E_I^{\mathrm{cav,p\cdot L}} + \Delta E_0^{\mathrm{cav,p\cdot L}} \tag{3.40}$$

Therefore, the combination of all cavitation terms in equation (3.35) is

$$\sum_I \Delta E_I^{\text{cav,P}\cdot\text{L}} + E^{\text{P}} + E^{\ell} - \sum_I \Delta E_I^{\text{cav,P}\cdot\ell} - \Delta E_0^{\text{cav,P}\cdot\ell} - \sum_I \Delta E_I^{\text{cav,p}\cdot\text{L}} - \Delta E_0^{\text{cav,p}\cdot\text{L}} = \\ -\sum_I \Delta E_I^{\text{cav,P}\cdot\text{L}} + E^{\text{P}} + E^{\ell} = -\sum_I \Delta\Delta E_I^{\text{cav}} \tag{3.41}$$

where after decomposing $E^{\text{P}} = \sum_{I\in P} \Delta E_I^{\text{cav,P}}$ and $E^{\ell} = \sum_{I\in L} \Delta E_I^{\text{cav,L}}$, one obtains

$$\Delta\Delta E_I^{\text{cav}} = \Delta E_I^{\text{cav,P}\cdot\text{L}} - \Delta E_I^{\text{cav,P}} \text{ for } I \in P;\ \text{for } I \in L,\ \Delta\Delta E_I^{\text{cav}} = \Delta E_I^{\text{cav,P}\cdot\text{L}} - \Delta E_I^{\text{cav,L}} \tag{3.42}$$

The reason why the es, disp, and rep terms have a tilde and the cav does not is because the former group includes interaction of fragment I with the dummy cavity (0), which the tilde signifies. The final expression for the bonding between the desolvated protein and ligand is

$$\Delta E^{\text{bon,P}\cdot\text{L}} = \sum_{I\in \text{P}\cdot\text{L}} \Delta E_I'' + \sum_{I\in \text{P}\cdot\text{L}} \Delta\Delta \tilde{E}_I^{\text{solv}} + \sum_{I>J\in P} \Delta\Delta E_{IJ}^{\text{int}} \\ + \sum_{I>J\in L} \Delta\Delta E_{IJ}^{\text{int}} + \sum_{I\in P, J\in L} \Delta E_{IJ}^{\text{int}} \tag{3.43}$$

It is made of (a) the protein–ligand interaction energies $\Delta E_{IJ}^{\text{int}}$ and (b) the differences in the fragment energies in the complex and in the standalone protein (ligand) in the complex cavity: internal energies $\Delta E_I''$, solvation energies $\Delta\Delta \tilde{E}_I^{\text{solv}}$, interaction energies $\Delta\Delta E_{IJ}^{\text{int}}$ within the protein (ligand). The reason why the last term has a single delta is because it is the absolute value of the interaction energy, whereas the $\Delta\Delta E_{IJ}^{\text{int}}$ terms are differences of $\Delta E_{IJ}^{\text{int}}$.

The expression can also be written as

$$\Delta E^{\text{bon,P}\cdot\text{L}} = \Delta E^{\text{PLd,P}} + \Delta E^{\text{PLd,L}} + \sum_{I\in P, J\in L} \Delta E_{IJ}^{\text{int}} \tag{3.44}$$

where the destabilization polarization (PLd) [12] terms are

$$\Delta E^{\text{PLd,P}} = \sum_{I\in P} \Delta E_I'' + \sum_{I\in P} \Delta\Delta \tilde{E}_I^{\text{solv}} + \sum_{I>J\in P} \Delta\Delta E_{IJ}^{\text{int}} \tag{3.45}$$

$$\Delta E^{\text{PLd,L}} = \sum_{I\in L} \Delta E_I'' + \sum_{I\in L} \Delta\Delta \tilde{E}_I^{\text{solv}} + \sum_{I>J\in L} \Delta\Delta E_{IJ}^{\text{int}} \tag{3.46}$$

So that the total binding energy, schematically shown in Figures 3.1 and 3.2, is

$$\Delta E = \Delta E^{\text{des,P}} + \Delta E^{\text{des,L}} + \Delta E^{\text{PLd,P}} + \Delta E^{\text{PLd,L}} + \Delta E^{\text{int,PL}} \tag{3.47}$$

where

$$\Delta E^{\text{int,PL}} = \sum_{I\in P, J\in L} \Delta E_{IJ}^{\text{int}} \tag{3.48}$$

Figure 3.2 also illustrates a very important point, that $\Delta E^{\text{int,PL}}$ is a much overestimated representation of the binding (ΔE). The bonding energy is the sum of the protein and ligand destabilization polarization and the protein–ligand pair interaction energies.

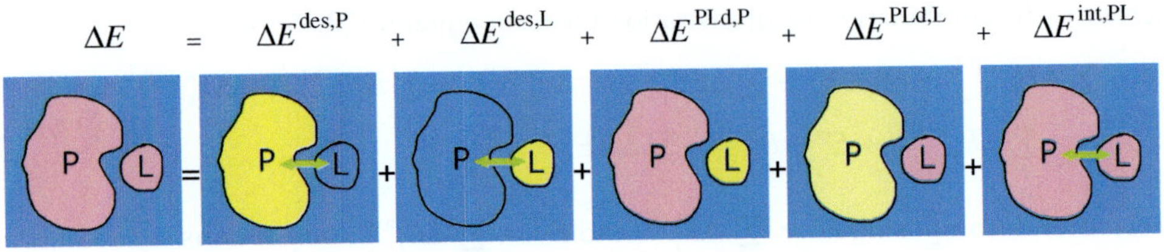

Figure 3.1 The total protein(P)–ligand(L) binding energy ΔE in solution is decomposed in SBA as the cost $\Delta E^{\mathrm{des,P}}$ to remove the water cluster filling the ligand L volume from protein P, the cost $\Delta E^{\mathrm{des,L}}$ to remove the water cluster filling the P volume from ligand L, the polarization destabilization $\Delta E^{\mathrm{PLd,P}}$ of P turning pink in the presence of L, the polarization destabilization $\Delta E^{\mathrm{PLd,L}}$ of L turning pink in the presence of P, and the interaction $\Delta E^{\mathrm{int,PL}}$ between polarized P and L.

Any polarization represents a change from a reference state to another (polarized) state; here, the reference state is the electronic state of the standalone protein (ligand) in the cavity of the complex; and the polarized state is the state of the protein and ligand in the complex. In other words, the polarization is of the protein by the ligand $\Delta E^{\mathrm{PLd,P}}$ and vice versa $\Delta E^{\mathrm{PLd,L}}$ (this protein and ligand polarization is done self-consistently, so that strictly speaking, both protein and ligand should be painted in pink for the $\Delta E^{\mathrm{PLd,P}}$ and $\Delta E^{\mathrm{PLd,L}}$ terms in Figure 3.1, but then the meaning will be hard to convey). The stabilization component of the polarization is implicitly included in $\Delta E_{IJ}^{\mathrm{int}}$ (in the electrostatic component in the PIE decomposition analysis, PIEDA [12, 13]). Roughly, one can expect that the stabilization component is about twice as large so that

$$\Delta E^{\mathrm{PLs,P}} \approx -2\Delta E^{\mathrm{PLd,P}} \tag{3.49}$$

$$\Delta E^{\mathrm{PLs,L}} \approx -2\Delta E^{\mathrm{PLd,L}} \tag{3.50}$$

So that the total polarization is

$$\Delta E^{\mathrm{PL,P}} = \Delta E^{\mathrm{PLd,P}} + \Delta E^{\mathrm{PLs,P}} \approx -\Delta E^{\mathrm{PLd,P}} \tag{3.51}$$

$$\Delta E^{\mathrm{PL,L}} = \Delta E^{\mathrm{PLd,L}} + \Delta E^{\mathrm{PLs,L}} \approx -\Delta E^{\mathrm{PLd,L}} \tag{3.52}$$

This relation was found [12] to hold quite well for water clusters.

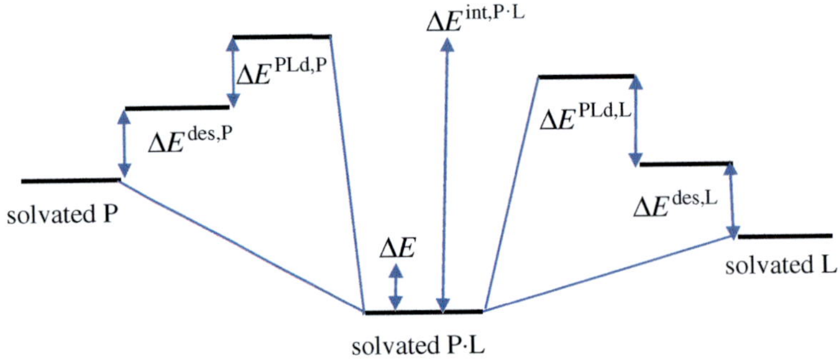

Figure 3.2 Schematic energy diagram for SBA, showing the protein (P) and ligand (L) binding energy $\Delta E = \Delta E^{\mathrm{des,P}} + \Delta E^{\mathrm{des,L}} + \Delta E^{\mathrm{PLd,P}} + \Delta E^{\mathrm{PLd,L}} + \Delta E^{\mathrm{int,PL}}$ (equation 3.47) and its components: the desolvation (des), destabilization polarization (PLd), and interaction (int) energies.

3.3.2.5 Fragment-Wise Elaboration of SBAC

The components in SBAC are obtained starting from SBA (equations 3.38, 3.45, and 3.46) and separating the cavitation energy. The bonding energy is

$$\Delta\bar{E}^{\mathrm{bon,P\cdot L}} = \Delta\bar{E}^{\mathrm{PLd,P}} + \Delta\bar{E}^{\mathrm{PLd,L}} + \sum_{I\in P,J\in L} \Delta E_{IJ}^{\mathrm{int}} \tag{3.53}$$

where

$$\Delta\bar{E}^{\mathrm{PLd,P}} = \sum_{I\in P} \Delta E_I'' + \sum_{I\in P} \Delta\Delta\bar{E}_I^{\mathrm{solv}} + \sum_{I>J\in P} \Delta\Delta E_{IJ}^{\mathrm{int}} \tag{3.54}$$

$$\Delta\bar{E}^{\mathrm{PLd,L}} = \sum_{I\in L} \Delta E_I'' + \sum_{I\in L} \Delta\Delta\bar{E}_I^{\mathrm{solv}} + \sum_{I>J\in L} \Delta\Delta E_{IJ}^{\mathrm{int}} \tag{3.55}$$

$$\Delta\Delta\bar{E}_I^{\mathrm{solv}} = \Delta\Delta\tilde{E}_I^{\mathrm{es}} + \Delta\Delta\tilde{E}_I^{\mathrm{disp}} + \Delta\Delta\tilde{E}_I^{\mathrm{rep}} \tag{3.56}$$

The cavitation energy is decomposed as

$$\Delta\Delta E^{\mathrm{cav}} = \sum_{I} \Delta\Delta E_I^{\mathrm{cav}} \tag{3.57}$$

3.3.3 Application of Analyses to Protein–Ligand Binding

The analysis is applied to the complexes of Trp-cage (PDB: 1L2Y) with two ligands. The structures are taken from a previous study [37]. The calculations are performed using FMO code [38, 39] in GAMESS [40] at the RHF-D/PCM<1>/6-31G* level of theory. The binding energy ΔE is computed as the difference between the energy of the complex and the energies of the protein and ligand separately, at their geometry in the complex. That is, the deformation energy was not added. The results are shown in Table 3.1. ABA and SBA bonding energies are related as

$$\Delta\tilde{E}^{\mathrm{bon,P\cdot L}} = \Delta E^{\mathrm{bon,P\cdot L}} + \Delta E^{\mathrm{des,L}} \tag{3.58}$$

Table 3.1 Trp-cage and ligand binding energies ΔE and components (kcal/mol) in the three analyses.

Ligand	ABA	SBA	SBAC
p-HOPhCOOH	$\Delta E^{\mathrm{des,P}} = -0.086$	$\Delta E^{\mathrm{des,P}} = -0.086$	$\Delta\bar{E}^{\mathrm{des,P}} = 10.62$
($\Delta E = -13.69$)		$\Delta E^{\mathrm{des,L}} = 7.88$	$\Delta\bar{E}^{\mathrm{des,L}} = 18.59$
	$\Delta\tilde{E}^{\mathrm{bon,P\cdot L}} = -13.60$	$\Delta E^{\mathrm{bon,P\cdot L}} = -21.49$	$\Delta\bar{E}^{\mathrm{bon,P\cdot L}} = -32.19$
			$\Delta\Delta E^{\mathrm{cav}} = -10.71$
p-HOPhCOO$^-$	$\Delta E^{\mathrm{des,P}} = -0.095$	$\Delta E^{\mathrm{des,P}} = -0.095$	$\Delta\bar{E}^{\mathrm{des,P}} = 10.20$
($\Delta E = -11.14$)		$\Delta E^{\mathrm{des,L}} = 8.88$	$\Delta\bar{E}^{\mathrm{des,L}} = 19.18$
	$\Delta\tilde{E}^{\mathrm{bon,P\cdot L}} = -11.04$	$\Delta E^{\mathrm{bon,P\cdot L}} = -19.93$	$\Delta\bar{E}^{\mathrm{bon,P\cdot L}} = -30.22$
			$\Delta\Delta E^{\mathrm{cav}} = -10.29$

The total binding energy in ABA, SBA, and SBAC, is decomposed as

$$\begin{aligned}\Delta E &= \Delta E^{\text{des,P}} + \Delta \tilde{E}^{\text{bon,P}\cdot\text{L}} \\ &= \Delta E^{\text{des,P}} + \Delta E^{\text{des,L}} + \Delta E^{\text{bon,P}\cdot\text{L}} \\ &= \Delta \bar{E}^{\text{des,P}} + \Delta \bar{E}^{\text{des,L}} + \Delta \bar{E}^{\text{bon,P}\cdot\text{L}} + \Delta\Delta E^{\text{cav}}\end{aligned} \tag{3.59}$$

3.4 GUI Development

3.4.1 Outline of FU

GUI software FU is developed for structure modeling, input data generation, and visualization of results (Figure 3.3). FU is designed to handle large and complex molecular systems like photosystem II (PSII). FU is open-source software [9] written in Python, although for acceleration of computation-intensive tasks it can also use FORTRAN modules replacing corresponding Python scripts. It can be executed on various platforms such as Microsoft Windows, Mac OS X, and Linux, which requires preinstalling Python and various libraries. FU comes with a Python shell window, and all functionality of FU can be accessed from either menu interactively or from a shell window by typing Python commands, so that automatic data processing becomes simple. The latter does require that one is familiar with the module names of FU commands, which can be learned from programming guide.

FU is designed to be a user-customizable program, allowing users to modify menus, add new functionality to suit their needs, and the license allows one to redistribute FU as

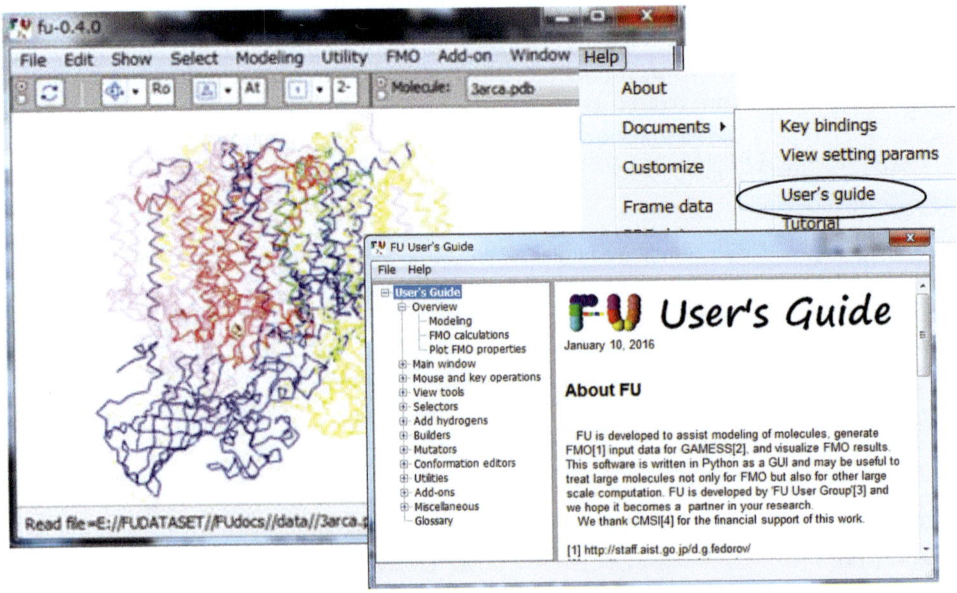

Figure 3.3 The main window of FU together with the user's guide panel.

well. FU has a special menu for executing Python scripts, so it is easy for the user to add a new script to the “Add-on” menu. FU allows general modeling, but at present, the functionality to prepare input files and display the results is limited to FMO and GAMESS. In future, support of other methods and programs can be easily added. FU has a system of “filters” for processing results from external programs. A filter is a script that reads an output and extracts necessary data. This way, extensions to future modifications of the output format of external programs are easy to accommodate. Each external program can have multiple filters to suit particular needs.

FU can read/write molecular structure files in the formats of protein data bank (PDB), Cartesian coordinates, MDL MOL, and Z-matrix. FU has a Z-matrix editor, which can use variables for equating variables, defining active, and freeze coordinates in geometry optimization, and so on. Fragmentation for FMO calculations can be done automatically for polypeptides and water, and manually for other molecules. A script in FU “fuplot” can be used to visualize pair interaction energies and draw density.

Protonation can be done in FU automatically for peptides or for general systems using either connectivity information from PDB (monomer files) or bond lengths. Experimental protein structures PDB can have missing atoms or unknown residues. FU can be used to repair structures using various tools, executed interactively under user’s careful guidance. A visual inspection of large molecular structures can be tiresome and time-consuming. FU has a tool called “one-by-one viewer” which draws individual residues.

3.4.2 Modeling and Result Visualization

As an example, a typical usage of FU for FMO is illustrated here, consisting of the following steps: 1) modeling of a protein–ligand complex, 2) creating FMO input for GAMESS, 3) running FMO in GAMESS, and 4) visualizing FMO results.

3.4.2.1 Modeling of an FKBP Protein Complex

The PDB structure file [41] of FKBP with its ligand FK506 (PDB ID: 1FKF) is read from the “File”-“Open” menu (Figure 3.4). Water molecules are removed with “Modeling”-“Delete”-“Waters”. Hydrogen atoms are added with “Modeling”-“Add hydrogen”, followed by “to polypeptide” (for the protein), and “to non-polypeptide(use frame data)” (for the ligand), see Figure 3.5. Note that adding hydrogen atoms also draws covalent bonds between all atoms.

In this example, HIS residues are protonated automatically, and a manual reprotonation can be done with “Modeling”-“Mutate amino acid residue”-“Open HIS form change” menu. A user-guided protonation can also be done with “Modeling”-”Open add hydrogen panel”. The added hydrogens are shown in green indicating they are in the “selected” state.

3.4.2.2 Creating FMO Input

The fragmentation of a protein for FMO is performed automatically with several choices of fragment sizes. Here, one residue per fragment is used, except that glycine residues are merged with a neighboring fragment. This is accomplished with “FMO”-”Fragment(auto)”-“AA residue/1resxGly” (Figure 3.6).

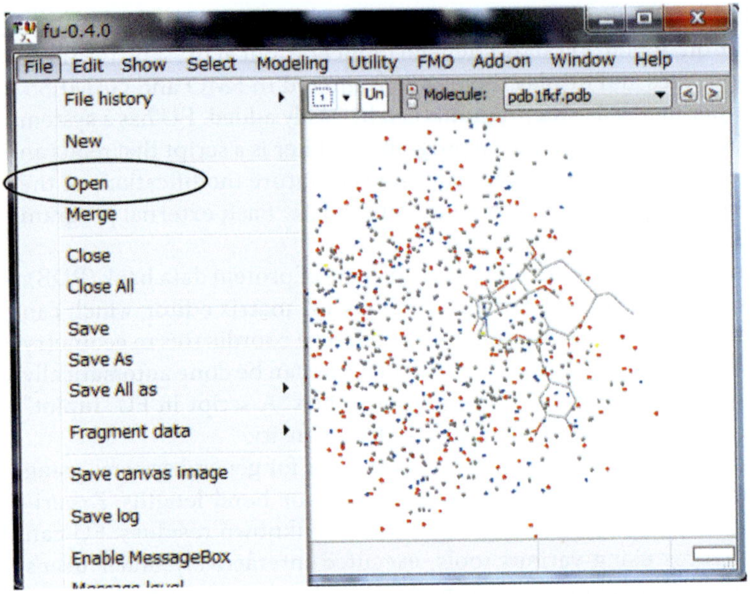

Figure 3.4 FKBP and its ligand FK506 after reading the PDB file from "File"-"Open". In the PDB, covalent bonds are given only for the ligand, so that other atoms are drawn as circles. All covalent bonds are added during consequent protonation.

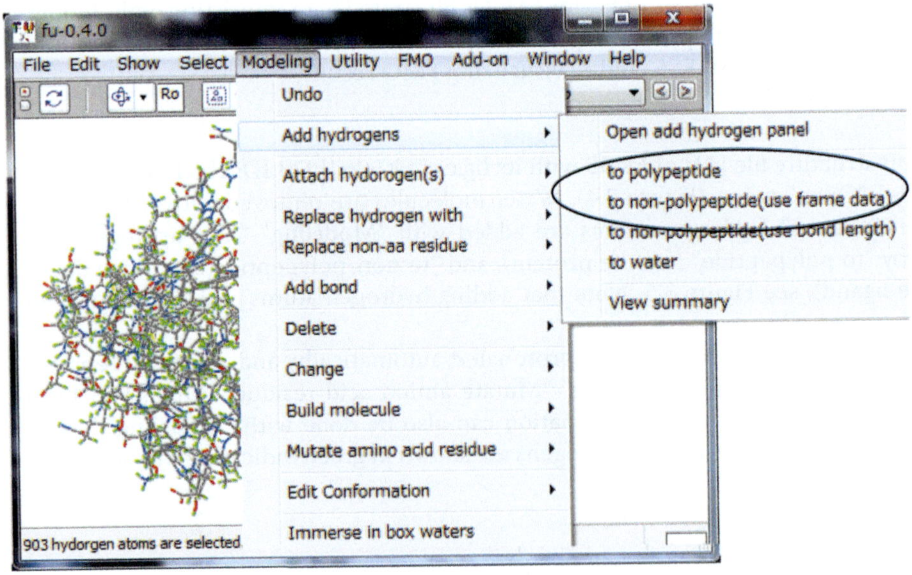

Figure 3.5 FKBP and its ligand FK506 after protonation with "Modeling"-"Add hydrogen(s)"-"to polypeptide" and "to non-polypeptide (use frame data)".

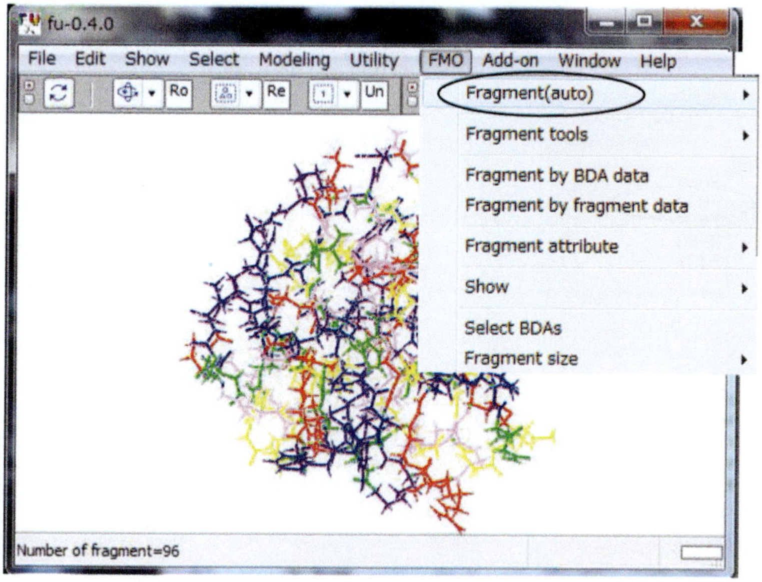

Figure 3.6 Fragmentation of FKBP and its ligand FK506 for FMO calculations, done with "FMO"-"Fragment(auto)"-"AA Residue/1resxGly".

One can check the location of detached bonds (the so-called bond detached atom (BDA) points) with "FMO"-"Show"-"BDA points" and display fragments in various colors with "FMO"-"Show"-"Paint fragment" menu. The ligand is not fragmented in the present example.

3.4.2.3 Running FMO in GAMESS

First, "Add-on"-"gamess-user.py" is used to open "GAMESS assist for beginner" panel (Figure 3.7). In "Method" box, "FMO" and then "FMO2" are selected from the pull-down list. The molecular coordinates can be read from a file or taken from the current FU modeling window. For the latter, "From FU" should be pressed.

A GAMESS/FMO input file has to be saved by pressing the "Save" button and then it can be executed by pressing "RunGMS" button. Before running GAMESS for the first time, the executable details have to be set up in "Help"-"Settings". For large jobs an external cluster can be used (FU at the moment has no support for running jobs remotely).

3.4.2.4 Visualizing FMO Results

After the computation is finished, "Add-on"-"fuplot.py" is used to start the FU plot interface. A GAMESS output file is read with "File"-"Open output file". The user has to choose a filter to process the output file, in this case "gamess-fmo.filter". The output file name displayed in the panel has to be selected and "Set for plot" pressed.

The properties collected by the filter are shown (Figure 3.8). Selecting "PIEDA" and pressing "Plot" will show the results in a new window. The strength of pair interaction energies can also be shown in the molecular structure window by selecting "MolView" (Figure 3.9).

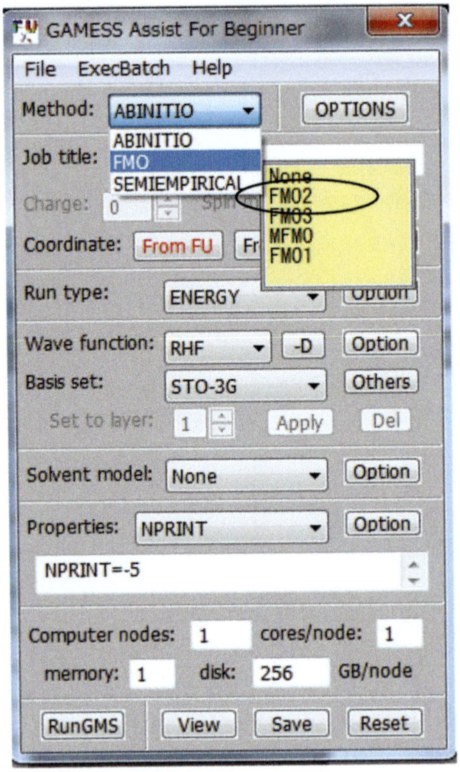

Figure 3.7 "GAMESS Assist For Beginner" panel, where details of FMO calculations are defined, such as wave function and basis set.

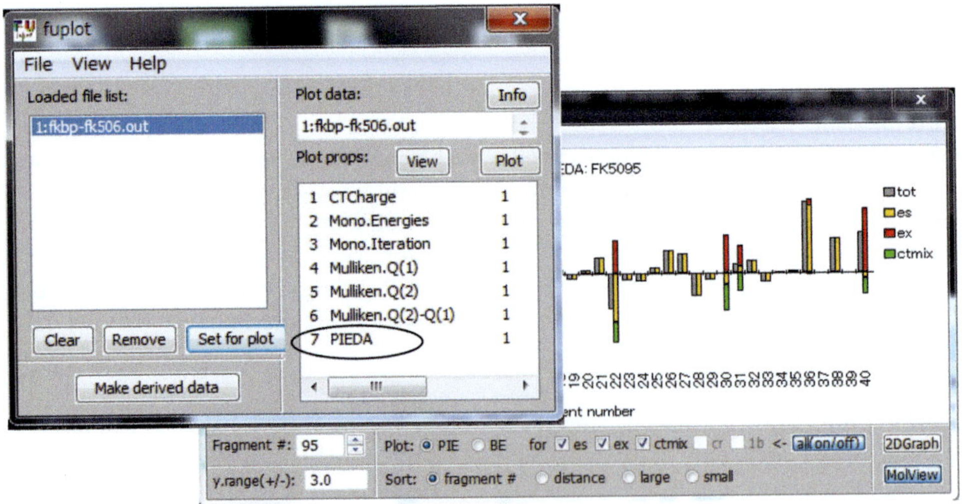

Figure 3.8 Visualization of FMO properties for FKBP + FK506 by "fuplot". The properties in the provided output file, found by the filter, are listed in the panel. PIEDA is selected in the list and the pair interaction energies between one fragment (#95, the ligand) and the rest are plotted on the right.

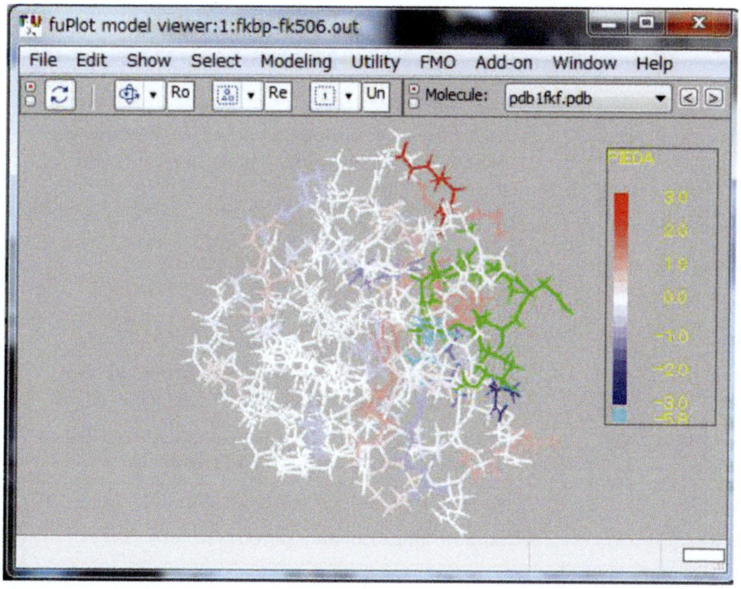

Figure 3.9 Fragment residues in FKBP are colored according to their pair interaction energy with the ligand (green). The color legend bar is shown on the right (repulsion and attraction are shown in red and blue, respectively).

3.4.3 An Overview of Using FU for a Complex System

PSII is an example of a complex system because it contains many kinds of molecules, and some missing atoms and residues in the PDB file (3ARC). Preparing this system for quantum-mechanical calculations requires many manipulations, which well illustrate the capabilities of FU.

PSII is a dimer. First, monomer *a* is prepared, then monomer *b*, and finally both are merged. The steps necessary for making a complete structure with FU are shown briefly in Table 3.2. FU can be used to make an FMO input file at the end.

3.4.4 Examples of Scripting in FU

FU has several thousands of subroutines which can be used in user scripts. Routine work should be done automatically by a script to reduce human fatigue and error. Two script examples of medium complexity are described: how to convert multiple PDB files into Z-matrix files, and how to draw dipole moments for fragments in FMO. Note that method in this subsection refers to the Python concept of a method.

3.4.4.1 Converting Multiple PDB Files into Z-matrix Files

The purpose of this example is to show how to process many files using a script. The Python code is shown in Figure 3.10. “GetFilesInDirectory” method in the “lib” module gets a list of files in a given directory. “ReadFiles” method, defined in “Model” class in “fumodel” module (“fum” is an instance of “Model” class), reads structure files and creates a “mol” object which is an instance of “Molecule” in “molec” module. The elements of the list attribute variables of “mol” are instances of “Atom” class in “molec” module,

Table 3.2 PSII (PDB: 3arc) modeling steps.

Step	Operation	Comments
1	Read comments in 3arc.pdb.	Check missing residues and atoms. Make a decision whether to repair missing parts or simply add hydrogen caps.
2	Run report unique residue tool.	Note 36 missing residues (UNL).
3	Run report missing atom tool.	Note 11 missing atoms in SQD:768:D.
4	Run split pdb tool on full PDB.	Save monomer *a* PDB.
5	Run split pdb on monomer *a*.	Split into peptide, non-peptide, and water parts.
6	Protonate peptide parts.	Use automatic tools, but some residues require manual corrections, done interactively.
7	Protonate non-peptide parts.	Use .mht files. For UNL use automatic tools with an interactive manual check and corrections.
8	Protonate water.	Use automatic tool for water protonation.
9	Run merge pdb tool.	Combine all protonated parts.
10	Run pick object tool.	Choose all HIS residues and interactively consider protonation of Hδ and Hε.
11	Use repair tools.	Repair missing parts by extracting similar units and putting them in the system.
12	Run add box water tool.	Add a shell or box of water.
13	Use run Tinker tool.	Use MM to optimize geometry.
14	Use extract residue tool.	If desired, reduce the system size by extracting the oxygen evolving center with some environment.

```
#!/bin/sh
# -*- coding: utf-8

# Script:pdbs2zmts.py
# Function: read multiple pdb files in a directory and save them as zmt files
# The same operation can be done in fu,
# 1) Drag-and-drop the "pdbdir" on the fu main window,
# 2) Execute the "File"-"Save All as"-"zmt files" menu.

import lib     # lib module in FU
import rwfile  # rwfile module in FU

pdbdir="c://FUDATASET//FUdocs//data//pdbdir" # pdb files directory
zmtdir="c://FUDATASET//FUdocs//data//zmtdir" # directory to save Z-matrix files
pdbfilelst=lib.GetFilesInDirectory(pdbdir) # get pdb files in directory
# "fum" and "mol" are instances of "Model" and "Molecule" classes, respectively
for pdbfile in pdbfilelst:
    fum.ReadFiles(pdbfile,True) # the second arg is True for draw molecule
    zelm,zpnt,zprm=lib.CCToZM(fum.mol) # convert to Z-matrix data
    head,tail=os.path.split(pdbfile); base,ext=os.path.splitext(tail)
    zmtfile=os.path.join(zmtdir,base+".zmt") # make zmt file name
    rwfile.WriteZMTFile(zmtfile,base,zelm,zpnt,zprm) # write zmt file
# end of script
```

Figure 3.10 Python script for FU, writing out a Z-matrix for each PDB file in a directory.

whose attribute variables are atomic data such as element and coordinates. "CCToZM" method in "lib" module converts Cartesian coordinates atoms in the "mol" object to Z-matrix data. "WriteZMTFile" method in "rwfile" module creates a Z-matrix file.

3.4.4.2 Drawing Dipole Moments with Arrows

The Python code is listed in Figure 3.11. There are four parameters, pdbfile, outfile, scale, and atmnam. The first two define file names, the third determines the scaling of an arrow

```
#!/bin/sh
# -*- coding: utf-8

# Script: draw-dipole.py
# Function: Draw FMO monomer dipoles with arrow
#
import rwfile
import lib

##### modify the following data ######
pdbfile="c://FUDATASET//FUdocs//data//ala10bet.pdb" # pdb file
outfile="c://FUDATASET//FUdocs//data//ala10bet.out"  #  FMO  output  file  with
Mulliken charges
scale=0.5       # scale factor of dipole moment
atmnam=" CA  " # atom name of arrow origin
######################################

def OriginOfArrow(atmlst,atmnam):
    x=None; y=None; z=None
    for i in atmlst:
        atom=fum.mol.atm[i]
        if atom.atmnam == atmnam:
            x=atom.cc[0];  y=atom.cc[1]; z=atom.cc[2] # cc: list of coordinates
            break
    return x,y,z

# read the pdb file
fum.ReadFiles(pdbfile,True)
# read Mulliken charges in the outfile (used for getting fragment info)
err,version,mulchg=rwfile.ReadFMOMulliken(outfile)
# read FMO one-body property
err,version,onebody=rwfile.ReadFMOOneBody(outfile)
natm=len(mulchg); frgdic={}
for i in range(natm): # assign atoms to fragment
    ifg=mulchg[i][1]-1 # FMO2 is assumed
    if not frgdic.has_key(ifg): frgdic[ifg]=[]
    frgdic[ifg].append(i)
# fragment loop
nfrg=len(frgdic); arrowpnts=[]
for i in range(nfrg):
    atmlst=frgdic[i]
    cx,cy,cz=OriginOfArrow(atmlst,atmnam)
    if cx is not None:
        mx=scale*onebody[i][3]; my=scale*onebody[i][4]; mz=scale*onebody[i][5]
        mx += cx; my += cy; mz += cz
        arrowpnts.append([[cx,cy,cz],[mx,my,mz]])
# draw arrows
fum.DrawAxisArrow(True,arrowpnts,True)
# end of script
```

Figure 3.11 Python script for FU, plotting fragment dipole moments in FMO.

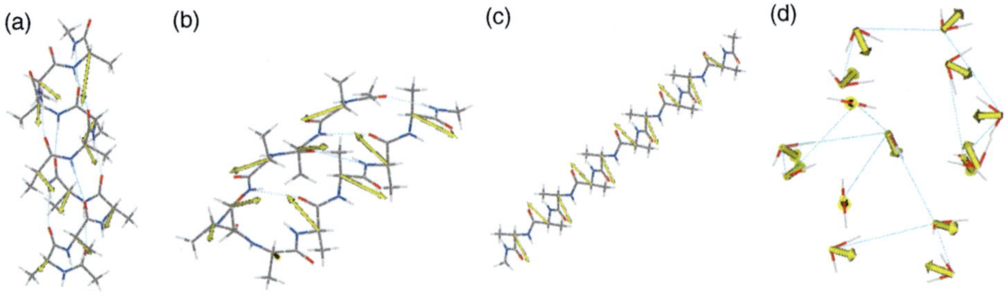

Figure 3.12 FMO monomer dipoles drawn as yellow arrows for (a) α-helix of alanine 10-mer, (b) β-turn of alanine 10-mer, (c) extended form of alanine 10-mer, and (d) cluster of 15 water molecules. In the α-helix dipole moments of fragment residues point toward a common direction, although they are tilted to some extent; in the extended form, dipole moments alternatively point up and down and the β-turn is a twisted extended form.

(the arrow length is given by the dipole moment times this constant), and the fourth is the PDB label for the atom, where the origin of an arrow is placed. Strictly speaking, for neutral fragments the origin can be anywhere, but for charged ones, the origin has to be at the center of charge, both when computed in GAMESS and in plotting, but neither is implemented at present.

This script reads a PDB file ("fum.ReadFiles"), fragmentation data, and Mulliken charges ("ReadFMOMulliken" method in "rwfile" class), and FMO one-body properties ("ReadFMOOneBody" method in "rwfile" module) from an output file of FMO in GAMESS, including fragment dipole moments. "OriginOfArrow" method is defined in the script to get the origin coordinates for a dipole moment arrow. "DrawAxisArrow" method in "Model" class draws arrows. The results of this script for several systems are shown in Figure 3.12.

3.5 Conclusions

New kinds of analyses of protein–ligand binding have been proposed, clearly defining the desolvation penalty. Their usefulness will be established in future applications, for example, by analyzing correlations between components proposed here and experimentally observed quantities. Clearly, the desolvation energies should also be considered when discussing the protein ligand binding, and not just the pair interaction energies in the complex.

A new GUI, FU, has been developed for facilitating modeling and result visualization in large and complex molecular systems. When the number of atoms reaches 100,000 (a current limit of PDB files), clearly, it is easy to make mistakes in failing to notice a missing atom or a dozen, or a clash in the geometry. A very efficient GUI is needed to aid in the preparation of structure for quantum-mechanical calculations, which are much more demanding to the quality of the molecular structures than, for example, force fields.

FU has been designed to be user-customizable, and an addition of new features to FU is straightforward. The ability to execute Python scripts using FU functionality should enable automation of many tedious manual operations and facilitate research for FMO and, in future, other methods.

Acknowledgments

We thank the Next Generation Super Computing Project, Nanoscience Program (MEXT, Japan), and Computational Materials Science Initiative (CMSI, Japan) for financial support.

References

1 Gordon, M. S.; Pruitt, S. R.; Fedorov, D. G.; Slipchenko, L. V. Fragmentation methods: a route to accurate calculations on large systems. *Chem. Rev.* **2012**, *112*, 632–672.

2 Otto, P.; Ladik, J. Investigation of the interaction between molecules at medium distances: I. SCF LCAO MO supermolecule, perturbational and mutually consistent calculations for two interacting HF and CH_2O molecules. *Chem. Phys.* **1975**, *8*, 192–200.

3 Gao, J. Toward a molecular orbital derived empirical potential for liquid simulations. *J. Phys. Chem. B* **1997**, *101*, 657–663.

4 Nakano, T.; Mochizuki, Y.; Fukuzawa, K.; Amari, S.; Tanaka, S. Developments and applications of ABINIT-MP software based on the fragment molecular orbital method, in *Modern Methods for Theoretical Physical Chemistry of Biopolymers*, E. B. Starikov, J. P. Lewis, S. Tanaka (Eds.), Elsevier, Amsterdam, 2006, pp. 39–52.

5 Suenaga, M. Development of GUI for GAMESS/FMO calculation. *J. Comput. Chem. Jpn.* **2008**, *7*, 33 (in Japanese).

6 http://www.paics.net/get_paics_view.html (accessed February 2, 2016).

7 Both FragIt and FMO portal are currently offline (accessed February 2, 2016).

8 https://staff.aist.go.jp/d.g.fedorov/fmo/fmoutil.html (accessed February 2, 2016).

9 http://sourceforge.net/projects/fusuite/ (accessed February 2, 2016).

10 Ishikawa, T.; Mochizuki, Y.; Amari, S.; Nakano, T.; Tokiwa, H.; Tanaka, S.; Tanaka, K. Fragment interaction analysis based on local MP2. *Theor. Chem. Acc.* **2007**, *118*, 937–945.

11 Mochizuki, Y.; Fukuzawa, K.; Kato, A.; Tanaka, S.; Kitaura, K.; Nakano, T. A configuration analysis for fragment interaction. *Chem. Phys. Lett.* **2005**, *410*, 247–253.

12 Fedorov, D. G.; Kitaura, K. Pair interaction energy decomposition analysis. *J. Comput. Chem.* **2007**, *28*, 222–237.

13 Fedorov, D. G.; Kitaura, K. Energy decomposition analysis in solution based on the fragment molecular orbital method. *J. Phys. Chem. A* **2012**, *116*, 704–719.

14 Kitaura, K.; Ikeo, E.; Asada, T.; Nakano, T.; Uebayasi, M. Fragment molecular orbital method: an approximate computational method for large molecules. *Chem. Phys. Lett.* **1999**, *313*, 701–706.

15 *The Fragment Molecular Orbital Method: Practical Applications to Large Molecular Systems*, D. G. Fedorov, K. Kitaura (Eds.), CRC Press, Boca Raton, FL, 2009.

16 Amari, S.; Aizawa, M.; Zhang, J.; Fukuzawa, K.; Mochizuki, Y.; Iwasawa, Y.; Nakata, K.; Chuman, H.; Nakano, T. VISCANA: visualized cluster analysis of protein–ligand

interaction based on the ab initio fragment molecular orbital method for virtual ligand screening. *J. Chem. Inf. Comput. Sci.* **2006**, *46*, 221–230.

17 Nakanishi, I.; Fedorov, D. G.; Kitaura, K. Molecular recognition mechanism of FK506 binding protein: an all-electron fragment molecular orbital study. *Proteins Struct. Funct. Bioinf.* **2007**, *68*, 145–158.

18 Sawada, T.; Fedorov, D. G.; Kitaura, K. Role of the key mutation in the selective binding of avian and human influenza hemagglutinin to sialosides revealed by quantum-mechanical calculations. *J. Am. Chem. Soc.* **2010**, *132*, 16862–16872.

19 Otsuka, T.; Okimoto, N.; Taiji, M. Assessment and acceleration of binding energy calculations for protein–ligand complexes by the fragment molecular orbital method. *J. Comput. Chem.* **2015**, *36*, 2209–2218.

20 Yoshida, T.; Yamagishi, K.; Chuman, H. QSAR study of cyclic urea type HIV-1 PR inhibitors using ab initio MO calculation of their complex structures with HIV-1 PR. *QSAR Comb. Sci.* **2008**, *27*, 694–703.

21 Mazanetz, M. P.; Ichihara, O.; Law, R. J.; Whittaker, M. Prediction of cyclin-dependent kinase 2 inhibitor potency using the fragment molecular orbital method. *J. Cheminf.* **2011**, *3*, 2.

22 Heifetz, A.; Chudyk, E. I.; Gleave, L.; Aldeghi, M.; Cherezov, V.; Fedorov, D. G.; Biggin, P. C.; Bodkin. M. J. The fragment molecular orbital method reveals new insight into the chemical nature of GPCR-ligand interactions. *J. Chem. Inf. Model.* **2016**, *56*, 159–172.

23 Alexeev, Y.; Mazanetz, M. P.; Ichihara, O.; Fedorov, D. G. GAMESS as a free quantum-mechanical platform for drug research. *Curr. Top. Med. Chem.* **2012**, *12*, 2013–2033.

24 Heifetz, A.; Trani, G.; Aldeghi, M.; MacKinnon, C. H.; McEwan, P. A.; Brookfield, F. A.; Chudyk, E. I.; Bodkin, M.; Pei, Z.; Burch, J. D.; Ortwine, D. F. Fragment molecular orbital method applied to lead optimization of novel interleukin-2 inducible T-cell kinase (ITK) inhibitors. *J. Med. Chem.*, **2016**, *59*, 4352–4363.

25 Murata, K.; Fedorov, D. G.; Nakanishi, I.; Kitaura. K. Cluster hydration model for binding energy calculations of protein–ligand complexes. *J. Phys. Chem. B* **2009**, *113*, 809–817.

26 Fedorov, D. G.; Kitaura, K. Theoretical development of the fragment molecular orbital (FMO) method, in *Modern Methods for Theoretical Physical Chemistry of Biopolymers*, E. B. Starikov, J. P. Lewis, S. Tanaka (Eds.), Elsevier, Amsterdam, 2006, pp. 3–38.

27 Fedorov, D. G.; Kitaura, K. Extending the power of quantum chemistry to large systems with the fragment molecular orbital method. *J. Phys. Chem. A* **2007**, *111*, 6904–6914.

28 Fedorov, D. G.; Asada, N. ; Nakanishi, I.; Kitaura, K. The use of many-body expansions and geometry optimizations in fragment-based methods. *Acc. Chem. Res.* **2014**, *47*, 2846–2856.

29 Nagata, T.; Fedorov, D. G.; Kitaura K. Mathematical formulation of the fragment molecular orbital method, in *Linear-Scaling Techniques in Computational Chemistry and Physics*, R. Zalesny, M. G. Papadopoulos, P. G. Mezey, J. Leszczynski (Eds.), Springer, New York, 2011, pp. 17–64.

30 Fedorov, D. G.; Nagata, T.; Kitaura, K. Exploring chemistry with the fragment molecular orbital method. *Phys. Chem. Chem. Phys.* **2012**, *14*, 7562–7577.

31 Tanaka, S.; Mochizuki, Y.; Komeiji, Y.; Okiyama, Y.; Fukuzawa, K. Electron-correlated fragment-molecular-orbital calculations for biomolecular and nano systems. *Phys. Chem. Chem. Phys.* **2014**, *16*, 10310–10344.

32 Mazanetz, M.; Chudyk, E.; Fedorov, D. G.; Alexeev. Y. Applications of the fragment molecular orbital method to drug research, in *Computer-Aided Drug Discovery*, W. Zhang (Ed.), Springer, New York, 2016, pp. 217–255.

33 Tomasi, J.; Mennucci, B.; Cammi, R. Quantum mechanical continuum solvation models. *Chem. Rev.* **2005**, *105*, 2999–3093.

34 Fedorov, D. G.; Kitaura, K.; Li, H.; Jensen, J. H.; Gordon, M. S. The polarizable continuum model (PCM) interfaced with the fragment molecular orbital method (FMO). *J. Comput. Chem.* **2006**, *27*, 976–985.

35 Li, H.; Fedorov, D. G.; Nagata, T.; Kitaura, K.; Jensen, J. H.; Gordon, M. S. Energy gradients in combined fragment molecular orbital and polarizable continuum model (FMO/PCM) calculation. *J. Comput. Chem.* **2010**, *31*, 778–790.

36 Nagata, T.; Fedorov, D. G.; Li, H.; Kitaura, K. Analytic gradient for second order Møller–Plesset perturbation theory with the polarizable continuum model based on the fragment molecular orbital method. *J. Chem. Phys.* **2012**, *136*, 204112.

37 Fedorov, D. G.; Kitaura, K. Subsystem analysis for the fragment molecular orbital method and its application to protein–ligand binding in solution. *J. Phys. Chem. A* **2016** *120*, 2218–2231.

38 Fedorov, D. G.; Kitaura, K. The importance of three-body terms in the fragment molecular orbital method. *J. Chem. Phys.* **2004**, *120*, 6832–6840.

39 Fedorov, D. G.; Olson, R. M.; Kitaura, K.; Gordon, M. S.; Koseki, S. A new hierarchical parallelization scheme: generalized distributed data interface (GDDI), and an application to the fragment molecular orbital method (FMO). *J. Comput. Chem.* **2004**, *25*, 872–880.

40 Schmidt, M. W.; Baldridge, K. K.; Boatz, J. A.; Elbert, S. T.; Gordon, M. S.; Jensen, J. H.; Koseki, S.; Matsunaga, N.; Nguyen, K. A.; Su, S.; Windus, T. L.; Dupuis, M.; Montgomery, J. A., Jr. General atomic and molecular electronic structure system. *J. Comput. Chem.* **1993**, *14*, 1347–1363.

41 http://www.rcsb.org/pdb/explore/explore.do?structureId=1fkf (accessed February 2, 2016).

4

Molecules-in-Molecules Fragment-Based Method for the Accurate Evaluation of Vibrational and Chiroptical Spectra for Large Molecules

K. V. Jovan Jose and Krishnan Raghavachari*

Department of Chemistry, Indiana University, USA
**Current address: School of Chemistry, University of Hyderabad, India*

4.1 Introduction

Vibrational spectroscopic techniques are among the most prevalent tools for studying the unique structural *fingerprints* of molecules [1]. In particular, infrared (IR) and Raman vibrational spectroscopic techniques yield information on the chemical composition and the internal structure of the molecule [2]. In addition to IR and Raman, several other complementary spectroscopic techniques are used to study important properties of molecules. For example, for determining the absolute configuration of chiral systems, vibrational circular dichroism [3–5] (VCD) and Raman optical activity [6–8] (ROA) spectroscopic techniques are widely used [9–15]. VCD characterizes the differential absorption of the left and right circularly polarized infrared (IR) radiation, while, ROA characterizes the differential inelastic scattering of the Stokes lines in the Raman spectra of chiral molecules. Due to the differences in the associated mechanisms, Raman and ROA spectroscopy yield complementary information to that in IR and VCD. The two enantiomeric spectra in VCD and ROA exhibit spectral intensities of the same magnitude but opposite signs, and the associated unique characteristics help in the determination of the absolute configurations of molecules by a careful comparison between theory and experiment.

There is a great need for the development of theoretical models to describe large molecular systems and the associated spectral complexity at a reasonable computational cost. In order to circumvent the high scaling associated with accurate computational methods, composite quantum chemical methods have been developed by many groups [16]. Composite methods reduce the computational cost by combining a series of calculations to extrapolate to the results from a direct calculation on a complex system that would otherwise be prohibitive. For example, the popular ONIOM composite method is widely used to study key regions of the molecule at highly accurate correlated levels of theory and large basis sets, and the rest of the molecule at a low level of theory.

Fragmentation: Toward Accurate Calculations on Complex Molecular Systems, First Edition.
Edited by Mark S. Gordon.

Fragment-based methods represent particularly effective composite models for the study of large molecules [17]. The basic idea in the fragment-based method is to spatially separate the whole molecule into computationally efficient units, and patch them together in a mathematically rigorous manner to evaluate accurate total energies. The different fragment-based methods differ in the way the fragments are constructed and the way the resulting energy is assembled [18]. The fragment-based methods bring down the computational dependence of the quantum chemical methods asymptotically to nearly linear scaling.

The calculation of the analytic higher energy derivatives of the total energy is the bottleneck for the evaluation of vibrational frequencies and spectroscopic intensities of large molecules [19–21]. Moreover, fairly large basis sets including diffuse and polarization functions are needed for a reliable assignment of experimental vibrational and chiroptical spectra [22, 23]. While a range of fragment-based methods have been developed by many different groups, only a few have the capability of determining the associated higher energy derivatives for the accurate calculation of the molecular spectroscopic properties [24]. The analytic expressions for gradient and higher energy derivatives are mathematically complicated and may be laborious to implement in many methods.

In order to simplify the evaluation of energies and higher energy derivatives, we have developed a composite, linear scaling, and parallelizable approach called Molecules-in-Molecules (MIM) [25–30]. MIM builds a large molecule from many small molecules, and employs a multilayer partitioning technique with multiple levels of theory, similar in spirit to the popular ONIOM method [17]. The use of a generalized hybrid energy expression makes it an efficient extrapolation method compared to many other fragment-based methods. In addition, MIM keeps the subsystem sizes reasonably small by having a second layer with a low level of theory to take into account the relevant long-range interactions needed to describe the electronic structure and properties of a large molecule. The multilayer approach makes MIM suitable for efficient evaluation of accurate higher energy derivatives. In addition to the energies [25], the performance of MIM has been carefully validated for accurate evaluations of IR [26], Raman [27], VCD [28], and ROA [29] spectra of large molecules. This chapter provides a review of the performance of MIM methods for the calculation of vibrational and chiroptical spectroscopic properties of large molecules.

The current chapter is organized as follows. Section 4.2 describes the theoretical background for evaluating MIM energies and their higher derivatives to evaluate the spectroscopic properties of interest. Section 4.3 presents an assessment of the performance of MIM for a range of benchmark systems as well as a comparison of MIM vibrational and chiroptical spectra with experiment for large molecules like α-D-cyclodextrin and cryptophane-A. Sections 4.4 and 4.5 provide an overall summary and the general conclusions.

4.2 Computational Methods and Theory

The evaluation of the higher energy derivatives is the bottleneck for the calculation of the vibrational spectra of large molecules. The MIM method provides an efficient path for the fast evaluation of higher energy derivative tensors. The following section gives a general strategy and implementation for the evaluation of MIM-vibrational and chiroptical spectra. All the actual and MIM calculations have been carried out using the Gaussian 09 program suite [31].

1. The input for carrying out the MIM algorithm [25–29] is the full molecular geometry, and the initial fragments are generated by cutting appropriate single bonds in the large molecule (typically C–C bonds). The interaction of neighboring fragments yields overlapping "primary subsystems." The overcounting due to the overlapping regions of the primary subsystems is accounted for through the "derivative subsystems" via the inclusion–exclusion principle [25]. We have followed the connectivity and number-based fragmentation schemes for the evaluation of vibrational spectra of large molecules. As in the standard ONIOM method [17], the dangling bonds of the subsystems are saturated with link hydrogen atoms that are placed along the vector connecting the supporting and host atoms at a distance defined using a scale factor that depends on the covalent radii of the different atoms [32]. Here the scale factor, g, and the link-atom Cartesian coordinates are related through a linear equation,

$$R_2 = R_1 + g(R_3 - R_1) \tag{4.1}$$

where R_2 is the Cartesian component of the link-atom, and R_1 and R_3 are those of the supporting- and host-atoms, respectively. MIM can have multiple layers that can be treated at different combinations of levels of theory and basis sets. The total energy of a two-layer MIM model (MIM2), can be constructed from the subsystem energies, through a generalized ONIOM-like expression

$$E_{\text{Total}} = E_{rl} - E_{ml}^{\text{total}} + E_{mh}^{\text{total}}, \tag{4.2}$$

where

$$E_{ml/mh}^{\text{total}} = \sum_i E_{ml/mh}^{i} - \sum_{ij} E_{ml/mh}^{i\cap j} + \cdots + (-1)^{n-1} \sum_{ij..n} E_{ml/mh}^{i\cap j\cap k\ldots\cap n} \tag{4.3}$$

Here, E_{ml} and E_{mh} in the standard ONIOM expression [17] are replaced by E_{ml}^{total} and E_{mh}^{total} to sum over each of the primary and derivative subsystems at the low and high levels of theory, and E_{rl} is the energy of the real (i.e., full) system calculated at the low level of theory.

2. The analytic MIM-Forces are evaluated through separate subsystem gradient calculations [26]. For a precise atomic force evaluation, the link-atom force components are projected back onto the supporting- and host-atoms through a general Jacobian projection method, as in the ONIOM method [32].

$$F_a = \left[\frac{\partial E_{\text{Total}}}{\partial X_\alpha}\right] = \frac{\partial E_{rl}}{\partial X_\alpha} - \sum_{L=1}^{M} \frac{\partial E_{ml}}{\partial X_\alpha} J\left(R_2; R_1, R_3\right) + \sum_{L=1}^{M} \frac{\partial E_{mh}}{\partial X_\alpha} J\left(R_2; R_1, R_3\right) \tag{4.4}$$

Here the summation runs over all the subsystems, and the index α corresponds to the Cartesian components of X, Y, or Z directions, and J the 3×3 Jacobi matrix for projecting the link-atom Cartesian components onto the corresponding supporting- and host-atoms. A complete geometry optimization is then carried out using the forces (*vide infra*) at the MIM2 level.

3. At this final optimized geometry, Hessian ($\boldsymbol{H}$) matrix for the whole molecule, is built in Cartesian coordinates from the subsystem Hessians. The mass-weighted Hessian matrix elements corresponding to the link atoms are projected back onto the real atoms through the general expression at the optimized geometry, R_e

$$H_{\alpha\beta} = \left[\frac{\partial^2 E_{\text{Total}}}{\partial X_\alpha \partial X_\beta}\right]_{R_e} = \frac{\partial^2 E_{rl}}{\partial X_\alpha \partial X_\beta} - \sum_{L=1}^{M} J^T \frac{\partial^2 E_{ml}}{\partial X_\alpha \partial X_\beta} J + \sum_{L=1}^{M} J^T \frac{\partial^2 E_{mh}}{\partial X_\alpha \partial X_\beta} J \tag{4.5}$$

Here, the $H_{\alpha\beta}$ elements, modified from the link-atom Jacobi-projection, are the blocks involving link-atom indices. The Hessian is a real- and symmetric-matrix, and is diagonalized to yield the harmonic frequencies and the corresponding normal vibrational modes. Also, this step allows us to evaluate the transformation matrix, $\boldsymbol{S}$, from Cartesian to normal modes, for the whole molecule.

4. The IR intensity of the vibrational normal mode can be derived from the square of the electric dipole transition moment, and this depends on the electric dipole derivative tensors or atomic polar tensors (APT), $P^{\lambda}_{\alpha\beta}$. The alpha and beta subscripts label the Cartesian component directions X, Y, or Z. The electronic part of the APT for the full molecule in Cartesian coordinates is built from the subsystem values, and is then used to derive the IR intensities in the MIM-IR method.

$$P^{\lambda}_{\alpha\beta,\text{Total}} = \frac{\partial^2 E_{\text{Total}}}{\partial X_{\lambda\alpha} \partial F_\beta} = P^{\lambda}_{\alpha\beta,rl} - \sum_{L=1}^{M} P^{\lambda}_{\alpha\beta,ml} J(R_2; R_1, R_3) + \sum_{L=1}^{M} P^{\lambda}_{\alpha\beta,mh} J(R_2; R_1, R_3) \tag{4.6}$$

Since the dipole-derivatives involve a first derivative in Cartesian coordinates, only the elements involving the host and supporting atoms are modified.

5. Similarly, the intensity of the VCD normal modes is determined from the rotational strength. The VCD intensity of the ith normal mode is proportional to the imaginary part of the scalar product of the electric dipole transition moment, $\langle 0|\mu_{\text{ele}}|1\rangle_i$ and the magnetic dipole transition moment, $\langle 1|\mu_{\text{mag}}|0\rangle_i$. The transition moments are defined by two tensors, representing the derivatives with respect to the Cartesian displacements for each atom, λ: the atomic polar tensor (APT) $P^{\lambda}_{\alpha\beta}$ considered above for MIM-IR, and the atomic axial tensor (AAT) $M^{\lambda}_{\alpha\beta}$. The AAT contributes to the electronic part of the magnetic transition dipole moment and is constructed in Cartesian coordinates from the corresponding subsystem tensors through the general expression,

$$M^{\lambda}_{\alpha\beta,\text{Total}} = M^{\lambda}_{\alpha\beta,rl} - \sum_{L=1}^{M} M^{\lambda}_{\alpha\beta,ml} J(R_2; R_1, R_3) + \sum_{L=1}^{M} M^{\lambda}_{\alpha\beta,mh} J(R_2; R_1, R_3) \tag{4.7}$$

The two tensors are transformed to the normal mode representation, and used to build the rotational strength and the VCD intensity of the vibrational mode.

6. The Raman and ROA intensities depend on the Raman and ROA invariants [9], which in turn are dependent on the higher order polarizability normal coordinate

derivatives: $(\partial\alpha_{\alpha\beta}/\partial Q_i)$, $(\partial G'_{\alpha\beta}/\partial Q_i)$ and $(\partial A_{\alpha\beta\gamma}/\partial Q_i)$. Here, $\alpha_{\alpha\beta}$ is the electric dipole–electric dipole polarizability tensor, $G'_{\alpha\beta}$ the electric dipole–magnetic dipole polarizability tensor, and $A_{\alpha\beta\gamma}$ is the electric dipole–electric quadrupole polarizability tensor. In MIM2, the individual $\alpha_{\alpha\beta}$, $G'_{\alpha\beta,}$ and $A_{\alpha\beta\gamma}$ tensors and their derivatives are constructed in Cartesian coordinates from the separate fragment calculations. The derivative of the electric dipole–electric dipole polarizability tensor is built from the subsystem tensors through the general expression

$$\left[\alpha^{\lambda}_{\alpha\beta,\mathrm{Total}}\right]_{R_e} = \alpha^{\lambda}_{\alpha\beta,rl} - \sum_{L=1}^{M} \alpha^{\lambda}_{\alpha\beta,ml} J(R_2; R_1, R_3) + \sum_{L=1}^{M} \alpha^{\lambda}_{\alpha\beta,mh} J(R_2; R_1, R_3) \tag{4.8}$$

The derivative of the electric dipole–magnetic dipole polarizability tensor is constructed as,

$$\left[G^{\lambda}_{\alpha\beta,\mathrm{Total}}\right]_{R_e} = G^{\lambda}_{\alpha\beta,rl} - \sum_{L=1}^{M} G^{\lambda}_{\alpha\beta,ml} J(R_2; R_1, R_3) + \sum_{L=1}^{M} G^{\lambda}_{\alpha\beta,mh} J(R_2; R_1, R_3) \tag{4.9}$$

Similarly, the derivative of the electric dipole–electric quadrupole moment polarizability tensor is constructed through the general expression,

$$\left[A^{\lambda}_{\alpha\beta\gamma,\mathrm{Total}}\right]_{R_e} = A^{\lambda}_{\alpha\beta\gamma,rl} - \sum_{L=1}^{M} A^{\lambda}_{\alpha\beta\gamma,ml} J(R_2; R_1, R_3) + \sum_{L=1}^{M} A^{\lambda}_{\alpha\beta\gamma,mh} J(R_2; R_1, R_3) \tag{4.10}$$

The electric dipole–electric dipole polarizability derivatives are used to form two Raman tensor invariants (a_i^2 and β_i^2), while all three higher order polarizability derivatives are required to construct the ROA invariants (aG'_i, $\beta^2_{G_i}$, and $\beta^2_{A_i}$). These invariants are then used to complete the calculation of the associated intensities of the Raman and ROA modes [9].

7. We have devised MIM models for incorporating the solvent effects on α-(alanine)$_{20}$ polypeptide (*vide infra*) through the IEFPCM implicit solvation model [33]. Here, each of the fragment energies and the higher energy derivatives are evaluated incorporating the IEFPCM implicit solvation model [33]. This "MIM-Implicit" model is then used to build the molecular tensors and to evaluate the spectral intensities incorporating implicit solvent effects.
8. Finally, we have developed the two-step-MIM procedure for evaluating the extrapolated Raman and ROA spectra of large molecules (*vide infra*). In this model, the geometry optimization, Hessian and the frequency evaluations are carried out using modest basis sets. Subsequently, the frequency-dependent polarizability derivative tensors for evaluating the chiroptical spectral intensities are evaluated using appropriate large basis sets.

MIM fragmentation schemes and all vibrational and chiroptical spectroscopic evaluations have been implemented through an external Perl script interface to Gaussian-09. More details on the evaluation of IR [26], Raman [27], VCD [28], and ROA [29] spectra with MIM2 have been given previously.

4.3 Results and Discussion

Carbohydrates and polypeptide molecules containing multiple chiral centers are attractive candidates for benchmarking the vibrational and chiroptical spectra. In order to showcase the accuracy and applicability of MIM vibrational spectroscopic methods, careful benchmark studies have been performed on a variety of calibration molecules. The agreement between MIM and actual (unfragmented) spectra are discussed for each of the methods. In addition, the performance of MIM method to evaluate the vibrational and chiroptical spectra of large systems such as α-D-cyclodextrin, cryptophane-A, and polypeptides is assessed by comparison with the available experimental spectra [34–38].

4.3.1 MIM Method for Geometry Optimization

A reliable geometry is an important prerequisite before evaluating accurate spectroscopic properties of large molecules. In particular, standard harmonic vibrational analysis is only applicable at the optimized geometry, rigorously maintaining the separation between rotations and vibrations. As discussed earlier, MIM provides an efficient way to evaluate the first derivatives (forces) and the second derivatives (Hessian) of the total energy, so that geometry optimization can be performed using any of the standard optimization methods. The accuracy of the optimized geometrical parameters has been compared with the actual full calculations for a range of test systems [26]. For a set of $(\text{Glycine})_5$ conformational isomers, the MIM2 geometrical parameters (x-axis) of benchmark systems have been compared with the actual optimized values (y-axis) in Figure 4.1. Figures 4.1a–f show the performance for bond lengths and bond- and dihedral angles using MIM1[B3LYP/6-311G(d,p)], MIM2[B3LYP/6-311G(d,p):HF/3-21G], and MIM2[B3LYP/6-311G(d,p):HF/6-311G(d,p)] levels of theory. In the notation used in MIM2, a colon (:) is used to separate the high and low levels of theory. The bond length correlation coefficient is ~0.99 in Figure 4.1 for all combinations (a, c, and e). However, the corresponding angular correlation coefficients show an improvement from 0.987 (intercept 1.560) to 0.997 (intercept 0.054), to 0.999 (y-intercept 0.000), respectively (b, d, and f). The principal source of error at the MIM1 level for bond and dihedral angles is due to the missing intramolecular weak interactions. The missing long-range interactions are taken into account at the second layer of theory in MIM2 model, leading to accurate geometrical parameters.

4.3.2 MIM Method for Evaluating IR Spectra (MIM-IR)

The calculation of the analytic harmonic force constant matrix (Hessian) and the dipole moment derivatives (with respect to Cartesian displacement coordinates) are the bottlenecks for vibrational IR spectra of large molecules. MIM-IR has been developed for the fast and efficient evaluation of the mass-weighted Hessian matrix to derive the harmonic vibrational frequencies and intensities [26]. The accuracy of the calculated force constants has been quantified by performing a one-to-one comparison of the Hessian matrix elements with those from the actual (unfragmented) calculation [26]. The contour distribution of the real and symmetric Hessian matrix elements of α- and β-$(\text{Glycine})_{10}$ are depicted as "heat diagrams" in Figures 4.2a and 4.2b, respectively. The

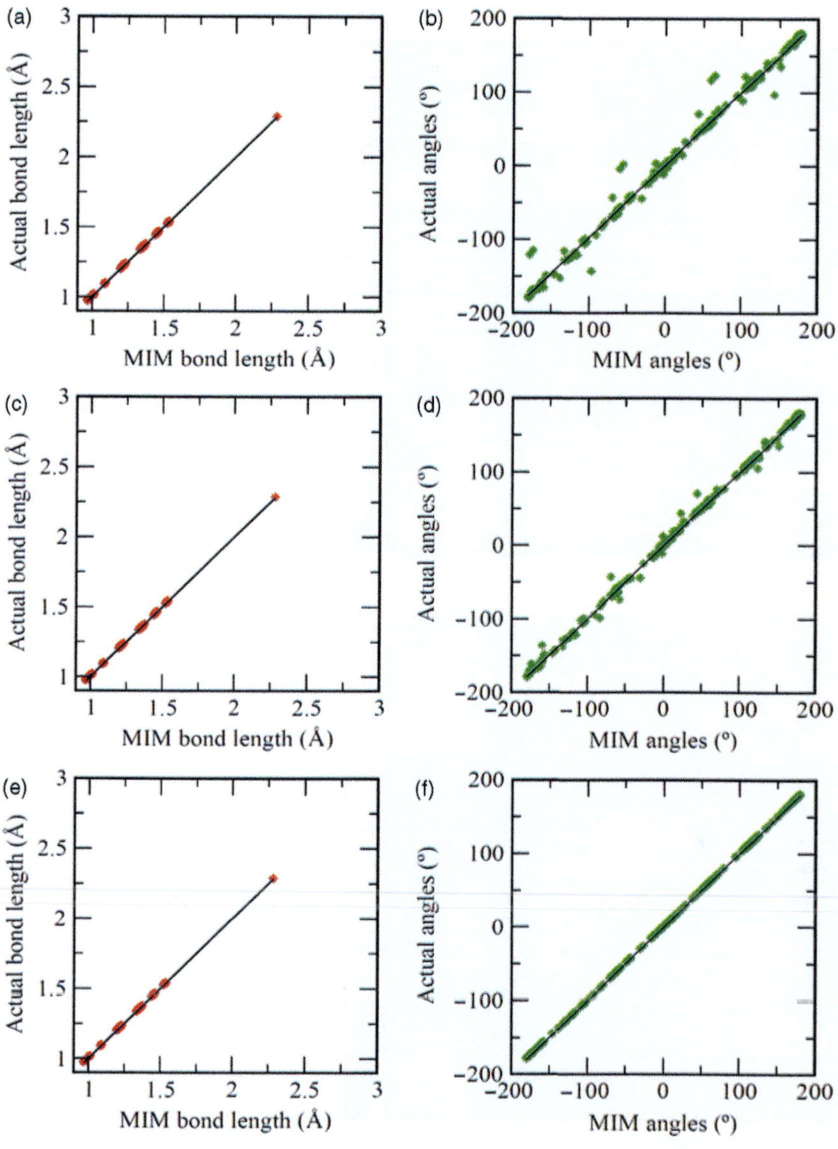

Figure 4.1 Comparison of the Actual[B3LYP/6-311G(d,p)] bond lengths (Å), bond angles (degrees), and dihedral angles (degrees) with (a, b) MIM1[B3LYP/6-311G(d,p)], (c, d) MIM2[B3LYP/6-311G(d,p):HF/3-21G], and (e, f) MIM2[B3LYP/6-311G(d,p):HF/6-311G(d,p)] levels of theory.

upper triangular matrices correspond to the actual force constant elements at B3LYP/6-311G(d,p) and the lower triangular matrices correspond to the corresponding values calculated at MIM2[B3LYP/6-311G(d,p):HF/3-21G]. Both α- and β-$(\text{Glycine})_{10}$ conformers have 82 atoms, and the Hessian matrix has a dimension 246×246. Each pixel in Figure 4.2 corresponds to an element of the full Hessian matrix. The highest valued Hessian matrix elements are the ones in red, and are distributed along the diagonal of this matrix.

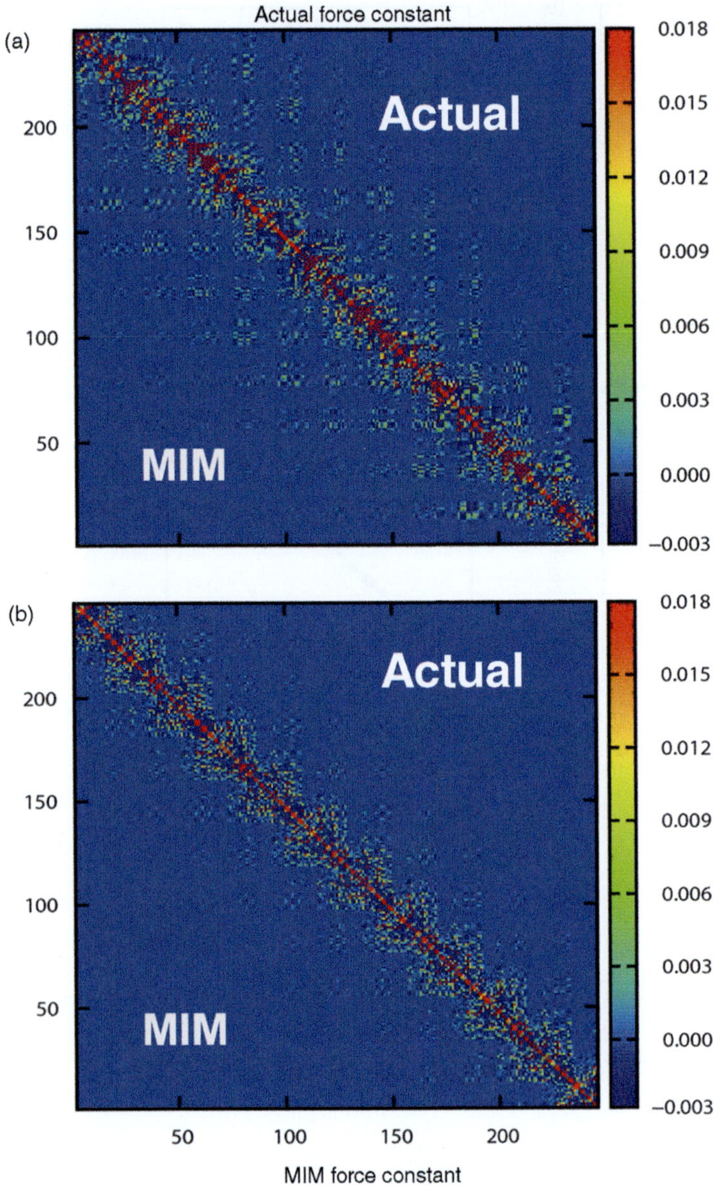

Figure 4.2 Heat diagram showing the distribution of the force constants of (a) α-(Glycine)$_{10}$ and (b) β-(Glycine)$_{10}$ at MIM2[B3LYP/6-311G(d,p):HF/3-21G] and Actual[B3LYP/6-311G(d,p)] levels of theory. In panels (a) and (b), the upper triangular matrix corresponds to the actual Hessian and the lower triangular matrix corresponds to the MIM Hessian. The Hessian matrix is 246×246 in dimension, and each pixel corresponds to one of the elements. The atomic numbering is according to the connectivity information from one end of the molecule to the other end.

As the interatomic distance increases, most of the Hessian elements are closer to zero and these matrix elements are depicted in progressive shades of blue. In the two-layer MIM model, some of the hydrogen bonding interactions are included in the first layer, and the rest of the long-range weak interactions are taken into account via the second low level of theory. The excellent agreement between the actual and MIM Hessian for both α- and β-(Glycine)$_{10}$ is only achieved at MIM2, since the second layer is important in capturing the long-range interactions and constructing accurate force constants (and vibrational frequencies). An associated important point is that this second layer correction at a low level of theory enables a reduction in the size of the primary overlapping subsystems. If the second layer is not included, large primary subsystems will be needed to incorporate all these long-range hydrogen bonding interactions, making the calculations very expensive, particularly for a helical system containing strong intramolecular interactions such as α-(Glycine)$_{10}$. However, for β-(Glycine)$_{10}$ conformer, off-diagonal elements in Figure 4.2b are relatively small in magnitude, compared to the dominant diagonal matrix elements. Hence, in such systems, with reasonably sized fragments, one layer of theory may suffice to give accurate structures and properties.

The Hessian matrix is real and symmetric, and can be diagonalized after mass-weighting to yield the eigenvalues and the associated harmonic vibrational frequencies. The corresponding eigenvectors yield the directions of the normal modes at the MIM2 level. The transformation of the dipole moment derivative matrices from the Cartesian to the normal mode framework subsequently yields the IR intensities. The calculated IR vibrational spectra of (Glycine)$_{10}$ and (Glycine)$_{15}$ at the MIM2[B3LYP/6-311G(d,p):PM6] level are depicted in Figure 4.3. For α-(Glycine)$_n$, the most intense peak in the spectrum corresponds to the amide I mode, and the second most intense band corresponds to the amide II mode. In case of β-(Glycine)$_n$, however, the most intense peak corresponds to the amide II band with a weaker amide I mode.

Vibrational frequencies are also important for the evaluation of the thermochemical properties of large molecules. Thus, the zero-point energy (ZPE) of a large molecule is another parameter to quantify the accuracy of the MIM fragmentation procedure. We have used the MIM2 frequencies to evaluate the zero-point energies (ZPE) of the (Glycine)$_{10}$ conformers using B3LYP/6-311G(d,p) as the primary level of theory, and PM6, HF/3-21G, and B3LYP/3-21G as the second low level of theory. On increasing the size of the primary subsystems from tripeptide to tetrapeptide for the (Glycine)$_{10}$ conformers, the accuracy of the calculated ZPE is improved substantially. The error in the zero-point energy is only ~0.5 kcal/mol using PM6 as the second layer, and less than 0.1 kcal/mol using HF/3-21G or B3LYP/3-21G. This accuracy suggests that the MIM2 model is a promising tool for predicting accurate thermochemistry of large molecules.

4.3.3 MIM Method for Evaluating Raman Spectra (MIM-Raman)

Raman and IR spectroscopic techniques are often complementary to each other and provide a wealth of structure-specific information. The Raman vibrational spectroscopic technique is widely used to characterize the structures, sizes, and shapes of large molecules. In the MIM-Raman method [27], the electric dipole–electric dipole polarizability derivative tensors from the individual subsystems are assembled in conjunction with the Jacobi link-atom treatment to yield the Raman intensities. MIM-Raman

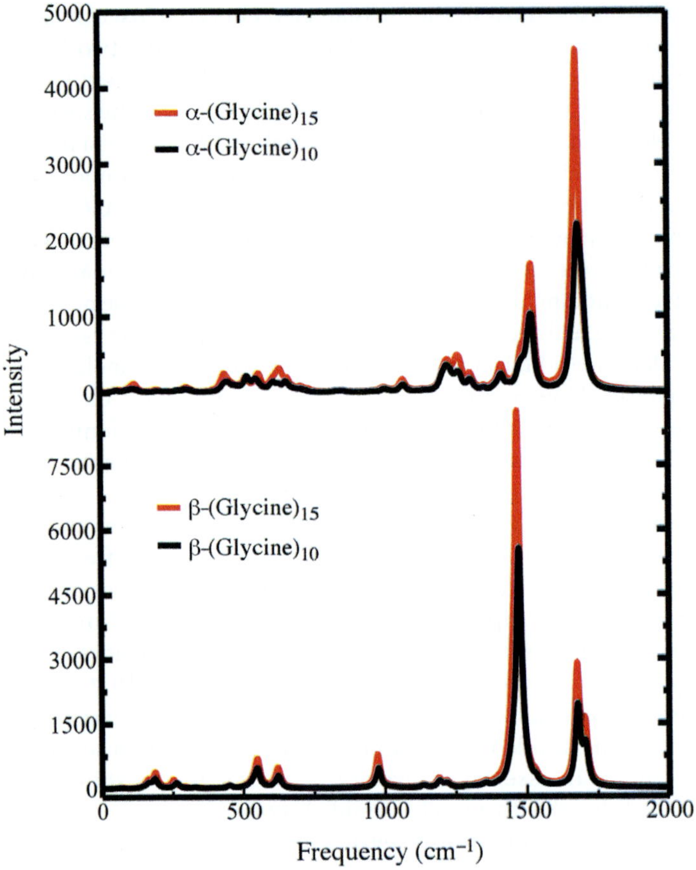

Figure 4.3 The intensity (in km/mol) vs frequency (in cm^{-1}) infrared spectrum of α-$(\text{Glycine})_{10;15}$ and β-$(\text{Glycine})_{10;15}$ evaluated at MIM2[B3LYP/6-311G(d,p):PM6].

method has been benchmarked on a set of inorganic, organic, or bio-oligomers containing boron, carbon, nitrogen, oxygen, fluorine, chlorine, and hydrogen atoms [27]. An initial benchmark study on 21 reference molecules shows 74% improvement in Raman frequencies and 41% in intensities, respectively, at MIM2 over MIM1 [27].

The MIM-Raman model has been employed for evaluating the Raman spectra of polystyrene [27]. Figure 4.4 depicts the most intense region in the actual and MIM vibrational Raman spectra of polystyrene evaluated at MIM2[B3LYP/6-311+G(d,p):HF/6-31G]. The MIM2 harmonic frequencies have been scaled uniformly by 0.967 in Figure 4.4, and compared with experiment. The calculated most intense Raman peak is at 982 cm^{-1}, and corresponds to the C–H bending mode. The MIM2 Raman spectral features in Figure 4.4b show excellent corroboration with the experimental spectrum in Figure 4.4a even though only a small oligomer (tetramer) has been used to represent polystyrene in these calculations [27]. This suggests that the positions of the vibrational modes in the Raman spectra converge very quickly with the size of the oligomers in polystyrene. A similar behavior has also been seen for the vibrational modes in the

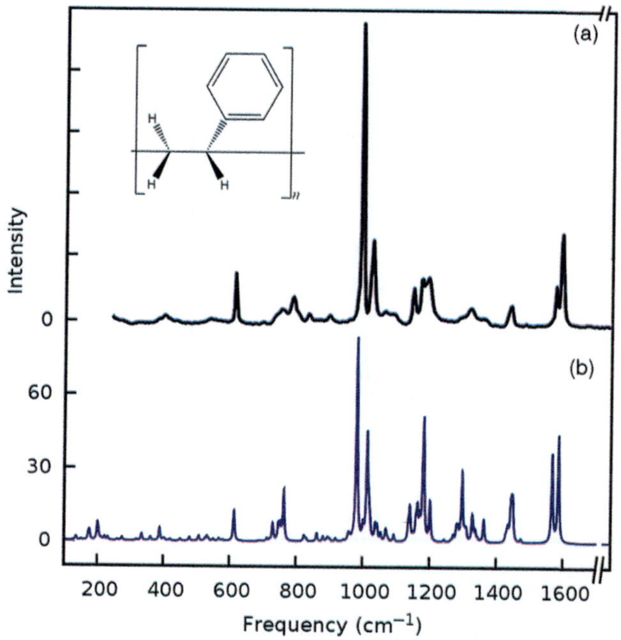

Figure 4.4 Comparison of experimental (black) with MIM-Raman (blue) spectra of polystyrene. The depicted frequencies are in cm^{-1}, experimental intensities are in arbitrary units and MIM-Raman intensities in $Å^4$/amu.

VCD spectra for poly(alanine)$_n$ converging quickly as the length of the peptide chain increases [30].

4.3.4 MIM Method for Evaluating VCD Spectra (MIM-VCD)

VCD spectroscopy is the chiroptical analogue of IR, and requires the evaluation of both electric- and magnetic-dipole transition moments. In the MIM-VCD implementation [28], the electronic contribution to the electric dipole transition moments (atomic polar tensors) and the corresponding magnetic terms (atomic axial tensors) for the large molecule are constructed from the subsystem calculations including the Jacobi link-atom projections. They are then transformed to the normal coordinate representation to yield the associated rotational strengths and the VCD intensities. Gauge-invariant atomic orbitals are used to ensure that the calculated spectra are origin-independent. The need to calculate the rotational strengths along with the Hessian matrix makes the accurate evaluation of *ab initio* VCD spectra for large molecules to be computationally expensive, particularly in conjunction with large basis sets. The MIM-VCD spectra have been compared with the actual (unfragmented) spectra for 10 carbohydrate enantiomeric pairs, and shown to be nearly superimposable, demonstrating the accuracy of the implementation [28]. In addition, MIM-VCD spectra have been determined for two larger molecules, *viz.* the (S)-enantiomer of D_3-anti-trans-anti-trans-anti-trans-perhydrotriphenylene (S-PHTP) and cryptophane-A, depicted in Figure 4.5 inset, and Figure 4.6b, respectively.

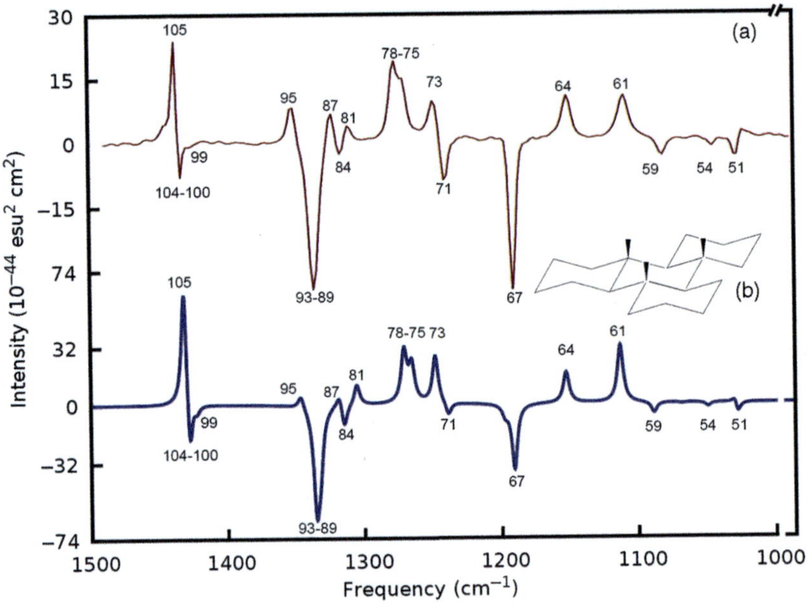

Figure 4.5 Comparison of (a) experimental (in brown) with (b) MIM2[MPW1PW91/aug-cc-pVTZ-f:MPW1PW91/6-31+G(d)] VCD (in blue) spectra of S-(+)-D3-anti-trans-anti-trans-anti-trans-perhydrotriphenylene.

Figure 4.5a depicts the experimental spectrum of S-PHTP that has been investigated by Stephens *et al.* [35]. Previous conformational analysis of perhydrotriphenylene has shown that all four cyclohexane rings in this molecule preferentially adopt a chair form over the other conformations. Hence, this stereoisomer is the most stable conformer, and is effectively a rigid molecule on the potential energy surface. This makes a theory-to-experiment comparison relatively straightforward as we can take into account only

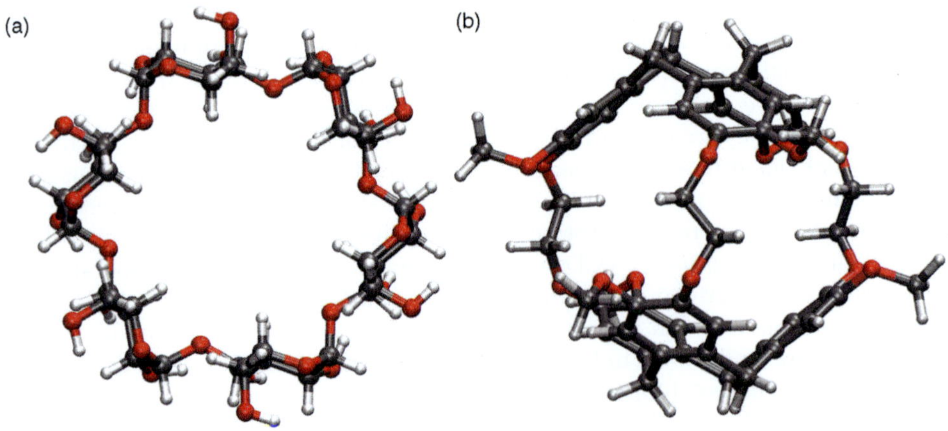

Figure 4.6 The optimized geometries of (a) α-cyclodextrin at MIM2[B3LYP/6-311++G(d,p):HF/3-21G] and (b) $T_1T_1T_1$ isomer of cryptophane-A at MIM2[MPW1PW91/6-31+G(d,p):MPW1PW91/6-31G].

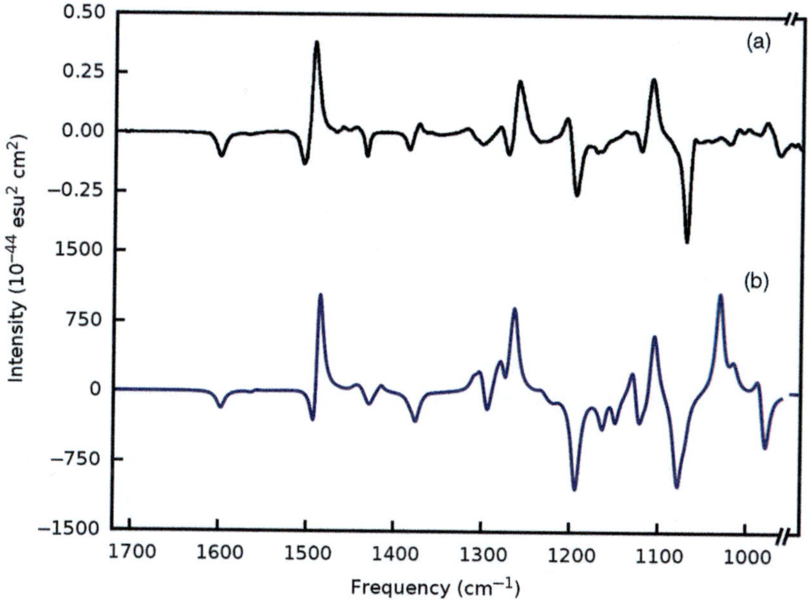

Figure 4.7 Comparison of (a) experimental with (b) MIM2[MPW1PW91/6-31+G(d,p):MPW1PW91/6-31G] VCD spectra of (+)-$T_1T_1T_1$ stereoisomer of cryptophane-A.

this most stable conformer. The MIM-VCD spectrum at [MIM2[MPW1PW91/aug-cc-pVTZ-f:MPW1PW91/6-31+G(d)] level of theory (Figure 4.5b) has been compared with the experimental spectrum in Figure 4.5a [28]. The calculated vibrational normal modes are scaled by a uniform factor of 0.971 to account for the systematic deviations, and the VCD spectrum is depicted with a Lorentzian line shape using a width of 3.5 cm^{-1} in Figure 4.5b. It is clear from Figure 4.5 that the calculated VCD spectrum of S-PHTP shows excellent agreement with experiment. The calculated most intense band at 1354 cm^{-1} [−57.5 esu^2 cm^2] corresponds to H—C—H wagging mode, and the second most intense normal mode at 1450 cm^{-1} [57.3 esu^2 cm^2] corresponds to the H—C—H scissoring mode. From the benchmark studies, it is clear that MIM-VCD calculations are appropriate to make peak assignments for large molecular systems such as perhydrotriphenylene.

As an additional benchmark system, MIM-VCD spectra have been evaluated for the stereoisomers of cryptophane-A, a large cage molecule with two cyclotriveratrylene bowls connected by three aliphatic linkers. Cryptophane molecules are among the first examples of synthetic "container molecules" that have been experimentally investigated [34, 36, 37], and there has been a significant interest in studying the VCD spectrum of cryptophane-A both experimentally and theoretically. Optically pure cryptophane-A isomers have been recently synthesized through an optical resolution method [38], and have been used for the experimental measurement of the VCD spectra in $CDCl_3$ solvent. It has previously been confirmed that the solvent molecules do not alter the VCD spectrum of the parent $T_1T_1T_1$ stereoisomer with D_3 symmetry [34]. The corresponding MIM-VCD spectra have been evaluated in the gas phase and compared with these experimental spectra directly. Figure 4.7 shows the experimental spectrum for cryptophane-A (top) with the MIM2 spectrum (bottom) using frequencies scaled by 0.952 and fitted with a Lorentzian function of half-width of 7.5 cm^{-1}. The MIM-VCD spectra for the D_3 stereoisomer in the 1000–1700 cm^{-1} region has been analyzed [34], and show a very good agreement with experiment. The overall performance on the set of calibration systems [28] suggests that MIM-VCD can be a powerful tool to assist in the assignment of VCD spectra for a variety of large molecules.

4.3.5 MIM Method for Evaluating ROA Spectra (MIM-ROA)

The ROA spectral intensities are a function of ROA-invariants [9] and the invariants are constructed from the higher order polarizability derivatives (*vide supra*). The calculation of ROA intensities requires the electric dipole–electric dipole, electric dipole–magnetic dipole, and electric dipole–electric quadrupole polarizability derivatives along the normal modes. As discussed earlier, the Cartesian components of the three molecular tensors are built from the corresponding subsystem tensors, and then used to build the ROA intensities. As in the case of MIM-VCD, gauge invariant atomic orbitals have been used to ensure that the resulting MIM-ROA intensities are origin-independent. The MIM-ROA method has been benchmarked on a set of 10 carbohydrate molecules in their cyclic forms at MIM2[MPW1PW91/aug-cc-pVDZ:MPW1PW91/6-31G(d)] level compared to the full calculations [29]. The Raman and ROA spectra can vary according to the polarity and angle of the incident and scattered electromagnetic radiation. The MAD at MIM2 shows ~71.6% improvement over MIM1 in frequencies and 48.2, 48.1 and 57.1% improvement in ROA intensities for scattering angle geometries of 0°, 90°, and 180°, respectively [29].

To validate the applicability of MIM-ROA in predicting experimental spectra, explicit comparison of the calculated MIM-ROA spectra have been compared with experiment for two larger test systems, α-D-cyclodextrin and cryptophane-A, depicted in Figure 4.6. α-D-cyclodextrins are cyclic oligosaccharides consisting of six D-glucopyranose residues, interconnected with glycosidic bonds [39]. The stereoisomer shown in Figure 4.6a is the most stable conformer, and MIM-ROA calculations have been carried out on this stereoisomer, enabling a careful theory-to-experiment comparison.

A comparison of the backscattered ROA experimental spectrum of α-D-cyclodextrin [40] with the calculated MIM-ROA spectrum at the MIM2[B3LYP/6-311++G(d,p):HF/3-21G] level between 400 and 1600 $cm^{-1}$is depicted in Figures 4.8a and 4.8b, respectively [29]. The frequency-dependent polarizability derivatives have been evaluated at an external frequency of 488.0 nm, same as that used in the backscattered experiment [40]. Each normal mode is scaled by 0.952 and fitted with a Lorentzian function of width 14.5 cm^{-1} (Figure 4.8).

As seen in Figure 4.8, the most intense peak in the experimental ROA spectrum is a couplet at 922 cm^{-1}. This is in agreement with a strong calculated couplet involving a motion of the glycosidic bond along with some contribution from C–O–H deformation. While some deviations are seen between theory and experiment, it should be noted that the backscattered experimental spectra for α-cyclodextrin have been measured in water. Despite the fact that hydrogen bonding interactions with water are missing in our calculations, the overall calculated ROA spectrum shows a very reasonable agreement with experiment.

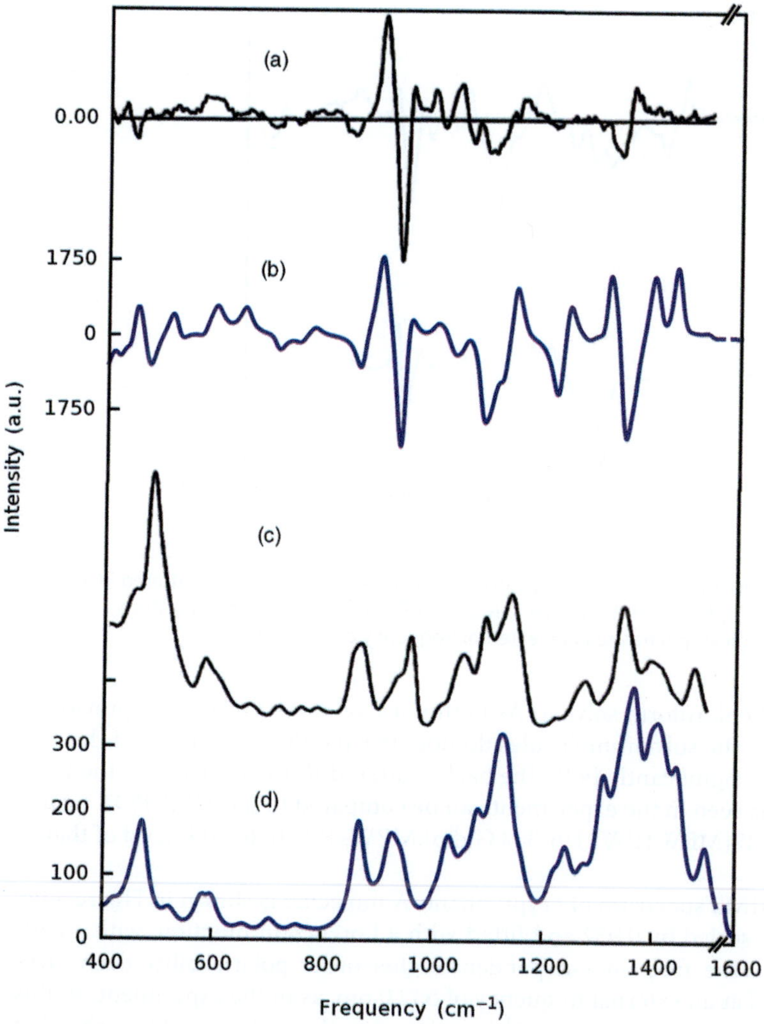

Figure 4.8 Comparison of α-cyclodextrin (a) experimental backscattered ROA spectrum (in black) with (b) MIM2[B3LYP/6-311++G(d,p):HF/3-21G] (in blue), and (c) experimental backscattered Raman spectrum (in black) with (d) MIM2[B3LYP/6-311++G(d,p):HF/3-21G] (in blue). Both experimental and theoretical spectra are at an external frequency of 488.0 nm.

The second benchmark molecule considered is the synthetic "container molecule", cryptophane-A whose ROA spectrum has also been measured experimentally [41]. A previous work has shown that the high symmetry D_3 isomer with a dihedral angle close to 180° (denoted as $T_1T_1T_1$) provides a very good agreement with experimental ROA. Thus, as in the case of MIM-VCD discussed earlier, the MIM-ROA spectrum has been determined only for this stereoisomer.

The most important factors in deciding whether a calculated ROA spectrum agrees well with experiment are relative peak location and magnitude, and the sign of each intensity. The gas phase theoretical spectra have been compared with the experiments

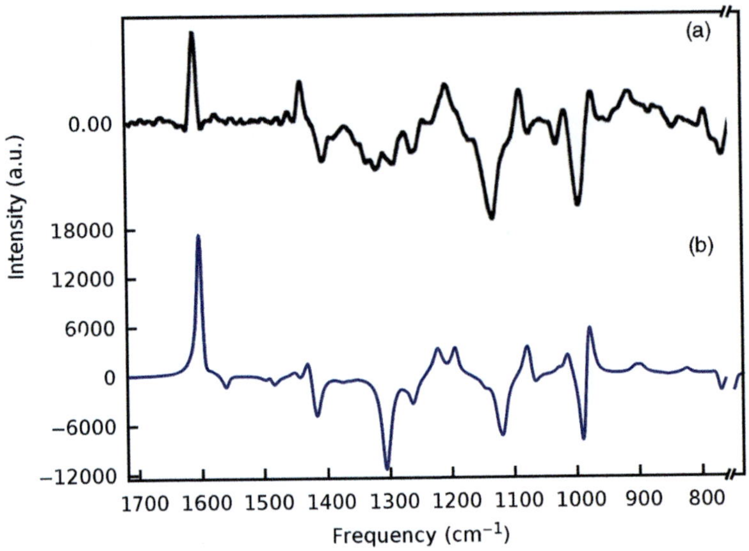

Figure 4.9 A comparison of $T_1T_1T_1$ isomer of cryptophane-A (a) experimental backscattered ROA spectrum (in black) with (b) MIM2[MPW1PW91/6-31+G(d,p):MPW1PW91/6-31G] (in blue). Both experimental and theoretical spectra are at an external frequency of 532.0 nm.

conducted in 0.1 M chloroform solvent. As in the case of MIM-VCD, it has previously been confirmed that the solvent molecules do not alter the ROA spectrum of the parent $T_1T_1T_1$ isomers significantly [42]. The back-scattered ROA spectrum in the 750–1700 cm^{-1} region as seen in the experiment can be compared to the MIM-ROA results obtained using MIM2[MPW1PW91/6-31+G(d,p):MPW1PW91/6-31G] level of theory (Figure 4.9) [29].

The gas-phase MIM2 spectrum of cryptophane-A molecule is shown in Figure 4.9b, the frequencies are scaled by 0.952 and fitted with a Lorentzian function, with a half-width of 7.5 cm^{-1}. The frequency-dependent higher-order polarizability derivatives have been evaluated at an external frequency of 532.0 nm, as in the experiment. In this spectrum, the normal mode at 1607 cm^{-1} is assigned to the C=C stretching vibration of the benzene rings [29]. The wagging and twisting of the CH_2 groups in the C–O–C linkers and bowls give rise to the three bands between 1350 and 1250 cm^{-1}, with the same sign. The region below 1200 cm^{-1} is more complex because the observed bands correspond to coupled modes involving several vibrations. Overall, the MIM2-ROA spectrum shows a good agreement with experiment for the $T_1T_1T_1$ stereoisomer.This suggests that MIM-ROA can be a powerful tool to assign ROA spectra for other large molecules.

4.3.6 Two-Step-MIM Scheme for Evaluating Raman and ROA Spectra

It is known that large basis sets containing polarization functions are needed to calculate accurate Raman and ROA spectral intensities.To minimize the associated computational expense, Cheeseman and Frisch have previously proposed a two-step model for evaluating Raman and ROA spectra of large molecules [43]. As per this model, a modest basis set is sufficient to derive the geometric parameters and the harmonic vibrational frequencies. However, a large basis set is essential to reproduce the ROA spectral intensities. Hence, in the two-step-MIM model, the geometry optimization, Hessian, and the frequency evaluations are carried out using modest basis sets. Subsequently, the frequency-dependent polarizability derivative tensors are evaluated using appropriate large basis sets and transformed to the normal mode representation from the original small basis set to derive the associated ROA spectral intensities.

This two-step-extrapolation method has also been adapted for evaluating MIM-Raman and MIM-ROA spectra of α-(alanine)$_{20}$ polypeptide as an illustrative application. In the MIM calculations, the optimization, Hessian, and frequency evaluations have been performed at a low level of theory, MIM2[MPW1PW91/6-31G:HF/3-21G] including IEFPCM continuum solvation model. In the second step of the two-step-MIM method, at the optimized geometry, the frequency-dependent polarizability derivatives are evaluated using larger basis sets at MIM2[MPW1PW91/6-311G(d,p):MPW1PW91/3-21G], again including the IEFPCM implicit solvation model. The calculations are performed under the same conditions used in the experiments: at an external frequency 514.5 nm, with dichloroacetic acid ($\varepsilon = 8.2$) IEFPCM solvation. This extrapolation step is performed separately on each fragment and the final higher-order polarizability tensors are then constructed to derive the chiroptical spectra.

From previous studies on (alanine)$_{20}$ polypeptide, α-helix is known to be a stable isomer in the solution phase [44]. The experimental spectrum for α(alanine)$_{20}$, shown in Figure 4.10a, and the MIM-ROA spectrum calculated at MIM2[MPW1PW91/6-311G(d,p):MPW1PW91/3-21G] level, shown in Figure 4.10b, are compared in the wavelength range 1100–1700 cm^{-1}. The MIM-ROA spectra could reproduce most of the intense experimental 1250 and 1290 cm^{-1} peaks in the amide III region, the couplet around 1460 cm^{-1} in the amide II region, and 1601 cm^{-1} amide I region peaks. However, the two most intense positive peaks observed experimentally around 1304 and 1340 are reproduced as a shoulder at 1300 and a strong peak at 1330 cm^{-1} in the MIM-ROA calculations. Explicit solvent interactions appear to be needed to reproduce a clear splitting between the modes [44]. Overall, this comparison shows that the incorporation of implicit solvation improves the characteristic intensities and agreement of the spectral peaks with experiment.

4.4 Summary

The MIM method has the following capabilities: (a) rapid evaluation of the Hessian matrix that yields the harmonic vibrational frequencies; (b) the evaluation of the electric dipole derivative matrix for the evaluation of IR spectral intensities; (c) electric and magnetic dipole derivative matrices for VCD intensities; (d) higher-order polarizabilities and polarizability derivative tensors for the calculation of Raman and ROA intensities; (e) the two-step MIM method for overcoming the bottlenecks for the evaluation of higher-order polarizability derivatives when used in conjunction with large basis sets; and (f) the incorporation of continuum solvation effects in the fragment calculations. Overall, the implementation of the MIM algorithms are general and will work with any

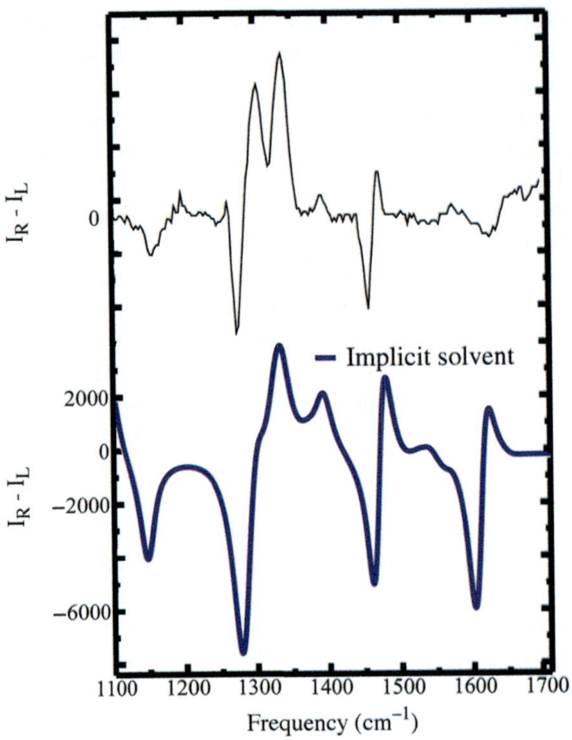

Figure 4.10 Comparison of (a) experimental and (b) MIM2[MPW1PW91/6-311G(d,p):MPW1PW91/6-31G] extrapolated ROA spectra of α-(Alanine)$_{20}$ with IEFPCM implicit solvation.

arbitrary fragmentation scheme at the primary level and any combination of theoretical methods for which the necessary derivatives are available.

The major bottleneck in the MIM-vibrational spectra calculations are the evaluation of the higher energy derivative tensors. The MIM method speed up the higher energy derivative calculations including the Hessian matrix, the dipole derivative matrices, and the polarizability derivative tensors. The calculation of higher energy derivatives on the overlapping primary subsystems at the first layer (i.e., high level of theory) is usually the rate-limiting step in the MIM calculations. As the size of the molecule increases, the number of primary subsystems increases linearly, while the sizes of the subsystems are relatively independent of the size of the molecule. Hence, the high level of theory scales asymptotically linearly with the size of the molecule. For normal systems, the time consumption at the second layer using a low-level of theory is usually small. However, as the molecule gets very large, the second layer may become rate-determining if the entire molecule is treated at this level. Three-layer models are being developed to make such calculations possible for large molecules containing a thousand atoms or more.

4.5 Conclusions

Direct evaluation of vibrational and chiroptical properties of large molecules using an accurate level of theory with large basis sets can be computationally expensive or prohibitive. The bottlenecks in such vibrational spectroscopic evaluations of large molecules can be overcome with MIM fragment-based methods. Efficient algorithms have been implemented for the evaluation of the gradients as well as the higher order energy derivatives to yield MIM-IR, MIM-Raman, MIM-VCD, and MIM-ROA methods. The incorporation of the Jacobian link atom projection method as well as the employment of multiple layers of theory to account for long-range weak interactions are important factors that contribute to the accuracy of the methods. The methods are linear-scaling and highly parallelizable, and have been benchmarked on a variety of systems to yield accurate results close to that from actual (unfragmented) calculations. In addition, MIM-chiroptical spectra on large molecules such as α-cyclodextrin and cryptophane-A show excellent correspondence with experiment. Overall, the MIM fragment-based approach opens up a wide range of systems to be studied through accurate *ab initio* methods.

Acknowledgments

This work was supported by funding from NSF Grant No. CHE-1266154 at Indiana University. The authors also thank the Indiana University Big Red II supercomputing facility for computing time.

References

1 Helgaker, T.; Coriani, S.; Jørgensen, P.; Kristensen, K.; Olsen, J.; Ruud, K. Recent advances in wave function-based methods of molecular-property calculations. *Chem. Rev.* **2012**, *112*, 543–631.

2 Hug, W., *Handbook of Vibrational Spectroscopy*, Chalmers J. M. and Griffiths P., Eds. John Wiley & Sons, New York, 2002, Vol. 1, pp. 745–758.

3 Nafie, L. A. *Vibrational Optical Activity Principles and Applications*. John Wiley & Sons: Chichester, West Sussex, 2011.

4 Stephens, P. J. *VCD Spectroscopy for Organic Chemists*. Devlin, F. J., Cheeseman, J. R., Eds., CRC Press, Boca Raton, FL, 2012.

5 Ruud, K. Ab initio methods for vibrational circular dichroism and Raman optical activity. In *Comprehensive Chiroptical Spectroscopy*, John Wiley & Sons, Inc., 2012; Vol. 1, pp. 699–727.

6 Barron, L. D. *Molecular Light Scattering and Optical Activity*. 2nd ed., Cambridge University Press: Cambridge, UK, 2004.

7 Fasman, G. Ed.; *Circular Dichroism and the Conformational Analysis of Biomolecules*. Plenum Press, New York, 1996.

8 Nafie, L. A. Infrared and Raman vibrational optical activity: Theoretical and experimental aspects. *Annu. Rev. Phys. Chem.* **1997**, *48*, 357–386.

9 Polavarapu, P. L. Ab initio vibrational Raman and Raman optical activity spectra. *J. Phys. Chem.* **1990**, *94*, 8106–8112.

10 Magyarfalvi, G.; Tarczay, G.; Vass, E. Vibrational circular dichroism. *Wiley Interdiscip. Rev. Comput. Mol. Sci.* **2011**, *1*, 403–425.

11 Stephens, P. J.; Lowe, M. A. Vibrational circular dichroism. *Annu. Rev. Phys. Chem.* **1985**, *36*, 213–241.

12 Polavarapu, P. L. Renaissance in chiroptical spectroscopic methods for molecular structure determination. *Chem. Rec.* **2007**, *7*, 125–136.

13 Bose, P. K.; Barron, L. D.; Polavarapu, P. L. Ab initio and experimental vibrational Raman optical activity in (+)-(R)-methylthiirane. *Chem. Phys. Lett.* **1989**, *155*, 423–429.

14 Pecul, M. New applications and challenges for computational ROA spectroscopy. *Chirality* **2009**, *21*, E98–E104.

15 Herrmann, C.; Reiher, M. First-principles approach to vibrational spectroscopy of biomolecules. In *Atomistic Approaches in Modern Biology*, Reiher, M., Ed., Springer, Berlin Heidelberg, 2007, Vol. 268, pp. 85–132.

16 Raghavachari, K.; Saha, A. Accurate composite and fragment-based quantum chemical models for large molecules. *Chem. Rev.* **2015**, *115*, 5643–5677.

17 (a) Chung, L. W.; Sameera, W. M. C.; Ramozzi, R.; Page, A. J.; Hatanaka, M.; Petrova, G. P.; Harris, T. V.; Li, X.; Ke, Z.; Liu, F.; Li, H.; Ding, L.; Morokuma, K. The ONIOM method and its applications. *Chem. Rev.* **2015**, *115*, 5678–5796. (b) Chung, L. W.; Hirao, H.; Li, X.; Morokuma, K. The ONIOM method: Its foundation and applications to metalloenzymes and photobiology. *WIREs: Comput. Mol. Sci.* **2012**, *2*, 327–350. (c) Collins, M. A.; Bettens, R. P. A. Energy-based molecular fragmentation methods. *Chem. Rev.* **2015**, *115*, 5607–5642. (d) Collins, M. A.; Cvitkovic, M. W.; Bettens, R. P. A. The combined fragmentation and systematic molecular fragmentation methods. *Acc. Chem. Res.* **2014**, *47*, 2776–2785. (e) Sahu, N.; Gadre, S. R. Molecular tailoring approach: A route for ab initio treatment of large clusters. *Acc. Chem. Res.* **2014**, *47*, 2739-2747. (f) Řezáč, J.; Salahub, D. R. Multilevel fragment-based approach (MFBA): A novel hybrid computational method for the study of large molecules. *J. Chem. Theory Comput.* **2009**, *6*, 91–99. (g) Vreven, T.; Byun, K. S.; Komáromi, I.; Dapprich, S.; Montgomery, J. A.; Morokuma, K.; Frisch, M. J. Combining quantum mechanics methods with molecular mechanics methods in ONIOM. *J. Chem. Theory Comput.* **2006**, *2*, 815–826. (h) Fedorov, D. G.; Ishida, T.; Kitaura, K. Multilayer formulation of the fragment molecular orbital method (FMO). *J. Phys. Chem. A* **2005**, *109*, 2638–2646. (i) Isegawa, M.; Wang, B.; Truhlar, D. G. Electrostatically embedded molecular tailoring approach and validation for peptides. *J. Chem. Theory Comput.* **2013**, *9*, 1381–1393. (j) Beran, G. J. O. Modeling polymorphic molecular crystals with electronic structure theory. *Chem. Rev.* **2016**, *116*, 5567–5613. (k) He, X.; Merz, K. M. Divide and conquer Hartree–Fock calculations on proteins. *J. Chem. Theory Comput.* **2010**, *6*, 405–411. (l) Nagata, T.; Fedorov, D. G.; Sawada, T.; Kitaura, K.; Gordon, M. S. A combined effective fragment potential–fragment molecular orbital method. II. Analytic gradient and application to the geometry optimization of solvated tetraglycine and chignolin. *J. Chem. Phys.* **2011**, *134*, 034110. (m) Mullin, J. M.; Roskop, L. B.; Pruitt, S. R.; Collins, M. A.; Gordon, M. S. Systematic fragmentation method and the effective fragment potential: An efficient method for capturing molecular energies. *J. Phys. Chem. A* **2009**, *113*, 10040–10049. (n) Guo, W.; Wu, A.; Zhang, I. Y.; Xu, X. XO: An extended ONIOM method for accurate and efficient modeling of large systems. *J. Comput. Chem.* **2012**, *33*, 2142–2160. (o) Pruitt, S. R.; Bertoni, C.; Kurt R. Brorsen, K. R.; Gordon, M. S. Efficient and accurate fragmentation methods. *Acc. Chem. Res.* **2014**, *47*, 2786–2794. (p) Li, S.; Li, W.; Ma, J. Generalized energy-based fragmentation approach and its applications to macromolecules and molecular aggregates. *Acc. Chem. Res.* **2014**, *47*, 2712–2720. (q) Merz, Jr.; K. M. Using Quantum mechanical approaches to study

biological systems. *Acc. Chem. Res.* **2014**, *47*, 2804–2811. (r) He, X.; Zhu, T.; Wang, X; Liu, J.; Zhang, J. Z. H. Fragment quantum mechanical calculation of proteins and its applications. *Acc. Chem. Res.* **2014**, *47*, 2748–2757. (s) Richard, R. M.; Lao, K. U.; Herbert, J. M. Aiming for benchmark accuracy with the many-body expansion. *Acc. Chem. Res.* **2014**, *47*, 2828–2836.

18 Gordon, M. S.; Fedorov, D. G.; Pruitt, S. R.; Slipchenko, L. V. Fragmentation methods: A route to accurate calculations on large systems. *Chem. Rev.* **2011**, *112*, 632–672.

19 Buckingham, A. D. Introductory lecture. The theoretical background to vibrational optical activity. *Farad. Discuss.* **1994**, *99*, 1–12.

20 Thorvaldsen, A. J.; Gao, B.; Ruud, K.; Fedorovsky, M.; Zuber, G.; Hug, W. Efficient calculation of ROA tensors with analytical gradients and fragmentation. *Chirality* **2012**, *24*, 1018–1030.

21 Helgaker, T.; Ruud, K.; Bak, K. L.; Jørgensen, P.; Olsen, J. Vibrational Raman optical activity calculations using London atomic orbitals. *Farad. Discuss.* **1994**, *99*, 165–180.

22 Zuber, G.; Hug, W. Rarefied basis sets for the calculation of optical tensors. 1. The importance of gradients on hydrogen atoms for the Raman scattering tensor. *J. Phys. Chem. A* **2004**, *108*, 2108.

23 Reiher, M.; Liégeois, V.; Ruud, K. Basis set and density functional dependence of vibrational Raman optical activity calculations. *J. Phys. Chem. A* **2005**, *109*, 7567–7574.

24 (a) Nakata, H.; Fedorov, D. G.; Yokojima, S.; Kitaura, K.; Nakamura, S. Simulations of Raman spectra using the fragment molecular orbital Method. *J. Chem. Theory Comput.* **2014**, *10*, 3689–3698. (b) Nakata, H.; Fedorov, D. G.; Yokojima, S.; Kitaura, K.; Nakamura, S. Efficient vibrational analysis for unrestricted Hartree–Fock based on the fragment molecular orbital method. *Chem. Phys. Lett.* **2014**, *603*, 67–74. (c) Nakata, H.; Nagata, T.; Fedorov, D. G.; Yokojima, S.; Kitaura, K.; Nakamura, S. Analytic second derivatives of the energy in the fragment molecular orbital method. *J. Chem. Phys.* **2013**, *138*, 164103.

25 Mayhall, N. J.; Raghavachari, K. Molecules-in-molecules: An extrapolated fragment-based approach for accurate calculations on large molecules and materials. *J. Chem. Theory Comput.* **2011**, *7*, 1336–1343.

26 Jovan Jose, K. V.; Raghavachari, K. Evaluation of energy gradients and infrared vibrational spectra through molecules-in-molecules fragment-based approach. *J. Chem. Theory Comput.* **2015**, *11*, 950–961.

27 Jovan Jose, K. V.; Raghavachari, K. Molecules-in-molecules fragment-based method for the evaluation of Raman spectra of large molecules. *Mol. Phys.* **2015**, *113*, 3057–3066.

28 Jovan Jose, K. V.; Beckett, D.; Raghavachari, K. Vibrational circular dichroism spectra for large molecules through molecules-in-molecules fragment-based approach. *J. Chem. Theory Comput.* **2015**, *11*, 4238–4247.

29 Jovan Jose, K. V.; Raghavachari, K. Raman optical activity spectra for large molecules through molecules-in-molecules fragment-based approach. *J. Chem. Theory Comput.* **2016**, *12*, 585–594.

30 Jovan Jose, K. V.; Raghavachari, K. Molecules-in-molecules fragment-based method for the calculation of chiroptical spectra of large molecules: Vibrational circular dichroism and Raman optical activity spectra of alanine polypeptides. *Chirality* **2016**, *28*, 755–768.

31 Frisch, M. J.; Trucks, G. W.; Schlegel, H. B.; Scuseria, G. E.; Robb, M. A.; Cheeseman, J. R.; Scalmani, G.; Barone, V.; Mennucci, B.; Petersson, G. A.; Nakatsuji, H.; Caricato, M.; Li, X.; Hratchian, H. P.; Izmaylov, A. F.; Bloino, J.; Zheng, G.; Sonnenberg, J. L.; Hada, M.; Ehara, M.; Toyota, K.; Fukuda, R.; Hasegawa, J.; Ishida, M.; Naka-jima, T.; Honda, Y.; Kitao, O.; Nakai, H.; Vreven, T.; Montgomery, J. A.; Peralta, J. E.; Ogliaro, F.; Bearpark, M.; Heyd, J. J.; Brothers, E.; Kudin, K. N.; Staroverov, V. N.; Kobayashi, R.; Normand, J.; Raghavachari, K.; Rendell, A.; Burant, J. C.; Iyengar, S. S.; Tomasi, J.; Cossi, M.; Rega, N.; Millam, J. M.; Klene, M.; Knox, J. E.; Cross, J. B.; Bakken, V.; Adamo, C.; Jaramillo, J.; Gomperts, R.; Stratmann, R. E.; Yazyev, O.; Austin, A. J.; Cammi, R.; Pomelli, C.; Ochterski, J. W.; Martin, R. L.; Morokuma, K.; Zakrzewski, V. G.; Voth, G. A.; Salvador, P.; Dannenberg, J. J.; Dapprich, S.; Daniels, A. D.; Farkas, Ã.; Foresman, J. B.; Ortiz, J. V.; Cioslowski, J.; Fox, D. J. *Gaussian 09*, Revision D.01, Gaussian, Inc., Wallingford, CT, 2016.

32 Dapprich, S.; Komáromi, I.; Byun, K. S.; Morokuma, K.; Frisch, M. J. A new ONIOM implementation in Gaussian98. Part I. The calculation of energies, gradients, vibrational frequencies and electric field derivatives. *J. Mol. Struct.: Theochem.* **1999**, *461–462*, 1–21.

33 Tomasi, J.; Mennucci, B; Cammi, R. Quantum mechanical continuum solvation models. *Chem. Rev.* **2005**, *105*, 2999–3093. Scalmani, G.; Frisch, M. J. Continuous surface charge polarizable continuum models of solvation. I. General formalism. *J. Chem. Phys.* **2010**, *132*, 114110.

34 Brotin, T.; Cavagnat, D.; Dutasta, J.-P.; Buffeteau, T. Vibrational circular dichroism study of optically pure cryptophane-A. *J. Am. Chem. Soc.* **2006**, *128*, 5533–5540.

35 Stephens, P. J.; Devlin, F. J.; Schürch, S.; Hulliger, J. Determination of the absolute configuration of chiral molecules via density functional theory calculations of vibrational circular dichroism and optical rotation: The chiral alkane D3-anti-trans-anti-trans-anti-trans-perhydrotriphenylene. *Theor. Chem. Acc.* **2008**, *119*, 19–28.

36 Brotin, T.; Dutasta, J.-P. Cryptophanes and their complexes – present and future. *Chem. Rev.* **2009**. *109*, 88–130.

37 Gabard, J.; Collet, J. Synthesis of a (D3)-bis(cyclotriveratrylenyl) macrocage by stereospecific replication of a (C3)-subunit. *Chem. Soc., Chem. Comm.* **1981**, 1137–1139.

38 Brotin, T.; Barbe, R.; Darzac, M.; Dutasta, J.-P. Novel Synthetic approach for optical resolution of cryptophanol-A: A direct access to chiral cryptophanes and their chiroptical properties. *Chem. Eur. J.* **2003**, *9*, 5784–5792.

39 Connors, K. A. The stability of cyclodextrin complexes in solution. *Chem. Rev.* **1997**, *97*, 1325–1358.

40 Barron, L. D.; Gargaro, A. R.; Wen, Z. Q.; MacNicol, D. D.; Butters, C. Vibrational Raman optical activity of cyclodextrins. *Tetrahedron: Asymmetry* **1990**, *1*, 513–516.

41 Brotin, T.; Daugey, N.; Vanthuyne, N.; Jeanneau, E.; Ducasse, L.; Buffeteau, T. Chiroptical properties of cryptophane-223 and -233 investigated by ECD, VCD, and ROA spectroscopy. *J. Phys. Chem. B* **2015**, *119*, 8631–8639.

42 Daugey, N.; Brotin, T.; Vanthuyne, N.; Cavagnat, D.; Buffeteau, T. Raman optical activity of enantiopure cryptophanes. *J. Phys. Chem. B* **2014**, *118*, 5211–5217.

43 Cheeseman, J. R.; Frisch, M. J. Basis set dependence of vibrational Raman and Raman optical activity intensities. *J. Chem. Theory Comput.* **2011**, *7*, 3323–3334.

44 (a) Luber, S. Solvent effects in calculated vibrational Raman optical activity spectra of α-Helices. *J. Phys. Chem. A* **2013**, *117*, 2760–2770. (b) Jacob, C. R.; Luber, S.; Reiher, M. Analysis of secondary structure effects on the IR and Raman spectra of polypeptides in terms of localized vibrations. *J. Phys. Chem. B* **2009**, *113*, 6558–6573. (c) Yamamoto, S.; Furukawa, T.; Bouř, P.; Ozaki, Y. Solvated states of Poly-l-alanine α-Helix explored by Raman optical activity. *J. Phys. Chem. A* **2014**, *118*, 3655–3662. (d) Sundaralingam, M.; Sekharudu, Y. Water-inserted alpha-helical segments implicate reverse turns as folding intermediates. *Science* **1989**, *244*, 1333–1337.

5

Effective Fragment Molecular Orbital Method

Casper Steinmann[1] and Jan H. Jensen[2]

[1] *Department of Physics, Chemistry and Pharmacy, University of Southern Denmark, Denmark*
[2] *Department of Chemistry, University of Copenhagen, Denmark*

5.1 Introduction

Modeling biological systems with modern computational approaches represents one of the most difficult tasks at hand because of the range, magnitude, and scope of the complex interactions involved and the sizes of the systems investigated. In a single system such as ligand binding to a protein, the full range of inter-molecular interactions can be at play: long-range effects such as electrostatics, many-body polarization, dispersion interactions or effects that are only short-range in nature such as exchange-repulsion or charge-transfer. To accurately describe all of these interactions, computational quantum mechanical methods that include electron correlation must be used such as MP2 or ideally coupled cluster [1] theory with perturbative triples, that is, CCSD(T). However, such correlated methods also come with unfathomable scaling problems rendering them useless for biologically relevant systems. An alternative and computationally feasible approach is to obtain reliable model potentials for the interactions listed above based on quantum mechanical formulations and calculations. Two such approaches exist in the form of the sum of interactions among fragments ab initio computed (SIBFA) [2] and the effective fragment potential (EFP) method [3–5]. EFP works by doing calculations of subsystems (fragments) of a large system and is able to accurately capture the complex inter-molecular interactions that are present between the fragments. While the EFP method is completely free of adjustable parameters, it does have frozen internal geometries. If one is interested only in interaction energies, the EFP method is a very efficient means to an end, but if you are interested in evaluating the reaction barrier of a protein-ligand complex, the setup is more complex involving QM-EFP buffer regions [6]. EFP has also successfully been used to determine pK_a values in proteins [7, 8].

On the other hand, obtaining total energies (or derivatives thereof) for a large system with modest computational cost, methods based on the many-body expansion of the energy [9] provide an attractive alternative to running full SCF calculations.

The fragment molecular orbital (FMO) method [10–13] is one such method. The strength of the FMO method is that it incorporates high order interactions into low order expansions in terms of fragments, that is, electrostatic interactions in FMO are included to all orders because it uses a mean-field approach where each fragment is

Fragmentation: Toward Accurate Calculations on Complex Molecular Systems, First Edition.
Edited by Mark S. Gordon.

evaluated using the self-consistent field procedure taking into account the electrostatic potential of all surroundings fragments. On the other hand, non-electrostatic effects (dispersion, charge-transfer, and exchange-repulsion) are only included by considering pairwise interactions explicitly (see Section 5.1.2 for more details).

The FMO method has been used to investigate many different chemical problems since its inception almost 20 years ago [14, 15].

The effective fragment molecular orbital (EFMO) method [16, 17] is a merger of the effective fragment potential method and the fragment molecular orbital method. But before we begin to discuss the EFMO method in detail in Section 5.2, it is instructive to briefly discuss how and why it came into existence in the context of inter-molecular interactions energies calculated through effective fragment potentials (EFPs) in Section 5.1.1 and obtaining total energies of large molecular systems through the fragment molecular orbital (FMO) method discussed in Section 5.1.2.

5.1.1 Effective Fragment Potentials

The effective fragment potential method was originally developed as an approach to give high accuracy interaction energies which would enable an accurate description of chemical process in solution [3]. The EFP energy is written as an approximation to the closed shell HF interaction energy between isolated molecules:

$$E^{\mathrm{EFP}} \approx E^{\mathrm{HF}} - \sum_{I=1}^{N} E_I^{\mathrm{HF}}. \tag{5.1}$$

Here, E^{HF} is the total energy of a system of N molecules in the gas phase and E_I^{HF} is the energy of the ith molecule in the gas phase. While there exists two versions of effective fragment potentials (oftentimes distinguished as EFP1 and EFP2 or general effective fragment potentials), only EFP2 is truly parameter-free in the sense that all required data to evaluate all the terms are derived from *ab initio* calculations on individual fragments. The total interaction energy, that is, the EFP energy, is written as a sum of five terms [4]:

$$E^{\mathrm{EFP}} = E^{\mathrm{coul}} + E^{\mathrm{pol}} + E^{\mathrm{disp}} + E^{\mathrm{exrep}} + E^{\mathrm{ct}}, \tag{5.2}$$

where E^{coul} is the Coulomb interaction energy. It is calculated as the interaction energy between two distributed multipole expansions with expansion points on atoms and bond midpoints. Typically, the multipole expansion is based on Stones distributed multipole expansion [18]. The polarization energy, E^{pol}, is described using induced electric dipoles that arise from distributed electric dipole–dipole polarizabilities. The polarizabilities are placed in the centroids of localized molecular orbitals. This term includes both the interaction with other induced dipoles on other fragments as well as the permanent multipole moments on other fragments. The induced dipoles are iterated to self-consistency which captures most of the electrostatic many-body effects. The first two terms in equation (5.2), E^{coul} and E^{pol}, are the same in both EFP1 and EFP2. The remaining terms are only present in EFP2 (which we will henceforth just call EFP). The third term in the EFP energy expression is the dispersion energy which is written as a sum of inverse powers of the separation distance R:

$$E^{\mathrm{disp}} = \sum_{n} C_n R^{-n}. \tag{5.3}$$

In EFP2, this term is truncated at the leading induced-dipole-induced-dipole term, R^{-6}, which yields good accuracy compared with SAPT calculations on dispersion bound complexes.

The remaining two terms are based on the inter-molecular overlap between fragments. The exhange-repulsion term, E^{exrep}, was derived as a series of three separate energy terms based on the explicit dependence on powers of the inter-molecular overlap in the molecular orbital basis, which, because of the localized molecular orbital nature of the EFPs, can be truncated at the quadratic level with satisfactory results. Since the exchange-repulsion term depends directly on the overlap between fragments, each fragment carries a basis set explicitly. Finally, the charge-transfer energy term, E^{ct}, considers the interaction between occupied orbitals on one fragment with unoccupied fragments on another fragment. The term is based on a perturbation expression of the inter-molecular interactions through second order using canonical molecular orbitals. The charge-transfer term represents the most expensive term to evaluate when the EFP interaction energy is evaluated due to the large number of virtual orbitals usually present when employing large basis sets. This has been addressed recently by Peng and Gordon by projecting the large basis set onto a minimum basis set that best represents the valence orbital space [19].

To generate the EFP potentials for a molecule, the `MAKEFP` step is invoked which generates a potential file. This potential file includes everything needed to run calculations with either EFP–EFP interactions or EFP–QM interactions in subsequent calculations. A bottleneck of using EFP is that the user must manually copy the potential files into a final input file for each unique fragment in the calculations. In Section 5.2, we will discuss the actions that was taken for the EFMO method.

In the original formulation of the EFMO method, only the first two terms, the Coulomb term and the polarization term, of equation (5.2) are included in the calculation of the energy as we will show in Section 5.2. In the fully integrated EFMO method (FIEFMO), all five terms of equation (5.2) are included in the evaluation of the energy. This is presented in Section 5.2.5.

5.1.2 Fragment Molecular Orbital Method

While also a fragment-based method, the scope, as already discussed, of the FMO method is very different from that of EFP. Here, the goal is to obtain the total energy of a system of N fragments that is not tractable by conventional approaches. The FMO method uses a many-body expansion of the energy and for our purposes in this chapter, we shall limit our discussion to the two-body fragment molecular orbital (FMO2) method. Here, fragments (monomers) and pairs of fragments (dimers) are explicitly considered:

$$E^{\text{FMO2}} = \sum_{I}^{N} E_I + \sum_{I<J}^{N} \Delta E_{IJ}, \tag{5.4}$$

where

$$\Delta E_{IJ} = E_{IJ} - E_I - E_J. \tag{5.5}$$

Here, E_I and E_{IJ} are energies of fragment I and fragment pair IJ, respectively. In FMO, the ESP of all other fragments are considered while evaluating the fragment energies. Neglecting this ESP would give energies E_I^0 and E_{IJ}^0 for fragments and pairs of fragments, respectively. While equation (5.4) allows for effective two-level parallelism, that is, coarse-grained parallelism on the fragment level where each fragment is distributed to different compute nodes and a fine-grained parallelism on each node where multiple CPU cores are used to evaluate the energy of a single fragment, the scaling is still $\sim \mathcal{O}(N^2)$ rendering to too costly for large values of N. To reduce the number of pairs that needs to be explicitly evaluated, a relative (unit-less) distance is introduced

$$R(I,J) = \min\left(\frac{|\vec{r}_i - \vec{r}_j|}{\vec{r}_i^{\,vdw} + \vec{r}_j^{\,vdw}}\right), \tag{5.6}$$

where $\vec{r}_i$ and $\vec{r}_i^{\,vdw}$ are the position vector and Van der Waals radius of atom i, respectively. If the monomers I and J that make up dimer IJ are separated by more than a user-specified cut-off, R_{cut}, the dimer is approximated as the QM electrostatic interaction energy between the two

$$\Delta E_{IJ}^{sep} = \mathrm{Tr}(D^I v^{I(J)}) + \mathrm{Tr}(D^J v^{J(I)}) + \sum_{\mu\nu\in I}\sum_{\lambda\sigma\in J} D_{\mu\nu} D_{\lambda\sigma}(\mu\nu|\lambda\sigma) + \Delta E_{IJ}^{nr} \tag{5.7}$$

Here, the first two terms are the interaction between the electron density of fragment $I(J)$ with the nuclear potential onto fragment $I(J)$ from fragment $J(I)$. The third term is the electron repulsion term between the two fragments with their respective density matrices and the fourth and final term is the nuclear repulsion between fragments I and J. Having introduced the cut-off distance in equation (5.6), the FMO2 energy can be written as

$$E^{FMO2} = \sum_I^N E_I + \sum_{I<J}^{N,R(I,J)\le R_{cut}} \Delta E_{IJ} + \sum_{I<J}^{N,R(I,J)>R_{cut}} \Delta E_{IJ}^{sep}, \tag{5.8}$$

where the dimer correction to the energy is split up into two contributions depending on the inter-fragment separation distance which can be adjusted by the user. The default value for this cut-off is $R_{cut} = 2.0$. The above formulation of the FMO2 equations is the form utilized in the EFMO energy that is presented in the next section.

5.2 Effective Fragment Molecular Orbital Method

The EFMO method is a hybrid between the FMO method and the EFP method. The EFMO method is based on the FMO molecular fragmentation scheme and the many-body energy expression but uses the EFP multipole-based energy expressions for long-range interactions and for evaluating the many-body polarization. The EFMO energy, E^{EFMO}, is written like the FMO energy as [16]

$$E^{EFMO} = \sum_I^N E_I^0 + \sum_{I<J}^{N,R(I,J)\le R_{cut}} \Delta\left(E_{IJ}^0 - E_{IJ}^{pol}\right) + \sum_{I<J}^{N,R(I,J)>R_{cut}} \Delta E_{IJ}^{coul} + E^{pol} \tag{5.9}$$

Here, instead of embedded fragment and fragment pair energies, E_I and E_{IJ}, the EFMO method uses gas phase energies for both fragments, E_I^0, and fragment pairs, $\Delta E_{IJ}^0 = E_{IJ}^0 - E_I^0 - E_J^0$. This skips the self-consistent charge iterations during the monomer evaluation of the energy which is one of the more expensive parts of an FMO calculation. Second, the approximate electrostatic interaction energy from FMO2 for separated dimers (equation 5.7) is replaced by the classical Coulomb interaction energy from EFP. The total polarization energy, E^{pol}, is added to take many-body effects into account. This term is also from EFP. Finally, because an *ab initio* gas phase dimer calculation (second term in equation 5.9) includes the mutual polarization between fragments I and J, this polarization has to be removed again through E_{IJ}^{pol}, so as to not double count the effect already included from the many-body polarization term, E^{pol}.

The classical terms (E_{IJ}^{coul}, E_{IJ}^{pol} and E^{pol}) in the EFMO energy (equation 5.9) all require parameters to be available to be evaluated. This leads directly to one of the very strong points of the EFMO method: After the wave function of each monomer has converged to obtain E_I^0, the `MAKEFP` step is invoked in code to derive the required EFP parameters for the classical calculations without needing the user to specify any input. In fact, with this step automated, the EFMO method can be considered to be an ideal force field because it (i) requires no input from the user to specify parameters (apart from basis set) and (ii) all parameters needed are derived from *ab initio* calculations for the current geometry on the fly.

For ease of understanding, we have outlined the computational approach needed to calculate the EFMO energy (equation 5.9) and as it is implemented in the GAMESS [20, 21] program.

1. Evaluate gas phase energies of fragments, E_I^0.
   (a) Evaluate the EFP parameters needed later in the calculation for the classical terms through the `MAKEFP` step automatically.
2. Evaluate total polarization energy, E^{pol}.
3. Evaluate dimer correction to the energy between all pairs of fragments either through the explicit dimer energy, ΔE_{IJ}^0 and classical pair polarization, E_{IJ}^{pol}, for close pairs of fragments or E_{IJ}^{coul} for separated pairs.
4. Sum up all contributions to obtain the EFMO energy.

Compared to the FMO energy (equation 5.4), the EFMO energy has considerable time savings at step 1) outlined above. For proteins and large molecular clusters, the FMO method requires roughly 12 to 17 self-consistent charge iterations before convergence. This step gives rise to the N-body polarization in FMO. In the EFMO method, this many-body polarization is handled by step (2) above, that is, the E^{pol} term in the EFMO energy equation. Since the user-defined cut-offs are the same for both methods, the time savings in EFMO over FMO at step (3) is dominated by two different terms: not evaluating the ESP from all other fragments and using a classical multipole-expansion to evaluate the interaction energies for separated pairs of fragments.

For large systems where fragmentation across covalent bonds are necessary, EFMO uses the adapted frozen orbital approach originally used in FMO [22]. When using the adapted frozen orbital approach, one specifies pairs of atoms which define fragment boundaries. A bond is said to be detached between these two atoms which are known as the bond-attached atom and the bond-detached atom. Each of the atoms belongs to a different fragment. A model system is then constructed around the bond-attached atom and the bond-detached atom on which an RHF calculation is carried out. The canonical HF orbitals are then localized according to the Edmiston–Ruedenberg localization scheme [23]. The localized occupied orbital which has the maximum overlap with the bond-attached and bond-detached atoms are stored for later use along with several virtual orbitals. These orbitals are then used in later monomer self-consistent field calculations to keep the electron density confined to the current fragment and not enter the surrounding environment.

5.2.1 Correlation Energies in the EFMO Method

Use of higher level methods is required to study the binding energies and reaction barriers with high accuracy. Introduction of correlation in the EFMO methods follows that of FMO closely [24–26]. We write the total correlated energy (here for MP2) as

$$E^{\text{EFMO-MP2}} = E^{\text{EFMO}} + E^{\text{MP2}}, \tag{5.10}$$

where E^{MP2} is given as the sum of monomer correlation energies calculated at the MP2 level of theory, E_I^{MP2} and pairwise corrections

$$E^{\text{MP2}} = \sum_{I}^{N} E_I^{\text{MP2}} + \sum_{I<J}^{N,R(I,J)\leq R_{\text{corsd}}} \Delta E_{IJ}^{\text{MP2}}. \tag{5.11}$$

Here, $\Delta E_{IJ}^{\text{MP2}} = E_{IJ}^{\text{MP2}} - E_I^{\text{MP2}} - E_J^{\text{MP2}}$, which is the same approach that was used at the SCF level in earlier equations. A new unitless distance parameter, R_{corsd}, is introduced. It determines which pairs of fragments are to be correlated. In most applications, $R_{\text{corsd}} = R_{\text{cut}}$ but can be set to a lower value of reduce cost. We note that any size-extensive post-HF scheme can be used to evaluate the correlation energies.

In Table 5.1, we show example calculations of EFMO-MP2 and FMO2-MP2 for water clusters with 20, 30, 40, and 50 water molecules using various basis sets. The table shows error per hydrogen bond compared to full MP2 for three different 20 water molecule water clusters that are generated randomly and minimized. For the analysis of how EFMO compared (and faired) to FMO, the molecular cluster analyses were done per hydrogen bond because EFMO only describes higher-order polarization effects. This comparison was done in the first EFMO paper by considering the Kitaura–Morokuma energy decomposition analysis following work by Chen and Gordon [27, 28]. It was found that the error per hydrogen bond in EFMO is a direct measure of neglect of non-electrostatic many-body effects [16]. In all calculations, the cut-offs $R_{\text{cut}} = R_{\text{corsd}} = 2.0$ were used [17]. Since MP2 is included for fragments and close pairs of fragments but not included through classical dispersion from EFP [29], there will be an added error. We note here that the fully integrated EFMO method described in detail in Section 5.2.5 does include this classical dispersion interaction for all pairs.

For the 20 water molecule clusters, EFMO-MP2/6-31G(d) results deviate a maximum of 0.78 per hydrogen bond which is slightly worse than what is obtained with FMO2-MP2/6-31G(d) which deviate a maximum of $-0.43\ \text{kcal mol}^{-1}$ per hydrogen bond. Increasing the basis set by including heavy atom diffuse functions (6-31+G(d)) or also on Hydrogen atoms (6-31++G(d)) gives much lower errors per hydrogen bond

Table 5.1 Error in the energy per hydrogen bond of several water clusters when using EFMO or FMO2 relative to an *ab initio* MP2 calculation.

N_{H_2O}	N_{HB}	EFMO	FMO2
6-31G(d)			
20	31	0.63	−0.43
	32	0.66	−0.38
	29	0.78	−0.38
6-31+G(d)			
20	31	0.02	−0.69
	32	0.01	−0.67
	29	0.02	−0.76
6-31G++(d)			
20	31	−0.05	−0.44
	32	−0.04	−0.43
	29	−0.05	−0.48
6-31G(d)			
30	51	0.59	−0.43
40	63	0.79	−0.41
50	86	0.74	−0.45

Source: Steinmann *et al.* 2012 [17]. CC-BY 4.0.

for EFMO with maximum errors of 0.02 and −0.05 kcal mol^{-1} per hydrogen bond. This lowering in error for EFMO with more diffuse basis sets was expected already at the HF level of theory [16] because of the EDA analysis yielded that the classical induction term recovered most of the error. Here it is also observed that FMO observes larger errors for the more diffuse basis sets of −0.76 and −0.48 kcal mol^{-1} per hydrogen bond for 6-31+G(d) and 6-31++G(d), respectively. This is a well-known problem with FMO which has been attempted to solve using screening [30] or by using auxiliary polarization basis sets [31]. We note that calculated errors per hydrogen bond for the larger (30, 40, and 50 molecules) water clusters are consistent with the smaller 20 water molecule clusters at the same level of theory. Table 5.2 shows the error to RHF and MP2 reference energies for various protein and protein-like structures. The first six rows are α and β sheets of varying length. For these systems, RHF-EFMO and both RHF-FMO2 and MP2-FMO2 show reasonable accuracy. For the α sheets with MP2-EFMO, the error seems to increase linearly with system size: 0.3, −9.7, and −18.9 kcal mol^{-1} for 10, 20, and 40 amino acids, respectively. Turning to the last three rows in Table 5.2, we observe that the error for EFMO is much more consistent with FMO. For the Trp-cage protein which has 20 residues, the error is −2.9 and −4.3 kcal mol^{-1} for EFMO using RHF and MP2, respectively. This calculated error lies between the errors for both α and β sheets of the same length. The error is also comparable in size to that obtained by FMO2, albeit with a sign difference. In conclusion, we generally obtain errors for EFMO when

Table 5.2 Error in energy compared to *ab initio* calculations on proteins and protein-like structures for values of $R_{cut} = R_{corsd} = 2.0$ using two residue per fragment and the 6-31G(d) basis set.

	EFMO		FMO2/AFO	
	$R_{cut} = 2.0$		$R_{cut} = 2.0$	
	RHF	MP2	RHF	MP2
α-$(ALA)_{10}$	−2.9	0.3	−0.8	−0.1
β-$(ALA)_{10}$	0.6	0.9	0.1	0.3
α-$(ALA)_{20}$	−2.8	−9.7	−2.3	−0.5
β-$(ALA)_{20}$	1.7	2.8	0.2	0.7
α-$(ALA)_{40}$	0.2	−18.9	−5.5	−1.6
β-$(ALA)_{40}$	4.1	6.5	0.5	1.6
Chignolin	1.8	1.5	0.4	1.5
Tryptophan cage	−2.9	−4.3	1.7	6.4
Crambine[a]	15.7	26.2	3.5	25.6

Source: Steinmann *et al.* 2012 [17].
All units in kcal mol^{-1}.
[a]Reference is FMO3-MP2/6-31G(d).

comparing to full *ab initio* calculations which are comparable to those of FMO2 for realistic protein structures.

5.2.2 The EFMO Gradient

In the original work on EFMO, an approximate expression for the gradient was derived [16]:

$$\frac{\partial E^{\rm EFMO}}{\partial X_a} = \sum_I^N \frac{\partial E_I^0}{\partial X_a} + \sum_{I<J}^{N,R(I,J)\le R_{\rm cut}} \left(\frac{\partial \Delta E_{IJ}^0}{\partial X_a} - \frac{\partial E_{IJ}^{\rm pol}}{\partial X_a} \right) + \sum_{I<J}^{N,R(I,J)>R_{\rm cut}} \frac{\partial E_{IJ}^{\rm coul}}{\partial X_a} + \frac{\partial E^{\rm pol}}{\partial X_a}. \tag{5.12}$$

Just as for the energy, the gradient consists of all individual terms added (or subtracted). In more detail, the approximate gradient takes the gradient and torque terms arising from the rigid EFP fragments and distributes them to the nearest atoms instead of to the center of mass. The EFP gradient terms have been presented previously [3, 4].

The EFMO gradient (equation 5.12) is approximate because it neglects the change in multipoles and dipole polarizability tensors with respect to the change in the internal structure of the monomers. It was argued that the effect of this approximation is ameliorated by the fact that (i) only long-range interactions are treated with multipoles and (ii) that multipoles and dipole polarizability tensors are recomputed at each step in a geometry optimization. Therefore, this approximation becomes more accurate as the changes in the internal geometries becomes smaller when geometry optimization converges.

Table 5.3 Timings in minutes from doing energies or gradients (∇) with or without MP2.

	R_{cut}	T(RHF)	T(∇RHF)	T(MP2)	T(∇MP2)
Chignolin					
EFMO	1.0	9.6	11.1	22.8	102.8
	1.5	13.2	13.1	28.7	106.4
	2.0	14.0	17.0	29.5	119.0
FMO2	2.0	38.5	59.7	58.6	149.9
Tryptophan-cage					
EFMO	1.0	24.2	23.5	42.7	161.0
	1.5	33.7	38.3	70.7	261.6
	2.0	37.6	43.0	78.9	314.0
FMO2	2.0	100.4	187.0	142.5	408.6

Source: Steinmann *et al.* 2012 [17]. CC-BY 4.0.

EFMO offers a gradient whose quality is similar to FMO2/AFO calculations but at a reduced cost [17]. The electronic structure part of the AFO gradient is analytical [32] but there are terms that are approximate in the EFMO part of the coming from the EFP terms as discussed above. The gradient is easily evaluated when using correlated methods too by including the required gas phase derivatives from equation (5.10). Work is in progress on deriving the fully analytical EFMO gradient.

5.2.3 Timings and Computational Efficiency

As outlined above, there are considerable time savings to be obtained at the various stages of an EFMO calculation over FMO. In Table 5.3, we show the absolute timing in minutes for various EFMO and FMO2 calculations with various settings for the different user-defined cutoffs. For RHF single-point energies, we see that EFMO takes 14 minutes before completion for $R_{cut} = 2.0$ whereas with the same settings but using FMO2, the time to completion is almost 39 minutes. In general, we find that the energies are evaluated around two and a half times faster with EFMO than FMO when using RHF but only a modest speedup of around two when evaluating the energy with MP2 due to the added computational cost of especially fragment pairs which both methods share for the same cutoffs. The costs of computing gradients with either EFMO and FMO2 are also presented in Table 5.3. Here we see for Chignolin that the added cost for EFMO is only 3 minutes (from 14 minutes to 17 minutes) whereas for FMO2, the cost increases from ~39 minutes to almost 60 minutes. The same timing pattern is observed for the Tryptophan-cage where the cost almost doubles (from 100 minutes to 187 minutes) using FMO but roughly 6 minutes are added for EFMO. We note that whenever MP2 is involved, and especially with the gradient, the differences in timings become less due to where the computational bottleneck is located.

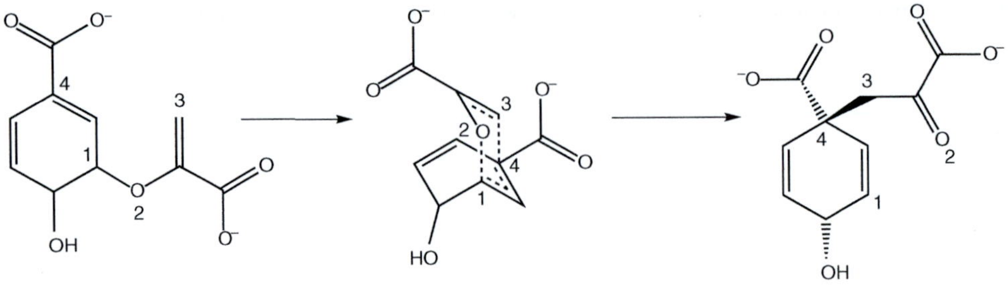

1 2 3

Figure 5.1 Illustration of the conversion of chorismate (left) to prephenate (right) through its transition state (center). *Source*: Steinmann *et al.* 2013 [33]. CC-BY 4.0.

5.2.4 Biochemistry with EFMO

An essential part of developing EFMO was to be able to treat systems of biological relevance. Having now seen its general applicability to both molecular clusters and to polypeptides, the EFMO was tested to see if it could capture the reaction barrier when converting chorismate to prephenate in chorismate mutase illustrated in Figure 5.1 [33]. The enzyme facilitates a binding pocket that gives rise to increased conversion rate of chorismate to prephenate. To map out the reaction, the following reaction coordinate was chosen

$$R = R_{21} - R_{43}, \tag{5.13}$$

The conversion from chorismate ($R = -2.0$, $R_{21} = 1.4$, $R_{43} = 3.4$) to prephenate ($R = 1.9$, $R_{21} = 3.3$, $R_{43} = 1.4$) was done by *adiabatic mapping* where the two bond lengths (R_{21} and R_{43}) were harmonically constrained and the surrounding structure is allowed to relax. After minimization, the distances R_{21} and R_{43} are changed 0.1 Å toward prephenate and everything is minimized again. This continues until the coordinates are those of prephenate. In order to keep the computational cost down, the EFMO method was adapted to utilize the frozen domain and frozen domain with dimers approach from FMO. Here, only a certain region of a large molecule is minimized. Basically, the EFMO energy is rewritten resulting in an energy expression that is very closely related to more conventional QM/MM energy expressions [33]:

$$E^{\text{EFMO}} = E_F^0 + E_b^0 + E_A^0 + E_{F/b}^0 + E_{F/A}^0 + E_{A/b}^0 + E^{\text{pol}}. \tag{5.14}$$

This energy expression allows a separation of the total system into different parts, the most important being the active region A that is to be optimized and a surrounding buffer region b which is frozen but which contribute to the gradient elements of A through the quantum mechanical treatment of pairs of fragments. The energy expression also includes interaction terms between the different regions. b is chosen large enough such that the rest of the protein, which is also frozen, F only interacts through the separated dimers expression with A and as a result is fast to compute. With this setup in place, only the fragments in A needs to have the classical parameters recomputed for every geometry step. At the start of the calculation, all fragments in all regions are evaluated once such that the EFP parameters are generated and stored for later use. We note that the polarization energy, E^{pol} for the entire system is evaluated in all steps

Table 5.4 Relative timings between different models using both the small (**S**) and large (**L**) with EFMO and the small model with FMO.

Model	E_{TS-R}	E_{P-R}	T_{rel}	T_{rel}^{full}
EFMO:**S**	46.3	−1.3	1.0	1.0
EFMO:**L**	46.4	−3.6	2.1	1.8
FMO:**S**	47.2	−1.4	6.7	7.5

Reaction barrier heights E_{TS-R} and reaction energies E_{P-R} are in kcal mol^{-1}. T_{rel} are relative timings to minimize a single reaction coordinate subject to harmonic constraints. T_{rel}^{full} are relative timings to minimize the entire reaction path.

but this step is comparably cheap to the SCF calculation on each fragment in the active region.

In the original work, two different sizes of the active region were attempted. The small model, EFMO:**S**, had every fragment which had at least one atom within a distance of 2 Å of chorismate in the active region. This amounted to 129 atoms in the active region to be minimized. The surrounding buffer region b contained all fragments which had at least one atom within 2.5 Å of any fragments in the active region. This amounted to 928 atoms. For the larger active region tested, EFMO:**L** consisted of all fragments which at at least one atom within 3 Å within chorismate which amounted to 241 atoms in the active region and 1151 atoms in the buffer region. The added flexibility by using EFMO:**L** over EFMO:**S** should give rise to better reaction energy profiles but that was not observed (see Figure 8 in Ref. [33]). In all tested cases using both EFMO and FMO and small and large regions, the computed reaction barriers using RHF/6-31G(d) was roughly 10 kcal mol^{-1} above of what was reported in the previous work [34]. We show the computed reaction barrier heights and reaction energies computed at the RHF/6-31G(d) level of theory in Table 5.4. It was reasoned that the computed barriers using EFMO and FMO were too high because there was no relaxation in either the b or F regions. This is confirmed with the reaction energy, E_{P-R} when using the large active region with EFMO, that is, EFMO:**L**, that decreases just over 2 kcal mol^{-1} compared to the EFMO:**S** model. The problem is most likely that the enzyme (apart from region A) is in effect fixed to the geometry of chorismate and does not have the flexibility to accommodate prephenate leading to large energies. Using an even larger active region could alleviate this but the computational cost could be prohibitive as shown in Table 5.4. Here, we observe that increasing the size of the active region from the small model to the large model doubles the time to completion for mapping out the reaction path. The time to completion for the smallest system for the entire reaction path was 4 days on 80 cores (on 20 compute nodes). We also observe that running FMO with the same settings as EFMO is vastly more expensive with the relative wall time to completion being more than seven times more expensive.

The computed reaction barrier was refined using the two-layer ONIOM [35, 36] approach where the energy at a high level of theory is approximated according to

$$E_{\text{real}}^{\text{high}} \approx E_{\text{real}}^{\text{low}} + E_{\text{model}}^{\text{high}} - E_{\text{model}}^{\text{low}}, \tag{5.15}$$

where the energy of the full system, that is, $E_{\text{real}}^{\text{low}} = E^{\text{EFMO}}$ is calculated from equation (5.14) at the RHF/6-31G(d) level of theory and the high-level correction is calculated at

Table 5.5 ONIOM reaction barriers calculated at the X:RHF-EFMO/6-31G(d) level of theory according to equation (5.15).

	E_{TS-R}	E_{P-R}
MP2/6-31G(d)	22.2	−3.2
MP2/cc-pVDZ	19.8	−5.5
MP2/cc-pVTZ	21.8	−1.1
MP2/cc-pVQZ	21.7	−0.8

Reaction barrier height E_{TS-R} and reaction energies E_{P-R} are in kcal mol^{-1}. Data taken from Ref. [33].

the MP2/cc-pVnZ (n = D, T and Q) level of theory. In fact, calculating the reaction barriers from equation (5.15) can be considered a special case of the more general ONIOM method based on FMO [37].

This is presented in Table 5.5. By using the ONIOM approach from equation (5.15), we see a large lowering in the energy from 46.3 kcal mol^{-1} using EFMO:**S** to 19.8 kcal mol^{-1} when correcting with MP2/cc-pVDZ. However, increasing the basis set size does not lead to a systematic increase in the reaction barrier. In fact, what we observe in Table 5.5 is that while the reaction barrier height converges toward some constant value, the overall reaction energy changes rather drastically from −5.5 kcal mol^{-1} at the MP2/cc-pVDZ level of theory while it decreases to −0.8 kcal mol^{-1} at the MP2/cc-pVQZ level of theory. This shows that the EFMO-RHF geometry with the ONION MP2 energy correction is not enough. To get around this problem, an extension was made to EFMO to allow for a single fragment to be optimized at the MP2 level of theory with a different basis set as well while the rest is treated at the RHF/6-31G(d) level of theory [38]. This is different from the case where everything in the active region, *A*, would be treated with MP2. Results from computing a single reaction barrier is shown in Table 5.6. These numbers are directly comparable to Table 5.5. We immediately observe that the computed barrier heights systematically decrease from 21.0, 19.2 down to 18.3 kcal mol^{-1} as the basis set is increased from 6-31G(d), cc-pVDZ to cc-pVTZ, respectively. The reaction energy is around −6 kcal mol^{-1} when either the cc-pVDZ or cc-pVTZ basis set is used which contrasts what was observed in Table 5.5. In conclusion, we note that the use of MP2 in the geometry optimization step substantially improves the calculated transition state compared to similar uncorrelated geometry optimization with a subsequent MP2 energy

Table 5.6 Calculated reaction barrier height and reaction energy with the geometry of chorismate optimized at the level of theory presented in the first column. All other fragments were treated using the RHF/6-31G(d) level of theory.

	E_{TS-R}	E_{P-R}
MP2/6-31G(d)	21.0	−4.8
MP2/cc-pVDZ	19.2	−6.8
MP2/cc-pVTZ	18.3	−6.2

Reaction barrier height E_{TS-R}, and reaction energies, E_{P-R}, are in kcal mol^{-1}. Data taken from Ref. [38].

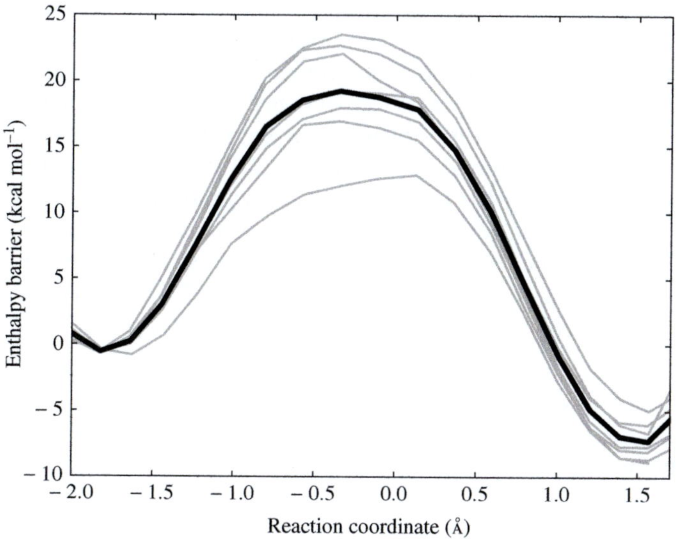

Figure 5.2 Reaction enthalpy profile from seven different reaction paths (shown in gray) computed at the ONIOM MP2/cc-pVDZ:EFMO-RHF/6-31G(d) level of theory. The resulting mean reaction enthalpy profile is shown in black. The figure is made with data from Ref. [33].

correction. The same is true for the product state. There is still an issue with the active region A being too small to accommodate the conformational changes needed in order to obtain results which are in agreement with the previous work.

To improve the computed reaction barrier, we carried out ensemble averaging starting from several different geometries extracted from a molecular dynamics simulation. We then apply the same treatment that has been discussed above where some snapshots from this trajectory are extracted, an active region is defined, and the rest is frozen. This is followed by a constrained geometry optimization along the same reaction coordinate (equation 5.13) for each starting geometry. Finally, an ONIOM energy correction is done to improve the energetics.

Results for such an MP2/cc-pVDZ:EFMO-RHF/6-31G(d) ONIOM refinement (equation 5.15) on the reaction barrier is shown in Figure 5.2 for a total of seven reaction paths shown in gray. The mean reaction enthalpy barrier is shown in black. We note in Figure 5.2 that the seven different reaction barriers gives rise to very different transition state energies varying from 12 kcal mol^{-1} to more than 23 kcal mol^{-1}. The barrier height computed for the mean reaction path can in general be computed in two different ways [39]:

$$\Delta H_1^{\ddagger} = \frac{1}{M}\sum_{i=1}^{M}(E_{\mathrm{TS},i} - E_{\mathrm{R},i}) - 1.6\,\mathrm{kcal\,mol^{-1}}, \tag{5.16}$$

where $M = 7$ or by [34]

$$\Delta H_2^{\ddagger} = \langle E_{\mathrm{TS}} \rangle - \langle E_{\mathrm{R}} \rangle - 1.6\,\mathrm{kcal\,mol^{-1}}. \tag{5.17}$$

In equation (5.16), one computes the barrier height for each trajectory ($E_{R,i}$ is the lowest energy with a negative reaction coordinate) and the computes the average. This is different from equation (5.17) where one computes the average reaction enthalpy barrier (black lines in Figure 5.2] and from that extracts the average transition state energy $\langle E_{TS} \rangle$, that is, the highest energy on the reaction path and the average lowest energy with a negative reaction coordinate, $\langle E_R \rangle$. The brackets in equation (5.17) means averaging over all seven reaction paths. The difference between equation (5.16) and equation (5.17) is the result of the non-communative sum and min/max operations over reaction coordinates. In both the equations above, the factor of 1.6 kcal mol^{-1} that is subtracted corrects for the change in zero point energy and thermal contributions [39]. The computed reaction barrier enthalpy from equation (5.16) is $\Delta H_1^{\ddagger} = 18.3 \pm 3.5$ kcal mol^{-1} whereas from equation (5.17), the computed barrier is $\Delta H_2^{\ddagger} = 18.2 \pm 6.9$ kcal mol^{-1}. The fact that the two computed reaction barrier enthalpies are so close within each other (around 0.1 kcal mol^{-1}) indicates that the transition state occurs at roughly the same value of reaction coordinate R for the majority of reaction paths.

5.2.5 Fully Integrated EFMO

The fully integrated EFMO method (FIEFMO) includes the rest of the terms from the EFP interaction energy from equation (5.2) [40]:

$$E^{\text{FIEFMO}} = \sum_{I}^{N} E_I^0 + \sum_{I<J}^{R(I,J)\leq R_{\text{cut}}} \left(\Delta E_{IJ}^0 - E_{IJ}^{\text{pol}}\right) + \sum_{I<J}^{R(I,J)>R_{\text{cut}}} \left(E_{IJ}^{\text{coul}} + E_{IJ}^{\text{ct}} + E_{IJ}^{\text{xr}} + E_{IJ}^{\text{disp}}\right) + E^{\text{pol}}, \tag{5.18}$$

Whereas correlation energies for the original EFMO method (equation 5.10) included only dispersion interactions for QM dimers, the inclusion of the dispersion term, E_{IJ}^{disp}, from EFP allows for the estimation of the full dispersion interaction for the entire system. Additionally, by including all intermolecular interactions from EFP, the user-defined cutoff value R_{cut} can be reduced to neglect any QM dimers. Thus, with the right set of user-defined cutoffs, the FIEFMO method can be considered to be the completion of a generalized EFP model. The FIEFMO method was tested on minimum energy water clusters with 8 and 16 water molecules [41] and on larger water clusters with 32 water molecules [42, 43] to gauge not only its accuracy but also timings when comparing FIEFMO to both the two- and three-body FMO methods, all MP2 calculations, and also the original EFMO formulation in equation (5.9).

As expected, the FIEFMO method and the EFMO method performed very similarly across all water clusters except when $R_{\text{cut}} \leq 0.6$ which is low enough to remove all QM dimers from a calculation. Surprisingly, this is where FIEFMO performed best with errors compared to full MP2 calculations of around 0.5 kcal mol^{-1}. At intermediate values of R_{cut}, that is, $R_{\text{cut}} = 0.8$, we observed that EFMO oftentimes did slightly better than FIEFMO. This is most likely due to an imbalance between the close-range EFP interactions and the polarization contribution from EFP needed for QM dimers plus the interactions captured with performing a full QM dimer.

For timings, it was found that the fragment-based approaches were indeed faster than the full MP2 calculation on the same system with the same basis sets. EFMO timings

Table 5.7 Tabulated differences between EFMO, FMO, and EFP.

	EFMO	FMO	EFP
Internal energy	QM	QM	0
Pair interaction	QM / classical	QM	Classical
Many-body pol	Classical	QM	Classical

were discussed in Section 5.2.3. FIEFMO timings were in general slightly slower because of the additional terms needed to calculate all interaction terms from the EFP mode. Not surprisingly, the fastest calculations were done for $R_{cut} = 0.6$ where all QM dimers are neglected. As already discussed, FMO includes polarization effects at the quantum mechanical level of theory through an iterative scheme on the monomer level which increases the computational cost.

The fully analytical gradients of the FIEFMO method was also published [44].

5.2.6 Remarks, Notes, and Comments

It is instructive to take a step back and compare and contrast the differences between EFP, FMO, and EFMO. We have summarized essential similarities and differences in Table 5.7. From the point of view of EFP, the EFMO method can be seen as the natural extension of EFP to allow the fragments to be flexible. This is done by recomputing the internal energies of all fragments at all required points during a geometry optimization, for instance. Compared to EFP, where the internal geometry is always fixed, this is very expensive. The difference in computational cost is related to obtaining the internal energy of each fragment. In FMO, the internal energy is also computed for each fragment as it is done for EFMO. However, in FMO, the fragments feel each other through the electrostatic potential of all the other fragments which results in the so-called self-consistent charge macro-iterations. Contrary to EFMO, this is expensive as the self-consistent field procedure must be evaluated for each fragment several times until all fragments are fully converged. The self-consistent charge iterations leads to the many-body polarization for the EFMO method. The many-body polarization is also computed in both EFMO and EFP, albeit classically from distributed anisotropic point–dipole polarizabilities which leads to induced dipoles that must also be iterated to self-consistency. This is, however, reasonably inexpensive compared to the many-body polarization obtained in FMO. All three methods are subject to the polarization catastrophe, however. In all three methods EFMO, EFP, and FMO, pair-wise corrections are computed. In EFP, this is done from perturbation expressions of the interaction energy between pairs of fragments. In EFMO and FMO, the energy for pairs of fragments which are close are evaluated by a regular SCF. The difference between EFMO and FMO is that in EFMO the SCF calculations is performed in vacuo whereas the FMO dimer SCF calculations is performed for a system embedded in the (now static) electrostatic potential of all other fragments. For both EFMO and FMO, this is very much more expensive than EFP. For large separation distances between pairs of fragments, the EFMO method approximates the interaction energy by expressions from EFP taking into consideration only the electrostatic interaction evaluated classically. In FMO, the interaction energy between largely separated dimers is also only computed through a Coulomb expression although the expression still includes terms which involves the electron densities and expensive two-electron integrals. This cost was further reduced in FMO by computing a multipole-expansion for each fragment and evaluating that electrostatic interaction energy for even larger separations of pairs of fragments making it much more like EFMO than EFP [45].

5.3 Summary and Future Developments

The research area of fragment-based methods is still very active. EFMO, as one of the fragment-based methods, can be expected to be useful in various biochemical studies, such as in enzymatic catalysis and protein-ligand binding where a compromise between computational speed and accuracy is needed for very large systems but without the associated problems of parameterizing the classical interactions from off-the-shelf force fields.

In addition to its parameter-free *ab initio*-based nature, EFMO, FIEFMO, as well as FMO offer chemical insight on the processes by providing subsystem information, such as the properties of individual fragments (e.g., the polarization energy) as well as the pair interaction energies between fragments.

References

1 Bartlett, R. J.; Musiał, M. *Rev. Mod. Phys.* **2007**, *79*, 291–352.

2 Gresh, N.; Cisneros, G. A.; Darden, T. A.; Piquemal, J.-P. *J. Chem. Theory Comput.* **2007**, *3*, 1960–1986.

3 Day, P. N.; Jensen, J. H.; Gordon, M. S.; Webb, S. P.; Stevens, W. J.; Krauss, M.; Garmer, D.; Basch, H.; Cohen, D. *J. Chem. Phys.* **1996**, *105*, 1968–1986.

4 Gordon, M. S.; Freitag, M. A.; Bandyopadhyay, P.; Jensen, J. H.; Kairys, V.; Stevens, W. J. *J. Phys. Chem. A* **2001**, *105*, 293–307.

5 Gordon, M. S.; Slipchenko, L.; Li, H.; Jensen, J. H. *Ann. Rep. Comp. Chem.* **2007**, 177–193.

6 Kairys, V.; Jensen, J. H. *J. Phys. Chem. A* **2000**, *104*, 6656–6665.

7 Minikis, R. M.; Kairys, V.; Jensen, J. H. *J. Phys. Chem. A* **2001**, *105*, 3829–3837.

8 Li, H.; Hains, A. W.; Everts, J. E.; Robertson, A. D.; Jensen, J. H. *J. Phys. Chem. B.* **2002**, *106*, 3486–3494.

9 Hankins, D. *J. Chem. Phys.* **1970**, *53*, 4544.

10 Fedorov, D. G.; Kitaura, K., Eds. *The Fragment Molecular Orbital Method: Practical Applications to Large Molecular Systems*; CRC Press: Boca Raton, FL, 2009.

11 Kitaura, K.; Ikeo, E.; Asada, T.; Nakano, T.; Uebayasi, M. *Chem. Phys. Lett.* **1999**, *313*, 701–706.

12 Nakano, T.; Kaminuma, T.; Sato, T.; Akiyama, Y.; Uebayasi, M.; Kitaura, K. *Chem. Phys. Lett.* **2000**, *318*, 614–618.

13 Fedorov, D. G.; Kitaura, K. *J. Phys. Chem. A* **2007**, *111*, 6904–6914.

14 Gordon, M. S.; Fedorov, D. G.; Pruitt, S. R.; Slipchenko, L. V. *Chem. Rev.* **2012**, *112*, 632–672.

15 Fedorov, D. G.; Nagata, T.; Kitaura, K. *Phys. Chem. Chem. Phys.* **2012**, *14*, 7562.
16 Steinmann, C.; Fedorov, D. G.; Jensen, J. H. *J. Phys. Chem. A* **2010**, *114*, 8705–8712.
17 Steinmann, C.; Fedorov, D. G.; Jensen, J. H. *PLoS ONE* **2012**, *7*, e41117.
18 Stone, A. *The Theory of Intermolecular Forces*, 2nd ed.; Oxford University Press, 2013.
19 Xu, P.; Gordon, M. S. *J. Chem. Phys.* **2013**, *139*, 194104.
20 Schmidt, M. W.; Baldridge, K. K.; Boatz, J. A.; Elbert, S. T.; Gordon, M. S.; Jensen, J. H.; Koseki, S.; Matsunaga, N.; Nguyen, K. A.; Su, S.; Windus, T. L.; Dupuis, M.; Montgomery, J. A. *J. Comput. Chem.* **1993**, *14*, 1347–1363.
21 Gordon, M. S.; Schmidt, M. W. In *Theory and Applications of Computational Chemistry, the First Forty Years*; Dykstra, C. E., Frenking, G., Kim, K. S., Scuseria, G. E., Eds.; Elsevier: Amsterdam, 2005; Chapter 41, pp 1167–1189.
22 Fedorov, D. G.; Jensen, J. H.; Deka, R. C.; Kitaura, K. *J. Phys. Chem. A* **2008**, *112*, 11808–11816.
23 Edmiston, C.; Ruedenberg, K. *Rev. Mod. Phys.* **1963**, *35*, 457–464.
24 Fedorov, D. G.; Kitaura, K. *J. Chem. Phys.* **2004**, *121*, 2483.
25 Fedorov, D. G.; Kitaura, K. *J. Chem. Phys.* **2005**, *123*, 134103.
26 Fedorov, D. G.; Ishimura, K.; Ishida, T.; Kitaura, K.; Pulay, P.; Nagase, S. *J. Comput. Chem.* **2007**, *28*, 1476–1484.
27 Kitaura, K.; Morokuma, K. *Int. J. Quantum Chem.* **1976**, *10*, 325–340.
28 Chen, W.; Gordon, M. S. *J. Phys. Chem.* **1996**, *100*, 14316–14328.
29 Adamovic, I.; Gordon, M. S. *Mol. Phys.* **2005**, *103*, 379–387.
30 Fedorov, D. G.; Slipchenko, L. V.; Kitaura, K. *J. Phys. Chem. A* **2010**, *114*, 8742–8753.
31 Fedorov, D. G.; Kitaura, K. *Chem. Phys. Lett.* **2014**, *597*, 99–105.
32 Fedorov, D. G.; Avramov, P. V.; Jensen, J. H.; Kitaura, K. *Chem. Phys. Lett.* **2009**, *477*, 169–175.
33 Steinmann, C.; Fedorov, D. G.; Jensen, J. H. *PLoS ONE* **2013**, *8*, e60602.
34 Ranaghan, K. E.; Ridder, L.; Szefczyk, B.; Sokalski, W. A.; Hermann, J. C.; Mulholland, A. J. *Org. Biomol. Chem.* **2004**, *2*, 968.
35 Svensson, M.; Humbel, S.; Froese, R. D. J.; Matsubara, T.; Sieber, S.; Morokuma, K. *J. Phys. Chem.* **1996**, *100*, 19357–19363.
36 Dapprich, S.; Komáromi, I.; Byun, K.; Morokuma, K.; Frisch, M. J. *J. Mol. Struct. Theochem.* **1999**, *461-462*, 1–21.
37 Asada, N.; Fedorov, D. G.; Kitaura, K.; Nakanishi, I.; Merz, K. M. *J. Phys. Chem. Lett.* **2012**, *3*, 2604–2610.
38 Christensen, A. S.; Steinmann, C.; Fedorov, D. G.; Jensen, J. H. *PLoS ONE* **2014**, *9*, e88800.
39 Claeyssens, F.; Harvey, J. N.; Manby, F. R.; Mata, R. A.; Mulholland, A. J.; Ranaghan, K. E.; Schütz, M.; Thiel, S.; Thiel, W.; Werner, H.-J. *Angew. Chem.* **2006**, *118*, 7010–7013.
40 Pruitt, S. R.; Steinmann, C.; Jensen, J. H.; Gordon, M. S. *J. Chem. Theory Comput.* **2013**, *9*, 2235–2249.
41 Day, P. N.; Pachter, R.; Gordon, M. S.; Merrill, G. N. *J. Chem. Phys.* **2000**, *112*, 2063.
42 Mullin, J. M.; Gordon, M. S. *J. Phys. Chem. B.* **2009**, *113*, 8657–8669.
43 Mullin, J. M.; Gordon, M. S. *J. Phys. Chem. B.* **2009**, *113*, 14413–14420.
44 Pruitt, S. R.; Bertoni, C.; Brorsen, K. R.; Gordon, M. S. *Acc. Chem. Res.* **2014**, *47*, 2786–2794.
45 Choi, C. H.; Fedorov, D. G. *Chem. Phys. Lett.* **2012**, *543*, 159–165.

6

Effective Fragment Potential Method: Past, Present, and Future

Lyudmila V. Slipchenko and Pradeep K. Gurunathan

Department of Chemistry, Purdue University, USA

6.1 Overview of the EFP Method

Non-covalent interactions play a governing role in the structure and properties of biological macromolecules such as proteins, membranes, and nucleic acids. Aggregation and self-assembly of natural and synthetic materials and physical properties of liquids and emulsions are also closely related to the interplay of non-covalent forces. However, theoretical modeling of non-covalent interactions is challenging in several aspects. First, non-covalent interactions are prevalent and consume a majority of computational time in the simulation of an extended system. Second, non-covalent interactions are typically small in magnitude and impose strict requirements on the accuracy of a computational method applied. Finally, non-covalent interactions originate from different underlying physical phenomena, such as electrostatic terms due to Coulomb forces between polar species, van der Waals terms due to quantum effects and electron correlation, and so on. Thus, a desired computational method should be able to describe different components of non-covalent interactions in a balanced fashion, and to do it efficiently.

Multiple computational approaches have been developed for describing and analyzing non-covalent interactions. Generally, they can be split into several partly overlapping categories. Force fields based on molecular mechanics (MM) are the simplest and most widely used approach for treating large systems. In force fields, all non-covalent interactions are parameterized based on a model Hamiltonian. The main advantage of force fields is a trivial computational cost. The disadvantages range from a complicated set-up of the calculations, to missing or inadequate parameters, and to an oversimplified form of the Hamiltonian that might not be sufficient for high accuracy, even if the best set of parameters were available. Fragmentation methods are on another pole of the spectrum [1–6], as they attempt to capture all the interactions in the system from a purely quantum-mechanical (QM) approach. While the computational demands of fragmentation methods are significantly higher as compared to classical force fields, they benefit from a linear scaling and can be applied to systems containing thousands of atoms. Being based on quantum-mechanical methods, fragmentation techniques share pros and cons of underlying QM models in reaching an accurate description of

Fragmentation: Toward Accurate Calculations on Complex Molecular Systems, First Edition.
Edited by Mark S. Gordon.

non-covalent interactions. For example, accurate description of dispersion interactions by computationally efficient QM models, such as density functional theory (DFT) methods, remains an active area of research and a universal solution is still to be found [7, 13]. Another concern in fragmentation methods is the treatment of many-body effects that sometimes converge slowly, which significantly increases computational cost. DFT and density functional tight binding (DFTB) methods, if based on functionals trained on non-covalent datasets, can approach a reference-value accuracy in describing weak forces [14–17].

Analysis of non-covalent interactions is important both for characterization of chemical systems and processes and for the development of next-generation computational models. Several fragmentation methods such as fragment molecular orbital (FMO) method [18–22] can provide decomposition of interaction energy [23, 24]. Symmetry adapted perturbation theory (SAPT) is also a powerful tool for the analysis of interaction energy components [22, 23, 25–27].

The effective fragment potential (EFP) method is an alternative approach for describing intermolecular interactions [28–32]. EFP is a quantum-mechanical based model potential that can be viewed either as a fragmentation method with pre-computed fragments or as a universal force field with a functional form and parameters extracted from QM calculations. Similar to force fields, the total non-covalent interaction energy in EFP is modeled as a sum of several interaction terms, namely, Coulomb (or electrostatic), polarization (or induction), dispersion, exchange-repulsion, and optional charge-transfer components. Each of these terms can be thought of as a truncated expansion. Different from force fields, the simplest unit in EFP is a fragment rather than an atom. The EFP method evolved from being a model for only water fragments, to general small-molecule fragments, and to arbitrary molecular fragments with a reasonably localized wave function.

Coulomb, polarization, and dispersion terms in EFP have their origins in long-range perturbation theory, with perturbation being a Coulomb interaction between involved chemical species [33]. Coulomb term appears in the first order, while polarization and dispersion originate from the second order in the perturbative series. Multipole expansion is employed throughout to bring these three terms to computationally efficient expressions, without compromising the overall accuracy. In order to account for charge-penetration effects and ensure correct asymptotic behavior at small inter-fragment distances, all long-range terms are augmented by short-range damping functions. The exchange-repulsion and charge-transfer terms are derived from the short-range perturbation series, in powers of an intermolecular overlap.

In order to ensure computational efficiency, all EFP terms are parameterized. However, in contrast to classical force fields, EFP parameters are obtained from a set of QM calculations on a gas-phase fragment. Thus, if a fragment's geometry is reasonably rigid (typically, if there are no free rotations around single bonds), these pre-computed parameters might be used for all consecutive calculations. Our recent work [34, 35], outlined in Section 6.4, suggests ways of extending the EFP method to flexible molecules.

The EFP parameters include the following information: for Coulomb term—multipole moments distributed at atoms and bond-mid-points, up to octopoles; for polarization term—static polarizability tensors distributed at localized molecular orbital (LMO) centroids; for dispersion term—dynamic polarizability tensors, computed at 12 different imaginary frequencies, distributed at LMO centroids; for exchange-repulsion and charge-transfer terms—localized wave function, Fock matrix, and atomic basis set. EFP parameters are dependent on the employed level of theory, that is, electronic structure method and basis set. One of the directions of the development of the EFP method pertains to the determination of relative accuracies of each EFP term and finding an optimal level of theory for computing EFP parameters [14].

The computational cost of the EFP method is determined by its cost-limiting terms. In the present implementation, these are exchange-repulsion and charge-transfer terms because computation of the interaction energies involves on-the-fly evaluation of the overlap integrals between each pair of fragments. As a result, the cost of the exchange-repulsion term at present consumes about 80–90% of the total computational time (if charge-transfer term is excluded), and charge-transfer term increases the cost by another order of magnitude. The next most expensive term after exchange-repulsion and charge-transfer is polarization. Thus, EFP (with omitted charge-transfer term) is about an order of magnitude more expensive than a polarizable classical force field and two orders more expensive than a classical non-polarizable force field. The cost difference becomes less severe with an increase in the system size when the amount of dimers, for which the exchange-repulsion term should be computed, diminishes due to the short-range nature of the exchange-repulsion.

The EFP method provides several benefits that enrich computational capabilities and enhance our understanding of non-covalent interactions. First of all, EFP provides QM-level treatment of non-covalent interactions for a fraction of the computational cost of QM methods. Second, the natural ability of EFP to provide decomposition of interaction energy makes the method valuable for understanding the make-up of non-covalent interactions, which is essential for a multitude of fundamental and practical problems such as guest–host interactions related to computational drug design, absorption on surfaces, phase transitions in crystals, etc. Additionally, EFP can be combined with QM methods in a QM/MM fashion, providing polarizable embedding for the description of chemical and photochemical phenomena in condensed phases.

The main focus of this chapter is on the use of the EFP method as a polarizable embedding scheme for describing electronic excitation, ionization, and attachment processes in condensed phase, interfaces, and biological systems.

6.2 Milestones in the Development of the EFP Method

6.2.1 EFP1 Water Model

The development of the EFP method was started in the beginning of the 1990s by Mark Gordon and coworkers [32]. The first implementation, later referred to as EFP1, was focused on the modeling of water and aqueous solutions [36]. The EFP1 water potential consists of electrostatic, polarization, and remainder terms:

$$E_{\mathrm{EFP1}} = E_{\mathrm{elec}} + E_{\mathrm{pol}} + E_{\mathrm{rem}} \tag{6.1}$$

The electrostatic term is based on a multipole expansion, with charges, dipoles, quadrupoles, and octopoles being computed at atoms and bond-mid-points using the distributed multipole analysis (DMA). Electrostatic energy for interactions between EFP fragments is composed of charge–charge, charge–dipole, charge–quadrupole, charge–octopole, dipole–dipole, dipole–quadrupole, and quadrupole–quadrupole terms. In QM/EFP1 treatment, electrostatic interactions between QM and EFP regions are described through one-electron multipole contribution V to the QM Hamiltonian:

$$V_k^{\text{Coul}}(x) = q_k T(r_{kx}) - \sum_{a}^{x,y,z} \mu_a^k T_a(r_{kx}) + \frac{1}{3}\sum_{a,b}^{x,y,z} \Theta_{a,b}^k T_{a,b}(r_{kx}) - \frac{1}{15}\sum_{a,b,c}^{x,y,z} \Omega_{a,b,c}^k T_{a,b,c}(r_{kx}) \tag{6.2}$$

where q, μ, Θ, and Ω are the net charge, dipole, quadrupole, and octopole, respectively, located at points k of effective fragments. Points k correspond to positions of atoms and bond-mid-points. T are electrostatic tensors of ranks 0–3; x is an electronic coordinate. Additionally, to account for short-range charge-penetration effects, the first (charge) term in equation (6.2) is augmented by a Gaussian damping function.

For polarization term, static anisotropic polarizability tensor of a fragment is pre-computed and is distributed at the centroids of LMOs. Polarization energy is then obtained as a product of induced dipoles and electric field acting on the dipoles:

$$E_{\text{pol}} = \frac{1}{2}\sum_{a}^{x,y,z}\sum_{k} \left(-\mu_a^k \left(F_a^{\text{mult},k} + F_a^{\text{nuc},k}\right) + \tilde{\mu}_a^k F_a^{ai,k}\right) \tag{6.3}$$

where μ^k and $\tilde{\mu}^k$ are the induced and conjugated induced dipoles at distributed point k (centroid of LMO on a fragment). F^{mult} is the electric field due to multipole moments of other fragments in the system; F^{nuc} and F^{ai} are electric fields due to nuclei and electrons of the QM subsystem, respectively. Induced dipoles at polarizability points k are computed self-consistently as a response to electric field due to static multipoles and induced dipoles of other fragments, as well as field due to nuclei and electrons of the QM region, namely:

$$\mu^k = \sum_k \alpha^k (F^{\text{mult},k} + F^{\text{ind},k} + F^{\text{nuc},k} + F^{ai,k}) \tag{6.4}$$

where F^{ind} is the field due to induced dipoles on other fragments.

Polarization contribution to the total ground state QM-EFP energy is also computed in a self-consistent fashion, where self-consistency is reached between QM wave function and induced dipoles on the fragments. Practically, the QM Hamiltonian is augmented by the induced dipole term (one-electron dipole integral):

$$V_k^{\text{pol}}(x) = \frac{1}{2}\sum_{a}^{x,y,z} \left(\mu_a^k + \tilde{\mu}_a^k\right) T_a(r_{kx}) \tag{6.5}$$

Induced dipoles in equations (6.3) and (6.5) are obtained in the following manner. For each iteration of the SCF procedure for the QM wave function, electric field due to the current wave function is computed at polarizability points, and induced dipoles are iterated until self-consistent with themselves, following equation (6.4). These induced dipoles enter equations (6.3) and (6.5) for the next iteration of the QM wave function and the procedure continues. As a result, upon convergence of the QM SCF, induced dipoles are fully consistent with themselves and the QM wave function as well.

The remainder term, which effectively captures exchange-repulsion, charge-transfer, and higher-order contributions, is parameterized against total QM interaction energies in a set of water dimers, that is,

$$E_{\text{rem}} = E_{\text{QM}} - E_{\text{elec}} - E_{\text{pol}} \tag{6.6}$$

For interactions among fragments, functional form of the remainder term is exponential. However, a Gaussian functional form is employed for QM-EFP interactions. The latter is done to simplify the computation of produced integrals that, in case of the Gaussian functional form of the remainder term, can be represented as one-electron three-center integrals.

As should be clear from the above discussion, all three interacting terms in QM/EFP1 appear as one-electron integrals in the QM Hamiltonian:

$$H_{\text{total}} = H_{\text{QM}} + \langle p|V^{\text{Coul}} + V^{\text{pol}} + V^{\text{rem}}|q\rangle \tag{6.7}$$

Thus, QM-EFP1 methods correspond to full-embedding schemes, in which a complete quantum-mechanical coupling exists between quantum and classical regions.

Two different versions of EFP1 exist, one is based on HF/DH(d,p) [32, 36] and another on DFT(B3LYP)/DH(d,p) [37] parameterizations. This means that parameters for all terms, electrostatic, polarization, and remainder, are precomputed at the given level of theory. As expected, EFP1-HF potential accurately reproduces HF/DH(d,p) results for water—both clusters and bulk, and EFP1-DFT matches B3LYP/DH(d,p) results. As follows, both from a functional form and from the underlying QM methods used for the EFP1 parameterization, dispersion interactions are not captured by the EFP1 water potentials, which might be not negligible in water solvation.

Analytic gradients for EFP1 water models, both for fragment–fragment and QM–fragment interactions [32], have enabled numerous applications exploring structure and property of water clusters, properties of solutes in aqueous solvation, chemical reactions and photochemistry in water [38–45].

6.2.2 EFP (EFP2) General Model

During 1996–1998, Jan Jensen developed a general form of the exchange-repulsion term that is derived from the exact HF interaction energy using spherical orbital overlap approximation [46, 47]. In this formulation, exchange-repulsion term depends only on the localized wave function of each fragment and is computed using pairwise interfragment overlap and kinetic integrals. The latter are evaluated on-the-fly for each energy calculation of the system. While numerous evaluations of the overlap integrals create a significant computational cost, this formulation of the exchange-repulsion energy provides a universal way for computing exchange-repulsion interactions in a system without introducing fitted parameters. Development of the exchange-repulsion energy and gradients between fragments has led to the creation of a general method for describing intermolecular interactions. To distinguish from the water models, this general formulation is sometimes called EFP2, but "EFP" is commonly used in the recent literature.

The next logical step was to develop a dispersion term, as dispersion contribution is expected to be significant or dominant in many molecular complexes, such as non-polar solvents and aromatic compounds. Adamovic and Gordon adapted the formulation of C6 dispersion coefficients by Amos *et al.* [48, 49] for EFP in 2005 [50], in which the C6 coefficients between a pair of distributed points k and j on different fragments are computed as integrals over imaginary frequencies ω of a product of distributed time-dependent polarizability tensors $\alpha_{\alpha\gamma}(i\omega)$:

$$C_6^{kj} = \int_0^{\infty} d\omega \alpha_{\alpha\gamma}^{k}(i\omega)\alpha_{\beta\delta}^{j}(i\omega) \tag{6.8}$$

The integral in equation (6.8) is computed using a quadrature, and dynamic polarizabilities on each fragment are pre-computed for a set of imaginary frequencies using time-dependent HF (TDHF) method. Similar to static polarizabilities, these dynamic polarizabilities are distributed over LMO centroids.

Using C6 coefficients, pairwise fragment–fragment dispersion energy is computed as:

$$E_{\text{EFP-EFP}}^{\text{disp}} = -\frac{4}{3}\sum_{k\in A}\sum_{j\in B}\frac{f_{\text{damp}}C_6^{kj}}{R_{kj}^6} \tag{6.9}$$

where f_{damp} is a damping function that accounts for short-range deviation of the dispersion energy from multipole representation. Practically, damping functions are evaluated using either Tang–Toennies formula [51] with a single predefined damping parameter, or overlap-based damping that is parameter-free which uses overlap integrals as a measure of damping strength [52].

Another unique energy term present in the general EFP method is the charge-transfer term developed and implemented in 2006 by Li, Gordon, and Jensen [53]. The goal of this term is to account for energy lowering due to resonance interaction between a closed-shell ground state of the dimer and a charge-transfer state in which an electron is transferred between the two monomers. The charge-transfer is derived as a function of the molecular overlap between occupied orbitals on one fragment and virtual orbitals on another. Later, the number of virtual orbitals used in the calculation of the overlap integrals was restricted to valence localized virtual orbitals obtained from quasiatomic orbitals (QUAMBO) [54, 55], which significantly lowered the computational cost of the charge-transfer term. However, even with these advances, the cost of the charge-transfer contribution is the highest among all the EFP terms.

Thus, at present, the general EFP method computes interaction energy as a sum of five terms:

$$E_{\text{EFP}} = E_{\text{elec}} + E_{\text{pol}} + E_{\text{disp}} + E_{\text{exrep}} + E_{\text{ct}}, \tag{6.10}$$

in which electrostatic and polarization terms are computed analogously to the EFP1 scheme. Analytic gradients of all these terms are implemented as well.

General EFP was exploited for predicting and rationalizing structures of molecular clusters and liquids [14, 56–59], using Monte Carlo (MC) and molecular dynamics (MD) simulations, combined with periodic boundary conditions. The strength of the EFP method is in the ability to describe systems with various type of bonding, including

H-bonds, hydrophobic hydration, π–π and π–H bonding, etc. For example, EFP was employed for analysis of the interplay of π–π, π–H, and H-bonding in water–benzene complexes [56]. Relative strengths of H-bonds and π–π stacking interactions in DNA base pairs are also accurately captured by EFP [29, 58]. Another type of interaction successfully modeled by EFP is a competition between H-bonding and hydrophobic attraction in water–alcohol systems, which were explored both at the level of small water–alcohol clusters and solutions [59, 60]. Affinity of halide anions to hydrophobic surfaces in aqueous solution is another example of heterogeneous environment where EFP reveals its strength as a universal well-balanced model potential [61].

Significant speed-up in EFP MD simulations was achieved by independent implementation of the EFP method in an open-source software library named *libefp* [62, 63]. *Libefp*, written in modular C, exploits both OpenMP and MPI parallelization. For example, our recent benchmarks show that EFP molecular dynamics simulations performed on one 16-processor node (2 × 8-Core Intel Xeon-E5 + 32 GB of RAM) achieves about 100,000 energy and gradient steps per day (i.e., ~0.1 ns with 1 fs timestep) for a system consisting of 256 water molecules. Additional speed-up in EFP MD can be achieved by introducing smart cut-offs for short-range charge-transfer and exchange-repulsion interactions, as well as by exploiting intrinsically different time-scales of van der Waals (exchange-repulsion and dispersion) and electrostatic (Coulomb, polarization, and potentially charge-transfer) interactions. These topics are currently being explored.

Along with the development of the EFP formalism, there is an ongoing effort to couple general EFP with QM in so called QM/EFP methods. As in traditional QM/MM schemes, the Hamiltonian of a combined QM/EFP system can be represented as:

$$H_{\text{QM/EFP}} = H_{\text{QM}} + H_{\text{EFP}} + H_{\text{QM-EFP}} \tag{6.11}$$

where H_{QM} and H_{EFP} are Hamiltonians of QM and EFP regions, respectively, and $H_{\text{QM-EFP}}$ is a coupling Hamiltonian that can be expressed as

$$H_{\text{QM-EFP}} = H_{\text{elec}} + H_{\text{pol}} + H_{\text{disp}} + H_{\text{ex-rep}} + H_{\text{ct}} \tag{6.12}$$

where H_{elec}, H_{pol}, etc. are coupling terms that couple QM and EFP subsystems through electrostatic, polarization, etc. interactions.

Polarization and electrostatic terms in general QM/EFP are computed analogously to QM/EFP1, as one-electron multipole integrals in QM Hamiltonian. Same strategy for a self-consistent treatment of polarization is employed. The QM/EFP exchange-repulsion energy term was developed in 2010 [64]. Similar to the EFP-EFP exchange-repulsion term, this formulation is based on approximations applied to the exact HF expression of the exchange-repulsion energy in the dimer. However, the QM-EFP exchange-repulsion implementation is more challenging than the EFP-EFP one, because not all two-electron integrals can be approximated as one-electron integrals without loss of accuracy. Despite that, the QM-EFP exchange-repulsion formulation provides means to account for repulsive interactions between quantum and classical subsystems in a rigorous self-consistent fashion because contributions from all the individual fragments are included in the QM Fock matrix.

Analytic gradients of the QM/EFP exchange-repulsion term are under development [65].

Initial implementation of the QM/EFP dispersion energy by Smith *et al.* [66] was done in the spirit of EFP-EFP dispersion, using multipole expansion on both EFP and QM sides (truncated after the first dipole term).

$$E^{\text{disp}}_{\text{QM-EFP}} = -\frac{6}{\pi}\sum_{j\in B}\sum_{k}^{\text{occ}}\sum_{r}^{\text{vir}}\frac{1}{R^6_{\text{kj}}}\langle k|\hat{\mu}|r\rangle\langle r|\hat{\mu}|k\rangle\int_0^\infty d\omega\frac{\omega^A_{\text{rk}}}{\left(\omega^A_{\text{rk}}\right)^2+\omega^2}\bar{\alpha}^j(i\omega) \tag{6.13}$$

where k and r are occupied and virtual orbitals in the QM region, respectively, j is the distributed point on fragment B at which dynamic polarizability tensors α^j are stored, ω_{rk} is the orbital energy difference between orbitals r and k, and R_{kj} is the distance between orbital k of the QM region and point j on the effective fragment. Anisotropic polarizability tensors are approximated by the corresponding isotropic polarizabilities $\bar{\alpha}$. For achieving sufficient accuracy, especially when larger QM subsystems are considered, occupied molecular orbitals of the QM subsystem are localized. Further, damping function and *ad hoc* factor of 4/3 are added, similar to the EFP-EFP dispersion formulation.

Recently, a more rigorous and computationally efficient treatment of the dispersion term was developed [67]. In this approach, electronic coordinate of the QM subsystem is not integrated out of the expression

$$E^{\text{disp}}_{\text{QM-EFP}} = -\frac{1}{\pi}\sum_{j\in B}\sum_{k}^{\text{occ}}\sum_{r}^{\text{vir}}\sum_{\alpha\beta}^{x,y,z}\left\langle k\middle|T^j_\alpha\middle|r\right\rangle\left\langle r\middle|T^j_\beta\middle|k\right\rangle\int_0^\infty\frac{\omega^A_{rk}}{\left(\omega^A_{rk}\right)^2+\omega^2}\alpha^j_{\alpha\beta}(i\omega)d\omega \tag{6.14}$$

where T^j_α and T^j_β are electric field integrals computed in occupied-virtual block. In this case, no orbital localization is required in the QM region, which saves computational cost and simplifies development of analytic gradients. Additionally, our benchmarks suggest that using anisotropic polarizabilities effectively increases values of dispersion energies, such that the *ad hoc* factor of 4/3 is not needed. Analysis of dispersion interactions on a number of dimers at different intermolecular separations suggests no notable necessity for introducing damping functions [67].

Figure 6.1 shows comparison of dispersion energies in sandwich and T-shaped isomers of the benzene dimer computed by SAPT and various EFP schemes, namely, EFP-EFP dispersion and two different implementations of QM-EFP dispersion. This and other benchmarks suggest that both QM-EFP dispersion formulations are comparable in accuracy and match SAPT dispersion energy within 10–15% relative errors. However, the formulation of equation (6.14) is more rigorous and more computationally efficient and thus should be preferred.

QM/EFP schemes were successfully applied for exploring excited state chemistry [68–75]. In terms of methodology, one needs to pay a close attention to polarization term in the QM-EFP Hamiltonian. This is because polarization requires self-consistent treatment and, in principle, each electronic excited state induces a different response of polarizable environment. In practice, however, state-specific changes in polarization, at least when energies are under question, are typically modest and might be either

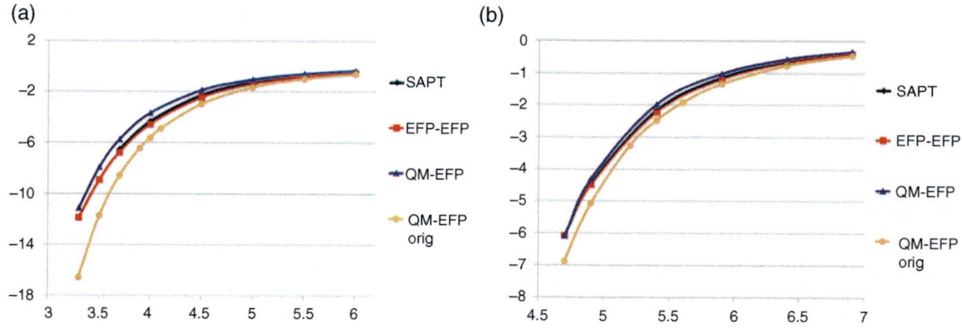

Figure 6.1 Dispersion energy in (a) sandwich and (b) T-shaped isomers of benzene dimer computed with SAPT (black), EFP-EFP (red), and two versions of QM-EFP, corresponding to equation (6.13) (orange) and equation (6.14) (blue).

ignored or approximated using perturbation theory [68, 69, 71]. An alternative solution to this problem is to find an average response of the environment to state-average wave function, as was implemented within state-average MCSCF and multiconfigurational quasidegenerate perturbation theory (MCQDPT) [72]. One more approach is to account for state-specific polarization through linear-response techniques [76]; however, it was not explored in the context of EFP.

For a consistent treatment of electronic excited states with QM/EFP formalism, QM-EFP dispersion term also needs to account for a change in the electron density upon excitation. This could be achieved either perturbatively, in the spirit of CIS(D) and related methods, in which perturbative dispersion correction is computed for each electronic state, or by computing dispersion correction for a transition density of a given electronic state. These ideas will be explored in future work.

Developments of the QM-EFP dispersion and exchange repulsion terms open unique opportunities for in-depth analysis of short-range solvatochromic effects. Indeed, classical QM/MM models are incapable of describing solvatochromic shifts in non-polar solvents or solutes, because QM and MM subsystems are not coupled through short-range van der Waals terms. Thus, QM wave function is not affected by changes in van der Waals interactions in the system. Another related problem appears in the description of electronic states with significant Rydberg character, where short-range cavity effects determine a spread of the diffuse state into the environment. These and other questions will be explored with QM/EFP models in the near future.

So far, QM/EFP schemes were used for describing electrostatic effects of the environment [68–75]. Initial applications targeted analysis of solvatochromic shifts and ionization energies in polar solvents, such as solvatochromic shifts on singlet and triplet excited states of para-nitroaniline chromophore in water, dioxane and cyclohexane solvents [68, 69], shifts of ionization energies in hydrated thymine [71], solvatochromic shifts of uracil in water [72], etc. Our recent work extends these ideas to two new directions; namely, characterization of solvent effects at interfaces and in biological systems. In both cases, solvent is not homogeneous anymore, such that QM/EFP polarizable embedding is expected to give an edge to electrostatic embedding of standard QM/MM.

The following section provides detailed description of two exemplary cases of QM/EFP usage.

6.3 Chemistry at Interfaces and Photobiology

6.3.1 OH Radical Solvated in Water

The hydroxyl radical (OH radical) is one of the most important oxidants in the atmosphere [77]. Yet, the dynamics of hydroxyl radical in atmospheric aerosols remain controversial. Even less is known about photochemical activity of the OH radical at the air–water interface. The absorption spectrum of OH radical in bulk water has two characteristic features: a narrow shoulder centered at around 310 nm and a broad peak at around 230 nm [78]. The narrow peak at 310 nm is due to the valence $n \rightarrow \pi^*$ transition, which can be observed in the gas phase excitation spectrum as well. The broad peak at 230 nm is believed to be due to the charge transfer excitations caused by the transfer of an electron from the surrounding water(s) to the hydroxyl radical [79]. This results in the formation of a positive hole on the surrounding water(s) and a hydroxide anion. However, it remains unclear how the photoactivity of a hydroxyl radical depends on its specific location and interaction with water. For example, do hydroxyl radicals at air–water surface and in bulk water exhibit significantly different absorption spectra? Which water molecules do contribute to the charge-transfer excitation and how delocalized is the charge-transfer state? We address these questions in the following section.

Excited state calculations of hydroxyl radical interacting with small water clusters generally agree with experimental assignment of two features in the absorption spectrum (Figure 6.2) [80]. Specifically, equation-of-motion coupled cluster method for ionized potentials (EOM-IP-CCSD) was employed in the 6-311++G** basis for a set of low-energy conformations of OH with one, two, three, and four waters. In EOM-IP-CCSD, coupled cluster equations are solved for a closed-shell anion, that is, OH^-–$(H_2O)_n$ and ionized states of the anion are computed after this step. As a result, the formalism is free of spin-contamination and provides a balanced description of excited states in radicals. Absorption spectra of these clusters, shown in Figure 6.2, reveal $n \rightarrow \pi^*$ transitions in the region of ~300 nm and charge-transfer states near 200 nm. The CT states of the lowest energy correspond to electron transfer from lone pairs of waters participating in H-bonding with OH radical. As expected, the number of CT states increases with an increase in the number of water molecules present. Intensity of some of these CT states also grows in larger clusters.

In order to explore the photochemistry of OH radical in bulk water and at air–water interface, we employed the following strategy. A classical MD NVE trajectory containing 864 water molecules and 4 hydroxyl radicals immersed in a rectangular periodic box of $3 \times 3 \times 10$ nm was obtained from the previous work by Roeselova *et al.* [81]. This setup ensures a layer of vacuum along the z-axis and emulates an air–water interface. Water molecules were described using POL3 water model, a polarizable water model developed by Caldwell and Kollman. The parameters corresponding to the hydroxyl radical were developed in-house by Roeselova *et al.* [82]. Ninty-four snapshots were extracted from the MD trajectory at equal intervals. In every snapshot, a single hydroxyl radical

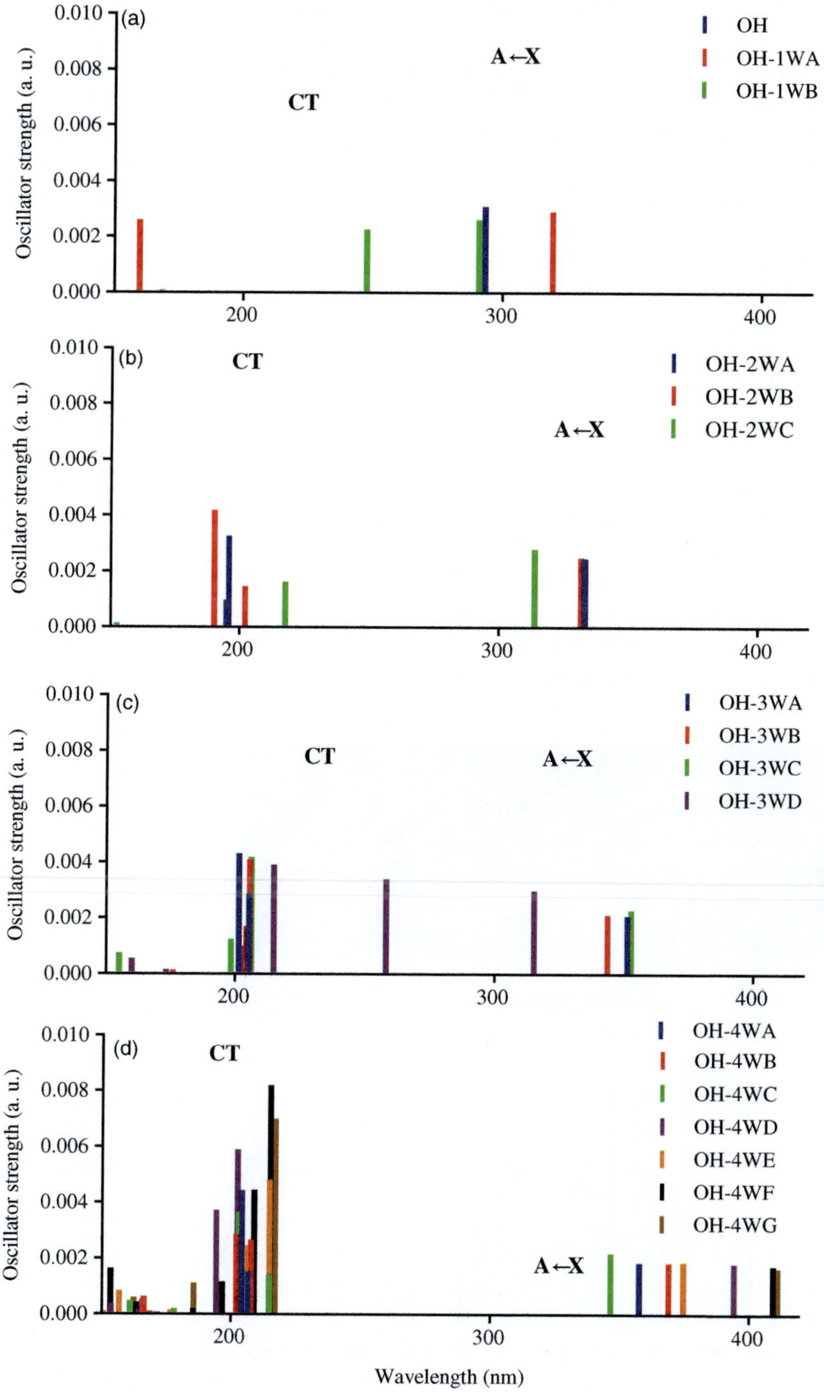

Figure 6.2 Excited state spectrum of OH radical solvated by (a) one, (b) two, (c) three, and (d) four water molecules. Several low-energy clusters are considered in each case. Valence A←X ($n \rightarrow \pi^*$) transitions are observed in the region between 300 nm and 400 nm; CT transitions are located around 200 nm. *Source*: Hoffman *et al.* [80]. Reproduced with permission of AIP Publishing.

was chosen at random for inclusion in the QM region. The remaining three hydroxyl radicals were described as EFP fragments. The geometries corresponding to the hydroxyl radical and the water molecules in the periodic box are shifted in such a way that the hydroxyl radical remains at the center of the box along the *Z*-axis. A sphere with radius of 0.6 nm is drawn from the bond midpoint of the hydroxyl radical. All water molecules falling within this range are described using QM methods, while all the other water molecules are described using EFP fragments. As mentioned earlier, other hydroxyl radicals are described using EFP fragments, irrespective of the distance from the QM hydroxyl radical. The choice of a cut-off radius of 0.6 nm is rationalized by a compromise of the size of the QM region that can be treated efficiently and a desire to explore energetics and intensity of long-range charge-transfer states. Generally, cut-off of 0.6 nm ensures inclusion of two shells of water molecules around hydroxyl radical.

Distribution of the number of waters within a sphere of radius 0.6 nm from an OH radical and representative snapshots are shown in Figure 6.3. We call snapshots in which OH radical is more exposed to the vacuum than to the bulk (generally when $n < 18$ waters are located within the 0.6 nm sphere around OH radical) as "OH in gas-phase." In a few other snapshots, the hydroxyl radicals are buried deep in the water layer ($n > 34$) and no water molecules within the 0.6 nm sphere are exposed to the vacuum layer. Such snapshots are termed as "OH in bulk water." In the bulk region, a hydroxyl radical is at least two shells of water underneath the surface. The region that lies in between, named as "OH in the surface region," comprises of snapshots where a hydroxyl radical is partially exposed to the surface or is not well buried inside the water box

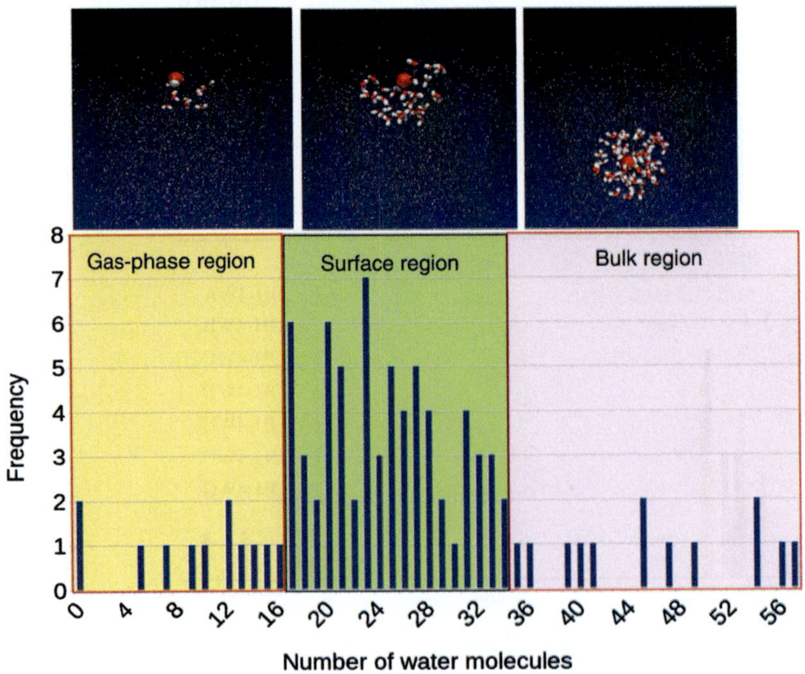

Figure 6.3 Distribution of the number of waters around an OH radical from representative snapshots of MD trajectory.

($17 < n < 35$). In these snapshots, there are at the most two hydration shells between the air–water interface and the hydroxyl radical. As is clear from Figure 6.3, and as was discussed in detail in Ref. [79], in majority of snapshots, the hydroxyl radical prefers the surface region. Using these snapshots, we analyze the charge-transfer states of hydroxyl radical in three different regions, surface, vacuum, and bulk.

It is not viable to compute the excited states of clusters including more than ~10 waters using equation-of-motion coupled cluster methods. To circumvent this issue, we have made use of Koopmans theorem applied to the anionic system. According to Koopmans theorem, the ionization potential (IP) from a given molecular orbital is equal to the negative of the orbital energy. Then, the excited state energies of a radical can be estimated as a difference in IPs from a HOMO orbital, which corresponds to a singly occupied molecular orbital (SOMO) in the radical and other occupied orbitals n:

$$E^{\text{neutral}}_{n\rightarrow\text{SOMO}} \approx IP^{\text{anion}}_{n} - IP^{\text{anion}}_{\text{HOMO}} \approx \varepsilon^{\text{anion}}_{n} - \varepsilon^{\text{anion}}_{\text{HOMO}} \tag{6.15}$$

This approximation is valid for singly excited states dominated by a single transition, which is the case for CT states in OH–water systems. Oscillator strengths of the excited states can be obtained using transition dipole moments of involved HF orbitals of the anion. Thus, Koopmans theorem can be considered as a frozen-orbital approximation to EOM-IP methodology for predicting excited states of radical species.

In order to confirm the accuracy of Koopmans theorem for computing excited states in OH–water system, we compared excitation energies and oscillator strengths of low-energy CT states obtained using Koopmans theorem and EOM-IP-CCSD in 6-31G* basis set for one of the MD snapshots in which only one OH radical and five closest water molecules were left (Figure 6.4).

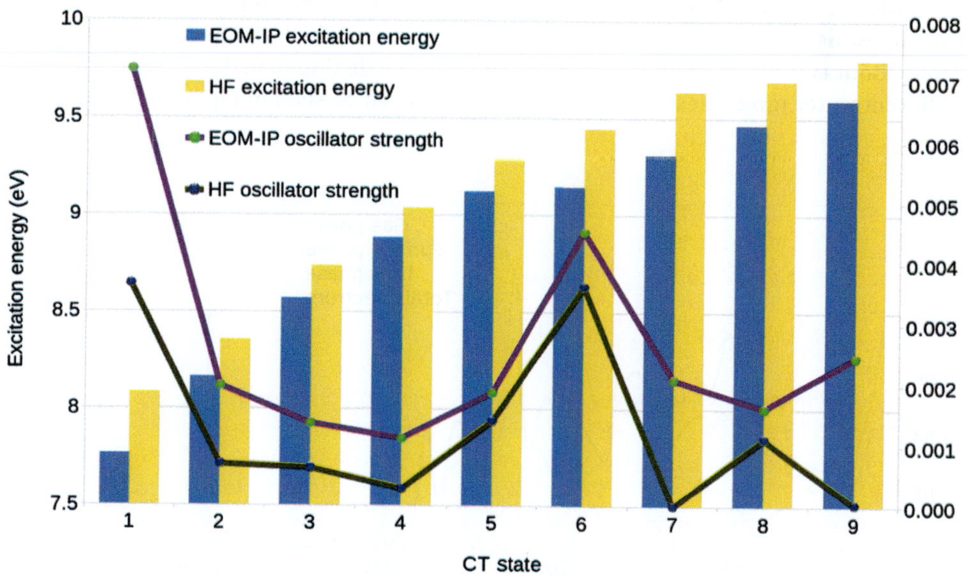

Figure 6.4 Excitation energies and oscillator strengths of low-energy CT states in a representative OH–$(H_2O)_5$ cluster computed using Koopmans theorem and EOM-IP-CCSD method.

As each surrounding water molecule possesses two lone-pair (LP) orbitals, 10 CT states of $H_2O_{LP}{\rightarrow}OH_{\pi}$ character, that is, excitations from water lone pairs to π orbital of OH radical, are observed. For the sake of clarity, the last CT state with energy higher than 10 eV is omitted from Figure 6.4. From the figure, it can be observed that the excited state energies obtained from Koopmans theorem systematically overestimate EOM-IP excited states, but do not differ by more than 0.3 eV. It is very promising that the ordering of the CT states coincides in both methods. On the other hand, the oscillator strengths obtained using HF orbitals are consistently underestimated, but general trends remain the same, that is, bright states are bright and dark states are dark in both methods. Thus, we conclude that Koopmans theorem is a viable alternative to qualitatively assess the nature of CT states in large clusters.

Figure 6.5 depicts CT absorption spectrum of OH solvated in water based on calculations at snapshot geometries extracted from the MD trajectory, using Koopmans theorem approach for QM/EFP system, that is, HF/EFP. Gaussians with FWHM of 3.73 nm were used to model the individual excitations, followed by which the individual spectra were combined together. The total CT spectrum is decomposed into contributions from snapshots corresponding to the location of OH radical in gas-phase, surface, and bulk water, following Figure 6.3. As can be seen in Figure 6.5, the contributions from the gas-phase and bulk phase hydroxyl radical to the total CT spectrum are lower compared to the contribution from the surface phase. This is primarily due to the higher number of snapshots representing the hydroxyl radical in the surface phase, as against the other two. The most striking feature in the CT spectrum is the presence of bright low-energy charge-transfer excitations that extend as far as 290 nm, bordering the near-UV region. These red-shifted transitions arise due to the hydroxyl radicals in bulk phase, in contrast to the rest of the spectrum dominated by snapshots from hydroxyl radical in the surface.

To understand the origin of such low-energy transitions, we considered a representative snapshot of OH radical in bulk water (Figure 6.6) in which the hydroxyl radical is located about two water shells beneath the surface. We distinguished the contributions of different water molecules to the CT spectrum. As can be seen in Figure 6.6, two QM

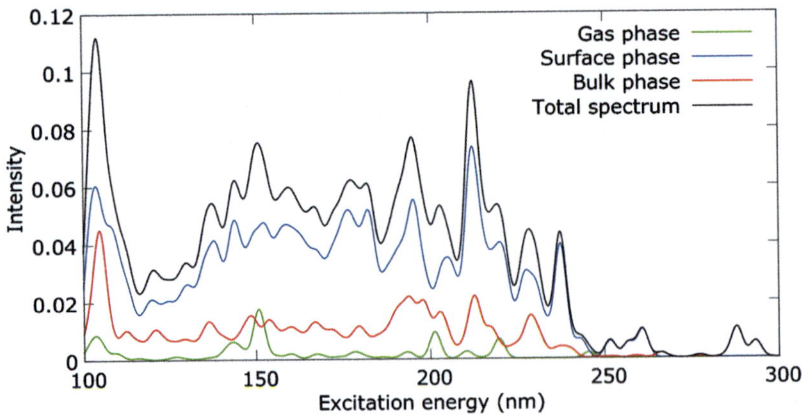

Figure 6.5 CT absorption spectrum of OH radical solvated in water.

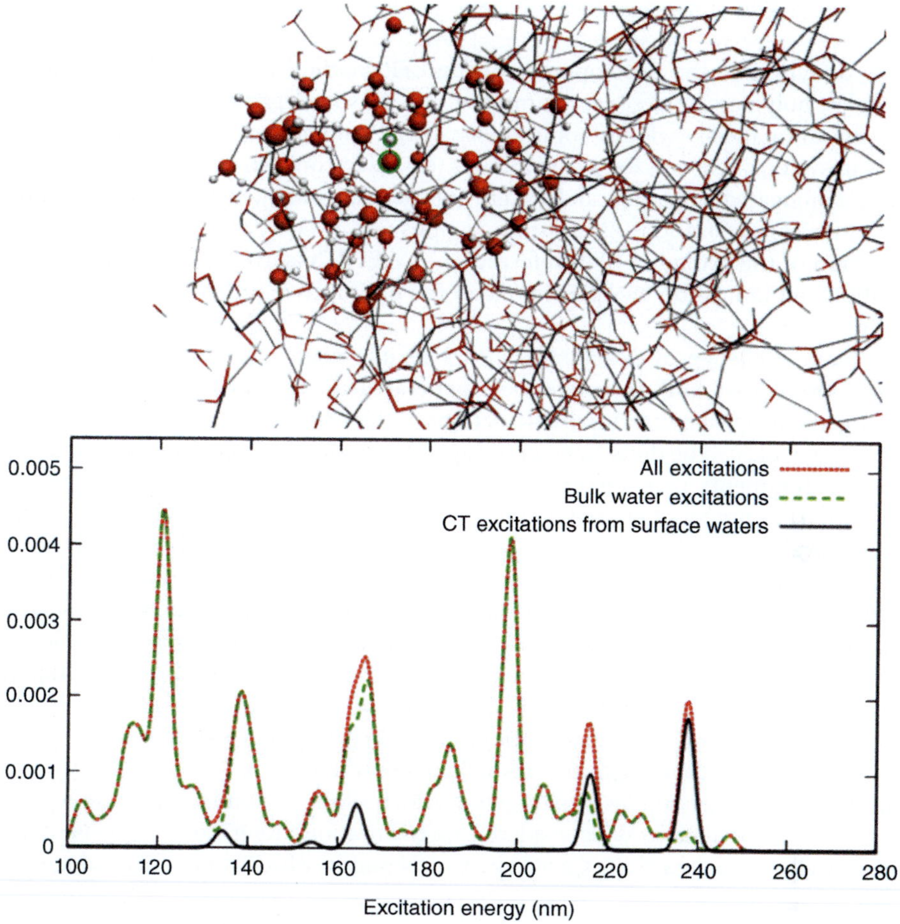

Figure 6.6 CT spectrum of a representative snapshot of OH radical in bulk water.

water molecules are present at the surface and have dangling H-bonds. We term these two water molecules as "surface waters." All the other water molecules in the QM region are located below the first layer of water in the box, and are considered as "bulk waters" for this purpose. It can be seen that the contribution of the surface waters to the CT spectrum is seen prominently in the low-energy region of the spectrum. This is because the water lone-pair orbitals that lack stabilization due to H-bonding to other waters are more prone to donate the electron to the hydroxyl radical, than the lone-pairs on bulk water molecules that are typically fully H-bonded with other water molecules. Thus, the interplay of H-bond water network and the position of the hydroxyl radical in the water play a major role in shaping the low-energy region of the CT absorption spectrum. It is also intriguing that such low-energy transitions are very bright suggesting that photoexcitation of solvated OH radical might lead to a long-range charge separation. It remains to be explored how these near-UV long-range CT states affect dynamics of OH–water systems.

6.3.2 EFP for Macromolecules and Polymers

A new direction in the development of the EFP method is an extension to macromolecules, polymers, and generally molecules with soft degrees of freedom [34]. This extension dramatically widens applicability of EFP and allows exploring environment effects in systems relevant for photovoltaics (e.g., electronic structure and dynamics in organic conjugated polymers and hetero-interfaces), energy storage (charge-transfer in Li batteries), separation (diffusion in zeolites and MOFs), etc. Here, we discuss an application of the extended EFP (referred to as mEFP, "macro-EFP") to the analysis of excitation and ionization energies in one of the photoactive protein, green fluorescent protein (GFP).

The main challenge of extending EFP to flexible molecules comes from the original constraint that EFP parameters are generated for a rigid fragment and remain unchanged during all consecutive calculations of the system. EFP parameterization of a flexible molecule can be done in several ways. One option is to split the macromolecule into small fragments that remain rigid during simulation. Then, the questions to be solved are (i) to prepare parameters for fragments with unsatisfied valences and (ii) to develop covalent links between the fragments to ensure stability of the macromolecule. The second option is to consider the macromolecule as a single fragment, but be able to adjust parameters upon geometrical changes of the molecule. One more possibility, to combine both ideas and to split the macromolecule into fragments with adjustable parameters, is exploited in a hybrid of EFP and FMO, called EFMO (effective fragment molecular orbital method) [6, 35, 83]. The following section will provide details of our recent implementation of the first option, that is, representing a system of interest as a set of small rigid fragments, on the example of peptide system.

Splitting of natural peptides into fragments can be done in a number of ways, as shown in Figure 6.7. Specifically, we considered fragmentation along Cα–C, Cα–N, peptide C–N, as well as two cuts along Cα–C and Cα–N bonds, which results in producing two

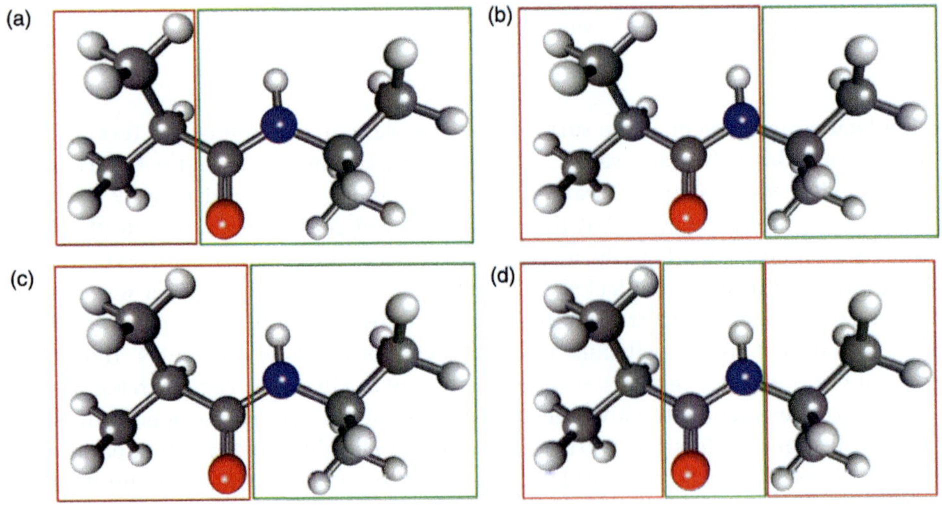

Figure 6.7 Possible fragmentation schemes for a peptide system.

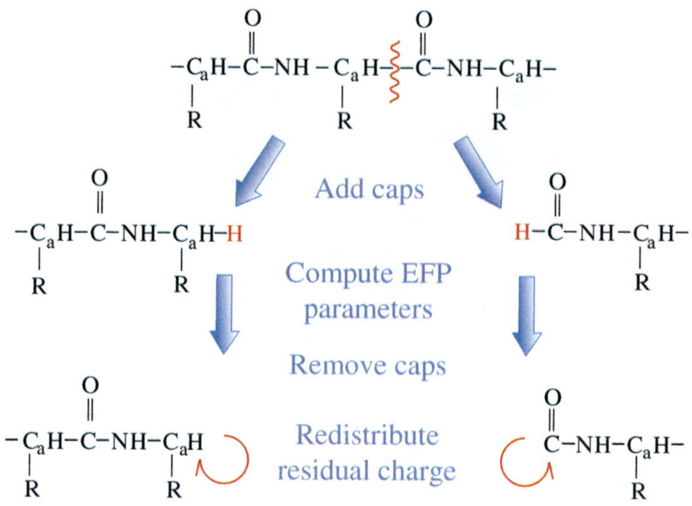

Figure 6.8 Preparing fragment parameters in mEFP.

fragments per amino acid. Generally, it is expected that fragmentation along peptide bond might be less favorable because this bond is the most polar and its breaking might result in an unphysical charge redistribution. Additionally, as fully flexible potential is a goal, fragmenting along the peptide bond does not give any advantage, as this bond typically remains stiff in proteins. On the other hand, three other fragmentation patterns have advantage of cutting a less polar bond that is better suited for capping, as well as allowing flexibility along either one or both ψ or φ torsional angles.

A general procedure for preparing EFP parameters in peptides, as shown in Figure 6.8, has the following steps:

- Peptide is cut into fragments following one of the fragmentation schemes;
- Open valencies are capped, typically with H atoms, positioned along cut bonds at 1.08 Å from C or at 1.00 Å from N;
- EFP parameters for the generated closed-shell fragments are computed;
- Capping atoms and parameters located at and around capping atoms (including parameters at bond-mid-points and LMO centroids associated with capping atoms) are removed from EFP potentials;
- Residual charges on fragments due to the removal of capping atoms are neutralized by adding the corresponding partial charge on the nearest C or N atom.

These steps were implemented in Python scripts and can be automatically executed, such that parameters for any natural peptide can be prepared efficiently without user intervention.

The above procedure can be generalized to other flexible molecules or macromolecules. However, in this case user needs to specify fragmentation pattern that might be unique for each considered system.

In the present implementation, covalent links between fragments are not fully tested, such that only static properties can be explored at the moment. However, the

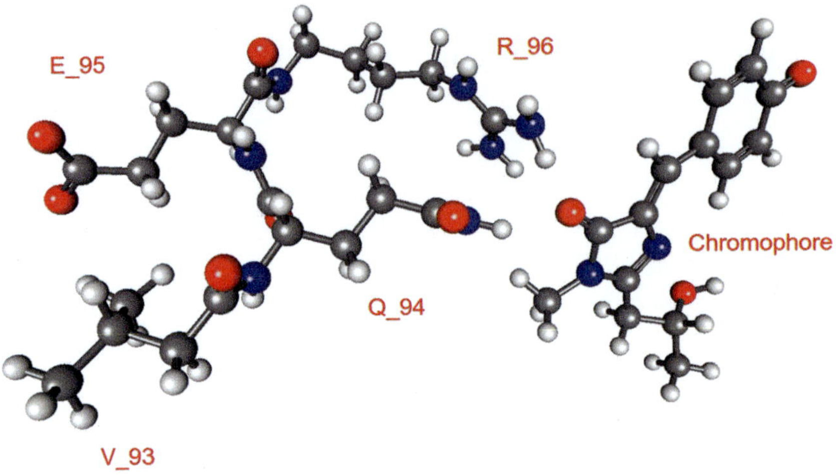

Figure 6.9 A model system containing GFP chromophore with 4-peptide VAL93-GLN94-GLU95-ARG96.

implementation of fully functional covalent links (using classical harmonic potentials) and analytic gradients, enabling MD simulations, is anticipated in near future.

Figure 6.9 shows a model system considered for validating mEFP approach in biology. The system consists of GFP chromophore and four amino acid sequence naturally present in enhanced GFP (pdb id:1EMA): VAL93, GLN94, GLU95, and ARG96. Chromophore as well as unsaturated valences in VAL93 and ARG96 were capped with hydrogens. GFP chromophore was considered in two forms, anionic (deprotonated) and neutral (protonated). GLU95 is deprotonated and ARG96 is protonated in case of the anionic chromophore, such that the total charge of the model system is −1 (negatively charged chromophore and GLU95, and positively charged ARG96). In case of neutral chromophore, GLU95 is protonated, resulting in the total system charge of +1.

Table 6.1 shows the performance of mEFP compared to full QM calculations, for vertical dissociation and ionization energies, VDE and VIE, respectively [34]. Several interesting effects are observed. For example, it is obvious that polarization of the environment is more significant in the anionic than in the neutral chromophore, as can be seen by comparing the VIE/VDE differences between polarization-on and -off values in the superfragment calculations, which are 0.18 eV and 0.11 eV for the anionic and neutral chromophores, respectively. Further, polarization is smaller in fragmented (mEFP) than in the superfragment (EFP) calculations. For example, for the anionic chromopore, polarization-induced change in VDE decreases from 0.18 eV in QM/EFP to 0.06–0.08 eV in QM/mEFP, and similarly for the neutral chromophore. This observation suggests that neighboring fragments effectively "depolarize" each other.

Electrostatic damping takes care of charge penetration energy, which was explored in detail in the context of interactions between two or more EFP-EFP fragments [36, 52, 84]. It is noteworthy to explore the importance of electrostatic damping for interactions between QM and EFP subsystems as well. As shown in Table 6.1, the QM-EFP electrostatic damping shifts energy differences by 0.01–0.05 eV and improves the agreement between the QM/EFP results and the reference QM data. Different fragmentation schemes produce similar shifts. To sum up, data shown in Table 6.1 suggest

Table 6.1 VDE and VIE (eV) in the anionic and neutral GFP chromophore surrounded by nearby four-peptide as a function of electrostatic and polarization damping in the EFP region.

	QM/EFP	Polarization off	Electrostatic damping off
Anionic GFP			
Full QM	4.03		
QM/EFP: superfragment	3.93	3.75	3.88
QM/mEFP: Cα–C	3.94	3.87	3.90
QM/mEFP: Cα–N	3.92	3.86	3.87
QM/mEFP: C–N	3.93	3.85	3.89
QM/mEFP: Cα–C and Cα–N	3.93	3.87	3.88
Neutral GFP			
Full QM	9.74		
QM/EFP: superfragment	9.67	9.56	9.66
QM/mEFP: Cα–C	9.68	9.67	9.67
QM/mEFP: Cα–N	9.81	9.79	9.80
QM/mEFP: C–N	9.69	9.66	9.68
QM/mEFP: Cα–C and Cα–N	9.66	9.65	9.65

ωB97X-D/6-31G(d) is used in all calculations.

that both electrostatic damping and polarization interactions are important for accurate description of VIEs/VDEs. For example, in the anionic chromophore, the best (with electrostatic damping and polarization) QM/mEFP calculations with Cα–C scheme underestimate VDE by 0.09 eV, while switching of electrostatic damping and polarization increases the errors by 0.04 eV and 0.07 eV, respectively. In the neutral chromophore, Cα–C scheme underestimates VIEs by 0.06 eV, but turning off electrostatic damping and polarization creates additional errors of 0.01 eV each.

This and other examples, discussed in detail in Ref. [81], suggest that QM/mEFP fragmentation approach for proteins accurately reproduces ionization as well as excitation and electron attachment energies, for electronic processes localized in the QM subsystem. Typically, discrepancies between the full QM calculations and QM/mEFP calculations do not exceed 0.1 eV. In most cases, the errors due to fragmenting a peptide into fragments (discrepancies between QM/mEFP and superfragment QM/EFP) are 0.01–0.02 eV. Fragmentation schemes with cutting Cα–C bond or cutting two bonds per amino acid residue (Cα–C and Cα–N) produce the most reliable results, even though other two fragmentation schemes also do not show dramatic failures. On the other hand, polarization interactions play a noticeable role when electronic properties of biological chromophores are under question. Short-range damping of electrostatic interactions between the QM and EFP subsystems, which corrects classical multipole expansion for charge-penetration energy, also brings the QM/mEFP results into a closer agreement with the reference full QM values.

Overall, we showed that the QM/mEFP approach provides a rigorous way to incorporate polarization embedding into studies of biological systems. mEFP algorithms will be generalized to other polymers and flexible molecules in future studies.

6.4 Future Directions and Outlook

While already being successful as a way to describe and characterize non-covalent interactions and solvent effects, EFP and QM-EFP schemes continue to be developed and enhanced. Our current work is focused along following directions: (i) development of an EFP-based force field for polymers and macromolecules; (ii) development of computationally efficient schemes that allow flexibility in fragments' geometries; (iii) development of fully-coupled (full-embedding) QM-EFP schemes capable of exploring dynamics of electronically excited species in various environments; and (iv) extending EFP to free energy calculations.

Development of the EFP-based force field that could be applicable for modeling flexible molecules and macromolecules is motivated by often-voiced concerns about the credibility of classical force fields, especially when dealing with chemical species containing non-standard bonding patterns or uncommon chemical elements. On the other hand, EFP parameters can be created automatically and from first principles for any chemical species, which solves the problem of missing parameters often experienced by classical force fields. Additionally, in all considered examples, EFP excels classical force fields when details of structural patterns and interaction energies are taken into account. As was mentioned earlier in this chapter, the two main bottlenecks on the way of employing EFP as a universal force field are the necessity to describe covalent interactions between the fragments and a high computational cost, when compared to classical force fields. Thus, while EFMO solves the first problem in an elegant and rigorous way, by treating covalent links between fragments analogous to the FMO method, the computational cost of this approach increases further compared to the cost of the original EFP method. An alternative scheme is to describe covalent links between fragments at the classical level, while all non-covalent interactions remain to be dealt with at the EFP level. Fragmentation step has already been explored using the example of electronic properties of GFP chromophore surrounded by a peptide, as described earlier in this chapter. This hybrid EFP-MM scheme has been implemented in *libefp* code and is currently undergoing testing and benchmarking. Toward diminishing the computational cost of EFP calculations, several scenarios can be considered. Adjustable cut-offs for short-range EFP terms, exchange-repulsion and charge-transfer, can solve part of the problem. Exploiting different time-scales of van der Waals and electrostatic interactions in MD simulations is another interesting possibility. Additionally, introducing efficient schemes for computing long-range parts of polarization and electrostatic terms might be of interest as well.

A question of flexibility of fragments' geometries arises in a context of modeling fragments that do not have flexible bonds, but still experience non-negligible geometrical changes in bond lengths and valence angles. One example is the changes in O–H and N–H bond lengths and involved valence angles upon formation of H-bond. While the EFP formalism assumes that fragments remain rigid, we found that these and other changes in internal geometries of fragments are significant enough and should be accounted for, implying that the usage of gas-phase geometries of fragments does not result in sufficiently accurate non-covalent energies. One solution is to recompute fragment parameters for each unique fragment geometry, assuming that the fragment geometries are obtained from a different QM or MM method. This approach, with underlying QM calculations of fragments' geometries, is an essence of the EFMO method. However, the resulting computational cost of EFMO makes it challenging for performing long-scale MD simulations, at least at typical in-house high-performance clusters. Another possibility is to adjust gas-phase parameters to a new geometry, by shifting and possibly rotating positions of multipoles, polarizabilities, and other parameters according to the changes in the fragment geometry. Our pilot studies suggest that the latter approach corrects main deficiencies of using gas-phase parameters and does not introduce any additional computational cost, and as such, can be combined with classical force fields (for describing bonded interactions within fragments) for efficient dynamics simulations.

Our recent work contributes to the development of full-embedding QM/EFP method capable of describing electronic structure and dynamics in condensed phases, with self-consistent treatment of all interacting terms between QM and EFP subsystems. Currently, ground state energy of the QM-EFP system can be computed with the inclusion of quantum-mechanically described electrostatic, polarization, dispersion, and exchange-repulsion terms. Implementation of dispersion and exchange-repulsion gradient terms need to be finalized for performing QM/EFP dynamics. Additionally, description of electronic excited states requires corresponding extension of the QM-EFP dispersion term. These method and code developments are work in progress.

Finally, the ability to compute free energies with the EFP method would provide new opportunities for enriching the field of computational drug design as well as other areas where the accurate prediction of binding energies is of importance. Computing free energies reliably is challenging by a necessity to properly sample configurational space that is often associated with high computational cost. Development of algorithms focused on minimizing computational cost of the EFP calculations, including hybrid EFP-MM schemes, will contribute to enabling EFP-based free energy calculations. Additionally, implementation of simple algorithms such as steered MD within *libefp* software library as well as interfacing *libefp* to free energy packages for more advanced ways of computing free energies are directions that will be undertaken in the near future.

References

1 Gordon, M. S., Fedorov, D. G., Pruitt, S. R. & Slipchenko, L. V. Fragmentation Methods: A Route to Accurate Calculations on Large Systems. *Chemical Reviews* **112**, 632–672, doi:10.1021/cr200093j (2011).

2 Mayhall, N. J. & Raghavachari, K. Molecules-in-Molecules: An Extrapolated Fragment-Based Approach for Accurate Calculations on Large Molecules and Materials. doi:papers3://publication/doi/10.1021/ct200033b (2011).

3 Fedorov, D. G. & Kitaura, K. Extending the Power of Quantum Chemistry to Large Systems with the Fragment Molecular Orbital Method. *The Journal of Physical Chemistry A* **111**, 6904–6914, doi:10.1021/jp0716740 (2007).

4 Wang, B. *et al.* Quantum Mechanical Fragment Methods Based on Partitioning Atoms or Partitioning Coordinates. *Accounts of Chemical Research* **47**, 2731–2738, doi:10.1021/ar500068a (2014).

5 Wesolowski, T. A. Embedding Potentials for Excited States of Embedded Species. *The Journal of Chemical Physics* **140**, doi:10.1063/1.4870014 (2014).

6 Steinmann, C., Fedorov, D. G. & Jensen, J. H. Effective Fragment Molecular Orbital Method: A Merger of the Effective Fragment Potential and Fragment Molecular Orbital

Methods. *The Journal of Physical Chemistry A* **114**, 8705–8712, doi:10.1021/jp101498m (2010).
7 Becke, A. D. & Johnson, E. R. Exchange-Hole Dipole Moment and the Dispersion Interaction Revisited. *The Journal of Chemical Physics* **127**, 154108, doi:10.1063/1.2795701 (2007).
8 Grimme, S., Antony, J., Ehrlich, S. & Krieg, H. A Consistent and Accurate Ab Initio Parametrization of Density Functional Dispersion Correction (DFT-D) for the 94 Elements H–Pu. *The Journal of Chemical Physics* **132**, 154104, doi:10.1063/1.3382344 (2010).
9 Burns, L. A., Mayagoitia, Á. V.-., Sumpter, B. G. & Sherrill, C. D. Density-Functional Approaches to Noncovalent Interactions: A Comparison of Dispersion Corrections (DFT-D), Exchange-Hole Dipole Moment (XDM) Theory, and Specialized Functionals. *The Journal of Chemical Physics* **134**, 084107, doi:10.1063/1.3545971 (2011).
10 Tkatchenko, A. & Scheffler, M. Accurate Molecular Van derWaals Interactions from Ground-State Electron Density and Free-Atom Reference Data. *Physical Review Letters* **102**, 073005 (2009).
11 Tkatchenko, A., DiStasio, R. A., Car, R. & Scheffler, M. Accurate and Efficient Method for Many-Body Van derWaals Interactions. *Physical Review Letters* **108**, 236402 (2012).
12 Chai, J.-D. & Head-Gordon, M. Long-range corrected hybrid density functionals with damped atom–atom dispersion corrections. *Physical Chemistry Chemical Physics* **10**, 6615–6620, doi:10.1039/B810189B (2008).
13 Xie, W. & Gao, J. Design of a Next Generation Force Field: The X-POL Potential. *Journal of Chemical Theory and Computation* **3**, 1890–1900, doi:10.1021/ct700167b (2007).
14 Flick, J. C., Kosenkov, D., Hohenstein, E. G., Sherrill, C. D. & Slipchenko, L. V. Accurate Prediction of Non-Covalent Interaction Energies with the Effective Fragment Potential Method: Comparison of Energy Components to Symmetry-Adapted Perturbation Theory for the S22 Test Set. *Journal of Chemical Theory and Computation* **8**, 2835–2843, doi:10.1021/ct200673a (2012).
15 Thanthiriwatte, K. S., Hohenstein, E. G., Burns, L. A. & Sherrill, C. D. Assessment of the Performance of DFT and DFT-D Methods for Describing Distance Dependence of Hydrogen-Bonded Interactions. *Journal of Chemical Theory and Computation* **7**, 88–96, doi:10.1021/ct100469b (2011).
16 Goerigk, L., Kruse, H. & Grimme, S. Benchmarking Density Functional Methods against the S66 and S66x8 Datasets for Non-Covalent Interactions. *ChemPhysChem* **12**, 3421–3433, doi:10.1002/cphc.201100826 (2011).
17 Yang, Yu, H., York, D., Cui, Q. & Elstner, M. Extension of the Self-Consistent-Charge Density-Functional Tight-Binding Method: Third-Order Expansion of the Density Functional Theory Total Energy and Introduction of a Modified Effective Coulomb Interaction. *The Journal of Physical Chemistry A* **111**, 10861–10873, doi:10.1021/jp074167r (2007).
18 Fedorov, D. G., Nagata, T. & Kitaura, K. Exploring Chemistry with the Fragment Molecular Orbital Method. *Physical Chemistry Chemical Physics* **14**, 7562–7577 (2012).
19 Kitaura, K., Ikeo, E., Asada, T., Nakano, T. & Uebayasi, M. Fragment Molecular Orbital Method: An Approximate Computational Method for Large Molecules. *Chemical Physics Letters* **313**, 701–706 (1999).
20 Fedorov, D. G. & Kitaura, K. In *Modern Methods for Theoretical Physical Chemistry of Biopolymers* (eds E. B. Starikov, J. P. Lewis, & S. Tanaka), 3–38. (Elsevier, 2006).

21 Fedorov, D. G. & Kitaura, K. Extending the Power of Quantum Chemistry to Large Systems with the Fragment Molecular Orbital Method. *The Journal of Physical Chemistry A* **111**, 6904–6914, doi: 10.1021/Jp0716740 (2007).

22 Fedorov, D. G. & Kitaura, K. *The Fragment Molecular Orbital Method: Practical Applications to Large Molecular Systems*. (CRC Press, 2009).

23 Fedorov, D. G. & Kitaura, K. Pair Interaction Energy Decomposition Analysis. *The Journal of Computational Chemistry* **28**, 222–237, doi: 10.1002/Jcc.20496 (2007).

24 Green, M. C. *et al.* Open-Shell Pair Interaction Energy Decomposition Analysis (PIEDA): Formulation and Application to the Hydrogen Abstraction in Tripeptides. *The Journal of Chemical Physics* **138**, 074111 (2013).

25 Hohenstein, E. G. & Sherrill, C. D. Density Fitting and Cholesky Decomposition Approximations in Symmetry-Adapted Perturbation Theory: Implementation and Application to Probe the Nature of Pi-Pi Interactions in Linear Acenes. *The Journal of Chemical Physics* **132**, 184111 (2010).

26 Hohenstein, E. G. & Sherrill, C. D. Efficient Evaluation of Triple Excitations in Symmetry-Adapted Perturbation Theory via Second-Order M[o-slash]ller–Plesset Perturbation Theory Natural Orbitals. *The Journal of Chemical Physics* **133**, 104107 (2010).

27 Hohenstein, E. G. *et al.* Large-Scale Symmetry-Adapted Perturbation Theory Computations via Density Fitting and Laplace Transformation Techniques: Investigating the Fundamental Forces of DNA-Intercalator Interactions. *The Journal of Chemical Physics* **135**, 174107 (2011).

28 Gordon, M. S., Slipchenko, L. V., Li, H. & Jensen, J. H. The Effective Fragment Potential: A General Method for Predicting Intermolecular Forces. *Annual Reports in Computational Chemistry* **3**, 177–193 (2007).

29 Ghosh, D. *et al.* Noncovalent Interactions in Extended Systems Described by the Effective Fragment Potential Method: Theory and Application to Nucleobase Oligomers. *The Journal of Physical Chemistry A* **114**, 12739–12754, doi:10.1021/jp107557p (2010).

30 Gordon, M. S., Smith, Q. A., Xu, P. & Slipchenko, L. V. Accurate First Principles Model Potentials for Intermolecular Interactions. *Annual Review of Physical Chemistry* **64**, 553–578, doi:10.1146/annurev-physchem-040412-110031 (2013).

31 Slipchenko, L. V. In *Many-Body Effects and Electrostatics in Biomolecules* (eds Q. Cui, M. Meuwly, & P. Ren) Ch. **5**, 147–190. (CRC Press, 2016).

32 Day, P. N. *et al.* An Effective Fragment Method for Modeling Solvent Effects in Quantum Mechanical Calculations. *The Journal of Chemical Physics* **105**, 1968–1986, doi:10.1063/1.472045 (1996).

33 Stone, A. J. *The Theory of Intermolecular Forces*. (Oxford University Press: Oxford, 1996).

34 Gurunathan, P. K. *et al.* Extension of the Effective Fragment Potential Method to Macromolecules. *The Journal of Physical Chemistry B* **120**, 6562–6574, doi:10.1021/acs.jpcb.6b04166 (2016).

35 Bertoni, C. & Gordon, M. S. Analytic Gradients for the Effective Fragment Molecular Orbital Method. *Journal of Chemical Theory and Computation*, doi:10.1021/acs.jctc.6b00337 (2016).

36 Gordon, M. S. *et al.* The Effective Fragment Potential Method: A QM-Based MM Approach to Modeling Environmental Effects in Chemistry. *Journal of Physical Chemistry A* **105**, 293–307, doi:10.1021/Jp002747h (2001).

37 Adamovic, I., Freitag, M. A. & Gordon, M. S. Density Functional Theory Based Effective Fragment Potential Method. *The Journal of Chemical Physics* **118**, 6725–6732, doi:10.1063/1.1559912 (2003).

38 Webb, S. P. & Gordon, M. S. Solvation of the Menshutkin Reaction: A Rigorous Test of the Effective Fragment Method. *The Journal of Physical Chemistry A* **103**, 1265–1273, doi:10.1021/jp983781n (1999).

39 Adamovic, I. & Gordon, M. S. Solvent Effects on the SN2 Reaction: ; Application of the Density Functional Theory-Based Effective Fragment Potential Method. *The Journal of Physical Chemistry A* **109**, 1629–1636, doi:10.1021/jp040665d (2005).

40 Yoo, S., Zahariev, F., Sok, S. & Gordon, M. S. Solvent Effects on Optical Properties of Molecules: A Combined Time-Dependent Density Functional Theory/Effective Fragment Potential Approach. *The Journal of Chemical Physics* **129**, 144112, doi:10.1063/1.2992049 (2008).

41 Arora, P. *et al.* Solvent-Induced Frequency Shifts: Configuration Interaction Singles Combined with the Effective Fragment Potential Method. *The Journal of Physical Chemistry A* **114**, 6742–6750 (2010).

42 Day, P. N., Pachter, R., Gordon, M. S. & Merrill, G. N. A Study of Water Clusters using the Effective Fragment Potential and Monte Carlo Simulated Annealing. *The Journal of Chemical Physics* **112**, 2063–2073, doi:10.1063/1.480775 (2000).

43 Kemp, D. D. & Gordon, M. S. Theoretical Study of the Solvation of Fluorine and Chlorine Anions by Water. *The Journal of Physical Chemistry A* **109**, 7688–7699, doi:10.1021/jp058086b (2005).

44 Kemp, D. D. & Gordon, M. S. An Interpretation of the Enhancement of the Water Dipole Moment due to the Presence of Other Water Molecules. *The Journal of Physical Chemistry A* **112**, 4885–4894, doi:10.1021/jp801921f (2008).

45 Minezawa, N. & Gordon, M. S. Optimizing Conical Intersections of Solvated Molecules: The Combined Spin-Flip Density Functional Theory/Effective Fragment Potential Method. *The Journal of Chemical Physics* **137**, 034116, doi:10.1063/1.4734314 (2012).

46 Jensen, J. H. & Gordon, M. S. An Approximate Formula for the Intermolecular Pauli Repulsion Between Closed Shell Molecules. *Molecular Physics* **89**, 1313–1325 (1996).

47 Jensen, J. H. & Gordon, M. S. An Approximate Formula for the Intermolecular Pauli Repulsion Between Closed Shell Molecules. II. Application to the Effective Fragment Potential Method. *The Journal of Chemical Physics* **108**, 4772–4782, doi:10.1063/1.475888 (1998).

48 Ioannou, A. G., Colwell, S. M. & Amos, R. D. The Calculation of Frequency-Dependent Polarizabilities using Current Density Functional Theory. *Chemical Physics Letters* **278**, 278–284, doi:10.1016/S0009-2614(97)00978-0 (1997).

49 Amos, R. D. *et al.* Ab Initio Prediction of Properties of Carbon dioxide, Ammonia, and Carbon dioxide...Ammonia. *The Journal of Physical Chemistry* **89**, 2186–2192, doi:10.1021/j100257a010 (1985).

50 Adamovic, I. & Gordon, M. S. Dynamic Polarizability, Dispersion Coefficient C6 and Dispersion Energy in the Effective Fragment Potential Method. *Molecular Physics* **103**, 379–387, doi:10.1080/00268970512331317246 (2005).

51 Tang, K. T. & Toennies, J. P. An Improved Simple Model for the Van der Waals Potential Based on Universal Damping Functions for the Dispersion Coefficients. *The Journal of Chemical Physics* **80**, 3726–3741, doi:10.1063/1.447150 (1984).

52 Slipchenko, L. V. & Gordon, M. S. Damping Functions in the Effective Fragment Potential Method. *Molecular Physics* **107**, 999–1016, doi:10.1080/00268970802712449 (2009).

53 Li, H., Gordon, M. S. & Jensen, J. H. Charge Transfer Interaction in the Effective Fragment Potential Method. *The Journal of Chemical Physics* **124**, 214108, doi:10.1063/1.2196884 (2006).

54 Lu, W. C. *et al.* Molecule Intrinsic Minimal Basis Sets. I. Exact Resolution of Ab Initio Optimized Molecular Orbitals in Terms of Deformed Atomic Minimal-Basis Orbitals. *The Journal of Chemical Physics* **120**, 2629–2637, doi:10.1063/1.1638731 (2004).

55 Xu, P. & Gordon, M. S. Charge Transfer Interaction using Quasiatomic Minimal-Basis Orbitals in the Effective Fragment Potential Method. *The Journal of Chemical Physics* **139**, 194104, doi:10.1063/1.4829509 (2013).

56 Slipchenko, L. V. & Gordon, M. S. Water–Benzene Interactions: An Effective Fragment Potential and Correlated Quantum Chemistry Study. *The Journal of Physical Chemistry A* **113**, 2092–2102, doi:10.1021/Jp808845b (2009).

57 Smith, Q. A., Gordon, M. S. & Slipchenko, L. V. Benzene–Pyridine Interactions Predicted by the Effective Fragment Potential Method. *The Journal of Physical Chemistry A* **115**, 4598–4609, doi:10.1021/jp201039b (2011).

58 Smith, Q. A., Gordon, M. S. & Slipchenko, L. V. Effective Fragment Potential Study of the Interaction of DNA Bases. *The Journal of Physical Chemistry A* **115**, 11269–11276, doi:10.1021/jp2047954 (2011).

59 Hands, M. D. & Slipchenko, L. V. Intermolecular Interactions in Complex Liquids: Effective Fragment Potential Investigation of Water–tert-Butanol Mixtures. *The Journal of Physical Chemistry B* **116**, 2775–2786, doi:10.1021/jp2077566 (2012).

60 Adamovic, I. & Gordon, M. S. Methanol–Water Mixtures: A Microsolvation Study using the Effective Fragment Potential Method. *The Journal of Physical Chemistry A* **110**, 10267–10273, doi:10.1021/jp060607n (2006).

61 Rankin, B. M. *et al.* Interactions between Halide Anions and a Molecular Hydrophobic Interface. *Faraday Discussions* **160**, 255–270, doi:10.1039/c2fd20082a (2013).

62 Kaliman, I. A. & Slipchenko, L. V. Software Library Implementation of the Effective Fragment Potential Method. (unpublished, 2012).

63 Kaliman, I. A. & Slipchenko, L. V. LIBEFP: A New Parallel Implementation of the Effective Fragment Potential Method as a Portable Software Library. *Journal of Computational Chemistry* **34**, 2284–2292, doi:10.1002/jcc.23375 (2013).

64 Kemp, D. D., Rintelman, J. M., Gordon, M. S. & Jensen, J. H. Exchange Repulsion between Effective Fragment Potentials and Ab Initio Molecules. *Theoretical Chemistry Accounts* **125**, 481–491, doi:10.1007/s00214-009-0660-7 (2010).

65 Xu, P. *The odd power of dispersion*, PhD thesis, Iowa State University (2014).

66 Smith, Q. A., Ruedenberg, K., Gordon, M. S. & Slipchenko, L. V. The Dispersion Interaction between Quantum Mechanics and Effective Fragment Potential Molecules. *The Journal of Chemical Physics* **136**, 244107, doi:10.1063/1.4729535 (2012).

67 Slipchenko, L. V., Ruedenberg, K. & Gordon, M. S. Dispersion interactions in QM/EFP. (in preparation, 2016).

68 Kosenkov, D. & Slipchenko, L. V. Solvent Effects on the Electronic Transitions of p-Nitroaniline: A QM/EFP Study. *The Journal of Physical Chemistry A* **115**, 392–401, doi:10.1021/jp110026c (2010).

69 Slipchenko, L. V. Solvation of the Excited States of Chromophores in Polarizable Environment: Orbital Relaxation Versus Polarization. *The Journal of Physical Chemistry A* **114**, 8824–8830, doi:10.1021/Jp101797a (2010).

70 DeFusco, A. *et al.* Modeling Solvent Effects on Electronic Excited States. *The Journal of Physical Chemistry Letters* **2**, 2184–2192, doi:10.1021/jz200947j (2011).

71 Ghosh, D., Isayev, O., Slipchenko, L. V. & Krylov, A. I. The Effect of Solvation on Vertical Ionization Energy of Thymine: from Microhydration to Bulk. *The Journal of Physical Chemistry A* **115**, 6028–6038, doi:10.1021/jp110438c (2011).

72 DeFusco, A., Ivanic, J., Schmidt, M. W. & Gordon, M. S. Solvent-Induced Shifts in Electronic Spectra of Uracil. *The Journal of Physical Chemistry A* **115**, 4574–4582, doi:10.1021/jp112230f (2011).

73 De Silva, N., Minezawa, N. & Gordon, M. S. Excited-State Hydrogen Atom Transfer Reaction in Solvated 7-Hydroxy-4-methylcoumarin. *The Journal of Physical Chemistry B* **117**, 15386–15394, doi:10.1021/jp404686c (2013).

74 Ghosh, D. *et al.* First-Principle Protocol for Calculating Ionization Energies and Redox Potentials of Solvated Molecules and Ions: Theory and Application to Aqueous Phenol and Phenolate. *The Journal of Physical Chemistry B* **116**, 7269–7280, doi:10.1021/jp301925k (2012).

75 Bose, S., Chakrabarty, S. & Ghosh, D. Effect of Solvation on Electron Detachment and Excitation Energies of a Green Fluorescent Protein Chromophore Variant. *The Journal of Physical Chemistry B* **120**, 4410–4420, doi:10.1021/acs.jpcb.6b03723 (2016).

76 List, N. H., Olsen, J. M. H. & Kongsted, J. Excited States in Large Molecular Systems through Polarizable Embedding. *Physical Chemistry Chemical Physics* **18**, 20234–20250, doi:10.1039/C6CP03834D (2016).

77 Yang, Y. *et al.* Towards a Quantitative Understanding of Total OH Reactivity: A Review. *Atmospheric Environment* **134**, 147–161, doi:10.1016/j.atmosenv.2016.03.010 (2016).

78 Chipman, D. M. Absorption Spectrum of OH Radical in Water. *The Journal of Physical Chemistry A* **112**, 13372–13381, doi:10.1021/jp807399b (2008).

79 Tsai, M.-K., Kowalski, K., Valiev, M. & Dupuis, M. Signature OH Absorption Spectrum from Cluster Models of Solvation: A Solvent-to-Solute Charge Transfer State. *The Journal of Physical Chemistry A* **111**, 10478–10482, doi:10.1021/jp074617f (2007).

80 Hoffman, G. J., Gurunathan, P. K., Francisco, J. S. & Slipchenko, L. V. Excited States of OH-$(H2O)_n$ Clusters for n = 1–4: An Ab Initio Study. *The Journal of Chemical Physics* **141**, 104315, doi:10.1063/1.4894772 (2014).

81 Roeselová, M. *et al.* Hydroxyl Radical at the Air–Water Interface. *Journal of the American Chemical Society* **126**, 16308–16309, doi:10.1021/ja045552m (2004).

82 Roeselová, M., Jungwirth, P., Tobias, D. J. & Gerber, R. B. Impact, Trapping, and Accommodation of Hydroxyl Radical and Ozone at Aqueous Salt Aerosol Surfaces. A Molecular Dynamics Study. *The Journal of Physical Chemistry B* **107**, 12690–12699, doi:10.1021/jp030592i (2003).

83 Pruitt, S. R., Steinmann, C., Jensen, J. H. & Gordon, M. S. Fully Integrated Effective Fragment Molecular Orbital Method. *Journal of Chemical Theory and Computation* **9**, 2235–2249, doi:10.1021/ct4001119 (2013).

84 Slipchenko, L. V. & Gordon, M. S. Electrostatic Energy in the Effective Fragment Potential Method: Theory and Application to Benzene Dimer. *Journal of Computational Chemistry* **28**, 276–291, doi:10.1002/jcc.20520 (2007).

7

Nucleation Using the Effective Fragment Potential and Two-Level Parallelism

Ajitha Devarajan,[1] Alexander Gaenko,[2] Mark S. Gordon,[3,4] and Theresa L. Windus[3,4]

[1] *Office of University Development, University of Michigan, USA*
[2] *Advanced Research Computing, University of Michigan, USA*
[3] *Ames Laboratory of United States Department of Energy, USA*
[4] *Department of Chemistry, Iowa State University, USA*

7.1 Introduction

Nucleation events are at the core of many important processes such as crystal growth, quantum dot fabrication, and the production of aerosols in the atmosphere. In the latter case, the nucleation and growth of aerosols forms one of the least-understood components of atmospheric forcing that directly affects our global climate (Seinfeld 1996). Understanding nucleation is a challenge for both experiment and computation due to the complex pathways for nucleation and the need for explicit understanding of the molecular interactions involved in the small clusters that form at the beginning of the nucleation cycle. While classical nucleation theory (CNT) (Abraham 1974; Pruppacher & Klett 1997) has been successful in examining long term nucleation effects, it relies on the use of bulk properties such as surface tension and equilibrium chemical potentials and, therefore, is not as effective at predicting the initial nucleation phases where small molecular clusters predominate. Below the critical cluster size—the point where the rate of association with the cluster is equal to the rate of dissociation from the cluster—another method is required.

One method that has proven to be effective for atmospheric processes (formation of liquid droplets in the gas phase) is the dynamical nucleation theory (DNT) using a Metropolis Monte Carlo (MC) sampling method (DNTMC) developed by Kathmann *et al.* (Schenter, Kathmann & Garrett 1999; Kathmann, Schenter & Garrett 2004; Kathmann, Schenter & Garrett 2005). DNTMC determines rates using variational transition state theory (VTST) (Garrett & Truhlar 1979) for each stage of the nucleation process. Since it is a molecular scale approach, DNTMC requires knowledge of the inter-molecular interaction potential and of ensembles of molecular configurations to compute the evaporation rate constant. Kathmann *et al.* suggested a practical implementation of DNT based on a Monte Carlo approach and studied water clusters using the Dang-Chang potential (Dang & Chang 1997). Crosby *et al.* (Crosby, Kathmann & Windus 2009; Windus, Kathmann & Crosby 2008) have extended the method to use

Fragmentation: Toward Accurate Calculations on Complex Molecular Systems, First Edition.
Edited by Mark S. Gordon.

ab initio potentials using an efficient sampling by using appropriate ensembles (described in Section 7.2.1), implemented it in an efficient parallel scheme into NWChem (Valiev *et al.* 2010), and applied it to study small water clusters. This initial work showed that the effect of the interaction potential has a significant impact on the evaporation rates for water dimer. Restricted Hartree-Fock (RHF) results showed a decrease in the evaporation rate of about a factor of 3 less than the previous Dang-Chang results. Using MP2, the rate was about a factor of 9 less than the previous Dang-Chang results. Most of these differences in rates could be attributed to the higher binding energies of MP2 and RHF.

The effective use of temperature scaling (TeS) was developed to take advantage of ensemble relationships to project properties of an ensemble derived at one temperature to other temperatures (Crosby & Windus 2009). The driving factor for TeS is to decrease the large computational expense of developing separate ensembles at multiple temperatures. In contrast to other work (Ferrenberg & Swendsen 1988; Frantz, Freeman & Doll 1990; Zhou & Berne 1997; Geyer & Thompson 1995; Swendsen & Wang 1986; Brown & Head-Gordon 2002), the TeS method performs a scaling on the original Markov chains, which potentially allows for multiple properties to be calculated from the ensemble. Since higher energy states will be more thermally accessible at higher temperatures, the scaling is such that ensembles developed at a high temperature can be scaled to produce an ensemble for a lower temperature. The maximum temperature change allowed before enthalpic and entropic contributions degrade the quality of the ensembles was determined using a relationship developed by Ferrenberg and Swendsen (1989). The method was applied and shown to produce excellent ensemble properties at various temperatures for an analytic potential known to be difficult to sample (Brown & Head-Gordon 2002) and to water clusters with 2–5 waters.

Devarajan *et al.* (Devarajan, Windus & Gordon 2011) have implemented the DNT method in the GAMESS software (Schmidt *et al.* 1993; Gordon & Schmidt 2005) using the *ab initio*-based effective fragment potential method (DNTEFP) for evaluating the inter-molecular interaction potential and applied it to compute evaporation rate constants of water from water clusters and for isoprene-water clusters. The results of these simulations produced evaporation rates that were similar to those of the previous *ab initio* DNTMC results, but with an order of magnitude decrease in the cost. These results showed that the evaporation rate of isoprene from a water hexamer is three times higher than that of a water molecule evaporating from a water hexamer. This result would have been very computationally expensive to produce using *ab initio* methods for evaluation of the molecular energies.

Since effective load balancing of the Markov chains in the MC method using *ab initio* energy evaluations is essential to performance, the use of static partitioning of processes into groups for multiple chain simulations was reexamined (Arafat *et al.* 2012). Load imbalance happens when process groups are in an idle state while waiting for other process groups to complete at a synchronization point within a calculation. When *ab initio* methods are used for energy evaluations, load imbalance plays a large role due to the different convergence rates of energies at different geometrical structures. Resource sharing barriers were implemented to put idle process groups to work. The main idea is that an idle process group will check to see if other process groups are still working and will initiate a merge with a working group to help compute the energies faster. Since adding too many process groups can actually slow down a computation, a user-defined upper limit for the number of total processes in a group can be set. This small code change to NWChem improved overall performance by up to 37%.

This chapter presents an overview of the components of the DNTEFP model, the use of the DNTEFP for three representative molecular systems of atmospheric importance, and implementation and performance of these methods using two-level parallelism on high performance computers.

7.2 Methods

In DNTEFP, several important concepts are combined to facilitate the computation of the evaporation rates—dynamical nucleation theory Monte Carlo (DNTMC), the effective fragment potential, and two-level parallelism. In this section, a brief introduction to each of the concepts is introduced to acclimate the reader to the ideas.

7.2.1 Brief Overview of DNTMC

Since the details of the DNTMC method have been given elsewhere, only a brief summary of the method is given here to give context to the rest of the chapter. One of the basic assumptions of the model is that the clusters grow by addition or loss of monomers:

$$N_{i+1} \underset{\beta_i}{\overset{\alpha_{i+1}}{\rightleftarrows}} N_i + N \tag{7.1}$$

where N_i and N_{i+1} is a cluster with i and $i + 1$ monomers, respectively. α_{i+1} is the evaporation rate constant for a monomer (N) leaving the N_{i+1} cluster and β_i is the condensation rate constant of a monomer adding to the N_i cluster. While this equation is written assuming that the monomers are all the same (a homogeneous system), the extension to multi-component systems is relatively straightforward. This monomer addition and loss approximation is very good for the low pressure situations in the upper atmosphere simulated in this work. Another assumption of the model is that all of the monomers are rigid, both outside and inside the cluster. For the molecules examined in this work, this assumption is not expected to have a large impact on the results. However, for molecules with low frequency vibrations, such as torsions, the rigid monomer assumption may not be valid.

DNT is used to calculate the evaporation rate for a cluster of i monomers, $\alpha_i(T, r_{\text{cut}})$ at a temperature T, using the derivative of the Helmholtz free energy, $A_i(T, r)$ with respect to a cluster radius, r:

$$\alpha_i(T, r_{\text{cut}}) = -(2\pi m_1 k_b T)^{-1/2} \left(\frac{dA_i(T,r)}{dr} \right) \Bigg|_{r=r_{\text{cut}}} \tag{7.2}$$

where k_b is Boltzmann's constant, m_l is the mass of the monomer, and a spherical volume is assumed as the central quantity in calculating the rate constant (thus the use of r). r_{cut} is variationally optimized as the spherical dividing surface for the evaporation (the zero flux surface).

The Helmholtz free energy is related to the canonical partition function through the equation

$$Q_i(t,r) = exp\left[\frac{-A_i(T,r)}{k_b T}\right] \tag{7.3}$$

$Q_i(t,r)$ can be determined by integration over the molecular configurational space, *i*. To reduce the number of molecular configuration evaluations required for the integral, a derivative canonical ensemble is used (Crosby, Kathmann & Windus 2009):

$$\frac{dQ_i(T,r')}{dr'} = \left(\frac{2\pi m_1 k_b T}{h^2}\right)^{3i/2}\left(\frac{1}{i!}\right)$$
$$\int dr^i \exp\left[\frac{-U_i\left(r^i\right)}{k_b T}\right]\sum_k^i \delta\left(r' - |R_i - r_k|\right)\prod_{j\neq k}^i \Theta\left(r' - |R_i - r_j|\right) \tag{7.4}$$

where Θ is the Heaviside step function, r' is the enclosing spherical volume centered on the center-of-mass, R_i, of the *i*th-cluster, r_j and r_k are monomer coordinates, δ is a delta function that restricts one or more monomers to lie on the surface of the spherical volume, *h* is Planck's constant, and U_i is the *i*th-cluster interaction potential. This derivative form of the canonical partition function is essentially a shell model that allows molecular configurations in a small volume to contribute to a larger volume without having to resample the whole volume. This savings in computational cost is especially important when *ab initio* methods are used for the intermolecular potential. The minimum in the derivative in equation (7.4) is also the point at which the minimum flux occurs and determines the value of r_{cut}. The cost of the potential energy function evaluation and the number of molecular configurations included in the integral determines the overall cost of the computation. When using *ab initio* potentials, of course, the cost can be prohibitive for large molecular clusters. This is where fragmentation methods can play an important role.

In addition to using fragmentation methods, other improvements in the software have been made in GAMESS to provide for more systematic and efficient sampling in the MC simulations. (i) Three different quaternion rotation algorithms were implemented, in addition to the quaternion implementation that was already available in GAMESS: (a) uniformly distributed unit quaternions, (b) uniformly distributed unit quaternions with rotation angles given as user input, and (c) uniformly distributed random rotation axes, with random uniformly distributed rotational angles within a user provided input. (ii) A dynamically adjustable stepping procedure is employed in which the step size is normally distributed and the variance is automatically adjusted to maintain a given range of translational acceptance rates. Similarly, the rotation angle limit is automatically adjusted to maintain a given range of rotational acceptance rates. (iii) To store the geometry and energy data of millions of configurations for any subsequent analysis, a compact binary data representation was chosen with some degree of redundancy to increase resilience to file corruption. To enable restarts of the multi-chain computation in case of an interrupted simulation (due to, e.g., a hardware failure), restart information for each chain is saved at regular intervals. (iv) A parallel-aware run-time database (RTDB) has been implemented to enable easy exchange of information in the Monte Carlo code.

7.2.2 Brief Overview of EFP

Due to the nature of the DNTMC model, treating each monomer separately and determining the interaction potentials between the monomers has a large computational advantage. While the use of classical potentials can decrease the cost of a computation at a point relative to the cost of an *ab initio* quantum mechanics calculation, the major challenge with the DNTMC approach is that the multidimensional potential must be developed for each unique chemical system that is examined. One effective approach to solve this problem is to use the effective fragment potential—a method that relies on the use of quantum mechanical quantities to determine a model potential for the intermolecular interactions (Day *et al.* 1996; Jensen & Gordon 1998; Jensen 2001; Gordon *et al.* 2001; Adamovic & Gordon 2005; Li & Gordon 2006; Adamovic, Freitag & Gordon 2003; Slipchenko & Gordon 2007; Slipchenko & Gordon 2009; Kemp *et al.* 2010). Again, only a brief overview of the EFP methodology is presented to provide context for the rest of the chapter. More details can be found in the EFP papers and also in Chapter 6 in this book. For the DNTEFP computations—a combination of DNTMC with EFP—described in this chapter, a monomer and a fragment are synonymous.

In the original EFP1 method for water clusters, the fragment interaction energy is

$$E = E^{\mathrm{Elec}} + E^{\mathrm{Pol}} + E^{\mathrm{Rem}} \tag{7.5}$$

The electrostatic term, E^{Elec}, is the Coulomb electrostatic contribution to the molecular interaction and is determined using a damped multipole expansion. The polarization term, E^{Pol}, also called the induction term, is a finite field dipole-induced dipole model in which the interaction is iterated to self-consistency. For each localized molecular orbital in the molecule (e.g., bonds and lone pairs), the polarizability is expressed as individual localized molecular orbital tensors over a set of points. The final remainder term, E^{Rem}, contains all of the rest of the interaction terms, such as exchange repulsion and charge transfer, and is fit to the total potential energy surface of water obtained using Hartree-Fock or density functional theory.

The more general EFP2 model can be used for any type of molecular system and has no empirically fitted parameters. In this model, the total fragment interaction energy is

$$E = E^{\mathrm{Elec}} + E^{\mathrm{Pol}} + E^{\mathrm{XR}} + E^{\mathrm{Disp}} + E^{\mathrm{CT}} \tag{7.6}$$

The first two terms are determined in the same manner as in EFP1 (equation 7.5), except that in the EFP2 model E^{Pol} is also damped. The inter-fragment exchange repulsion term, E^{XR}, is related to intermolecular overlaps and kinetic energy integrals. The dispersion term, E^{Disp}, is expressed in terms of an inverse expansion in the distance between the two fragments, where the coefficients of the expansion are derived from frequency dependent polarizabilities, and each term is damped. The charge transfer term, E^{CT}, depends on the interactions between the occupied valence molecular orbitals on one fragment with the virtual orbitals of the other fragment in a supermolecule approach.

The charge transfer term is the most expensive of the EFP2 terms and, therefore, is the bottleneck in DNTEFP calculations. To decrease the cost, an EFP-CT cut-off scheme (CTcut) was used to evaluate the distance at which the CT contribution to the pair interaction energy is reduced to near-zero values and can be ignored. In this statistical approach, E^{CT} is computed at many geometries using a Monte Carlo simulation, and the resulting values are binned according to the distance (r) between the monomers and by the E^{CT} value, forming a two-dimensional grid of bins. From the number of hits in each bin, the probability density $p(E^{CT},r)$ is computed. Also, the maximum contribution of the charge transfer energy for each monomer pair at the distance r ($E^{CT}{}_{max}(r)$) is recorded. The graphs for water dimer are presented in Figure 7.1. The two-dimensional

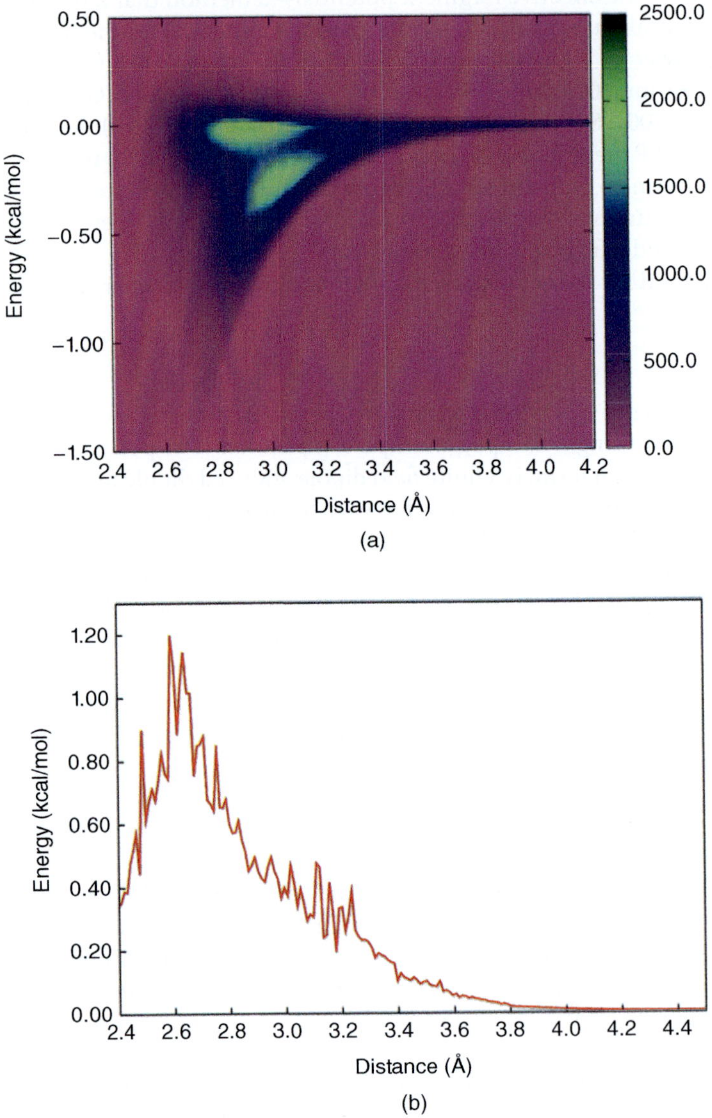

Figure 7.1 Statistical evaluation of the charge transfer interaction between two waters: (a) the probability density of the energy in kcal/mol as a function of the distance between the waters in Å; (b) the maximum charge transfer energy for a given distance between the waters.

graph of the probability density (Figure 7.1a) shows that (i) the most probable configurations correspond to small distances between monomers; (ii) even at small distances, the configurations with high charge transfer energy contributions are relatively rare; (iii) as inter-monomer distance increases, the charge transfer contribution becomes small. The latter statement is also corroborated by the ${E^{CT}}_{max}(r)$ graph (Figure 7.1b). Comparisons of results with and without this CTcut scheme are given in Section 7.3.1.

7.2.3 Overview of the Two-Level Parallelism Approach

One of the challenges in DNTEFP calculations is the large number of potential energy evaluations that must be completed—on the order of millions of structures—to obtain the necessary probability distributions to calculate the evaporation rates and other properties of the clusters. One method to obtain faster throughput for any calculation is to break the problem into multiple levels of parallelism. In this scheme, the original problem is broken down into course grained tasks that can be independently computed (highest level of parallelism). Each of these coarse grained tasks, may be further broken down to operations (finer grained tasks) on different processors (or groups of processors) to accomplish the coarse grained task (the second level of parallelism). By continuing this process, one can obtain multiple levels of parallelism. Since the EFP calculations are much less computationally expensive than an *ab initio* method, only two levels of parallelism are used for the DNTEFP simulations.

As mentioned earlier, the configurational sampling required in the integral of equation (7.4) is completed using a Metropolis MC method. To obtain results faster and to reduce the statistical noise, multiple independent MC Markov chains are started in different parts of the configurational space. Usually, the starting geometric configuration for each chain is determined by randomly placing the monomers within a specific size sphere, where each chain is given a different radius for the sphere. Each chain then uses one of the quaternion methods described above to move in a random translation and rotation direction. If the energy of the new configuration is lower than the energy of the original configuration, the new configuration is added to the ensemble and the next step is taken, assigning the new configuration to be the "original" configuration. If the energy of the new configuration is higher than the energy of the original configuration, then acceptance of the new configuration into the ensemble is determined by comparing a random number to the Boltzmann weighted energy difference between the two configurations. If the new configuration is rejected, then another step is taken starting from the original configuration. As each of these chains completes a certain number of steps (determined by the user), the information on the probability distributions of each chain is combined into a global probability distribution (GPD) to ascertain how well the configuration space has been represented. When the GPD does not change significantly between these communication steps, or after a user-defined total number of steps in the simulation, the simulation finishes.

For the DNTEFP method, a two-level parallel algorithm has been developed that is represented in Figure 7.2. At the first level of parallelism (the highest level), the processes are divided into *groups* of processes, where each group runs a Markov chain. The division of processes is accomplished using the generalized distributed data interface (GDDI) within GAMESS (Olson *et al.* 2003; Fedorov *et al.* 2004). While the processes do not have to be divided evenly, they usually are divided as evenly as possible since the

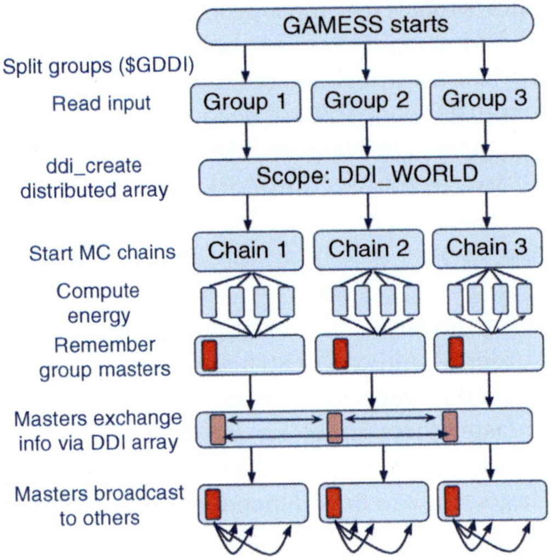

Figure 7.2 Two-level parallel DNTEFP algorithm. See text for details.

size of the molecular clusters are the same in each Markov chain. Each group sets up its own data structures for computing energies (ddi_create in Figure 7.2). Within each group (the lower level of parallelism), the EFP energy is cooperatively computed for each step using a parallel implementation of the EFP method (Netzloff & Gordon 2004). The basic premise of this level of parallelism is that each processor is assigned a certain number of fragment molecules that it "owns". Each processor then only calculates interactions for the fragment that it owns (avoiding double counting), mostly at a loop level of parallelism. The code also uses nonblocking communication to expedite the calculation of the induced dipoles for the polarization term.

A user-defined number of MC steps is taken within each group independent of the others. At the specified number of steps, statistics must be gathered from each group and combined to determine if the simulation should stop or continue. This represents a synchronization point within the calculation, since all of the groups have to finish the required number of steps before the communication between the groups can be completed. In each group, a "master" process is identified (this is the same master that MPI provides in its subgroup implementation) that coordinates communication between groups to collect necessary statistics and to decide whether the simulation should continue. It is also the process that writes out information to disk with the configurations and energies of the ensemble associated with the Markov chain. Once the masters have completed their communication, they communicate with the rest of their own group to determine whether to continue computing or to stop the simulation.

The number of Markov chains and fragments in each chain can be used to guide judicious choices of the number of processors to apply to any given challenge. More information on the performance of this two-level parallel scheme is given in the results section.

All of the calculations discussed in this chapter were obtained on the now-retired Kraken machine at the National Institute for Computational Sciences. Kraken was a Cray XT5 machine with two 2.6 GHz six-core AMD Opteron processors per node and 9408 compute nodes, 147 TB of memory, 3.3 PB of raw disk storage, and a Cray SeaStar2+ router.

7.3 Results

This section presents the results of the use of the DNTEFP method to study several chemical systems. An assessment of the performance of the software is also discussed. The chemical systems investigated here are representative of different types of calculations that can be performed using the DNTEFP method—homogeneous water clusters, heterogeneous water clusters with ions, and homogeneous dehydrated systems. More complete DNTEFP results of calculations on homogenous water clusters are given in Devarajan, Windus and Gordon (2011).

7.3.1 Evaporation Rate of Water Hexamer Cluster at 243K

One of the common test cases for the DNTMC method is a water hexamer. It is simple enough to do extensive calculations, yet complex enough to determine whether the intermolecular potential is reasonable and the MC sampling is complete. In that context, the water hexamer was used as a test case to examine the use of the CTcut method described earlier. EFP2 potentials were determined using the RHF method and the 6-311++G(d,p) basis set (Hehre, Ditchfield & Pople 1972; Clark *et al.* 1983; Hariharan & Pople 1973). Three different simulations were run: (1) full EFP2 including all contributions and full computation of all CT terms (full EFP2) for a total of 20 million configurations to provide a good reference; (2) full EFP2 including all contributions, but using a cut-off for the CT terms when the distance between the monomers was greater than 3.8 Å (EFP2CTcut) for a total of 4.7 million configurations—this cut-off was determined using the data shown in Figure 7.1; (3) EFP2 without any CT contribution at all (EFP2noCT) for a total of 60 million configurations since the calculations are so fast. For each of the three types of simulations, 10 chains were run on a total of 80 processors.

An interesting feature of the water hexamer simulations is that different numbers of hydrogen bonds are found in different configurations in the ensemble. As in previous work (Devarajan, Windus & Gordon 2011), a hydrogen bond is defined by a distance of 1.4-2.5 Å between a hydrogen and an oxygen on two different fragments and an O–H–O (where the H is the hydrogen bonding hydrogen) angle in the range 120–180°. Based on this definition, the total ensemble can be broken down into configurations that contain five to eight hydrogen bonds in the water hexamer clusters. This then can be used to individually plot the energy versus probability distribution plots shown in Figure 7.3.

As can been seen in Figure 7.3, the CTcut scheme recovers the total distribution, as well as the distribution of the hydrogen bonded structures. In contrast, totally ignoring the CT term gives significantly different results. When the CT term is omitted, the contributions from the five and six hydrogen bond structures dominate the ensemble. The r_{cut} value and the evaluated evaporation rates are also in good agreement for the full EFP2 and CTcut schemes, while the values using noCT are significantly different. In this simulation, the use of the CTcut scheme resulted in a threefold decrease in the

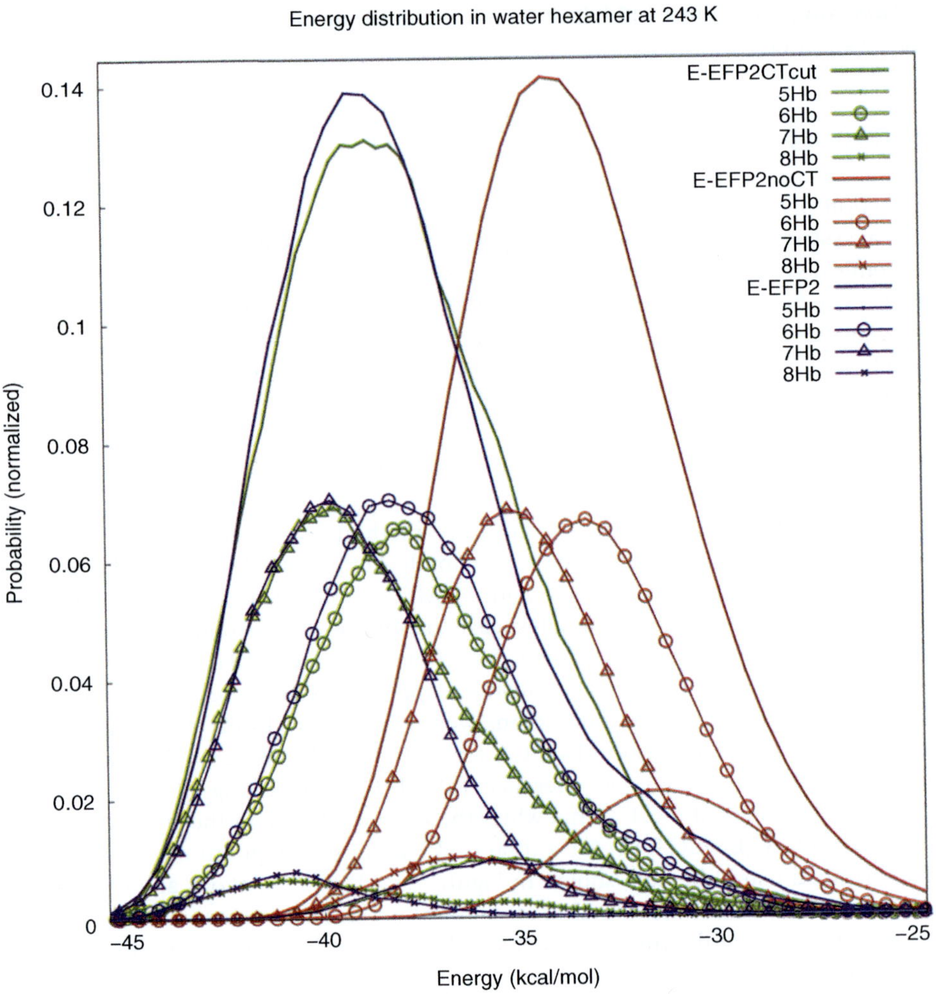

Figure 7.3 Energy distribution in water hexamer clusters for EFP2CTcut (green lines), EFP2noCT (red lines), and the full EFP2 (purple lines). The solid lines with the peaks around 14% probability show the full distribution using the full EFP2 and EFP2-CTcut potentials, while the lines with symbols show the breakdown of the distributions based on the number of hydrogen bonds (Hb) in each ensemble.

cost to solution without noticeable loss of accuracy. Of course, for larger clusters, this performance gain is expected to be much larger.

7.3.2 Ion Mediated Nucleation

Galactic cosmic rays generate up to 20–30 ion pairs per cubic centimeter in the upper troposphere (Usoskin *et al.* 2008). Simple primary ions quickly react with atmospheric trace gases such as water vapor, sulfuric acid, nitric acid and ammonia, and various organic species forming molecular clusters including the hydronium ion series and sulfuric acid aggregates (Arnold 1980). As might be expected, the effect of charge on the

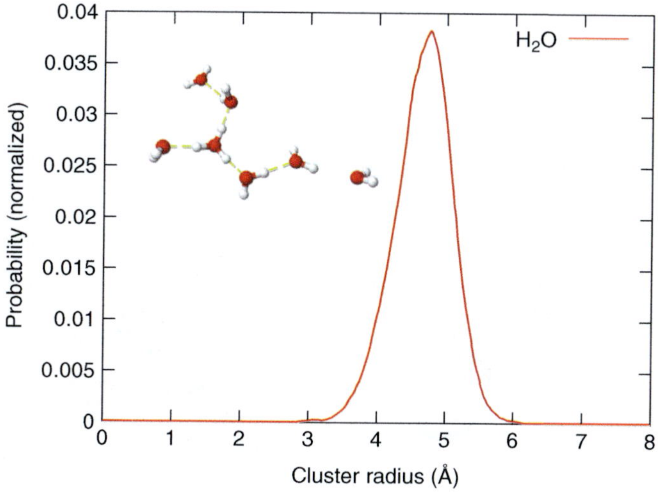

Figure 7.4 Plot of the probability distribution for water versus the cluster radii (Å) for the hydronium ion and hexamer water cluster. The inset figure shows a representative structure from the transition state region for a water to leave the cluster.

nucleating kernel is very strong for small clusters. In the present work, the DNTEFP method is applied for the first time to study ion-mediated nucleation and the effect of hydronium ion on the evaporation reaction of water clusters.

Using the same level of theory as discussed in Section 7.3.2 for the EFP fragment, a total of 36 MC chains were used to obtain 5.8 million configurations of a cluster of six water molecules and one hydronium ion at 243 K. The distribution of the radii of the water molecule clusters versus the normalized probability is shown in Figure 7.4. The most probable radius for the cluster is 4.8 Å, and the value of r_{cut} for a water molecule to leave the cluster is determined to be 6.2 Å. A representative cluster structure at this value of r_{cut} is shown in the inset in Figure 7.4. As with most of the clusters associated with this chemical system, the hydronium is near the center of the cluster with waters surrounding it. The departing water is the most likely pathway for evaporation of these clusters. The rate of evaporation of a water monomer from the hexamer cluster in the presence of hydronium ion is predicted to be $1.5 \times 10^9\ s^{-1}$, while the evaporation rate of water from the neutral hexamer cluster is $16 \times 10^9\ s^{-1}$. These results suggest that the hydronium ion decreases the evaporation rate of water from the cluster and hence favors nucleation of water. This is not an unexpected result given the charge on the hydronium, but the simulation results confirm this and show that the effect is about an order of magnitude in the evaporation rate.

7.3.3 Evaporation Rate of Sulfuric Acid from Neutral Sulfuric Acid Dimer Clusters

Sulfuric acid is the most common nucleating species in the atmosphere because of its low vapor pressure. The vapor pressure is reduced further in the presence of water due to the large mixing enthalpy of water and sulfuric acid (Weber *et al.* 1999). Experimental measurements by Weber *et al.* (1999) revealed that gaseous sulfuric acid concentrations

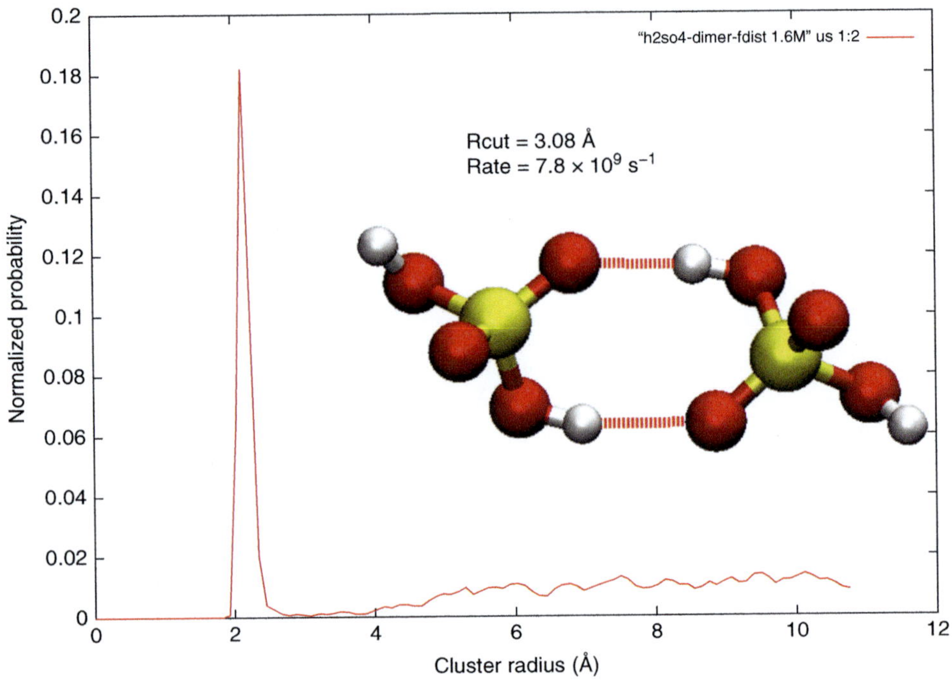

Figure 7.5 Probability distribution of cluster radii for sulfuric acid dimer clusters. The r_{cut} value (Rcut) shows the minimum in the flux surface from which the rate is calculated. The inset figure is the most probable structure in the transition state region.

in excess of 100,000 molecules per cubic centimeter are necessary to observe new particle formation in the atmosphere. Suggested mechanisms of nucleation in the presence of sulfuric acid include both neutral and ion-assisted pathways. In this work, two-level DNTEFP2 calculations were performed on the sulfuric acid dimer as an initial probe into the neutral mechanism.

The EFP2 potential for sulfuric acid was obtained with the 6-311++G(d,p) basis set, using the minimized geometry available on the NIST website (NIST n.d.). 1.7 million configurations were sampled using 10 Markov chains and employing 80 processor cores. A plot of the distribution of cluster radii versus the normalized probability is shown in Figure 7.5. A desirable feature in these types of simulations is sampling in the binding, transition state, and entropic regions. The binding region in Figure 7.5 is the high probability region between about 2 and 3 Å, where the fragments are close to one another and interacting. The transition state region is the low probability region between 3 and 4 Å, where the leaving water molecule has a kinetic bottleneck and where r_{cut} and the rate is determined. The entropic region is the region that exhibits slowly increasing probability for cluster radii greater than 4 Å, where the interaction potential between the monomers is small and the configurational space is increasing in size. Only through effective and sufficient sampling of these regions, will the evaporation rate be determined accurately.

For sulfur dimer, the r_{cut} value is 3.08 Å and the evaporation rate is 7.8×10^9 s^{-1}. By examining the transition state region, one can determine the transition state topology, which is shown as an inset in Figure 7.5. Interestingly, this structure still contains two hydrogen bonds, suggesting that the orientation of the evaporating monomers is still represented by the “face on” interaction of the hydrogen bonds—this motif is common in the binding region of the ensemble as well. One of the complicating features of sulfuric acid that has not been considered in these simulations, since EFPs are internally rigid, is that the barrier for rotation around the S–O–H bond of a sulfuric acid is less than about 1.5 kcal/mol. A second concern is the anharmonicity in the low frequency dimer vibrational modes that is not accounted for here. These concerns will be addressed in future work.

7.3.4 Two-Level Parallel DNTEFP Performance Analysis

To assess the performance of the two-level parallel DNTEFP code, a series of simulations were conducted with varying numbers of chains (N), numbers of processes, and numbers of steps (m) before synchronization between the group masters occurred. All of the computations were carried out for a 32-water cluster using a 6-31G(d,p)-based EFP2 simulation with all of the EFP contributions included. All of the level two energy evaluations in a Markov chain were completed using two 12-core nodes on the Kraken computer. Twenty-two of the processors were used for computing the energy, while the other two processors (one per node) were used to manage the data server. The number of chains (N) and the number of steps (m) between synchronizations were varied. So, in these simulations, the number of processors used changes as $N * 22$—a weak scaling test of the parallelism as the number of chains is increased. In all simulations, each chain took a total of $M = 5000$ steps, with the total number of steps for a particular simulation being $M * N$. A total of 30 simulations using up to 768 CPU cores were completed.

Figure 7.6 shows the relative speed-up of the sampling as the number of chains is increased for various numbers of steps before synchronization. The relative sampling speed-up, $x(N,m)$, is defined as the ratio of the sampling speed $s(N,m)$ of the given computation to the sampling speed, $s(1,M)$, of a single-chain computation with a single synchronization: $x(N,m) = s(N,m) / s(1,M)$. The sampling speed, $s(N,m)$ is defined as the total number of potential energy surface (PES) points sampled by all of the chains in a given computation per unit time: $s(N,m) = N * M / t(N,M,m)$, where $t(N,M,m)$ is the wall-clock time of the computation of M steps, N chains, m steps between interchain synchronizations.

Ideally, the speed of scanning the PES should increase proportionally with the number of chains; however, the difference in the speed of individual chains (due to slightly different computation times per energy, and varying operating system processes on the processors), chain creation and synchronization overhead, may decrease the speed-up. As seen in Figure 7.6a, the speed-up is practically linear (with coefficient of about 0.75) with respect to the number of chains. Increasing the synchronization frequency results in some decrease in speed-up, with the effect becoming more pronounced as the number of chains increases.

If it is assumed that the slow-down is caused exclusively by the overhead of synchronization operations, the total computation time should depend linearly on the number of synchronizations, $K = M / m$ during the run: $t(N,M,m) = t_0 + t_1(N) * K = t_0 + t_1 * M / m$, where t_0 is the computation with no synchronizations and t_1 is the synchronization time. Since the synchronization time should increase as the number

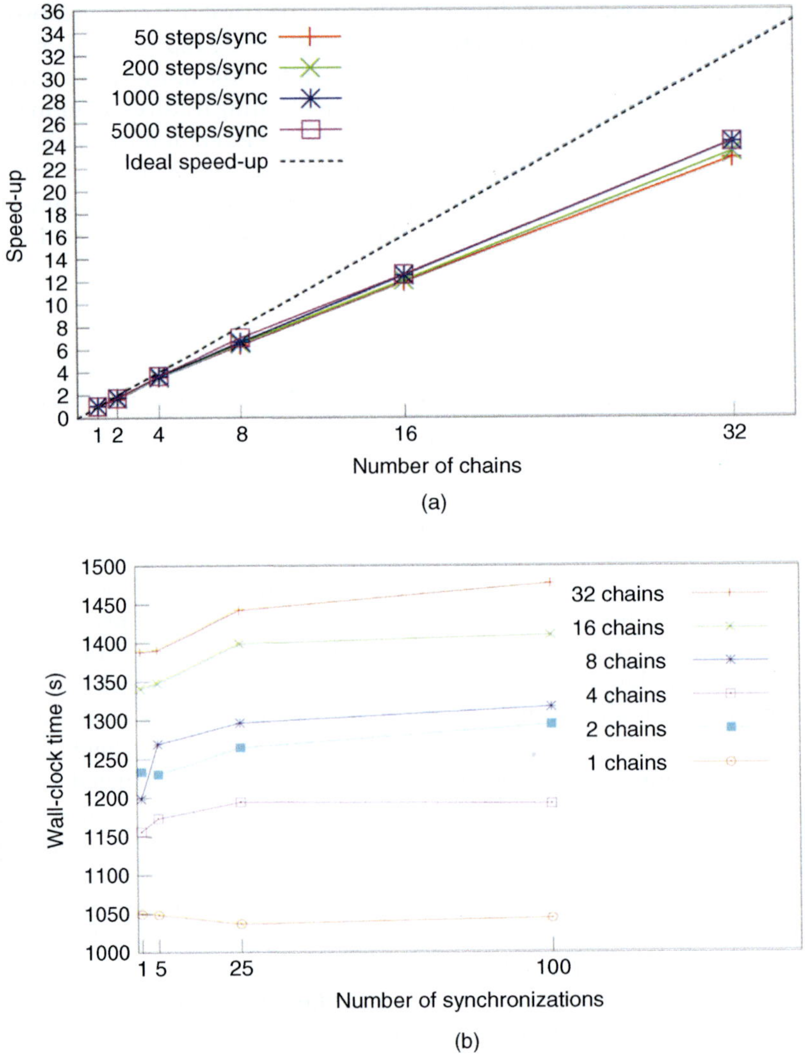

Figure 7.6 Performance analysis of the DNTEFP code: the effect of the number of chains and group synchronization frequency. (a) Sampling speed-up (with respect to a single-chain computation) as a function of the number of chains; (b) wall-clock time for different numbers of chains as a function of the number of synchronizations per simulation.

of communicating chains increases, it can be further expected that $t_1(N)$ is an increasing function.

To verify this hypothesis, a plot of the wall-clock time versus the number of synchronizations, K, is shown in Figure 7.6b. Although the wall-clock time does in general increase with both the number of chains and the number of synchronizations, the dependence is not linear. As has been shown previously (Crosby, Kathmann & Windus 2009), if one assumes that the sampling speed of the chains is not constant through the simulation, the increase in the frequency of synchronization will result in an increase in the overall waiting time, meaning more load imbalance. Thus, the results in Figure 7.6b suggest that the increase in computation time is due to the time delay undergone by the faster chains while they wait for the slower ones at the synchronization points. These different sampling speeds for the chains also explains the lower than unity coefficient of the linear dependence shown in Figure 7.6a. Unlike previous work on the use of resource sharing to decrease the overall load imbalance (Arafat 2012), the energy computations discussed in this chapter are short enough that the overhead of implementing load balancing techniques may outweigh the improvement in the load imbalance. Therefore, it is suggested that one should add extra chains in an efficient way to increase the overall sampling speed in order to get the desired throughput. One should also keep the number of steps between synchronizations as large as possible. Keeping the number of steps large is fairly straightforward at the early stages of the computation, but may not be as practical in the later stages of a simulation since one does not want to evaluate many more energies than are needed to obtain converged probability distributions.

7.4 Conclusions

In this chapter, the use and implementation of a two-level parallel DNT method using the *ab initio*-based EFP method are presented. A charge-transfer cut-off scheme that reduces the time to evaluate the molecular interaction potential was employed and resulted in a threefold increase in the performance for the water hexamer cluster, without noticeable loss of accuracy. DNTEFP was applied for the first time to study ion-induced nucleation mediated by a hydronium ion. The presence of the hydronium ion decreases the evaporation rate of a water monomer by an order of magnitude over that of a water molecule leaving a water hexamer and is predicted to be $1.5 \times 10^9\ s^{-1}$. Using DNTEFP, the evaporation rate of neutral sulfuric acid dimer cluster, a critical first step in the nucleation of neutral sulfuric acid clusters, was predicted to be $7.8 \times 10^9\ s^{-1}$. Performance analysis of the MP-DNTEFP was carried out, and the total number of Monte Carlo steps performed per unit time was seen to increase linearly with the number of chains (up to 72 parallel chains), thus indicating that the synchronization overhead is small. In addition, the performance of the algorithm can be enhanced by increasing the number of steps between synchronizations.

Acknowledgments

The authors acknowledge Ames Laboratory and Iowa State University for providing funding for computational resources. In addition, computing resources using the now-retired Kraken computing system at the National Institute for Computational Sciences were invaluable for assessing the parallel performance of the methods. This work was supported by the U. S. Department of Energy, Office of Basic Energy Sciences, Geosciences, and Biosciences through the Ames Laboratory. The Ames Laboratory is operated for the U. S. Department of Energy by Iowa State University under Contract No. DE-AC02-07CH11358.

References

Abraham, FF, 1974, *Homogeneous Nucleation Theory*, Academic Press, New York.

Adamovic, I, Freitag, MA, & Gordon, MS, 2003, "Density functional theory based effective fragment potential method," *Journal of Chemical Physics*, vol. 118, pp. 6725–6732.

Adamovic, I & Gordon, MS, 2005, "Dynamic polarizability, dispersion coefficient C_6 and dispersion energy in the effective fragment potential method," *Molecular Physics*, vol. 103, pp. 379–387.

Arafat, H, Dinan, J, Krishnamoorthy, S, Windus, TL, & Sadayappan, P, 2012, "Load Balancing of Dynamical Nucleation Theory Monte Carlo Simulations through Resource Sharing Barriers," *Proceedings of the 26th International Parallel and Distributed Processing Symposium (IPDPS)*, Shanghai, China.

Arnold, F, 1980, "Multi-ion complexes in the stratosphere—implications for trace gases and aerosol," *Nature*, vol. 284, pp. 610–611.

Brown, S & Head-Gordon, T, 2002, "Cool walking: a new Markov chain Monte Carlo sampling method," *Journal of Computational Chemistry*, vol. 24, pp. 68–76.

Clark, T, Chandrasekhar, J, Spitznagel, GW, & Schleyer, PVR, 1983, "Efficient diffuse function-augmented basis sets for anion calculations. III. The 3-21+G basis set for first-row elements, Li–F," *Journal of Computational Chemistry*, vol. 4, pp. 294–301.

Crosby, LD, Kathmann, SM, & Windus, TL, 2009, "Implementation of dynamical nucleation theory with quantum potentials," *Journal of Computational Chemistry*, vol. 30, pp. 743–749.

Crosby, LD & Windus, TL, 2009, "A temperature scaling method for Markov chains," *Journal of Physical Chemistry A*, vol. 113, pp. 607–616.

Dang, LX & Chang, TM, 1997, "Molecular dynamics study of water clusters, liquid, and liquid—vapor interface of water with many-body potentials," *Journal of Chemical Physics*, vol. 106, pp. 8149–8159.

Day, PN, Jensen, JH, Gordon, MS, Webb, SP, Stevens, WJ, Krauss, M, Garmer, D, Basch, H, & Cohen, D, 1996, "An effective fragment method for modeling solvent effects in quantum chemistry," *Journal of Chemical Physics*, vol. 105, pp. 1968–1986.

Devarajan, A, Windus, TL, & Gordon, MS, 2011, "Implementation of dynamical nucleation theory effective fragment potentials (DNTEFP) method for modeling aerosol chemistry," *Journal of Physical Chemistry A*, vol. 115, pp. 13987–13996.

Fedorov, DG, Olson, RM, Kitaura, K, Gordon, MS, & Koseki, S, 2004, "A new hierarchical parallelization scheme: generalized distributed data interface (GDDI), and an application to the fragment molecular orbital method (FMO)," *Journal of Computational Chemistry*, vol. 25, pp. 872–880.

Ferrenberg, AM & Swendsen, RH, 1988, "New Monte Carlo technique for studying phase transitions," *Physical Review Letters*, vol. 61, pp. 2635–2638.

Ferrenberg, AM & Swendsen, RH, 1989, "Optimized Monte Carlo data analysis," *Physical Review Letters*, vol. 63, pp. 1195–1198.

Frantz, DD, Freeman, DL, & Doll, JD, 1990, "Reducing quasi-ergodic behavior in Monte Carlo simulations by J-walking: applications to atomic clusters," *Journal of Chemical Physics*, vol. 93, pp. 2769–2784.

Garrett, BC & Truhlar, DG, 1979, "Generalized transition state theory. Classical mechanical theory and applications to collinear reactions of hydrogen molecules," *Journal of Physical Chemistry*, vol. 83, pp. 1052–1079.

Geyer, CJ & Thompson, EA, 1995, "Annealing Markov chain Monte Carlo with applications to ancestral inference," *Journal of the American Statistics Association*, vol. 90, pp. 909–920.

Gordon, MS, Freitag, MA, Bandyopadhyay, P, Jensen, JH, Kairys, V, & Stevens, WJ, 2001, "The effective fragment potential method: a QM-Based MM approach to modeling environmental effects in chemistry," *Journal of Physical Chemistry A*, vol. 105, pp. 293–307.

Gordon, MS & Schmidt, MW, 2005, "Advances in electronic structure theory: GAMESS a decade later," in *Theory and Applications of Computational Chemistry, The First Forty Years*, edited by C. E. Dykstra, G. Frenking, K. S. Kim, and G. E. Scuseria, Elsevier, Amsterdam, pp. 1167–1189.

Hariharan, PC & Pople, JA, 1973, "The influence of polarization functions on molecular orbital hydrogenation energies," *Theoretica Chimica Acta*, vol. 28, pp. 213–222.

Hehre, WJ, Ditchfield, R, & Pople, JA, 1972, "Self—consistent molecular orbital methods. XII. further extensions of Gaussian—type basis sets for use in molecular orbital studies of organic molecules," *Journal of Chemical Physics*, vol. 56, pp. 2257–2261.

Jensen, JH, 2001, "Intermolecular exchange-induction and charge transfer: derivation of approximate formulas using nonorthogonal localized molecular orbitals," *Journal of Chemical Physics*, vol. 114, pp. 8775–8783.

Jensen, JH & Gordon, MS, 1998, "An approximate formula for the intermolecular Pauli repulsion between closed shell molecules. II. Application to the effective fragment potential method," *Journal of Chemical Physics*, vol. 108, pp. 4772–4782.

Kathmann, SM, Schenter, GK, & Garrett, BC, 2002, "Understanding the sensitivity of nucleation kinetics: a case study on water," *Journal of Chemical Physics*, vol. 116, pp. 5046–5057.

Kathmann, SM, Schenter, GK, & Garrett, BC, 2004, "Multicomponent dynamical nucleation theory and sensitivity analysis," *Journal of Chemical Physics*, vol. 120, pp. 9133–9141.

Kemp, DD, Rintelman, J, Gordon, MS, & Jensen, JH, 2010, "Exchange repulsion between effective fragment potentials and ab initio molecules," *Theoretical Chemical Accounts*, vol. 125, pp. 481–491.

Li, H & Gordon, MS, 2006, "Charge transfer interaction in the effective fragment potential method," *Journal of Chemical Physics*, vol. 124, art. 214108.

National Institute of Standards and Technology, n.d., *Sulfuric Acid*, NIST, USA, viewed April 2015, http://webbook.nist.gov/cgi/cbook.cgi?ID=7664-93-9

Netzloff, HM & Gordon, MS, 2004 "Fast fragments: the development of a parallel effective fragment potential method," *Journal of Computational Chemistry*, vol. 25, pp. 1926–1936.

Olson, RM, Schmidt, MW, Gordon, MS, & Rendall, AP, 2003, "Enabling the efficient use of SMP clusters: the GAMESS/DDI model," *Proceedings of Supercomputing*.

Pruppacher, HR & Klett, JD, 1997, *Microphysics of Clouds and Precipitation*, Kluwer.

Schenter, GK, Kathmann, SM, & Garrett, BC, 1999, "Variational transition state theory of vapor phase nucleation," *Journal of Chemical Physics*, vol. 110, 7951–7959.

Schmidt, MW, Baldridge, KK, Boatz, JA, Elbert, ST, Gordon, MS, Jensen, JJ, Koseki, S, Matsunaga, N, Nguyen, KA, Su, S, Windus, TL, Dupuis, M, & Montgomery, JA, 1993, "General atomic and molecular electronic structure system," *Journal of Computational Chemistry*, vol. 14, pp. 1347.

Seinfeld, JH, Charlson, R, Durkee, PA, Hegg, D, Huebert, BJ, Kiehl, J, McCormick, MP, Ogren, JA, Penner, JE, Ramaswamy, V, Slinn, WGN, Slade, DH, & Bouadjemi, D, 1996, *Aerosol Radiative Forcing and Climate Change*, National Academy Press, Washington, D.C.

Slipchenko, LV & Gordon, MS, 2007, "Electrostatic energy in the effective fragment potential method: theory and application to benzene dimer," *Journal of Computational Chemistry*, vol. 28, pp. 276–291.

Slipchenko, LV & Gordon, MS, 2009, "Damping functions in the effective fragment potential method," *Molecular Physics*, vol. 107, pp. 999–1016.

Swendsen, RH & Wang, JS, 1986, "Replica Monte Carlo simulation of spin-glasses," *Physical Review Letters*, vol. 57, pp. 2607–2609.

Usoskin, I, Desorgher, L, Velinov, P, Storini, M, Fluckiger, E, Butikofer, R, & Kovaltsov, G, 2008, "Ionization of the earth's atmosphere by solar and galactic cosmic rays," *Acta Geophysica*, vol. 57, pp. 88–101.

Valiev, M, Bylaska, EJ, Wang, D, Kowalski, K, Govind, N, Straatsma, TP, Nieplocha, J, Aprà, E, Windus, TL, & de Jong, WA, 2010, "NWChem: a comprehensive and scalable open-source solution for large scale molecular simulations," *Computer Physics Communications*, vol. 181, pp. 1477–1489.

Weber, RJ, McMurry, PH, Mauldin III, RL, Tanner, DJ, Eisele, FL, Clarke, AD, & Kapustin, VN, 1999, "New particle formation in the remote troposphere: the observations at various sites," *Geophysical Research Letter*, vol. 26, pp. 307–310.

Windus, TL, Kathmann, SM, & Crosby, DL, 2008, "High performance computations using dynamical nucleation theory," *Journal of Physics: Conference Series*, vol. 12, art. 012017.

Zhou, R & Berne, BJ, 1997, "Smart walking: a new method for Boltzmann sampling of protein conformations," *Journal of Chemical Physics*, vol. 107, pp. 9185–9196.

8

Five Years of Density Matrix Embedding Theory

Sebastian Wouters,[1,2] Carlos A. Jiménez-Hoyos,[2] and Garnet K.L. Chan[2]

[1] Center for Molecular Modelling, Ghent University, Belgium
[2] Frick Chemistry Laboratory, Department of Chemistry, Princeton University, USA

Density matrix embedding theory (DMET) describes finite fragments in the presence of a surrounding environment. In contrast to most embedding methods, DMET explicitly allows for quantum entanglement between both. In this chapter, we discuss both the ground-state and response theory formulations of DMET, and review several applications. In addition, a proof is given that the local density of states can be obtained by working with a Fock space of bath orbitals.

8.1 Quantum Entanglement

In this section, we review nomenclature and several concepts from quantum information theory, which are necessary to follow the discussion on density matrix embedding theory. Quantum many-body theory can be formulated in a Fock space of single-particle states. In quantum chemistry, the single-particle states are orbitals, each with d possible occupations. For spin-orbitals, there are $d = 2$ possible occupations: empty or occupied. For spatial orbitals, there are $d = 4$ possible occupations: empty, occupied with one electron with spin-projection $s^z = \pm\frac{1}{2}$, or occupied with two electrons.

Consider a bipartition of the total Hilbert space (the Fock space of all orbitals) into two subsystems A and B (the Fock spaces of orbital groups A and B). One can think, for example, about the left and right halves of a polyene, or the split-up into active and external orbital spaces in a complete active space (CAS) calculation. Let L_A (L_B) denote the number of orbitals in subsystem A (B). The Hilbert space of subsystem A (B) has size $N_\mathrm{A} = d^{L_\mathrm{A}}$ ($N_\mathrm{B} = d^{L_\mathrm{B}}$). Let $\{|i\rangle_\mathrm{A}\}$ ($\{|j\rangle_\mathrm{B}\}$) denote a particular many-body basis of size N_A (N_B) for subsystem A (B). One can think, for example, about all possible Slater determinants formed by considering all d possible occupations of all L_A (L_B) orbitals in subsystem A (B). Any unitary rotation of this many-body basis is of course as good a choice.

The total Hilbert space $\mathcal{H}$ is the tensor product of the Hilbert spaces of subsystems A and B: $\mathcal{H} = \mathcal{H}_\mathrm{A} \otimes \mathcal{H}_\mathrm{B}$. $\mathcal{H}$ is spanned by $d^{L_\mathrm{A}+L_\mathrm{B}} = N_\mathrm{A}N_\mathrm{B}$ many-body basis states. One particular choice is $\{|i\rangle_\mathrm{A}|j\rangle_\mathrm{B}\}$. Any state in $\mathcal{H}$ can be expressed as

$$|\Psi\rangle = \sum_{i}^{N_\mathrm{A}} \sum_{j}^{N_\mathrm{B}} C_{ij}|i\rangle_\mathrm{A}|j\rangle_\mathrm{B}. \tag{8.1}$$

Fragmentation: Toward Accurate Calculations on Complex Molecular Systems, First Edition.
Edited by Mark S. Gordon.

By considering the singular value decomposition of the coefficient tensor

$$C_{ij} = \sum_{\alpha}^{\min(N_A,N_B)} U_{i\alpha} \lambda_{\alpha} V^{\dagger}_{\alpha j}, \tag{8.2}$$

this state can be rewritten as

$$|\Psi\rangle = \sum_{i}^{N_A} \sum_{j}^{N_B} \sum_{\alpha}^{\min(N_A,N_B)} U_{i\alpha} \lambda_{\alpha} V^{\dagger}_{\alpha j} |i\rangle_A |j\rangle_B = \sum_{\alpha}^{\min(N_A,N_B)} \lambda_{\alpha} |\alpha\rangle_A |\alpha\rangle_B. \tag{8.3}$$

The unitary transformation $U_{i\alpha}$ ($V^*_{j\beta} = V^{\dagger}_{\beta j}$) rotates the many-body basis $\{|i\rangle_A\}$ ($\{|j\rangle_B\}$) to the new many-body basis $\{|\alpha\rangle_A\}$ ($\{|\beta\rangle_B\}$). The particular form of $|\Psi\rangle$ in equation (8.3) is called the Schmidt decomposition. When $\mathcal{H}_A$ and $\mathcal{H}_B$ are of different sizes, the Schmidt decomposition allows for a compact representation of $|\Psi\rangle$. When, for example, $N_A < N_B$, only N_A many-body basis states in $\mathcal{H}_B$ are needed to represent $|\Psi\rangle$ exactly.

When only one of the Schmidt or singular values λ_{α} is nonzero, the state $|\Psi\rangle$ is factorizable with respect to the given bipartition, and the two subsystems are said to be unentangled. When several of the Schmidt values are nonzero, the state $|\Psi\rangle$ is not factorizable with respect to the given bipartition, and the two subsystems are said to be entangled.

In CAS wavefunctions, the occupations of the external orbitals are invariant over all determinants with nonzero coefficients. Such a wavefunction is factorizable with respect to the bipartition into active and external orbital spaces, and these two subsystems are therefore unentangled. In a polyene, the π-conjugation requires a multireference description. The occupations of the orbitals in the left (or right) half of the polyene differ over the determinants with nonzero coefficients. The multireference ground-state wavefunction of a polyene is therefore not factorizable with respect to the bipartition into left and right halves, and these subsystems are therefore entangled.

8.2 Density Matrix Embedding Theory

Suppose one is interested in the properties of a small subsystem of the whole quantum system. Synonyms for this subsystem are the *impurity, cluster*, or *fragment*. While the impurity is most often a specific region in space, it can be any subsystem of the total Hilbert space. Other examples are a band in momentum space, a particular non-localized set of orbitals in a molecule, or even an incomplete many-body basis which cannot be rewritten as a Fock space of orbitals. In this chapter, we will limit ourselves to a Fock space of impurity orbitals. The region outside of the impurity is called the *environment*.

The general goal in embedding theories is to obtain the properties of interest of the impurity, without doing expensive calculations on the whole quantum system. (Part of) the environment can for example be replaced with a solvent model, static charges, or a mean-field description which is unentangled with the impurity. For many systems, such a description gives accurate results. However, when there is static correlation between the impurity and the environment, these methods will fail.

In DMET, the wavefunction for the impurity plus environment is written as in equation (8.3). If the exact wavefunction $|\Psi\rangle$ is known, the decomposition from the previous section can be followed to construct a Schmidt basis for the environment. But this requires a priori knowledge of $|\Psi\rangle$, and then we would not have to use an embedding theory in the first place. The idea of DMET is to embed the impurity A in an approximate bath B. The bath can be thought of as N_A approximate many-body basis states for the environment. Solving the impurity plus bath system is called the *embedded* problem. Various possibilities exist to obtain a bath space. In this chapter, we will limit ourselves to a Fock space of bath orbitals, obtained from a *low-level* particle-number conserving mean-field wavefunction [1, 2]. This is, however, not the only possibility. One can also work with single-particle states from Hartree-Fock-Bogoliubov theory [3, 4] or with many-body basis states from a Schmidt decomposition of anti-symmetrized geminal power (AGP) wavefunctions for electrons [5], coherent state wavefunctions for phonons [6], or block-product states for spins [7]. Although there are real benefits to using the most accurate feasible wavefunctions for the bath construction, it is also convenient to recycle the large number of existing quantum many-body solvers when solving the embedded problem. They rely on Fock spaces of orbitals, as they are constructed for second-quantized Hamiltonians.

Once initial bath states have been obtained from a so-called *low-level* wavefunction, they can be left unoptimized. We call this *single-shot* DMET embedding. Alternatively, once an Ansatz for the bath states is chosen, its parameters can be variationally optimized with a *high-level* method [7]. Another option, which is most often used when the entire system is tiled with impurities, see Figure 8.1, is to introduce a DMET correlation potential $\hat{u}$ to link a single low-level wavefunction, which produces the bath states for the different impurities, with the high-level wavefunctions of these impurities [1, 2]. The Hamiltonian which yields the low-level wavefunction is augmented with $\hat{u}$, and $\hat{u}$ itself is optimized in order to match properties in the low- and high-level wavefunctions. In DMET, as the name indicates, (parts of) the density matrices of the low- and high-level wavefunctions are matched. In this chapter, we will limit ourselves to the latter strategy.

Dynamical mean-field theory (DMFT) [8–11] is another successful method to embed an impurity in an environment. Instead of setting up a self-consistent scheme to match density matrices, the Green's function of the impurity is determined self-consistently by fitting the frequency-dependent hybridization function. The fit parameters turn up as bath orbitals which are coupled to the impurity. The result of DMFT is a correlated

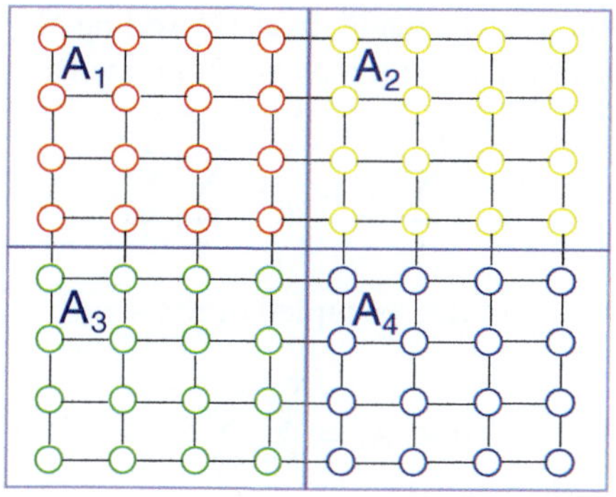

Figure 8.1 The total system is tiled with impurities. When the circles denote orbitals, each impurity is a Fock space of impurity orbitals.

frequency-dependent Green's function, which should still be integrated over a contour to obtain ground-state energies. For ground-state properties, DMET provides a computationally simpler and cheaper alternative with similar accuracy. But DMET is not at all limited to ground-state properties, as will be discussed in Section 8.6.

8.3 Bath Orbitals from a Slater Determinant

Consider a Slater determinant approximation $|\Phi_0\rangle$ for the ground-state of the full system. In second quantization, it can be written as

$$|\Phi_0\rangle = \prod_{\mu}^{N_{\text{occ}}} \hat{a}_{\mu}^{\dagger}|-\rangle. \tag{8.4}$$

The N_{occ} occupied spin-orbitals are denoted by $\mu\nu$, the L orthonormal spin-orbitals for the impurity and its environment by $klmn$, and the orthonormal impurity and bath orbitals by $pqrs$. There are L_{A} orbitals in the impurity A. In what follows, we initially assume that $N_{\text{occ}} \geq L_{\text{A}}$.

The occupied orbitals can be written in terms of the impurity and environment orbitals:

$$\hat{a}_{\mu}^{\dagger} = \sum_{k}^{L} \hat{a}_{k}^{\dagger} C_{k\mu}. \tag{8.5}$$

The physical wavefunction represented by equation (8.4) does not change when the occupied orbitals are internally rotated. Ref. [2] discusses how this freedom can be used to split the occupied orbital space into two parts: orbitals with and without overlap on the impurity. This construction can be understood by means of a singular value decomposition. Consider the occupied orbital coefficient block with indices on the impurity: $k \in \text{A}$. The singular value decomposition of the $L_{\text{A}} \times N_{\text{occ}}$ coefficient block $C_{k\mu}$ yields an occupied orbital rotation matrix $V_{\mu p}$:

$$C_{k\mu}(k \in \text{A}) = \sum_{p}^{L_{\text{A}}} U_{kp} \lambda_p V_{p\mu}^{\dagger}, \tag{8.6}$$

which can be made square by adding $N_{\text{occ}} - L_{\text{A}}$ extra columns: $W = [V\widetilde{V}]$. The occupied orbital space can now be rotated with the $N_{\text{occ}} \times N_{\text{occ}}$ matrix W:

$$\hat{a}_{p}^{\dagger} = \sum_{\mu}^{N_{\text{occ}}} \hat{a}_{\mu}^{\dagger} W_{\mu p} = \sum_{\mu}^{N_{\text{occ}}} \sum_{k}^{L} \hat{a}_{k}^{\dagger} C_{k\mu} W_{\mu p} = \sum_{k}^{L} \hat{a}_{k}^{\dagger} \widetilde{C}_{kp}. \tag{8.7}$$

Of the rotated occupied orbitals, only L_{A} have nonzero overlap with the impurity:

$$\widetilde{C}_{kp}(k \in \text{A}) = \sum_{\mu}^{N_{\text{occ}}} \sum_{q}^{L_{\text{A}}} U_{kq} \lambda_q V_{q\mu}^{\dagger} W_{\mu p} = \begin{cases} U_{kp}\lambda_p & \text{if } p \leq L_{\text{A}} \\ 0 & \text{otherwise} \end{cases}. \tag{8.8}$$

The N_A bath states $\{|\alpha\rangle_B\}$ with nonzero Schmidt number can be found by diagonalizing

$$\hat{\rho}_B = \sum_i^{N_A} \langle i |_A \Phi_0\rangle \langle \Phi_0 \mid i\rangle_A . \tag{8.9}$$

The occupied orbitals without weight on the impurity will always be fully occupied for all projections $\langle \Phi_0 \mid i\rangle_A$. In contrast, the occupied orbitals with weight on the impurity can become partially occupied. The DMET bath construction then yields a CAS Ansatz:

1. The external core orbitals are the $N_{occ} - L_A$ occupied orbitals without overlap on the impurity.
2. The active orbitals consist of the impurity and bath orbitals. The bath orbitals are the L_A-occupied orbitals with an overlap on the impurity, after projection onto the environment:

$$\hat{a}_r^\dagger(r \leq L_A) = \sum_{k>L_A} \hat{a}_k^\dagger \frac{\widetilde{C}_{kr}}{\sqrt{\sum_{l>L_A} | \widetilde{C}_{lr} |^2}} = \sum_{k>L_A} \sum_{\mu}^{N_{occ}} \hat{a}_k^\dagger \frac{C_{k\mu} V_{\mu r}}{\sqrt{1-\lambda_r^2}}. \tag{8.10}$$

3. The external virtual orbitals are the $L - N_{occ} - L_A$ unoccupied orbitals, after projection onto the environment.

Note that for this CAS Ansatz, the impurity and the environment can be entangled, as the latter is represented by L_A bath orbitals in the active space. If the active space comprises solely of the impurity orbitals, this is not the case.

To find the bath orbitals, the overlap of the occupied orbitals with the impurity is diagonalized in Ref. [2]:

$$S_{\mu\nu} = \sum_k^{L_A} C_{\mu k}^\dagger C_{k\nu} = \sum_p^{L_A} V_{\mu p} \lambda_p^2 V_{p\nu}^\dagger. \tag{8.11}$$

At most, L_A eigenvalues of $S_{\mu\nu}$ are nonzero. The L_A corresponding eigenvectors yield the bath orbitals in equation (8.10). A completely equivalent construction of the bath orbitals is to consider the environment block of the mean-field density matrix:

$$D_{kl}(kl > L_A) = \langle \Phi_0 \mid \hat{a}_l^\dagger \hat{a}_k \mid \Phi_0\rangle = \sum_\mu^{N_{occ}} C_{k\mu} C_{\mu l}^\dagger = \sum_p^{N_{occ}} \widetilde{C}_{kp} \widetilde{C}_{pl}^\dagger. \tag{8.12}$$

MacDonald's theorem [12] states that when a row and the corresponding column of a Hermitian matrix are removed, the eigenvalues of the new matrix lie in between the original eigenvalues. Therefore, at most L_A eigenvalues of the $(L - L_A) \times (L - L_A)$ environment subblock $D_{kl}(kl > L_A)$ will lie in between 0 and 1. The corresponding eigenvectors are the orthonormal bath orbitals from equation (8.10). The $N_{occ} - L_A$ eigenvectors with eigenvalue 1 are the external core orbitals. The $L - N_{occ} - L_A$ eigenvectors with eigenvalue 0 are the external virtual orbitals.

The overlap matrix $S_{\mu\nu}$ in equation (8.11) is a projector of the occupied orbitals onto the impurity. Analogously, $D_{kl}(kl > L_A)$ is a projector of the environment orbitals onto the occupied orbitals. Eigenvectors with partial weight signal occupied orbitals with support on both the impurity and the environment, that is, they are entangled orbitals.

We have assumed that $N_{\text{occ}} \geq L_{\text{A}}$. However, for very large basis sets this will not be the case. Because bath orbitals arise from the decomposition of occupied orbitals with support on both the impurity and the environment, there can be, at most, N_{occ} bath orbitals. Even if $N_{\text{occ}} \gtrsim L_{\text{A}}$, the DMET bath orbital selection will try to add low-lying core electrons corresponding to atoms in the environment into the bath space, an undesired effect. One way to circumvent this problem is to define for the impurity a core, valence, and virtual orbital space (in the molecular sense) and to try to find bath orbitals only for the valence orbitals. In Ref. [13], we show that Knizia's intrinsic atomic orbitals [14] are particularly suited for this strategy. Another way would be to take the most correlated orbitals, typically one bath orbital per chemical bond. This boils down to Sun's optimal QM/MM boundary scheme [15]. In the remainder, we will use L_{B} to denote the number of bath orbitals. All other environment orbitals are restricted to be fully occupied or empty. The deficit in electron number between the external core orbitals and N_{occ} is the number of electrons in the active space N_{act}.

8.4 The Embedding Hamiltonian

The Hamiltonian for the total system can be written as:

$$\hat{H} = E_{\text{nuc}} + \sum_{kl}^{L} t_{kl}\hat{a}_k^\dagger \hat{a}_l + \frac{1}{2}\sum_{klmn}^{L} (kl|mn)\hat{a}_k^\dagger \hat{a}_m^\dagger \hat{a}_n \hat{a}_l, \tag{8.13}$$

where t_{kl} and $(kl|mn)$ are diagonal in the spin indices of spin-orbitals k and l, and $(kl|mn)$ is diagonal in the spin indices of m and n as well. We assume fourfold permutation symmetry $(kl|mn) = (mn|kl) = (lk|nm)$ for the electron repulsion integrals.

Suppose the total system is tiled with impurities, see Figure 8.1. For each impurity A_x, a Hermitian one-particle operator $\hat{u}^x$ is introduced which acts solely within the impurity A_x:

$$\hat{u}^x = \sum_{kl}^{L_{\text{A}_x}} u_{kl}^x \hat{a}_k^\dagger \hat{a}_l. \tag{8.14}$$

The sum of all these one-particle operators forms the DMET correlation potential:

$$\hat{u} = \sum_x \hat{u}^x. \tag{8.15}$$

With $\vec{u}$ we denote all of its independent variables. The mean-field low-level wavefunction $|\Phi_0(\vec{u})\rangle$ is obtained as either the mean-field solution of $\hat{H} + \hat{u}$, or as the eigenfunction of a one-particle operator $\hat{h} + \hat{u}$. For local electron repulsion integrals, for example, in the Hubbard model, the one-particle operator $\hat{h}$ is the hopping matrix. For nonlocal quartic electron repulsion integrals, which arise in quantum chemistry, the one-particle operator $\hat{h}$ is the Fock operator corresponding to $\hat{H}$.

All variables $\vec{u}$ influence the bath orbitals of each impurity A_x. Together with the impurity orbitals of A_x, these bath orbitals form an active space. The N^x_{act} active space electrons interact with the external core electrons as well, and therefore it is important to take them into account via their density matrix:

$$D^{\text{core},x}_{kl} = \sum_{p \in \text{core}} \widetilde{C}_{kp} \widetilde{C}^{\dagger}_{pl}. \tag{8.16}$$

The one-electron integrals of the embedding Hamiltonian

$$\hat{H}^{\text{emb},x} = \sum_{pq}^{L_{A_x}+L_{B_x}} h^x_{pq} \hat{a}^{\dagger}_p \hat{a}_q + \frac{1}{2} \sum_{pqrs}^{L_{A_x}+L_{B_x}} (pq|rs) \hat{a}^{\dagger}_p \hat{a}^{\dagger}_r \hat{a}_s \hat{a}_q - \mu_{\text{glob}} \sum_{r}^{L_{A_x}} \hat{a}^{\dagger}_r \hat{a}_r \tag{8.17}$$

contain the Coulomb and exchange contributions due to the external core electrons:

$$h^x_{kl} = t_{kl} + \sum_{mn}^{L} [(kl|mn) - (kn|ml)] D^{\text{core},x}_{mn} \overset{\widetilde{C}_{kp}}{\rightarrow} h^x_{pq}. \tag{8.18}$$

This embedding Hamiltonian has an *interacting bath*, as there are electron repulsion integrals for the bath. The DMET correlation potential appears only indirectly through its effect on the form of the bath and external core orbitals. To ensure that the total number of electrons in all impurities A_x adds up to N_{occ}, it becomes necessary to introduce a global chemical potential μ_{glob} for the impurity orbitals. This global chemical potential is independent of the specific impurity A_x.

For local electron repulsion integrals, for example, in the Hubbard model, the core electrons do not have Coulomb or exchange interactions with the impurity orbitals. In this case, a simpler version of DMET is commonly used, where we replace the electron repulsion in the bath, due to the external core and bath electrons, by the DMET correlation potential rotated to the bath orbitals:

$$\hat{H}^{\text{emb},x} = \sum_{pq}^{L_{A_x}+L_{B_x}} t_{pq} \hat{a}^{\dagger}_p \hat{a}_q + \sum_{pq}^{L_{B_x}} u_{pq} \hat{a}^{\dagger}_p \hat{a}_q + \frac{1}{2} \sum_{pqrs}^{L_{A_x}} (pq|rs) \hat{a}^{\dagger}_p \hat{a}^{\dagger}_r \hat{a}_s \hat{a}_q. \tag{8.19}$$

On the impurity, the original electron repulsion integrals act. This embedding Hamiltonian has a *noninteracting bath*, as there are no electron repulsion integrals for the bath. The DMET correlation potential then has a triple role: it determines the form of the bath and external core orbitals, represents the electron repulsion in the bath, and ensures that the total number of electrons in all impurities A_x adds up to N_{occ}.

The ground state $|\Psi_x\rangle$ of the embedding Hamiltonian $\hat{H}^{\text{emb},x}$ for impurity A_x is calculated with a high-level method, typically full configuration interaction [1, 2], the density-matrix renormalization group [3, 16], or coupled-cluster theory [17]. In order to calculate expectation values of interest, it should be possible to obtain one- and two-particle density matrices of the active space with the high-level method:

$$D^x_{rs} = \langle \Psi_x \mid \hat{a}^{\dagger}_r \hat{a}_s \mid \Psi_x \rangle , \tag{8.20}$$

$$P^x_{pq|rs} = \langle \Psi_x \mid \hat{a}^{\dagger}_p \hat{a}^{\dagger}_r \hat{a}_s \hat{a}_q \mid \Psi_x \rangle . \tag{8.21}$$

For *local* operators, when all orbital indices correspond to one particular impurity, the density matrices of that impurity are used to obtain the expectation values. For *non-local* operators, a democratic partitioning of the Hermitian expectation values is used. Suppose, for example, that orbital i (j) is part of impurity A_x (A_y):

$$\langle \hat{a}_i^\dagger \hat{a}_j + \hat{a}_j^\dagger \hat{a}_i \rangle = \langle \Psi_x | \hat{a}_i^\dagger \hat{a}_j | \Psi_x \rangle + \langle \Psi_y | \hat{a}_j^\dagger \hat{a}_i | \Psi_y \rangle . \tag{8.22}$$

By convention, the density matrix of the impurity corresponding to the first index is used. For an embedding Hamiltonian with interacting bath, this gives rise to the following formula for the total energy:

$$E_{\text{tot}} = E_{\text{nuc}} + \sum_x E_x, \tag{8.23}$$

$$E_x = \sum_p^{L_{A_x}} \left(\sum_q^{L_{A_x}+L_{B_x}} \frac{t_{pq} + h_{pq}^x}{2} D_{pq}^x + \frac{1}{2} \sum_{qrs}^{L_{A_x}+L_{B_x}} (pq|rs) P_{pq|rs}^x \right). \tag{8.24}$$

The one-electron integrals in equation (8.24) avoid the double counting of Coulomb and exchange contributions of the external core electrons when they arise in the active spaces of other impurities. The factor $\frac{1}{2}$ is similar to the difference between the Fock operator and energy expressions in HF theory.

Note that DMET energies are not variational, because density matrices of different high-level calculations, each in their own active space consisting of impurity and bath orbitals, enter in equations (8.23) and (8.24).

8.5 Self-Consistency

The DMET correlation potential $\hat{u}$ is determined by matching (parts of) the low-level and high-level one-particle density matrices as closely as possible. These parts can be: for each impurity, the full one-particle density matrix in its active space; for each impurity, the impurity block of the one-particle density matrix; for each impurity, the diagonal of the impurity block of the one-particle density matrix; or just the total number of electrons in all impurities. In order to have a well-posed optimization problem, the number of variables in $\vec{u}$ should not exceed the number of independent expectation values which are matched. For the four cases discussed above, respectively, the full DMET correlation potential; the full DMET correlation potential; the diagonal of the DMET correlation potential; or just the global chemical potential μ_{glob} are optimized. The DMET correlation potential $\hat{u}$ and the global chemical potential μ_{glob} can be obtained by least-squares minimization of:

$$\Delta_{pq}^x(\vec{u}) = D_{pq}^{\text{low},x}(\vec{u}) - D_{pq}^{\text{high},x}, \tag{8.25}$$

$$\Delta_N(\mu_{\text{glob}}) = N_{\text{tot}}(\mu_{\text{glob}}) - N_{\text{occ}}. \tag{8.26}$$

Note that the high-level density matrix $D_{pq}^{\text{high},x}$ also depends on $\vec{u}$, either solely through the form of the bath and external core orbitals for an interacting bath, or also directly through the Hamiltonian matrix elements in equation (8.19) for a noninteracting bath. However, thus far the DMET correlation potential has always been optimized while keeping the high-level density matrix $D_{pq}^{\text{high},x}$ fixed. The appendix in Ref. [13] discusses how to obtain analytic gradients of $D_{pq}^{\text{low},x}(\vec{u})$ with respect to $\vec{u}$.

As discussed extensively in Ref. [5], trying to match (a part of) a mean-field density matrix with (a part of) a high-level correlated density matrix is not always possible because the former is idempotent while the latter does not have to be. Thus, obtaining a correlation potential for which the cost function becomes zero is therefore not always possible.

$\vec{u}$ can be optimized per impurity A_x by projecting the mean-field problem into its active space. Only the elements u_{kl}^x corresponding to the specific impurity A_x are then optimized. The low-level wavefunction should then be calculated in a small orbital space, but as all elements $\vec{u}$ can influence all bath spaces, this approach is prone to limit cycles and slow convergence due to overshooting. It might therefore be better to first solve the mean-field problem in the total system, and to subsequently project its density matrix to the different active spaces [13]. The desired parts of the density matrices are then matched simultaneously for all impurities. Stationary points of the latter approach will also be stationary points of the former.

In Ref. [13], we discuss how the function minimization of the difference of the one-particle density matrices with respect to $\hat{u}$ can be recast into a functional optimization problem with respect to $|\Phi_0\rangle$.

The DMET algorithm with interacting bath is summarized in Figure 8.2.

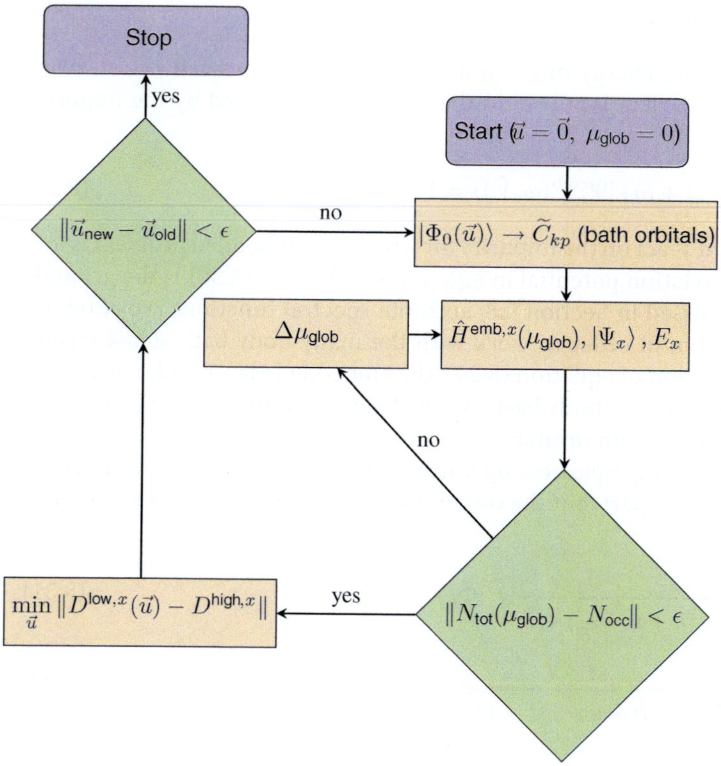

Figure 8.2 Flow chart of the DMET algorithm.

8.6 Green's Functions

The DMET algorithm is not limited to ground-state properties, but can be extended to calculate response properties as well. In this section, we review Ref. [18] in which the ground-state algorithm is extended to calculate Green's functions. In Ref. [18], a bath space of many-body states was constructed by Schmidt decomposition of an approximate first-order response. In this section, we will show that for the local density of states (LDOS), only one additional bath orbital is required. In other words, for the LDOS, the many-body bath space of Ref. [18] can be rewritten as a Fock space of bath orbitals.

Suppose one is interested in Green's functions of the form

$$\mathcal{G}(\omega, \hat{X}, \hat{V}) = \langle \Psi_0 \mid \hat{X}^\dagger \frac{1}{\omega - (\hat{H} - E_0) + i\eta} \hat{V} \mid \Psi_0 \rangle . \tag{8.27}$$

In the full Hilbert space, they can be calculated by first solving the linear problem

$$(\omega - (\hat{H} - E_0) + i\eta)\, |\Psi_1(\omega, \hat{V})\rangle = \hat{V}\, |\Psi_0\rangle , \tag{8.28}$$

which subsequently yields $\mathcal{G}(\omega, \hat{X}, \hat{V}) = \langle \Psi_0 \mid \hat{X}^\dagger \mid \Psi_1(\omega, \hat{V})\rangle$. The same strategy is followed in Ref. [18]. First, an approximate bath space is constructed by Schmidt decomposition of

$$|\Phi_1(\omega, \hat{V})\rangle = \frac{1}{\omega - (\hat{h} + \hat{u} - \epsilon_0) + i\eta} \hat{V} |\Phi_0\rangle, \tag{8.29}$$

where ϵ_0 is the ground-state energy of $\hat{h} + \hat{u}$ associated with the wavefunction $|\Phi_0\rangle$. Subsequently, the linear problem is solved in the active space formed by the impurity and bath space:

$$\left(\omega - \left(\hat{H^{\text{emb}}} - E_0^{\text{emb}}\right) + i\eta\right) |\Psi_1^{\text{emb}}(\omega, \hat{V})\rangle = \hat{V}\, |\Psi_0^{\text{emb}}\rangle . \tag{8.30}$$

In this formalism, $\hat{X}$ and $\hat{V}$ act on the impurity for which the embedded problem is formulated. The DMET correlation potential in equations (8.29) and (8.30) is the ground-state one. As will be discussed in Section 8.8, accurate spectral functions are obtained with this method. However, in order to work with the many-body bath states arising in the Schmidt decomposition of equation (8.29), the embedded Hamiltonian has to be constructed explicitly in the many-body basis. We will now show that for the LDOS, one can work with a Fock space of bath orbitals.

For LDOS, $\hat{X}$ and $\hat{V}$ are single-particle operators acting on a specific orbital. Here, we will only treat the addition part, but the discussion of the removal part is analogous. Equation (8.29) becomes:

$$|\Phi_1(\omega, \hat{a}_k^\dagger)\rangle = \frac{1}{\omega - (\hat{h} + \hat{u} - \epsilon_0) + i\eta} \hat{a}_k^\dagger |\Phi_0\rangle, \tag{8.31}$$

$$= \sum_{\mu}^{\text{virt}} \frac{1}{\omega - (\hat{h} + \hat{u} - \epsilon_0) + i\eta} \hat{a}_\mu^\dagger C_{\mu k}^\dagger |\Phi_0\rangle, \tag{8.32}$$

$$= \sum_{\mu}^{\text{virt}} \frac{1}{\omega - \epsilon_\mu + i\eta} \hat{a}_\mu^\dagger C_{\mu k}^\dagger |\Phi_0\rangle, \tag{8.33}$$

where ϵ_μ is the single-particle energy associated with virtual orbital μ of the single-particle operator $\hat{h} + \hat{u}$. We can now augment the ground-state bath orbital space with one additional orbital, arising from the added electron. This electron has weight on both the impurity and the environment:

$$\sum_{\mu}^{\text{virt}} \frac{C^{\dagger}_{\mu k}}{\omega - \epsilon_\mu + i\eta} \hat{a}^{\dagger}_{\mu} = \sum_{l}^{L_A} \sum_{\mu}^{\text{virt}} \frac{C_{l\mu} C^{\dagger}_{\mu k}}{\omega - \epsilon_\mu + i\eta} \hat{a}^{\dagger}_{l} + \sum_{l > L_A} \sum_{\mu}^{\text{virt}} \frac{C_{l\mu} C^{\dagger}_{\mu k}}{\omega - \epsilon_\mu + i\eta} \hat{a}^{\dagger}_{l}. \tag{8.34}$$

The part on the environment can be added to the ground-state bath orbital space, after normalization similar to equation (8.10). If real-valued instead of complex-valued orbitals are desired, two additional bath orbitals should be added, for the real and imaginary parts. The response (8.29) is then spanned exactly in the Fock space of the augmented set of bath orbitals.

8.7 Overview of the Literature

Ground-state DMET has been applied to a variety of condensed matter systems. It has been used to study the one-dimensional Hubbard model [1, 19], the one-dimensional Hubbard–Anderson model [5], the one-dimensional Hubbard–Holstein model [6], the two-dimensional Hubbard model on the square [1, 3, 4] as well as the honeycomb lattice [16], and the two-dimensional spin-$\frac{1}{2}$ J_1-J_2-model [7].

Within the context of quantum chemistry, the method has been used to study hydrogen rings and sheets [2, 13], beryllium rings [13], an S_N2 reaction [13], polymers [17], boron-nitride sheets [17], and crystalline diamond [17].

The DMET bath orbital construction from Section 8.3 can also be used to construct optimal QM/MM boundaries [15] and to contract primitive Gaussians into adaptive atomic basis sets for correlated calculations [20].

The DMET formalism is not limited to ground-state properties. By augmenting the ground-state bath space with additional correlated many-body states from a Schmidt decomposition of the response wavefunction, accurate spectral functions have been obtained [16, 18]. We have shown in Section 8.6 that for the local density of states, it is sufficient to augment the bath orbital space with one additional response orbital.

8.8 The One-Band Hubbard Model on the Square Lattice

In this section, we review the DMET calculations of Refs. [1], [3], and [18] on the one-band Hubbard model on the square lattice. This model contains sufficient physics to exhibit d-wave superconductivity. For this reason, many groups have invested considerable numerical effort to map its rich phase diagram, which contains a Mott metal–insulator transition, d-wave superconductivity, and magnetism. For a detailed overview, we refer the reader to Refs. [21] and [4].

The Hubbard Hamiltonian

$$\hat{H}_{\text{Hubbard}} = -t \sum_{<ij>,\sigma} \hat{a}^{\dagger}_{i\sigma} \hat{a}_{j\sigma} - t' \sum_{\ll ij \gg,\sigma} \hat{a}^{\dagger}_{i\sigma} \hat{a}_{j\sigma} + U \sum_{i} \hat{n}_{i,\uparrow} \hat{n}_{i,\downarrow} \tag{8.35}$$

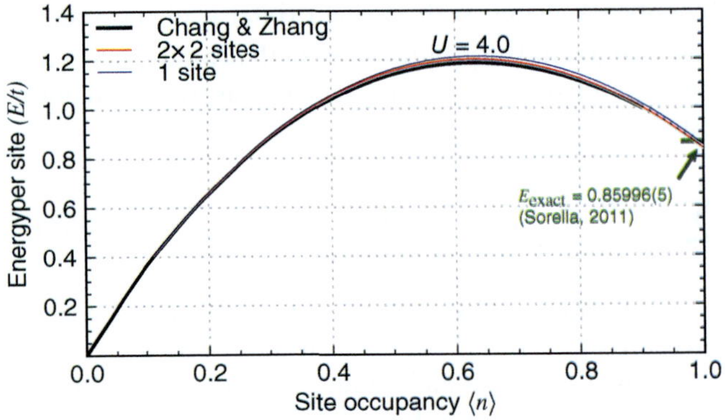

Figure 8.3 Ground-state energy per site for the one-band Hubbard model on the square lattice with $U/t = 4$ and nearest-neighbour hopping only ($t' = 0$). DMET calculations with impurity sizes 1×1 and 2×2 are in very good agreement with earlier QMC results from Chang and Zhang [22] and Sorella [23]. *Source*: Knizia and Chan 2012 [1]. Reprinted with permission of American Physical Society.

is expressed in terms of spatial orbitals and contains nearest-neighbour (t) and next-nearest-neighbour (t') hopping, as well as on-site electron repulsion (U).

For half-filling ($n = 1$) and nearest-neighbour hopping only ($t' = 0$), the model has particle-hole symmetry. Quantum Monte Carlo (QMC) becomes exact in this regime, because there is no fermion sign problem. It is therefore instructive to compare DMET calculations with QMC. Figure 8.3 compares DMET results with QMC [22, 23] both at and away from half-filling. The results improve as the impurity size becomes larger, and local bulk properties such as the energy per site will eventually saturate with increasing impurity size. Table 8.1 contains DMET results which are extrapolated with respect to impurity size, and compares them with AFQMC and DMRG [4]. The DMET ground-state energies are in very good agreement with the QMC and DMRG results.

Table 8.1 Ground-state energy per site for the one-band Hubbard model on the square lattice with nearest-neighbour hopping only ($t' = 0$).

U/t	Filling	DMET	AFQMC	DMRG
2	1.0	−1.1764(3)	−1.1763(2)	−1.176(2)
4	1.0	−0.8604(3)	−0.8603(2)	−0.862(2)
6	1.0	−0.6561(5)	−0.6568(3)	−0.658(1)
8	1.0	−0.5234(10)	−0.5247(2)	−0.5248(2)
12	1.0	−0.3686(10)	−0.3693(2)	−0.3696(3)
4	0.8	−1.108(2)	−1.110(3)	−1.1040(14)
4	0.6	−1.1846(5)	−1.185(1)	
4	0.3	−0.8800(3)	−0.879(1)	

Source: Zheng and Chan 2016 [3]. Reprinted with permission of American Physical Society.
The DMET results are extrapolated with respect to impurity size. For filling $n = 1.0$ there is no fermion sign problem and the AFQMC results are exact. The DMET, AFQMC, and DMRG results [4] are in very good agreement.

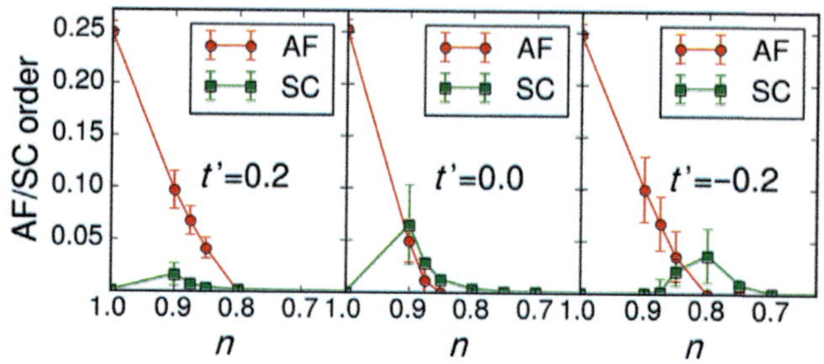

Figure 8.4 DMET antiferromagnetic (AF) and superconducting (SC) order parameters for the one-band Hubbard model on the square lattice as a function of lattice filling n for $U/t = 4$. Near half-filling ($n = 1.0$) antiferromagnetism is observed, and below half-filling d-wave superconductivity. *Source*: Zheng and Chan 2016 [3]. Reprinted with permission of American Physical Society.

This gives confidence to study not only ground-state energies, but various properties as well. Figure 8.4 shows the antiferromagnetic and superconducting order parameters as a function of lattice filling n for $U/t = 4$. Near half-filling antiferromagnetism is observed, and below half-filling d-wave superconductivity.

The paramagnetic phase of the one-band Hubbard model on the square lattice can be studied by using *restricted* Hartree-Fock theory as the low-level method. For this phase, the DMET correlation potential $\hat{u}$ opens a single-particle energy gap in $\hat{h} + \hat{u}$ with increasing U/t. This gap is a qualitative signature of the metal-insulator transition. With the Green's function DMET method explained in Section 8.6, the local density of states can be calculated, which yields a quantitatively accurate gap. Figure 8.5 shows

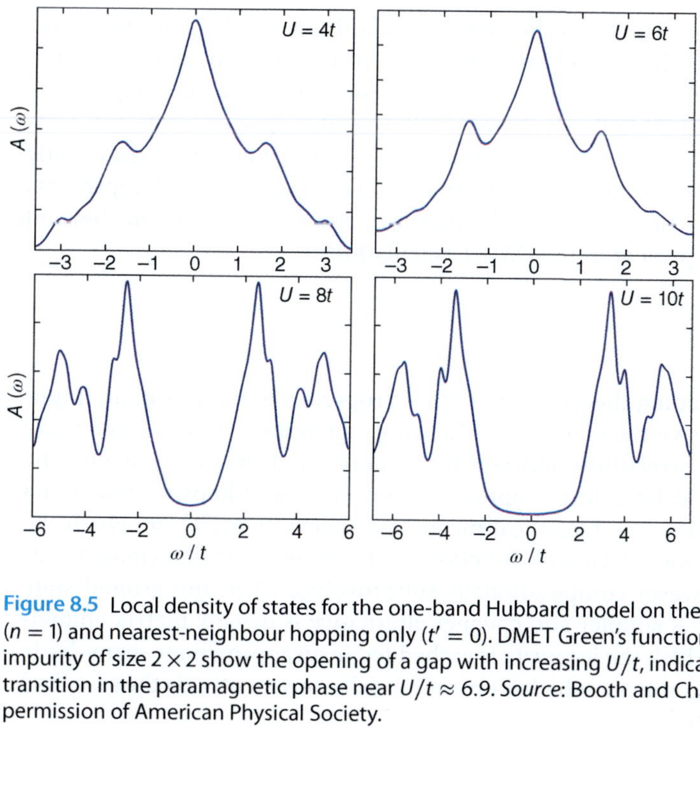

Figure 8.5 Local density of states for the one-band Hubbard model on the square lattice at half-filling ($n = 1$) and nearest-neighbour hopping only ($t' = 0$). DMET Green's function calculations on an impurity of size 2×2 show the opening of a gap with increasing U/t, indicating a metal-insulator transition in the paramagnetic phase near $U/t \approx 6.9$. *Source*: Booth and Chan 2015 [18]. Reprinted with permission of American Physical Society.

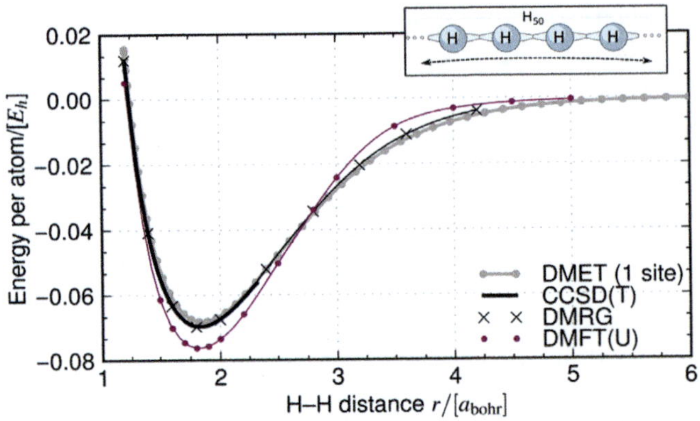

Figure 8.6 Ground-state energy per atom for the symmetric stretch of a linear hydrogen chain with 50 atoms in the STO-6G basis. CCSD(T), DMFT [24], and DMET calculations with one atom per impurity are compared with the numerically exact DMRG energies [25]. *Source*: Knizia and Chan 2013 [2]. Reprinted with permission of American Chemical Society.

the opening of the gap with increasing U/t for the one-band Hubbard model at half-filling ($n = 1$) and nearest-neighbour hopping only ($t' = 0$). A metal-insulator transition occurs near $U/t \approx 6.9$.

8.9 Dissociation of a Linear Hydrogen Chain

The dissociation of hydrogen chains has become a standard test case in quantum chemistry for strong correlation. The ground-state energy per atom for the symmetric stretch of a linear hydrogen chain with 50 atoms in the STO-6G basis is shown in Figure 8.6. CCSD(T), DMFT [24], and DMET calculations with one atom per impurity are compared with the numerically exact DMRG energies [25]. DMET performs significantly better than DMFT for this case, most likely due to the interacting bath which represents long-range electron interactions between the impurity and the environment beyond mean-field, which are not present in the DMFT calculations.

8.10 Summary

The general goal of embedding theories is to obtain properties of interest of an impurity, without doing expensive calculations on the whole quantum system. When there is entanglement (strong correlation) between the impurity and the environment, the embedding method should be able to capture it. The most well-known embedding method for strong correlation is dynamical mean-field theory (DMFT), in which the correlated frequency-dependent Green's function of the impurity is determined self-consistently. DMFT is, however, computationally quite involved when only ground-state properties are of interest. A simpler and cheaper alternative is density matrix embedding theory (DMET), in which the impurity is embedded in an approximate many-body Schmidt basis for the environment, which is optimized self-consistently via the so-called DMET correlation potential.

In Section 8.1, we have reviewed quantum entanglement and the Schmidt decomposition, two concepts from quantum information theory which are necessary to understand DMET. In Section 8.2, a general introduction to DMET is given. For mean-field low-level wavefunctions, the many-body Schmidt basis for the environment is a Fock space of bath orbitals. In Section 8.3, we have outlined how these bath orbitals can be calculated. This leads to a DMET active space which consists of the impurity and bath orbitals. The embedding Hamiltonian for DMET can be constructed similar to the CASCI effective Hamiltonian, that is, the one-electron integrals contain Coulomb and exchange terms from the external core orbitals. In Section 8.4, the construction of the embedded Hamiltonian and the calculation of DMET properties are discussed. The DMET correlation potential allows to fine-tune the bath orbital space and is optimized self-consistently as described in Section 8.5. While DMET was introduced as a simpler and cheaper alternative to DMFT for ground-state properties, it is not limited to ground-state properties. In Section 8.6, the extension of DMET to linear response and Green's functions is outlined. For the local density of states, the many-body Schmidt basis for the approximate response in the environment can be obtained by augmenting the ground-state bath orbital space with one additional response orbital. A general overview of the literature is given in Section 8.7. DMET calculations on the one-band Hubbard model on the square lattice and the symmetric stretch of a hydrogen chain are reviewed in Sections 8.8 and 8.9, respectively.

DMET has shown success in describing condensed matter and quantum chemical systems where strong correlation is present. Although in principle, the method can also handle weak correlation, we have not discussed how well it can be handled with the outlined algorithm. We refer the reader to Ref. [13] for a brief discussion.

Acknowledgments

S. W. gratefully acknowledges a Gustave Boël - Sofina - B.A.E.F. postdoctoral fellowship from the King Baudouin Foundation and the Belgian-American Educational Foundation for the academic year 2014–2015, and a postdoctoral fellowship from the Research Foundation Flanders (Fonds Wetenschappelijk Onderzoek Vlaanderen) for the academic years 2015–2018. G. K.-L. C. acknowledges support from the U. S. Department of Energy through DE-SC0010530. Additional support was provided from the Simons Foundation through the Simons Collaboration on the Many-Electron Problem.

References

1 Knizia, G. and Chan, G.K.L. (2012) Density matrix embedding: A simple alternative to dynamical mean-field theory. *Phys. Rev. Lett.*, **109** (18), 186404. http://dx.doi.org/10.1103/PhysRevLett.109.186404

2 Knizia, G. and Chan, G.K.L. (2013) Density matrix embedding: A strong-coupling quantum embedding theory. *J. Chem. Theory Comput.*, **9** (3), 1428–1432. http://dx.doi.org/10.1021/ct301044e

3 Zheng, B.X. and Chan, G.K.L. (2016) Ground-state phase diagram of the square lattice Hubbard model from density matrix embedding theory. *Phys. Rev. B*, **93** (3), 035126. http://dx.doi.org/10.1103/PhysRevB.93.035126

4 LeBlanc, J.P.F., Antipov, A.E., Becca, F., Bulik, I.W., Chan, G.K.L., Chung, C.M., Deng, Y., Ferrero, M., Henderson, T.M., Jiménez-Hoyos, C.A., Kozik, E., Liu, X.W., Millis, A.J., Prokof'ev, N.V., Qin, M., Scuseria, G.E., Shi, H., Svistunov, B.V., Tocchio, L.F., Tupitsyn, I.S., White, S.R., Zhang, S., Zheng, B.X., Zhu, Z., and Gull, E. (2015) Solutions of the two-dimensional Hubbard model: Benchmarks and results from a wide range of numerical algorithms. *Phys. Rev. X*, **5** (4), 041041. http://dx.doi.org/10.1103/PhysRevX.5.041041

5 Tsuchimochi, T., Welborn, M., and Van Voorhis, T. (2015) Density matrix embedding in an antisymmetrized geminal power bath. *J. Chem. Phys.*, **143** (2), 024107. http://dx.doi.org/10.1063/1.4926650

6 Sandhoefer, B. and Chan, G.K.L. (2016) Density matrix embedding theory for interacting electron-phonon systems. *Phys. Rev. B*, **94** (8), 085115. http://dx.doi.org/10.1103/PhysRevB.94.085115

7 Fan, Z. and Jie, Q.L. (2015) Cluster density matrix embedding theory for quantum spin systems. *Phys. Rev. B*, **91** (19), 195118. http://dx.doi.org/10.1103/PhysRevB.91.195118

8 Metzner, W. and Vollhardt, D. (1989) Correlated lattice fermions in $d = \infty$ dimensions. *Phys. Rev. Lett.*, **62** (3), 324–327. http://dx.doi.org/10.1103/PhysRevLett.62.324

9 Georges, A. and Krauth, W. (1992) Numerical solution of the $d = \infty$ Hubbard model: Evidence for a Mott transition. *Phys. Rev. Lett.*, **69** (8), 1240–1243. http://dx.doi.org/10.1103/PhysRevLett.69.1240

10 Georges, A., Kotliar, G., Krauth, W., and Rozenberg, M.J. (1996) Dynamical mean-field theory of strongly correlated fermion systems and the limit of infinite dimensions. *Rev. Mod. Phys.*, **68** (1), 13–125. http://dx.doi.org/10.1103/RevModPhys.68.13

11 Zgid, D. and Chan, G.K.L. (2011) Dynamical mean-field theory from a quantum chemical perspective. *J. Chem. Phys.*, **134** (9), 094115. http://dx.doi.org/10.1063/1.3556707

12 MacDonald, J.K.L. (1933) Successive approximations by the Rayleigh-Ritz variation method. *Phys. Rev.*, **43** (10), 830–833. http://dx.doi.org/10.1103/PhysRev.43.830

13 Wouters, S., Jiménez-Hoyos, C.A., Sun, Q., and Chan, G.K.L. (2016) A practical guide to density matrix embedding theory in quantum chemistry. *J. Chem. Theory Comput.*, **12** (6), 2706–2719. http://dx.doi.org/10.1021/acs.jctc.6b00316

14 Knizia, G. (2013) Intrinsic atomic orbitals: An unbiased bridge between quantum theory and chemical concepts. *J. Chem. Theory Comput.*, **9** (11), 4834–4843. http://dx.doi.org/10.1021/ct400687b

15 Sun, Q. and Chan, G.K.L. (2014) Exact and optimal quantum mechanics/molecular mechanics boundaries. *J. Chem. Theory Comput.*, **10** (9), 3784–3790. http://dx.doi.org/10.1021/ct500512f

16 Chen, Q., Booth, G.H., Sharma, S., Knizia, G., and Chan, G.K.L. (2014) Intermediate and spin-liquid phase of the half-filled honeycomb Hubbard model. *Phys. Rev. B*, **89** (16), 165134. http://dx.doi.org/10.1103/PhysRevB.89.165134

17 Bulik, I.W., Chen, W., and Scuseria, G.E. (2014) Electron correlation in solids via density embedding theory. *J. Chem. Phys.*, **141** (5), 054113. http://dx.doi.org/10.1063/1.4891861

18 Booth, G.H. and Chan, G.K.L. (2015) Spectral functions of strongly correlated extended systems via an exact quantum embedding. *Phys. Rev. B*, **91** (15), 155107. http://dx.doi.org/10.1103/PhysRevB.91.155107

19 Bulik, I.W., Scuseria, G.E., and Dukelsky, J. (2014) Density matrix embedding from broken symmetry lattice mean fields. *Phys. Rev. B*, **89** (3), 035140. http://dx.doi.org/10.1103/PhysRevB.89.035140

20 Sorella, S., Devaux, N., Dagrada, M., Mazzola, G., and Casula, M. (2015) Geminal embedding scheme for optimal atomic basis set construction in correlated calculations. *J. Chem. Phys.*, **143** (24), 244112. http://dx.doi.org/10.1063/1.4938089

21 Scalapino, D.J. (2007) Numerical Studies of the 2D Hubbard Model, in *Handbook of High-Temperature Superconductivity: Theory and Experiment* (eds. J.R. Schrieffer and J.S. Brooks), Springer, New York, chap. 13, pp. 495–526. http://dx.doi.org/10.1007%2F978-0-387-68734-6_13

22 Chang, C.C. and Zhang, S. (2008) Spatially inhomogeneous phase in the two-dimensional repulsive Hubbard model. *Phys. Rev. B*, **78** (16), 165101. http://dx.doi.org/10.1103/PhysRevB.78.165101

23 Sorella, S. (2011) Linearized auxiliary fields Monte Carlo technique: Efficient sampling of the fermion sign. *Phys. Rev. B*, **84** (24), 241110. http://dx.doi.org/10.1103/PhysRevB.84.241110

24 Lin, N., Marianetti, C.A., Millis, A.J., and Reichman, D.R. (2011) Dynamical Mean-Field Theory for Quantum Chemistry. *Phys. Rev. Lett.*, **106**, 096402. http://dx.doi.org/10.1103/PhysRevLett.106.096402

25 Hachmann, J., Cardoen, W., and Chan, G.K.L. (2006) Multireference correlation in long molecules with the quadratic scaling density matrix renormalization group. *J. Chem. Phys.*, **125** (14), 144101. http://dx.doi.org/10.1063/1.2345196

9

Ab initio Ice, Dry Ice, and Liquid Water

So Hirata,[1] Kandis Gilliard,[1] Xiao He,[2,3] Murat Keçeli,[1,4] Jinjin Li,[1,5,6] Michael A. Salim,[1] Olaseni Sode,[1,7] and Kiyoshi Yagi[1,8]

[1] Department of Chemistry, University of Illinois at Urbana-Champaign, USA
[2] School of Chemistry and Molecular Engineering, East China Normal University, China
[3] NYU-ECNU Center for Computational Chemistry, NYU Shanghai, China
[4] Argonne National Laboratory, USA
[5] National Key Laboratory of Science and Technology on Micro/Nano Fabrication, Shanghai Jiao Tong University, China
[6] Department of Micro/Nano Electronics, Shanghai Jiao Tong University, China
[7] Department of Chemistry, Biochemistry, and Physics, The University of Tampa, USA
[8] Theoretical Molecular Science Laboratory, RIKEN, Japan

9.1 Introduction

Computer simulations of condensed matter have long relied on empirical force fields as a trade-off between their feasibility and fidelity. They were revolutionized by density-functional theory (DFT) [1, 2], which enabled fast quantum-mechanical calculations of potential energy surfaces (PES) and other electronic properties of condensed matter. While DFT made immeasurable contributions to quantitative computational condensed matter physics and materials science, its applicability and the trustworthiness of its prediction are fundamentally limited by nonsystematic approximations to the elusive exact exchange-correlation functional.

In the meantime, computational chemistry for atomic and molecular electronic structures advanced on a much broader front, including both DFT and *ab initio* wave function theory (WFT) [3–5]. The latter includes several hierarchies of systematic approximations converging toward the exact solution of the Schrödinger equation or its relativistic extension. They are, for instance, the configuration-interaction (CI), many-body perturbation (MP), and coupled-cluster (CC) theories, which are systematic expansions of a many-electron wave function. For one- and two-electron basis expansions, systematic Gaussian-type-orbital (GTO) basis sets [6–8] and the explicitly correlated ('R12' or 'F12') methods [9–11] have been developed. Combining these, chemists have demonstrated predictive (i.e., controlled-error) calculations of virtually any observable of atoms and molecules [12–14]. For example, structures and thermochemistry of combustion intermediates and interstellar molecules may now be most reliably determined computationally with *ab initio* WFT, but not necessarily with DFT.

Today, many consider the extension of *ab initio* WFT to condensed matter as a major frontier of physics, chemistry, and computing; many aspire to replicate the triumph of

Fragmentation: Toward Accurate Calculations on Complex Molecular Systems, First Edition.
Edited by Mark S. Gordon.

predictive computing for atoms and molecules, marked by Pople's Nobel Prize in Chemistry [15], in condensed matter physics and materials science.

This may, at first glance, appear to be a straightforward exportation of well-established molecular electronic structure theories and algorithms to infinite systems on ever faster computers. This is not the case. When the system has Avogadro's number of particles, one must integrate quantum mechanics and statistical thermodynamics to describe it. The mathematical theories for condensed matter should thus have temperature, pressure, chemical potential, etc. as essential parameters. One also needs new *ab initio* WFT that can describe phases unique to condensed matter, that is, metals and superconductors. They should be able to treat condensed-phase properties such as dispersion, transport, and phase transitions. They also have to be rigorously (i.e., diagrammatically) size-consistent for both extensive and intensive observables [16, 17]. On top of these, their algorithms ought to be extremely fast and scalable to be routinely feasible for these infinite systems; to achieve this, molecular algorithms often have to be discarded and completely new ones need to be invented [18–20].

For a certain class of condensed matter, however, such a novel algorithm has recently come into existence, enabling routine *ab initio* WFT calculations to predict a variety of condensed-phase properties at finite temperature and pressure. The systems in question are the so-called molecular solids and liquids, which hold immense importance in a range of scientific fields. They consist of weakly interacting molecules and can be viewed as a superposition of overlapping dimers, trimers, etc. of the constituent molecules, which are embedded in the electrostatic environment of the whole system. These "fragments" can then be computationally characterized by sophisticated molecular *ab initio* WFT and their computer programs, independently of one another, that is, in a massively parallel, linear-scaling manner. This embedded-fragment method [21–27] realizes an extremely efficient, strongly size-consistent, systematically accurate, versatile, and easily implemented algorithm of virtually any *ab initio* WFT for molecular solids and liquids. Chemical interactions that bind molecular solids and liquids together are diverse (covalent to van der Waals bonds), whose strengths span 3 orders of magnitude, whereas energy differences between polymorphic phases can be tiny. Therefore, *ab initio* WFT with hierarchical approximations such as MP and CC with systematic GTO basis sets is essential to make consistent progress toward predictive simulation fidelity.

Molecular solids and liquids include nature's most abundant and important condensed-phase systems such as ice [28, 29] and liquid water [30–32]. This is because the surface of a planet, including that of the Earth, tends to be covered by condensation of its atmospheric molecules, which are invariably molecular solids and liquids. For instance, notwithstanding the temperature reaching 700K at the equator, Mercury is observed [33] to have water ice in its permanently dark craters at the north pole. On Venus, it rains sulfuric acid [34], a molecular liquid. On Titan, the largest satellite of Saturn, seas of liquid methane are believed to exist [35, 36]. Uranus and Neptune are the so-called "ice giants," whose mantles are made of a huge amount of dense molecular liquids ("ices") of water, ammonia, and methane [37]. The presence of ices on Mars is well documented; its polar ice caps are clearly visible and consist of permanent water ice covered by permanent and/or seasonal layers of solid carbon dioxide (CO_2) or dry ice. Liquid water is believed to exist in the recent past, forming Martian gullies [38], or even today transiently at night [39]. Others [40] hypothesized that the gullies could be caused by sublimation of dry ice in today's climate rather than in the past.

Of course, it is on the Earth that the most decisive influence the properties of ice and liquid water have on everything from climate to geology to biology. What may then be surprising is the fact that there are still a number of unsolved mysteries and unsettled controversies surrounding these properties. For instance, to explain two hydrogen-bond-stretch peaks at 28.4 and 37.9 meV in the inelastic neutron scattering (INS) spectra of ice-Ih, Li and Ross hypothesized [41, 42] the existence of two types of hydrogen bonds differing in strength by a factor of 2. This hypothesis remained controversial for the next nearly two decades, until it was disproved by two independent quantitative simulations [43, 44]. Unlike most other solids, ice-Ih contracts thermally at low temperatures before it expands at higher temperatures [45, 46]. Ice-Ih expands upon heavy-isotope (deuterium) substitution, which is the opposite behavior from normal solids and is said to have the anomalous volume isotope effect (VIE) [45, 46]. They mirror similar unusual thermodynamic behavior of liquid water, underscoring the importance of their quantitative understanding. Furthermore, ice-Ih undergoes a pressure-induced amorphization [47] to become high-density amorphous (HDA) ice, which is distinct from low-density amorphous (LDA) ice. This led to a hypothesis [48] that there are also two distinct phases of liquid water—high-density and low-density liquids—between which a liquid–liquid phase transition is possible. This remains to be a highly controversial issue [49–51]. There seems perennial disagreement among researchers about the average number of hydrogen bonds per molecule [52] and the mean dipole moment [53–58] in liquid water. The fact that liquid water exists only at excessively high temperature and pressure in DFT simulations has puzzled researchers [59], also casting a shadow of doubt on the value of some of their predictions.

When it comes to less common ices, their structural and thermodynamic properties are even less established. Take solid CO_2 as an example. The most well-characterized phase of CO_2 besides dry ice (phase I) is phase III, which can be reached only by low-temperature compression of phase I (Figure 9.1) [61]. The reported transition pressures range widely from 2.5 GPa at 80K (Ref. 62) to above 18 GPa at room temperature [63], with no consensus. It is speculated to be a metastable state with underlying stable phase II [64–66], whose structure and energy are not well understood, either. Even for dry ice (phase I), which is perhaps the most familiar ice besides water ice, quantitative prediction of thermal expansion and sublimation temperature has not been achieved until recently [67].

In this chapter, we summarize our recent applications of *ab initio* WFT to a variety of properties of water ices, liquid water, and solid CO_2 as well as the embedded-fragmentation techniques that enable such applications at *ab initio* theory levels of second-order MP (MP2) and CC with singles and doubles (CCSD). They mark the beginning of a new era of condensed matter simulations based on systematic *ab initio* WFT.

9.2 Computational Method

In this section, we review the embedded-fragment methodology taking an application to a solid as an example (except in the last subsection where we consider applications to liquids mainly). It has enabled *ab initio* calculations of a wide variety of properties of molecular crystals, amorphous solids, and liquids at finite temperature and pressure.

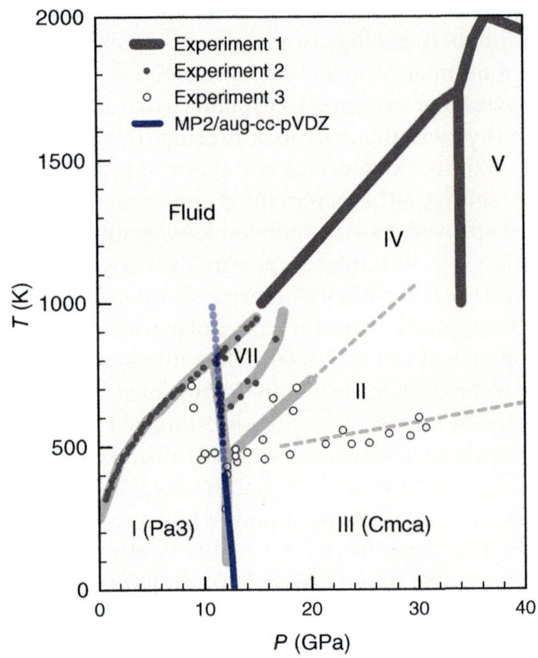

Figure 9.1 Phase diagram of CO_2. *Source*: Li *et al.* 2013 [60]. Reproduced with permission of Nature.

The formalisms and algorithmic strategies described here are by no means limited to condensed H_2O and CO_2 or to MP2 and CCSD.

9.2.1 Internal Energy

In the embedded-fragment method truncated after the two-body level (the binary-interaction method), the internal energy per unit cell of a molecular crystal is approximated [68] as

$$E_e \approx \sum_i E_{i(\mathbf{0})} + \frac{1}{2} \sum_{\mathbf{n}} \sum_{i,j} \{E_{i(\mathbf{0})j(\mathbf{n})} - E_{i(\mathbf{0})} - E_{j(\mathbf{n})}\} + E_{\mathrm{LR}}, \tag{9.1}$$

where $E_{j(\mathbf{n})}$ is the energy of the jth monomer in the $\mathbf{n}$th unit cell, $E_{i(\mathbf{0})j(\mathbf{n})}$ is the energy of the dimer consisting of the ith monomer in the central ($\mathbf{0}$th) unit cell and the jth monomer in the $\mathbf{n}$th unit cell, and E_{LR} is a long-range electrostatic energy correction. The first summation runs over all monomers in the unit cell, whereas the second goes over all unit cells and all pairs of monomers excluding the self interaction ($i = j$ and $\mathbf{n} = \mathbf{0}$) and those whose contributions are negligible because of a great distance between the $i(\mathbf{0})$th and $j(\mathbf{n})$th monomers. These energies are computed with an *ab initio* electron-correlated theory, MP2 or higher, independently of one another and therefore in parallel, at a computational cost scaling only quadratically with the number of molecules in the unit cell and linearly with global system size (i.e., when one views a solid as a giant cluster). Basis-set superposition errors (BSSE) can be removed [69] by a function-counterpoise method, when evaluating the second summation of equation (9.1), sharing the same advantage as local-basis, linear-scaling methods [70–72].

Each of these monomers and dimers is embedded in an electrostatic field created by the rest of the crystal. In our implementations, the field is represented by molecular dipole moments [73] or atomic partial charges [69], the latter reproducing the short-range electrostatic field of the molecular charge density in the linear-least-squares sense [74, 75]. More importantly, the field is determined self-consistently through the embedded-fragment calculation at the one-body truncation level using, e.g., a less expensive electronic structure theory such as Hartree–Fock (HF), prior to evaluating equation (9.1); the calculation of the dipole moment or atomic partial charges of a monomer is cyclically repeated in the presence of the field created by all the other monomers until the field ceases to change.

There are many other valid ways of defining the field. In fact, our method is a direct simplification of and differs only in the definition of the field from the pair interaction molecular orbital method of Kitaura *et al.* [76], the predecessor of their extremely successful fragment molecular orbital (FMO) method [77, 78]. In Kitaura *et al.* [76], an HF electron density is used as the embedding field, necessitating molecular integrals to be evaluated during electron-correlated calculations of monomers and dimers in the field. Our embedding field, in contrast, can be represented by point charges, which many existing molecular electronic structure programs can handle with no modifications. Other choices of the field definition include the one based on the Mulliken population analysis [79], the one involving an HF calculation for the entire system [80, 81], or a force field used in classical molecular dynamics [82]. We argue that our definition has the advantage of being polarizable because of self-consistency, being systematically improvable (both in the order of multipole moments and the level of electronic structure theory that computes them), not requiring any modification to existing molecular programs, and strictly maintaining linear scaling of cost by fragmenting a solid in every step of the calculation. Other implementations of the embedded-fragment method have been proposed [83–90]. For molecular crystals, a systematic effort by Beran's group is most notable [67, 82, 91–100].

The energy in the two-body truncation approximation, equation (9.1), includes two-body electrostatic, exchange, and correlation (dispersion) interactions explicitly and all higher-order electrostatic interactions in the dipole or atomic-partial-charge approximation, thus including the polarizability (induction) effect. The missing interactions, such as three-body dispersion or long-range quadrupole polarization effect, can be systematically recuperated [73] by including embedded trimers in the many-body expansion or self-consistent molecular quadrupole moments in the embedding field. In other words, the approximation embodied by equation (9.1) is the lowest member of a systematic hierarchy converging toward exactitude in much the same spirit as systematic *ab initio* many-body theories for electronic structures such as MP and CC are designed. It may also be noticed that it can be combined with virtually any electron-correlation theory, even one that is not diagrammatically size-consistent, yielding size-consistent and meaningful results for an infinitely extended system. This is because equation (9.1) fundamentally (but implicitly) alters the structure of the wave function in a solid from that of its constituent monomer or dimer. In this sense, we call the embedded-fragmentation strongly size-consistent.

The long-range electrostatic correction, E_{LR}, sums classical electrostatic interactions between dipole moments or atomic partial charges, which make up the embedding field, up to a large distance. It is, therefore, a part of the Madelung constant, not accounted for by other terms in equation (9.1). Great care needs to be exercised not to double count electrostatic interactions across the dimer energy sum, interactions between a monomer or a dimer and the embedding field, and this long-range correction [101].

9.2.2 Structure and Phonons

The energy in our method being additive, its derivative with respect to a perturbation is also calculated as a sum of monomer and dimer contributions in a linear-scaling, parallel fashion. For example, the gradients with respect to atomic coordinate x or lattice constant a are approximated as

$$\frac{\partial E_{\mathrm{e}}}{\partial x} \approx \sum_{i} \frac{\partial E_{i(\mathbf{0})}}{\partial x} + \frac{1}{2} \sum_{\boldsymbol{n}} \sum_{i,j} \left\{ \frac{\partial E_{i(\mathbf{0})j(\boldsymbol{n})}}{\partial x} - \frac{\partial E_{i(\mathbf{0})}}{\partial x} - \frac{\partial E_{j(\boldsymbol{n})}}{\partial x} \right\} + \frac{\partial E_{\mathrm{LR}}}{\partial x}, \tag{9.2}$$

$$\frac{\partial E_{\mathrm{e}}}{\partial a} \approx \frac{1}{2} \sum_{\boldsymbol{n}} \sum_{i,j} \left\{ \frac{\partial E_{i(\mathbf{0})j(\boldsymbol{n})}}{\partial a} - \frac{\partial E_{j(\boldsymbol{n})}}{\partial a} \right\} + \frac{\partial E_{\mathrm{LR}}}{\partial a}, \tag{9.3}$$

which can be evaluated analytically if their components are. The gradients with respect to a in equation (9.3) can be variously expressed in terms of those with respect to the x-coordinates of the jth molecule, depending on how one places each atom relative to the unit cell dimensions. The contribution from the long-range electrostatic correction is essential especially in the gradient with respect to a lattice constant because even a small change in the unit cell size is greatly amplified in the positions of distant atoms. Neglected here are the derivatives of the magnitudes (not the positions) of the dipole moments and atomic partial charges in the self-consistent embedding field.

The elements of a Hessian are also calculable efficiently and analytically (assuming that molecular Hessian elements are also available analytically) as

$$\frac{\partial^2 E_{\mathrm{e}}}{\partial x \partial y} \approx \sum_{i} \frac{\partial^2 E_{i(\mathbf{0})}}{\partial x \partial y} + \frac{1}{2} \sum_{\boldsymbol{n}} \sum_{i,j} \left\{ \frac{\partial^2 E_{i(\mathbf{0})j(\boldsymbol{n})}}{\partial x \partial y} - \frac{\partial^2 E_{i(\mathbf{0})}}{\partial x \partial y} - \frac{\partial^2 E_{j(\boldsymbol{n})}}{\partial x \partial y} \right\} + \frac{\partial^2 E_{\mathrm{LR}}}{\partial x \partial y}, \tag{9.4}$$

if x and y are in-phase collective coordinates that maintain periodicity of the crystal. Alternatively, x and y can be individual atomic coordinates. That such non-periodic coordinates can be used is because the embedded-fragment method does not rely strongly on the periodic boundary conditions unlike crystal orbital theory or planewave methods that use periodic basis functions (Bloch orbitals) and k-space integrations. One can therefore evaluate a full set of force constants across different unit cells as

$$\frac{\partial^2 E_{\mathrm{e}}}{\partial x^{\gamma}_{i(\mathbf{0})} \partial y^{\eta}_{j(\boldsymbol{n})}} \approx \frac{\partial^2 E_{i(\mathbf{0})j(\boldsymbol{n})}}{\partial x^{\gamma}_{i(\mathbf{0})} \partial y^{\eta}_{j(\boldsymbol{n})}}, \tag{9.5}$$

in the case of $\boldsymbol{n} \neq \mathbf{0}$, where $y^{\eta}_{j(\boldsymbol{n})}$ denotes the y-coordinate of atom η in the jth monomer in the $\boldsymbol{n}$th unit cell. With these, one obtains phonon dispersion curves in the entire reciprocal unit cell (not just at the Γ point) and, therefore, phonon density of states (DOS), which is difficult or inefficient to extract from crystal-orbital or planewave calculations. Phonon dispersion curves in the entire reciprocal space are crucial when temperature effects are to be included in the calculated properties of solids (see Section 9.2.5). In the same token, the embedded-fragment method works just as accurately and efficiently for nonperiodic solids (such as proton-disordered ice-Ih and amorphous ice; see Section 9.3.2) and liquids (see Section 9.3.4) as for periodic crystals.

9.2.3 Spectra

The harmonic frequency, ω_{nk}, of a fundamental lattice vibration in the nth phonon dispersion branch with wave vector $\boldsymbol{k}$ is obtained by solving

$$\mathbf{D}(\boldsymbol{k})\mathbf{L}_{nk} = \mathbf{L}_{nk}\omega_{nk}^2, \tag{9.6}$$

where $\mathbf{L}_{nk}$ is the corresponding mass-weighted normal-mode vector. $\mathbf{D}(\boldsymbol{k})$ is the dynamical force-constant matrix defined by

$$\mathbf{D}(\boldsymbol{k}) = \sum_{\boldsymbol{n}} \mathbf{F}(\boldsymbol{n}) \exp(i\boldsymbol{k} \cdot \boldsymbol{R}_{\boldsymbol{n}}), \tag{9.7}$$

where $\boldsymbol{R}_{\boldsymbol{n}} = (n_x a, n_y b, n_z c)$ is the lattice vector and $\mathbf{F}(\boldsymbol{n})$ is the mass-weighted force-constant matrix of the form,

$$\{\mathbf{F}(\boldsymbol{n})\}_{x_i^\gamma y_j^\eta} = m_\gamma^{-1/2} m_\eta^{-1/2} \frac{\partial^2 E_\mathrm{e}}{\partial x_{i(\mathbf{0})}^\gamma \partial y_{j(\boldsymbol{n})}^\eta}, \tag{9.8}$$

where γ and η label atoms (with masses m_γ and m_η) in the ith and jth monomer, respectively. As discussed above, the Hessian matrix element for individual atomic coordinates can be evaluated efficiently and often analytically in our method.

The plot of ω_{nk} with respect to $\boldsymbol{k}$ gives the phonon dispersion curves in the entire reciprocal space. Three of them are acoustic branches that converge at zero frequencies at the Γ point, whereas the rest are optical branches with nonzero frequencies at Γ. Owing to numerical errors caused by truncation of long-range force constants, the acoustic branches often need to be forced to satisfy the asymptotic condition at Γ by extrapolation of long-range force constants. One such method is given in an appendix of Ref. [102]. A phonon DOS is obtained as the histogram of frequencies:

$$g(\omega) \propto \sum_n \int \delta(\omega - \omega_{nk}) d\boldsymbol{k}. \tag{9.9}$$

The INS cross section (Sections 9.3.1 and 9.3.3) is given by [103]

$$S(Q, \omega) \propto g_\mathrm{H}(\omega) \exp\{-2W(Q)\}, \tag{9.10}$$

where Q is the neutron momentum and $g_\mathrm{H}(\omega)$ is the hydrogen-amplitude-weighted phonon DOS defined by

$$g_\mathrm{H}(\omega) \propto \sum_n \sum_\gamma^\mathrm{H} \int |L_{nk}^\gamma|^2 \delta(\omega - \omega_{nk}) d\boldsymbol{k}, \tag{9.11}$$

with L_{nk}^γ being the γth element of the normal-mode vector $\mathbf{L}_{nk}$, where γ runs over only hydrogen atoms. The last factor in equation (9.10) is the Debye–Waller factor, which accounts for the temperature effect and is often treated phenomenologically [104].

Only the Γ-point phonons are infrared- and/or Raman-active because only such zero-momentum phonons can conserve momentum with photons. Hence, the positions of infrared or Raman bands are determined by the Γ-point frequencies of the fundamental optical branches. Band intensities are related to dipole derivatives and polarizability derivatives [44], which are also sums of the corresponding monomer and dimer quantities (Section 9.3.3). In the Cartesian coordinates, they read

$$\frac{\partial \boldsymbol{\mu}}{\partial x} \approx \sum_i \frac{\partial \boldsymbol{\mu}_{i(0)}}{\partial x} + \frac{1}{2}\sum_{\boldsymbol{n}}\sum_{i,j}\left\{\frac{\partial \boldsymbol{\mu}_{i(0)j(\boldsymbol{n})}}{\partial x} - \frac{\partial \boldsymbol{\mu}_{i(0)}}{\partial x} - \frac{\partial \boldsymbol{\mu}_{j(\boldsymbol{n})}}{\partial x}\right\}, \tag{9.12}$$

$$\frac{\partial \boldsymbol{\alpha}}{\partial x} \approx \sum_i \frac{\partial \boldsymbol{\alpha}_{i(0)}}{\partial x} + \frac{1}{2}\sum_{\boldsymbol{n}}\sum_{i,j}\left\{\frac{\partial \boldsymbol{\alpha}_{i(0)j(\boldsymbol{n})}}{\partial x} - \frac{\partial \boldsymbol{\alpha}_{i(0)}}{\partial x} - \frac{\partial \boldsymbol{\alpha}_{j(\boldsymbol{n})}}{\partial x}\right\}, \tag{9.13}$$

where $\boldsymbol{\mu}$ and $\boldsymbol{\alpha}$ are the dipole moment and polarizability tensor, respectively, and x is an in-phase collective coordinate, which is associated with an infrared- and/or Raman-active Γ-point phonon. It should be recalled that dipole moment and polarizability are themselves derivatives of energies with respect to local electric-field strengths, justifying their many-body expansions underlying the above expressions. Remarkably, this formalism effortlessly circumvents a conundrum concerning the unbounded nature of the dipole operator in an infinite system [105–108].

9.2.4 Pressure Effects

The effect of pressure (P) can be included easily by evaluating enthalpy per unit cell, H_e, instead of internal energy:

$$H_\mathrm{e} = E_\mathrm{e} + PV, \tag{9.14}$$

where V is the unit cell volume. Gradients and Hessian of H_e can be obtained in the same way as above, allowing the determination of the structures and phonons of a solid under pressure. For example, in an orthorhombic unit cell ($V = abc$), the equilibrium value of lattice constant a is the one that equates the lattice force along a with the external force applied to the bc-plane by pressure,

$$-\frac{\partial E_\mathrm{e}}{\partial a} = P(bc), \tag{9.15}$$

which can be solved by a trivial modification to any geometry optimizer.

Using this method, one can determine the equation of state (the P-V curve) at 0K, from which the bulk modulus can be obtained, either by evaluating the following expression,

$$B = -V\left(\frac{\partial P}{\partial V}\right)_{T=0}, \tag{9.16}$$

semi-analytically (not used in our previous studies so far) or by linear-least-square fitting the equation of state over a range of V and then differentiating it analytically (Section 9.3.7). The third-order Birch–Murnaghan equation is sometimes used for the fitting.

It should be cautioned that P here is the so-called *virial* pressure, originating from the internal energy only. It does not include the *phonon* pressure, which arises from quantized nuclear motion, or the ideal-gas contribution in a liquid at finite temperature. The phonon pressure is nonzero even at 0K in an anharmonic crystal. These can, however, be included (see Section 9.2.5).

9.2.5 Temperature Effects

Inclusion of temperature effects is also surprisingly straightforward. In an insulating solid, heat is stored exclusively in phonons. Hence, Helmholtz and Gibbs energies per unit cell, A and G, of a crystal at temperature T is written as

$$A = E_{\mathrm{e}} + U_{\mathrm{v}} - TS_{\mathrm{v}}, \tag{9.17}$$

$$G = H_{\mathrm{e}} + U_{\mathrm{v}} - TS_{\mathrm{v}}, \tag{9.18}$$

where E_{e} and H_{e} are the electronic energy and enthalpy per unit cell at 0K, respectively, U_{v} is the vibrational energy per unit cell at T, and S_{v} is the vibrational entropy per unit cell. They are, in turn, derived from the vibrational partition function, Z_{v}, which, in the harmonic approximation, can be written as

$$Z_{\mathrm{v}} = \prod_n \prod_{\boldsymbol{k}} \frac{\exp(-\beta\omega_{n\boldsymbol{k}}/2)}{1 - \exp(-\beta\omega_{n\boldsymbol{k}})}, \tag{9.19}$$

where $\beta = (k_{\mathrm{B}}T)^{-1}$. Since

$$U_{\mathrm{v}} = \frac{T}{\beta K}\frac{\partial \ln Z_{\mathrm{v}}}{\partial T}, \tag{9.20}$$

$$S_{\mathrm{v}} = \frac{1}{\beta K}\frac{\partial \ln Z_{\mathrm{v}}}{\partial T} + \frac{1}{\beta K}\frac{\ln Z_{\mathrm{v}}}{T}, \tag{9.21}$$

where K is the number of evenly-spaced wave-vector ($\boldsymbol{k}$) sampling points in the discretized reciprocal unit cell, one finds

$$A = E_{\mathrm{e}} + \frac{1}{K}\sum_n \sum_{\boldsymbol{k}} \frac{\omega_{n\boldsymbol{k}}}{2} + \frac{1}{\beta K}\sum_n \sum_{\boldsymbol{k}} \ln\{1 - \exp(-\beta\omega_{n\boldsymbol{k}})\}, \tag{9.22}$$

$$G = A + PV, \tag{9.23}$$

which can be readily evaluated with the phonon dispersion curves in the entire reciprocal space available from the embedded-fragment method. G as a function of T and P can be used directly to draw computational phase diagrams (Section 9.3.6).

Other observable thermodynamic properties can also be calculated. For instance, the unit cell isochoric heat capacity is obtained (Section 9.3.1) by evaluating

$$C_V \equiv \left(\frac{\partial U_{\mathrm{v}}}{\partial T}\right)_V = \frac{k_{\mathrm{B}}}{K}\sum_n \sum_{\boldsymbol{k}} c_{n\boldsymbol{k}}, \tag{9.24}$$

$$c_{n\boldsymbol{k}} = (\beta\omega_{n\boldsymbol{k}})^2 \frac{\exp(\beta\omega_{n\boldsymbol{k}})}{\{\exp(\beta\omega_{n\boldsymbol{k}}) - 1\}^2}, \tag{9.25}$$

of which the latter may be called a mode heat capacity.

Another important thermodynamic property of a solid is the rate of thermal expansion, which can be evaluated by our method in at least three different ways. In

Grüneisen's theory [109, 110] (Section 9.3.7), the volumetric thermal expansion coefficient, α_V, at 0K is written as

$$\alpha_V \equiv \frac{1}{V}\left(\frac{\partial V}{\partial T}\right)_P \tag{9.26}$$

$$= -\frac{1}{V}\left(\frac{\partial V}{\partial P}\right)_T\left(\frac{\partial P}{\partial T}\right)_V = \frac{1}{B}\left(\frac{\partial P}{\partial T}\right)_V = \frac{1}{B}\left(\frac{\partial S_v}{\partial V}\right)_T \tag{9.27}$$

$$= \frac{C_V\gamma}{BV}, \tag{9.28}$$

where γ is the Grüneisen parameter defined as

$$\gamma \equiv \frac{V}{C_V}\left(\frac{\partial S_v}{\partial V}\right)_T. \tag{9.29}$$

The first equality of equation (9.27) uses a cyclic relation, the second identifies a factor as the reciprocal of B, and the third uses a Maxwell relation. Substituting equation (9.21) into equation (9.29), one finds

$$\gamma = -\frac{1}{C_V K}\sum_n \sum_k \frac{V}{\omega_{nk}}\frac{\partial \omega_{nk}}{\partial V}c_{nk} \tag{9.30}$$

$$= \frac{1}{C_V K}\sum_n \sum_k \gamma_{nk} c_{nk}, \tag{9.31}$$

with

$$\gamma_{nk} = -\frac{V}{\omega_{nk}}\frac{\partial \omega_{nk}}{\partial V} = -\frac{\partial \ln \omega_{nk}}{\partial \ln V}, \tag{9.32}$$

which is known as the mode Grüneisen parameter, directly calculable from the phonon dispersion curves as a function of pressure. The mode heat capacity, c_{nk}, defined by equation (9.25) is calculated similarly.

While Grüneisen's theory may not be the most accurate method of evaluating α_V, it nicely captures the essence of physics of thermal expansion. Thermal expansion is entirely an anharmonic effect; if the potential were exactly harmonic, the curvature of the potential would be the same wherever it is measured and the phonon frequency would therefore be independent of V. Then, all of the mode Grüneisen parameters would be zero and thus $\alpha_V = 0$. In reality, the potential is always anharmonic, and as the solid expands, phonons explore the potential at longer atom–atom distances, where it is usually softer, and their frequencies tend to become lower (corresponding to *positive* mode Grüneisen paramaters). Lower frequencies mean more thermally accessible vibrational states and an entropy gain, energetically favoring the expansion. In this sense, solids expand or contract for essentially the same entropic reason underlying the elasticity of rubber [111], although the change in entropy is caused by anharmonicity in the former while it originates from polymer conformations in the latter.

The discussion in the previous paragraph implies that according to Grüneisen's theory, the volume change by anharmonic phonons is null at 0K. This is not the case in reality because of the existence of zero-point vibrations. Not only do they make the volumes with zero-point vibration corrections differ from the ones without, but they also render their volumes isotope-dependent, an effect known as the volume-isotope effect (VIE), which persists even at 0K. Grüneisen's theory completely misses VIE.

The quasiharmonic approximation (QHA) [112] can account for both thermal expansion and VIE by directly computing equation of state (P–V curve) in which the quantum anharmonic effects of lattice vibrations are included in both P and V. A vibrationally corrected volume of a solid at temperature T and pressure P is obtained by solving

$$\left(\frac{\partial A}{\partial V}\right)_T = -P, \tag{9.33}$$

where A is calculated by equation (9.22), that is, in the harmonic approximation, hence the name, QHA. The anharmonicity of the potential is taken into account by the parametrical dependence of the harmonic frequencies on V, at essentially the same approximate level as in Grüneisen's theory. The quantum nature of lattice vibrations and thus the zero-point effect are included through the quantum-mechanical formula of the partition function, equation (9.19). Equation (9.33) can be written as

$$\frac{\partial E_{\mathrm{e}}}{\partial V} + \frac{\partial (U_{\mathrm{v}} - TS_{\mathrm{v}})}{\partial V} = -P, \tag{9.34}$$

the left-hand side of which sums the *virial* pressure (the first term) and the *phonon* pressure (the second term). They are both included in QHA, whereas only the virial pressure is accounted for in Grüneisen's theory.

The third and most accurate method of describing quantum anharmonic lattice vibrations is path-integral molecular dynamics (PIMD) [113, 114] (not used in our previous studies so far). It is exact for quantum anharmonic effect in the limit of an infinite number of beads and can be applied equally to a solid and a liquid.

9.2.6 Born–Oppenheimer Molecular Dynamics

An *ab initio* molecular dynamics (MD) simulation for a molecular liquid can be performed using on-the-fly atomic forces evaluated by the embedded-fragment method. In the canonical ensemble, for instance, Newton's equation of motion,

$$m_\gamma \frac{\partial^2 x_i^\gamma}{\partial t^2} = -\frac{\partial E_{\mathrm{e}}}{\partial x_i^\gamma} \tag{9.35}$$

is integrated while N (the number of molecules in a unit cell), V, and T are held fixed. The right-hand side is computed at an *ab initio* electron-correlated level in each MD step by equation (9.2). This is facilitated by the embedded-fragment method as the differentiating coordinate can be either a collective in-phase atomic coordinate (useful in geometry optimization) or the individual atomic coordinate of the γth atom in the ith molecule (necessary in this case). The instantaneous pressure exerted outward on the yz-plane is

$$P = -\frac{1}{V}\sum_i \sum_\gamma x_i^\gamma \frac{\partial E_{\mathrm{e}}}{\partial x_i^\gamma} + \frac{N}{\beta V}, \tag{9.36}$$

of which the first term is the virial pressure and the second is the ideal-gas (kinetic) contribution [115].

Infrared and Raman spectra can be simulated as Fourier transform of time-correlation functions of the unit cell's dipole moment and polarizability, respectively [116, 117]. The latter are, in turn, the sums of the corresponding quantities of monomers embedded in the electrostatic field of the liquid in our method. The infrared absorption intensity as a function of frequency (ω) is written [118, 119] as

$$I_{\mathrm{IR}}(\omega) \propto \beta\omega^2 \int_{-\infty}^{\infty} \langle \boldsymbol{\mu}(0) \cdot \boldsymbol{\mu}(t) \rangle \exp(-i\omega t)\, dt, \tag{9.37}$$

where $\boldsymbol{\mu}(t)$ is the vector sum of dipole moments of all constituent molecules in a unit cell at time t, and $\langle \cdots \rangle$ denotes an ensemble average. Likewise, the isotropic and anisotropic Raman scattering cross sections are obtained [119, 120] as

$$I^{\mathrm{iso.}}_{\mathrm{Raman}}(\omega) \propto \frac{1}{9}\, \beta\omega \int_{-\infty}^{\infty} \langle \mathrm{tr}\, \boldsymbol{\alpha}(0)\, \mathrm{tr}\, \boldsymbol{\alpha}(t) \rangle \exp(-i\omega t)\, dt, \tag{9.38}$$

$$I^{\mathrm{aniso.}}_{\mathrm{Raman}}(\omega) \propto \frac{2}{15}\, \beta\omega \int_{-\infty}^{\infty} \langle \mathrm{tr}\, \boldsymbol{\beta}(0) \cdot \boldsymbol{\beta}(t) \rangle \exp(-i\omega t)\, dt, \tag{9.39}$$

where $\boldsymbol{\alpha}(t)$ is the polarizability tensor of the unit cell at time t and $\boldsymbol{\beta}(t) = \boldsymbol{\alpha}(t) - \boldsymbol{I}\, \mathrm{tr}\, \boldsymbol{\alpha}(t)/3$, where $\boldsymbol{I}$ is the identity tensor. The phenomenon in which the isotropic and anisotropic components of a Raman band have different centers is known as the Raman noncoincidence effect [121–123], which reflects the inhomogeneity of a local environment surrounding the oscillator responsible for the band. INS spectra can also be obtained as Fourier transform of time-correlation functions [103] (not considered in our previous studies).

Although atoms experience anharmonic forces in these *ab initio* simulations, the anharmonic effects on energies, structures, and spectra are included only partially or sometimes to an arbitrary degree. This is because atomic motions are not quantized in these simulations, and quantization is essential for a correct inclusion of anharmonicity. For instance, the O–H stretching band position in the infrared or Raman spectra of liquid water determined as Fourier transform of time-correlation functions captures little to no anharmonic effect. This is ascribed to the fact that at room temperature ($\approx 200\ \mathrm{cm}^{-1}$), classical atoms explore a far smaller, near-harmonic portion of the O–H stretching potential, whereas the true quantum zero-point vibration has 10 times as high an effective temperature ($\approx 2000\ \mathrm{cm}^{-1}$). Here, PIMD again offers a rigorous path to including quantum anharmonic effects in liquids on spectra [124, 125].

9.3 Case Studies

9.3.1 Ice-Ih

For ice-Ih, we consider three controversial or mysterious properties: (i) the two-hydrogen-bond hypothesis of Li and Ross, (ii) the anomalous heat capacity at low temperatures, and (iii) the anomalous temperature- and isotope-dependence of volume.

The red traces in Figure 9.2 reproduce the INS spectrum of ice-Ih obtained by Li and Ross [41]. It shows two distinct peaks at $229\ \mathrm{cm}^{-1}$ (28.4 meV) and $306\ \mathrm{cm}^{-1}$ (37.9 meV) (labeled "3" and "4," respectively, in the figure) in the hydrogen-bond stretching region.

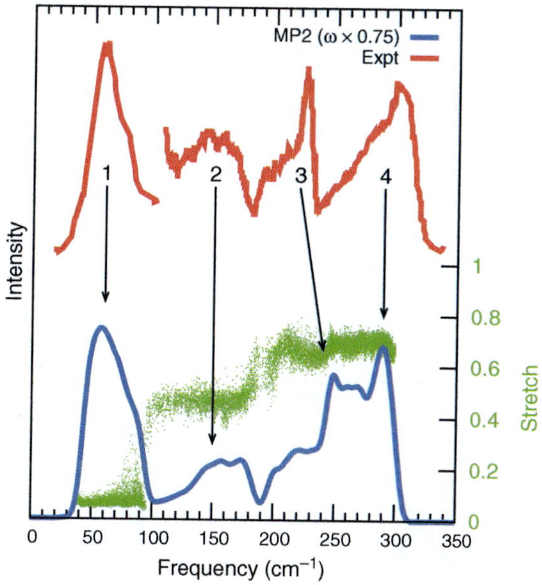

Figure 9.2 Inelastic neutron scattering from ice-Ih. The frequencies of the MP2/aug-cc-pVDZ phonon DOS are scaled by 0.75. The green dots plot the mode stretching parameters of Tanaka [126]. The arrows show the correspondence of peaks between theory and experiment [41].

On this basis, Li and Ross put forth a hypothesis that there were two types of hydrogen bonds with strengths differing by a factor of approximately two (which roughly accounted for the frequency ratio of 1.33). While this hypothesis was never widely accepted, the origin of the two peaks and their differences remained unknown.

We thus performed *ab initio* calculations of the infrared, Raman, and INS spectra of an infinite crystal of ice-Ih [44]. The blue trace in Figure 9.2 is the INS spectrum calculated by MP2/aug-cc-pVDZ for a fully proton-disordered structure with 64 molecules in a unit cell (unpublished). It has the same features as the calculated spectrum in Ref. [44] obtained with a quasi-proton-disordered structure. The peak positions in the calculated spectrum are systematically blue-shifted as compared with our earlier calculated spectrum or the experiment (because of the uncorrected BSSE and a slightly stronger atomic-partial-charge embedding field used in its calculation), but the four major peaks (labeled "1" through "4") are unmistakably assignable as indicated by the arrows. Hence, the calculated INS spectrum displays two peaks in the hydrogen-bond stretching region, while the calculated O–H and O···O distances have narrow distributions. This unequivocally disproves the hypothesis of Li and Ross; it is not necessary for there to be two hydrogen-bond types to explain the two distinct hydrogen-bond stretching peaks in INS. The same conclusion was reached by Li and coworkers using DFT [43].

The green dots show the "mode stretching parameters" of Tanaka [126]. They indicate that peak 1 has nearly zero mode stretching parameter values and is characterized as hydrogen-bond *bending* modes. Peak 2 has the parameter values of ≈ 0.5 and consists of mixed stretching-bending modes. Peaks 3 and 4 have high mode stretching parameter values; indeed both 229- and 306-cm^{-1} modes are of the hydrogen-bond stretching type, as Li and Ross correctly identified. The difference between these two modes turns out to exist in the polarization of the transition dipole moments (although INS does not obey the selection rules dictated by them). The lower-frequency (229-cm^{-1}) peak originates from a hydrogen-bond stretching motion that modulates dipole moment along the *c*

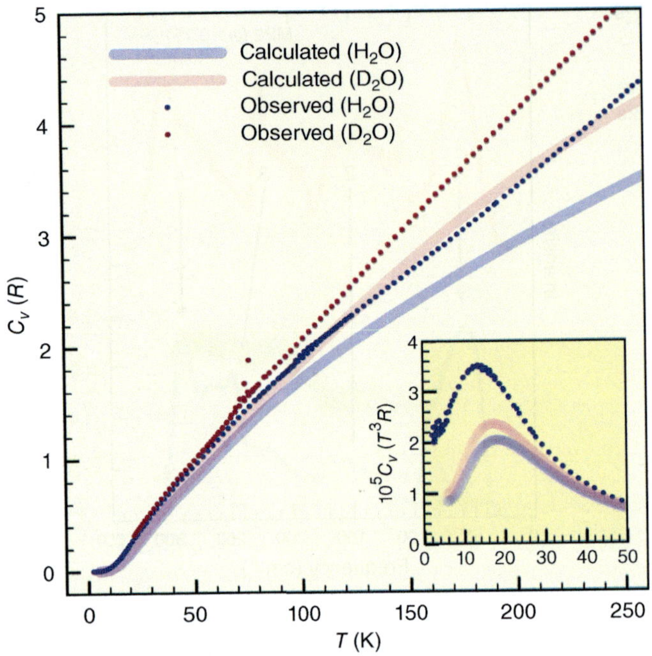

Figure 9.3 Isochoric molar heat capacity of ice-Ih in units of gas constant (R) calculated by MP2/aug-cc-pVDZ. *Source*: He *et al.* 2012 [44]. Reprinted with permission of AIP Publishing.

axis, whereas the higher-frequency (306-cm^{-1}) peak along the b axis. Since the b and c axes of ice-Ih are not equivalent, these two modes have different frequencies.

Figure 9.3 plots the isochoric heat capacity (C_V) of H_2O and D_2O ice-Ih as a function of temperature. The BSSE-corrected MP2/aug-cc-pVDZ curves are in excellent agreement with the observed up to $\approx$ 100 K, after which they begin to underestimate C_V, which is ascribed to the neglect of anharmonicity in the calculations. The inset in the figure shows the deviation of the observed C_V from Debye's T^3 behavior at low temperatures (< 30 K). This is reproduced by MP2 semi-quantitatively. This anomalous increase in C_V at low temperatures is caused nearly exclusively by the hydrogen-bond bending (transverse acoustic) modes centered at 60 cm^{-1} (peak 1 in Figure 9.2), which serve as a reservoir of heat.

Ice-Ih, unlike other solids, contracts upon heating at low temperatures (< 70 K), and then expands at higher temperatures [45, 46] (as seen in the solid curves in Figure 9.4 that initially decrease then turn to increase). Upon deuterium substitution for hydrogens, it expands by 0.09% at 0 K [45, 46], which is the opposite behavior for most other solids and is called the anomalous VIE (as seen in the solid red curve lying above the solid blue curve in Figure 9.4). Our study [127] shows that MP2/aug-cc-pVDZ in QHA can reproduce the thermal contraction at low temperatures quantitatively and the thermal expansion at higher temperatures qualitatively, the rate of which is underestimated (dots in Figure 9.4). It confirms the origin of the thermal contraction as the negative mode Grüneisen parameters of hydrogen-bond bending (transverse acoustic) modes (peak 1 in Figure 9.2). These modes coincide with the volume-collapsing motion of hollow

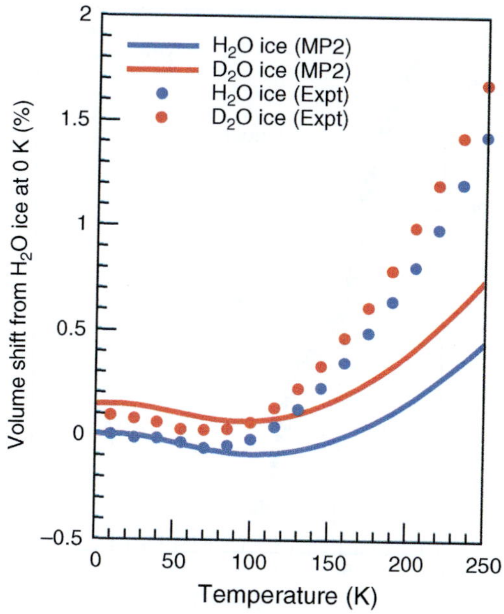

Figure 9.4 Change in the vibrationally corrected volumes of H_2O and D_2O ice-Ih relative to that of H_2O ice-Ih at 0 K observed [45, 46] or calculated by MP2/aug-cc-pVDZ. *Source*: Salim *et al.* 2016 [127]. Reprinted with permission of AIP Publishing.

hydrogen-bond cages in ice-Ih and, therefore, decrease their frequencies (i.e., increase entropy and lower Gibbs energy) upon compression. On the other hand, the subsequent thermal expansion is caused primarily by the hydrogen-bond stretching modes (peaks 2–4 in Figure 9.2) and to a much lesser extent by the librational modes. The underestimation of the expansion coefficient is due to the overestimation of the bulk modulus B (implicit in QHA) typical of MP2/aug-cc-pVDZ (see Section 9.3.7 for a fuller discussion). Recall that B appears as a factor in the denominator of Grüneisen's expression of thermal expansion coefficient α_V (equation 9.27). MP2/aug-cc-pVDZ without a BSSE correction systematically overbinds molecular crystals and predict too stiff an ice-Ih crystal, which therefore cannot expand computationally as much as it does in reality.

The same MP2/QHA study gives qualitatively varied predictions about the anomalous VIE, depending on the basis set and embedding field employed. This extreme sensitivity of even the sign of VIE is traced to the fact that net VIE at 0 K is the result of delicate cancellation of competing (i.e., volume-expanding and contracting) effects of *all* fundamental phonons that exist in the zero-point state. No one group of phonons determines the sign (normality or anomaly) of VIE. Despite the good agreement with the observed VIE obtained by MP2/aug-cc-pVDZ testified in Figure 9.4, we believe that a truly reliable prediction cannot be obtained without going to a much higher electron-correlation level (such as CC with singles, doubles, and noniterative triples or CCSD(T)) in the complete-basis-set (CBS) limit plus a more accurate embedding field. The anomalous VIE of ice-Ih, therefore, remains to be an open question.

9.3.2 Ice-HDA

Compression of ice-Ih at slightly below 0°C results in a phase transition to liquid water. In the temperature range of 160–250 K, upon pressure loading, ice-Ih transforms to

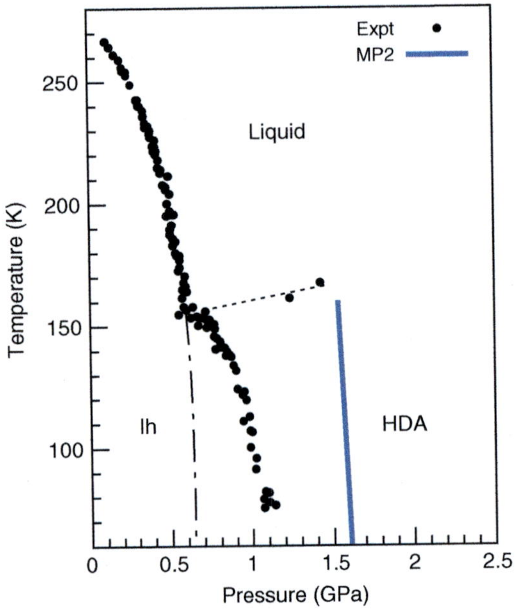

Figure 9.5 Phase diagram of water (small). Experimental data were taken from Mishima [129]. The dashed line is the crystallization line and the dotted-dashed curve is an extrapolation of the melting curve. The solid blue curve is the thermodynamic phase boundary estimated by MP2/aug-cc-pVDZ. *Source*: Salim *et al.* 2016 [127]. Reprinted with permission of AIP Publishing.

metastable supercooled liquid water. These are a manifestation of Le Chatelier's principle because liquid water is denser than ice-Ih. The thermodynamic phase boundary then has a negative slope according to the Clausius–Clapeyron relationship (Figure 9.5). In 1984, Mishima *et al.* [47] extrapolated the negative-slope melting curve to lower temperatures, predicting and then observing a pressure-induced transition from ice-Ih to liquid-like disordered high-density amorphous (HDA) phase at 1.0 GPa and 77 K.

However, this is not a thermodynamic phase transition or thermal melting; rather, it is mechanical melting of ice-Ih caused by the violation of Born's stability condition [128]. In other words, under the relevant conditions, ice-Ih is metastable and ice-HDA has a lower Gibbs energy, but the former lacks a thermal energy to cross an energy barrier for thermal melting to the latter. The qualitative change in the nature of the transition between lower and higher temperatures is reflected by the kink in the melting curve at 160 K (Figure 9.5).

In the course of the *ab initio* MP2/aug-cc-pVDZ calculations of thermodynamic properties of ice-Ih described in the previous subsection [127], we observed what could only be considered as computational detection and characterization of the pressure-induced amorphization to ice-HDA. The following are the observations [127].

First, with increasing pressure, geometry optimization of ice-Ih starting from an initial crystalline structure becomes more difficult at > 2.35 GPa. By 3 GPa, it never converges to a minimum-energy crystalline structure, but instead leads to a loss of symmetry and long-range order with a 15% reduction in volume relative to the ambient-pressure value. Figure 9.6 contrasts the optimized crystalline structure at 0 GPa and a nonconverged structure at 3 GPa, which may be viewed as being in an initial stage of pressure-induced amorphization via mechanical melting, crushing the hollow spaces of ice-Ih and densifying the solid. The volume reduction of 15% is in good agreement with the value of 15% at 2.5 GPa predicted by the lattice-dynamical model based on the INS data [130]. Our calculated transition pressure (2.35–3 GPa) is roughly consistent with the observed

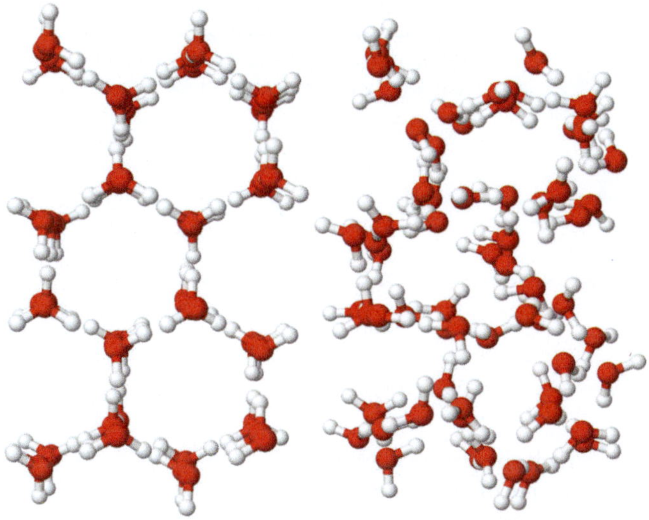

Figure 9.6 The MP2/aug-cc-pVDZ structures of H_2O ice-Ih at 0 GPa (optimized structure) and 3 GPa (nonconverged snapshot structure). *Source*: Salim *et al.* 2016 [127]. Reprinted with permission of AIP Publishing.

value of 1.0 GPa by Mishima *et al.* [47] and the aforementioned semi-observed value of 2.5 GPa by Strässle *et al.* [130].

Second, concomitant with the difficulty in optimizing the geometry, we observe pressure-induced red-shifting of phonon DOS due to acoustic branches, starting at around 2 GPa. An acoustic phonon near the edge of the Brillouin zone is shown to soften rapidly, with its frequency becoming imaginary at 2.35 GPa, coinciding with the onset of the geometry optimization failure. This pressure is, therefore, an upper bound for the mechanical instability, according to our MP2/aug-cc-pVDZ calculations. Clearly, the order of Gibbs energies between ice-Ih and ice-HDA switch at an even lower pressure (thermodynamic transition pressure) than 2.35 GPa, where ice-Ih is not even in a local minimum of its PES. This is consistent with the currently accepted view that below 160 K, there are two types of melting of ice-Ih to ice-HDA.

Third, an *ab initio* MD simulation using on-the-fly MP2/aug-cc-pVDZ atomic forces confirms that the amorphization-like geometrical change during the optimization process at > 2.35 GPa is not some computational artifact. At 2.5 GPa and 75 K, starting from an initial crystalline structure, the MD simulation records a rapid loss of long-range order and a volume reduction, which constitutes a real-time, computational observation of the pressure-induced amorphization. Without the latter, a relaxation to the stable crystalline structure accompanying lattice vibrations should be seen instead.

Finally, using the MP2 internal energies of ice-Ih and ice-HDA as well as a crude estimate of configuration entropy in the latter, we calculate the *thermodynamic* transition pressure between the two phases to be 1.6 GPa at 0 K with little temperature-dependence: the blue curve in Figure 9.5. This corresponds to the experimental thermodynamic phase boundary, dotted-dashed curve [129] in the same figure, inferred by the extrapolation of the melting curve of ice-Ih. The experimental *mechanical* phase

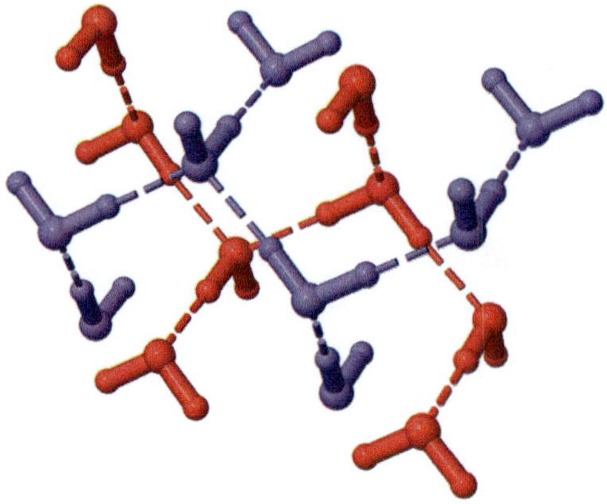

Figure 9.7 Crystal structure of ice-VIII. Two hydrogen-bonded sublattices are colored differently. *Source*: Gilliard *et al.* 2014 [131]. Reprinted with permission of AIP Publishing.

boundary is the dots in this figure, which should correspond to 2.35 GPa at 0 K according to MP2. Therefore, the phase behavior of ice-Ih and ice-HDA predicted by MP2 is fully consistent with experimentally inferred knowledge, if not quantitatively accurate.

Overall, we have a significant amount of evidence showing that MP2 computationally and spontaneously detected the pressure-induced amophization of ice-Ih to ice-HDA [127].

9.3.3 Ice-VIII

Ice-VII, VIII, and X form a group of phases sharing the same oxygen lattice, which consists of two interpenetrating cubic sublattices known as self-clathrate [132] (Figure 9.7) and is twice as dense as ice-Ih for the same O–O distances. These three phases differ from one another only by proton positions: ice-VII and VIII are proton-disordered and ordered, respectively, while ice-X is proton-shared (hence not a molecular crystal). They occur at wide ranges of temperature and pressure [133–135] and occupy a large area of the known phase diagram of ice (Figure 9.8) [29, 136]. Some of these phases are suspected to exist in Uranus [37], Neptune [37], Jovian, and Saturnian satellites [137] as well as in the deep ocean floor or upper mantle of the Earth [138].

Discontinuous changes in structural parameters in ice-VIII were recorded upon pressure loading of 2–3 GPa [139]. Concomitantly, nonlinear pressure-dependence of the Raman [140], infrared [141], and far-infrared spectra [142, 143] was also observed. On these basis, an isostructural-isosymmetric phase transition to a new phase (VIII′) was proposed by Tse and Klug [144], involving a movement of the interpenetrating sublattices in opposite directions [141]. At 10–14 GPa, a more dramatic change was observed by Yoshimura *et al.* [145]. A pseudo-translational band in the Raman spectra disappears and reappears at the same time the unit-cell axial ratio (c/a) shifts. This was interpreted as a precursor of the VIII-X phase transition. Whether these are real [146] and if so, what the exact nature of these transitions is have not been established.

Our MP2/aug-cc-pVDZ calculation of ice-VIII does not observe an isostructural-isosymmetric transition at 2–3 GPa in the fractional coordinate of oxygen, z(O), which has been said to display a stepwise change and testify to the sliding motion of sublattices (Figure 9.9). It instead shows a monotonic decrease with increasing volume, which is in agreement with the experimental data when the volume difference is adjusted. When

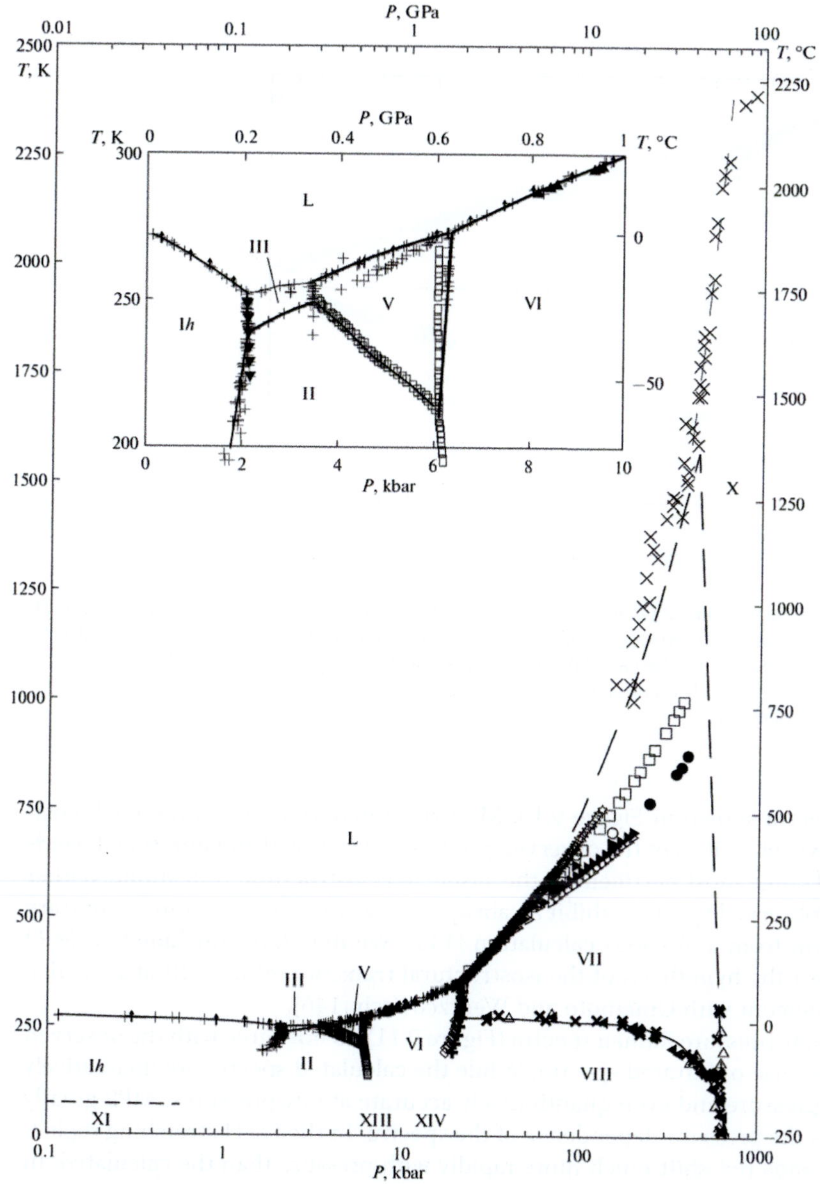

Figure 9.8 Phase diagram of water (large). The points denote the experimental data, and the lines show the approximation by the least squares method [136]. *Source*: Dunaeva *et al.* 2010 [136]. Reprinted with permission of Springer.

overlaid with the calculated results, it is in fact hard to see a stepwise change either in experimental $z(\mathrm{O})$ or in the related tetragonal distortion parameter, $\epsilon = 2c\{1/8 - z(\mathrm{O})\}$. We therefore find no supporting evidence of an isostructural transition in the MP2 structures of ice-VIII, although such a transition should be within the domain of applicability of embedded-fragment MP2. This conclusion is reinforced by the calculated

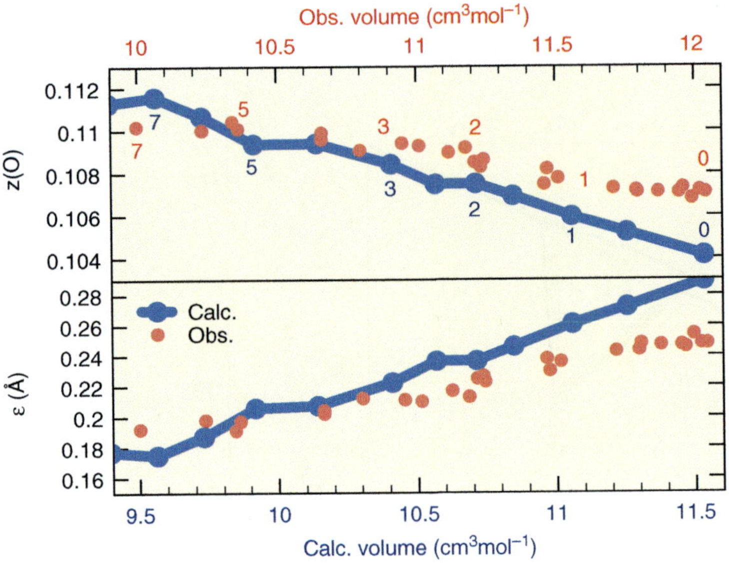

Figure 9.9 Fractional coordinate z(O) and tetragonal distortion parameter ϵ of ice-VIII as a function of calculated (lower axis) or observed (upper axis) molar volume. Integers in the upper panel indicate the corresponding pressures in GPa. The calculation was based on MP2/aug-cc-pVDZ and the experimental data were taken from Besson *et al.* [139]. *Source*: Gilliard *et al.* 2014 [131]. Reprinted with permission of AIP Publishing.

far-infrared spectra shown in Figure 9.10. MP2 accurately reproduces the peak positions and relative intensities of the observed pseudo-translational and librational bands in the pressure range most pertinent to the suspected isostructural transition. Neither calculated nor observed spectra exhibit an abrupt change in their appearance, contrary to the suggestion from a previous calculation [143]. We therefore conclude that MP2 does not support the hypothesis of the isostructural transition of ice-VIII at 2–3 GPa, which is in agreement with Umemoto and Wentzcovitch [146].

Calculated high-pressure Raman spectra (Figure 9.11) do not agree with the observed as well as in the case of infrared spectra. While the calculated spectra are qualitatively correct at any pressure and even quantitatively accurate at low pressures, MP2 greatly underestimates the pressure-dependence of the spectra. In the O–H stretching region, the observed bands red-shift much more rapidly with pressure than the calculated. In other words, the potential for the O–H stretching modes is extraordinarily anharmonic, so much so that QHA or Grüneisen's theory will miss a significant portion of its effect on their band positions. One plausible mechanism for this strong anharmonicity is that the potential for concerted proton transfers undergoes a transition from a double well to a single well (proton shared) upon a large pressure loading of tens of gigapascals. Not only the harmonic approximation but also the embedded-fragmentation employed in our calculations breaks down at some point during the transition to proton-shared ice-X. It may sound paradoxical, but the quantitative *disagreement* between the harmonic, embedded-fragment MP2 and observed spectra supports the interpretation of Yoshimura *et al.* [145] that the observed pressure-dependence of spectra is a precursor

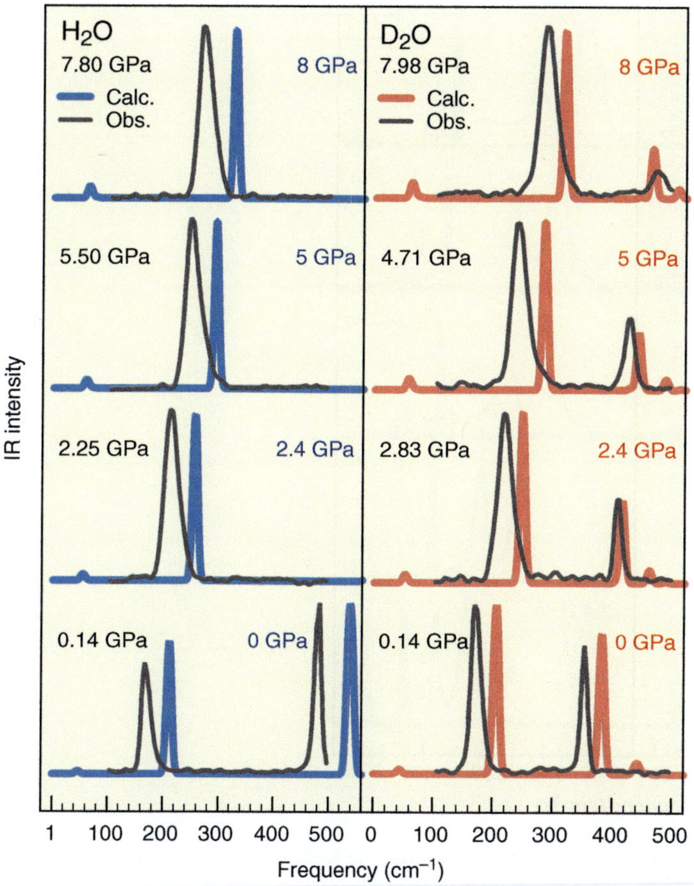

Figure 9.10 Far-infrared spectra of H_2O and D_2O ice-VIII. The calculation was based on MP2/aug-cc-pVDZ and the experimental data (85 K) were taken from Klug *et al.* [143]. *Source*: Gilliard *et al.* 2014 [131]. Reprinted with permission of AIP Publishing.

to the VIII-X phase transition. In the same token, the new bands that appear at pressures of 14.5 and 18.9 GPa are completely missing in the MP2 spectra and may therefore be related to the VIII-X phase transition also, because such bands cannot be described by embedded-fragment MP2. The assignments of these bands, whose frequencies blue-shift with pressure rather than red-shift, have not been made and are another open question.

Figure 9.12 illustrates the rather involved nature of the theory-experiment comparison of vibrational spectra of ice-VIII. Nearly all observed band positions fall nicely on the pressure-dependence curves of the fundamentals, overtones, or combinations calculated by MP2. The major discrepancies are highlighted by the solid black curve (diagonal to the graph), indicating the rapid softening of the O–H stretching Raman bands, and the solid grey line (near the bottom), corresponding to the mysterious blue-shifting Raman bands reported by Yoshimura *et al.* [145]. Our calculations do not contradict the interpretation of these bands by Yoshimura *et al.* [145] in terms of the VIII-X phase

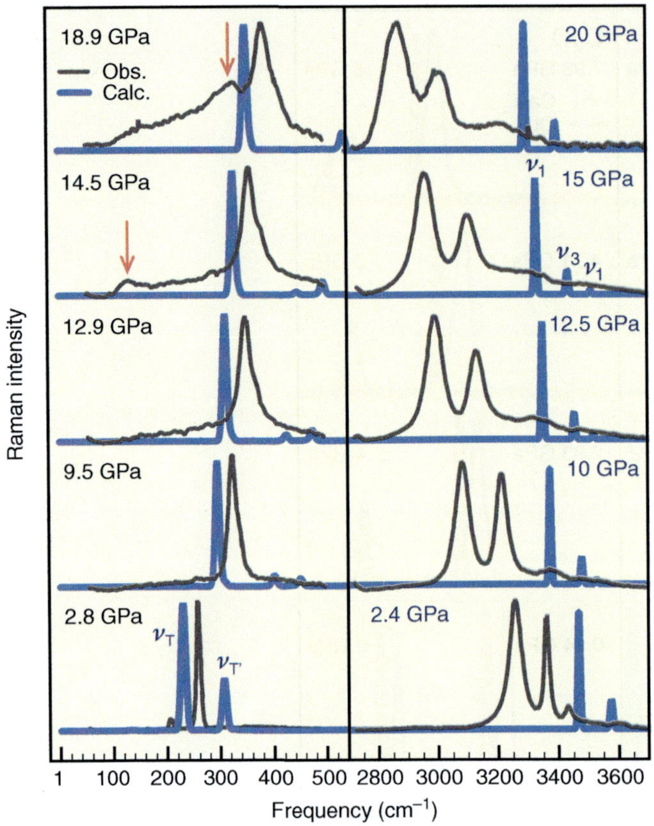

Figure 9.11 Raman spectra of ice-VIII. The calculation was based on MP2/aug-cc-pVDZ and the experimental data (80 K) were taken from Yoshimura *et al.* [145]. The red arrows indicate the appearance of a new band. *Source*: Gilliard *et al.* 2014 [131]. Reprinted with permission of AIP Publishing.

transition. On the other hand, a spectral anomaly in the bending-mode region ($\approx 1500\ \text{cm}^{-1}$) at 2–3 GPa (grey-shaded circle), which could be mistaken as a proof for the isostructural transition, is merely a consequence of anharmonic mixing between the bending fundamentals and librational overtones, as first pointed out by Song *et al.* [141].

9.3.4 Liquid Water

The Car–Parrinello method revolutionized thermodynamic simulations of liquids by allowing atomic forces in MD to be computed on the fly by DFT [53]. The field entered a yet new phase when Del Ben *et al.* [151–153] performed *ab initio* MP2 Monte Carlo and MD simulations of liquid water. The necessary speedup of MP2 was achieved by an ingenious algorithm [154]; they evaluate each of the two-electron integrals (that enter the MP2 energy expression) directly on a Cartesian grid using the electrostatic potentials exerted by a pair of orbitals (a factor in the integrals) determined by solving the Poisson equation with a fast Fourier transform. It is highly scalable with the number of

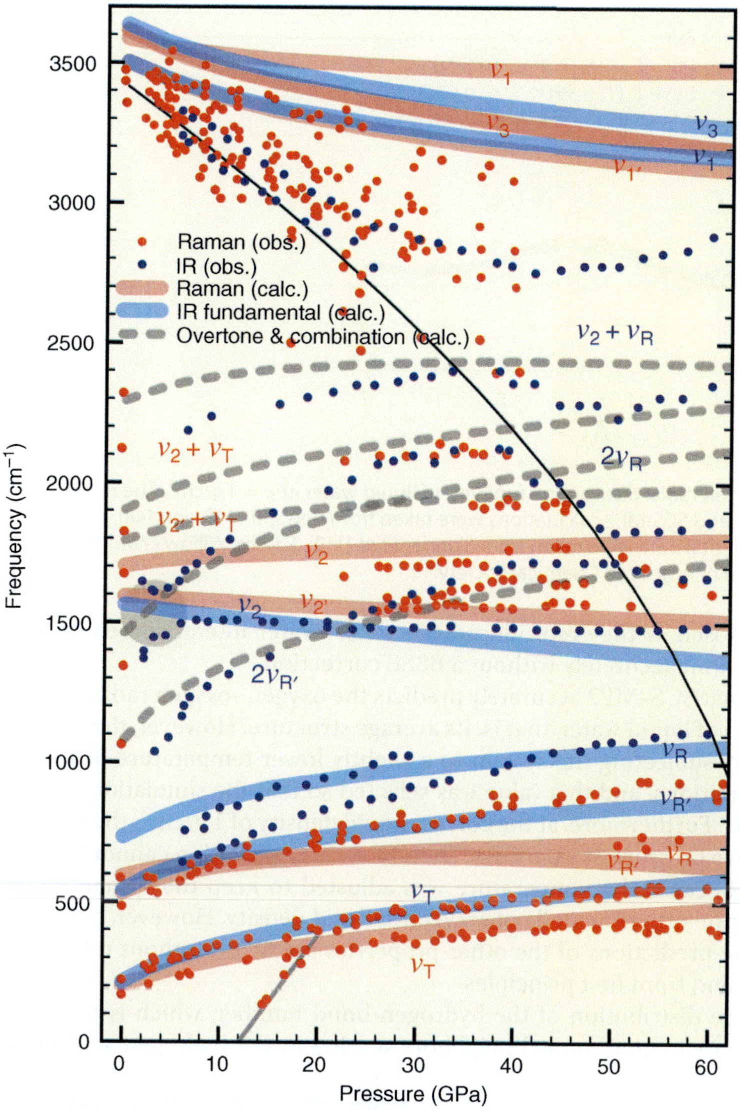

Figure 9.12 Pressure-dependence of the infrared and Raman band positions of ice-VIII. See the original paper [131] for details. *Source*: Gilliard *et al.* 2014 [131]. Reprinted with permission of AIP Publishing.

processors, broadly applicable to any insulating solids and liquids, but more difficult to implement or extend to higher-order methods or other properties.

We took a different approach for *ab initio* MP2 simulations of liquid water: the embedded-fragmentation. It is also highly scalable, easy to implement for a variety or properties, easy to raise the level of theory, but limited in applicability. Capitalizing upon the pioneering work by the FMO group [155–160], we performed *ab initio* MD (*NVT*) for liquid water using on-the-fly atomic forces calculated by the embedded-fragment spin-component-scaled (SCS) MP2/aug-cc-pVDZ method [101]. This level of theory

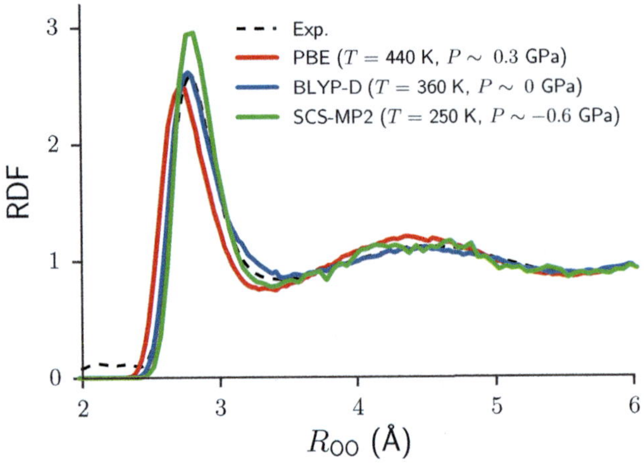

Figure 9.13 Oxygen–oxygen radial distribution function of liquid water at $\rho = 1$ g/cm^3. The results of the DFT-PBE, DFT-BLYP-D, and SCS-MP2 calculations were taken from Refs. [101, 148, and 148], respectively. The experimental data were taken from Skinner *et al.* [149]. *Source*: Willow *et al.* 2016 [150]. Reprinted with permission of American Chemical Society.

and basis set were chosen as they reproduced the water–water interaction energy of CCSD(T) in the CBS limit accurately without a BSSE correction.

Figure 9.13 shows that SCS-MP2 accurately predicts the oxygen–oxygen radial distribution function (RDF) of liquid water, that is, its average structure. However, this agreement was obtained by subjecting the system to a slightly lower temperature of 250 K than the ambient conditions, and this value was selected so that the simulation reproduced the correct RDF. Furthermore, at the correct liquid density of 1 g/cm^3, the system exerted a large negative pressure of −0.6 GPa. Therefore, this calculation cannot be said to be truly *ab initio* because the temperature was adjusted to keep the system be in the liquid phase with the experimentally observed RDF and density. However, once the conditions are set, the predictions of the other properties are made without reference to experimental data and from first principles.

Figure 9.14 plots the distribution of the hydrogen-bond number, which is strongly peaked at 4 with two hydrogen-bonded neighbors serving as the hydrogen acceptors

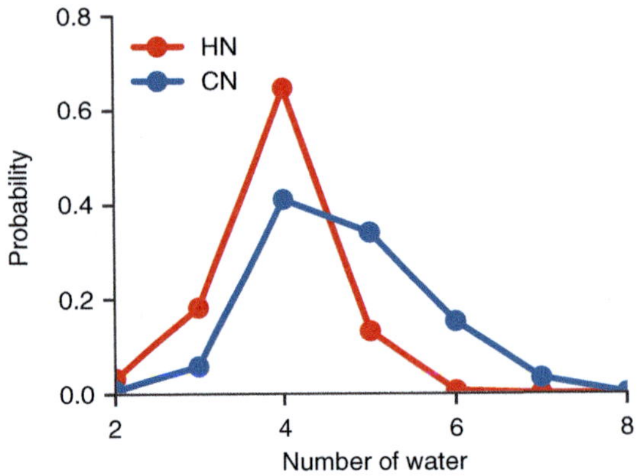

Figure 9.14 The coordination number (CN), the number of water molecules in the first shell ($0 < R_{OO} \leq 3.36$ Å) and the hydrogen-bond number (HN), the number of water molecules in the first shell with ∠HOO ≤ 40° in liquid water.
Source: Willow *et al.* 2015 [101]. Reproduced with permission of Nature.

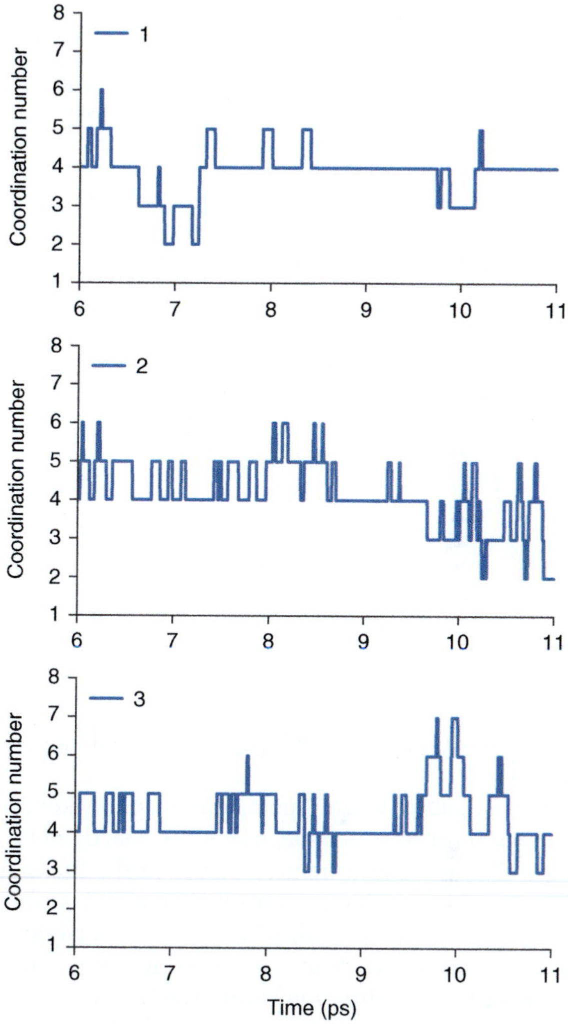

Figure 9.15 Fluctuation of the coordination number for three random water molecules in liquid water. *Source*: Willow *et al.* 2015 [101]. Reproduced with permission of Nature.

and the other two as the hydrogen donors. Hence, MP2 confirms "the standard picture" of liquid water [31] with an average hydrogen-bond number of 3.8, and disagrees with "the string theory" of liquid water [52], which asserts that each water molecule has an average of only two hydrogen bonds, forming a chain or a ring. The distribution of the coordination number has a long tail in the higher number side, with an average of 4.7. There are almost as many penta-coordinated water molecules as tetra-coordinated ones, accounting for the greater density of liquid water than that of ice-Ih.

These penta-coordinated structures are transient. See Figure 9.15. A 5-4-5 hetero-exchange is a dynamical process in which a penta-coordinated water molecule loses its non-hydrogen-bonded neighbor to become tetra-coordinated, and a different water molecule takes its place, restoring the penta-coordinated state. It occurs approximately every 1.0 ps and lasts on average for 100 fs, and is a chief mechanism of inter-shell exchange and therefore of self-diffusion. We note that the SCS-MP2-calculated

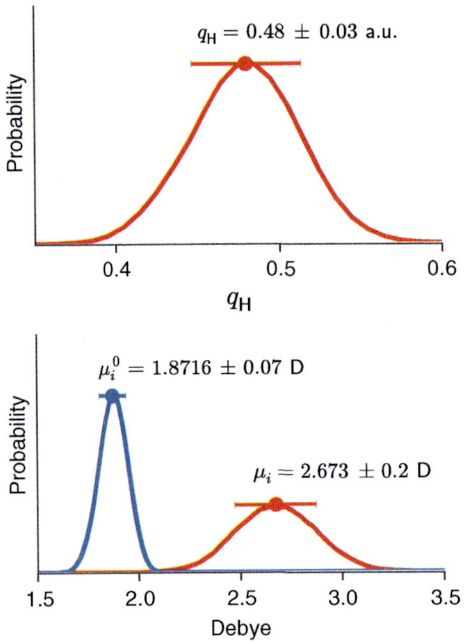

Figure 9.16 Permanent (μ_i^0) and total (μ_i) dipole moments and hydrogen partial charges (q_H) of water molecules in liquid water. *Source*: Willow *et al.* 2015 [101]. Reproduced with permission of Nature.

self-diffusion rate (0.27 Å^2/ps) agrees reasonably well with experiment (0.23 Å^2/ps) [161].

The magnitude of the dipole moment of a water molecule in the liquid phase has been another controversial issue, partly because there is no experimental means of measuring it directly. The computed values in the literature range widely from 2.6 to 3.0 D [54–57]. The gas-phase water dipole moment calculated by SCS-MP2/aug-cc-pVDZ is 1.88 D, which agrees well with the observed value of 1.86 D. The liquid-phase water dipole moment calculated by embedded-fragment SCS-MP2 can be divided into the permanent and induced portions, the former being obtained with the embedding field turned off. As shown in Figure 9.16, the permanent portion forms a narrow distribution centered at 1.87 D, which is essentially unchanged from the gas-phase value. The distribution is caused by geometry fluctuation in the liquid phase. The total (permanent plus induced) dipole ranges more widely from 1.8 to 3.5 D with an average of 2.67 D, which is near the lower end of the range spanned by the previously calculated values. The broader distribution reflects the diversity of local electrostatic fields as well as the geometry fluctuation. The calculated hydrogen partial charge of 0.48 ± 0.03 a.u. is consistent with the experimentally suggested value of 0.5 ± 0.1 a.u. [58].

The SCS-MP2 MD simulations (Figure 9.17) reproduce the general intensity profiles of the observed infrared and parallel- and perpendicular-polarized Raman spectra extremely well (the maximum intensity in each calculated spectrum was equated to the corresponding experimental value). Unlike most other calculated spectra, band widths in the calculated spectra are meaningful, and they reflect the diversity of local environments surrounding the oscillators. The observed O–H stretching infrared band is centered at 3400 cm^{-1} and has an asymmetric shape with a skirt in the lower frequency side and a full-width-at-half-maximum (FWMH) of 375 cm^{-1}. The calculated band is

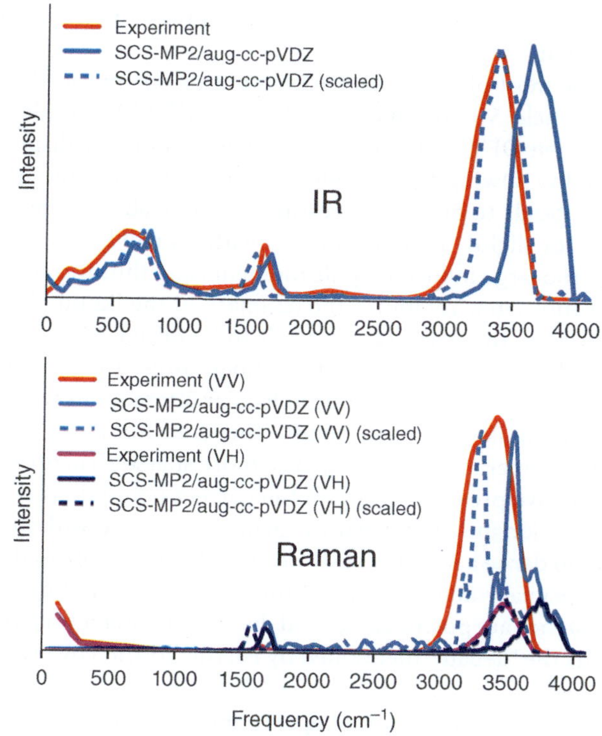

Figure 9.17 The infrared and Raman spectra of liquid water. The dashed curves are obtained by scaling the calculated frequencies by 0.93. The parallel-polarized (VV) and perpendicular-polarized (VH) Raman spectra are related to the isotropic and anisotropic components (equations 9.38 and 9.39) by $I^{VV}_{Raman}(\omega) = I^{iso.}_{Raman}(\omega) + I^{aniso.}_{Raman}(\omega)$ and $I^{VH}_{Raman}(\omega) = 0.75\, I^{aniso.}_{Raman}(\omega)$. The experimental data were taken from Refs. [162, 163]. *Source*: Willow *et al.* 2015 [101]. Reproduced with permission of Nature.

blue-shifted at 3700 cm^{-1} for the reasons given in the last paragraph of Section 9.2.6, but it reproduces a similar asymmetric band shape and a FWHM of 375 cm^{-1}, which is expected to include the motional narrowing effect by as much as 30% [164]. It also appears to recover the far-infrared band at 200 cm^{-1}, which has been ascribed to the first-shell dynamics [165], but owing to a rather large time interval of 1 fs used in our simulations, it is unclear whether the agreement is accidental. The simulations also predict correctly a lower peak position of the isotropic component than the anisotropic one in the O–H stretching Raman band, that is, the noncoincident effect [166], which also reflects the inhomogeneity of the local environments surrounding the O–H oscillators.

Finally, we address [150] the different thermodynamic conditions (temperature and pressure) required to keep "MP2-water" in the liquid phase from the ambient conditions. As mentioned above, a slightly lower temperature of 250 K needed to be applied for MP2-water to stay liquid [101, 151–153], which furthermore exerts a negative pressure of −0.6 GPa at the density of 1 g/cm^3. On the other hand, it is well-known that water simulated by DFT with a dispersionless gradient-corrected-approximation (GGA) functional (e.g., PBE functional) tends to be liquid at a much higher temperature (440 K) under a large positive pressure (0.3 GPa) (see the legends of Figure 9.13) [147]. A dispersion correction [167] (e.g., BLYP-D functional) resolves the wrong pressure (to 0 GPa), but falls short of fixing the temperature problem; the temperature (360 K) required to prevent DFT-water from being supercooled is still too high and close to the boiling temperature [148]. As Parrinello *et al.* [59] put it in 2013, "For the case of neat water, most standard GGA functionals are known to overstructure water at ambient conditions compared to what is known experimentally. There is currently an ongoing lively debate on the origins of these discrepancies which remains an unsolved problem."

The most plausible explanation of this was put forth by a comparison of polarizable-force-field MD results with DFT and MP2 MD results [150]. The primary differences between DFT and MP2 are the descriptions of dispersion interaction and of polarizability. Owing to the spurious self-interaction in virtually all exchange functionals, DFT tends to grossly overestimate polarizability, whereas MP2 should be more accurate (or slightly underestimate it with a small basis set). As a consequence, DFT-water has too large induced dipole moments and thus too strong hydrogen bonds, which hinder molecular rotations, the hallmark of the liquid phase. For DFT-water to be liquid, therefore, the temperature needs to be raised. MP2-water, on the other hand, exists as liquid at lower, more realistic temperatures. The lack of dispersion in some DFT functionals leads to weaker intermolecular interactions with little to no orientational dependence, unlike hydrogen bonds. This does not affect much the molecular rotation and thus temperature, but predicts a more sparse structure that needs to be compressed by a positive pressure to regain the correct liquid density of 1 g/cm^3. MP2 overestimates dispersion interaction with a small basis set and without a BSSE correction, which leads to too dense a structure that needs to be decompressed by a negative pressure. In short, the primary determinants of the temperature and pressure of the liquid water's stable existence are polarizability and dispersion interaction, respectively [150]. This hopefully ends the "debate" mentioned by Parrinello and coworkers [59].

9.3.5 CO_2-I: Pressure Tuning of Fermi Resonance

The Fermi resonance of the CO_2 molecule is a textbook example [168] of strong anharmonic mode–mode coupling. The destination states of its symmetric stretch (ω_1) and bending overtone ($2\omega_2$) transitions are nearly degenerate and have the same symmetry, thus undergoing resonance to create the ν_+ and ν_- states as their linear combinations (Figure 9.18). In the pertinent frequency region of the Raman spectra of gaseous CO_2, therefore, one finds two peaks due to ν_+ and ν_- separated by 107 cm^{-1} with an intensity ratio of 2.1, instead of just one due to ω_1 as the harmonic approximation may predict ($2\omega_2$ is a dark state).

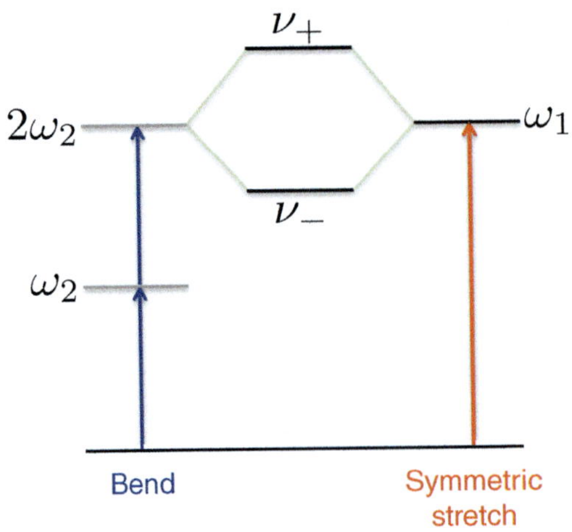

Figure 9.18 A schematic vibrational-energy-level diagram of CO_2. *Source*: Sode *et al.* 2013 [169]. Reprinted with permission of AIP Publishing.

The Fermi resonance persists in condensed phases, which is more interesting than that in the gas phase. This is because one can apply a large pressure on condensed-phase CO_2 and thereby modulate the energy levels of the resonating states and influence its resonance. As we will see shortly in this section, ω_1 increases with pressure, while ω_2 decreases; one can therefore bring the two states out of near-degeneracy by applying pressure and eventually turn off the resonance. This phenomenon is known as the pressure tuning of Fermi resonance, and has an immense technological implication in geoscience. Geochemists have developed one of the most accurate geobarometers by utilizing the pressure tuning of Fermi resonance exhibited by minerals with CO_2 inclusions (see Figure 9.19) [170–174]. The residual pressure experienced by CO_2 trapped in a mineral can be measured directly and accurately from the Fermi dyad splitting and intensity ratio and informs us about the depths at which the mineral was entrained by

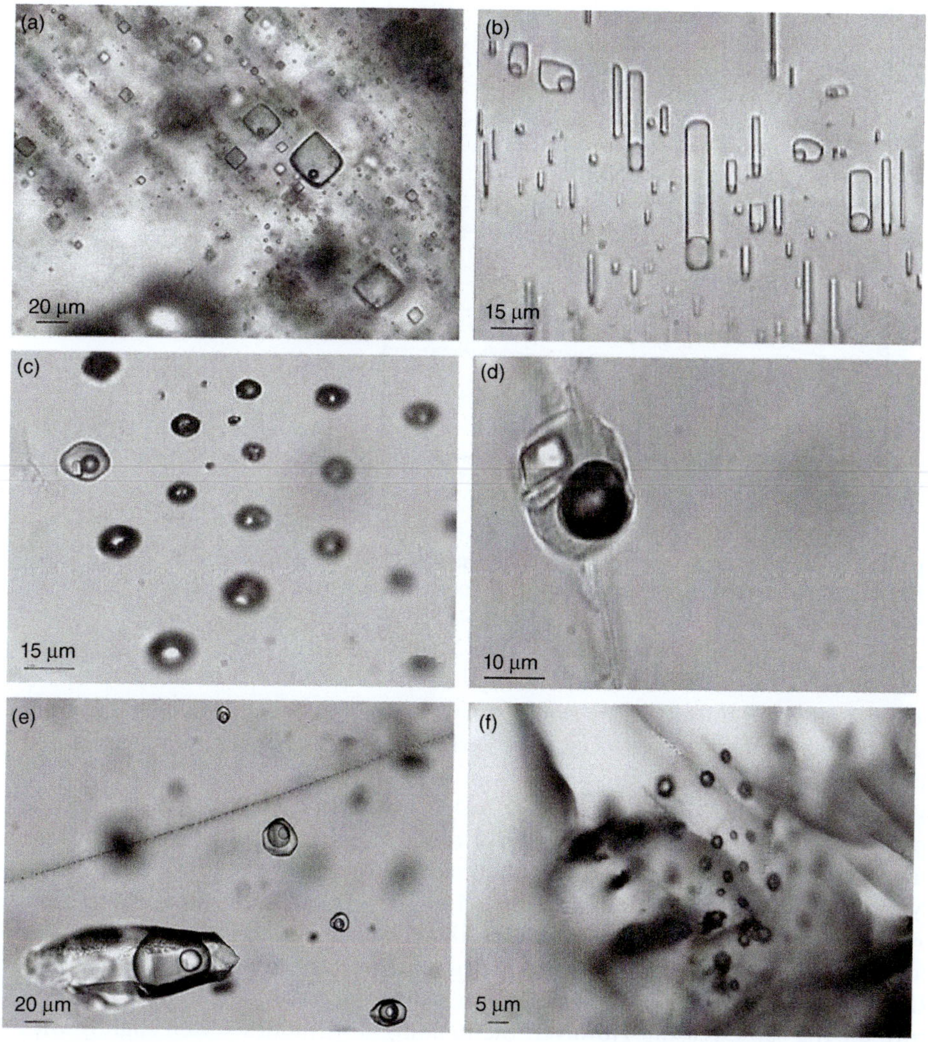

Figure 9.19 Photomicrographs of minerals with H_2O and/or CO_2 fluid inclusions. *Source*: Frezzotti *et al.* 2012 [174]. Reprinted with permission of Elsevier.

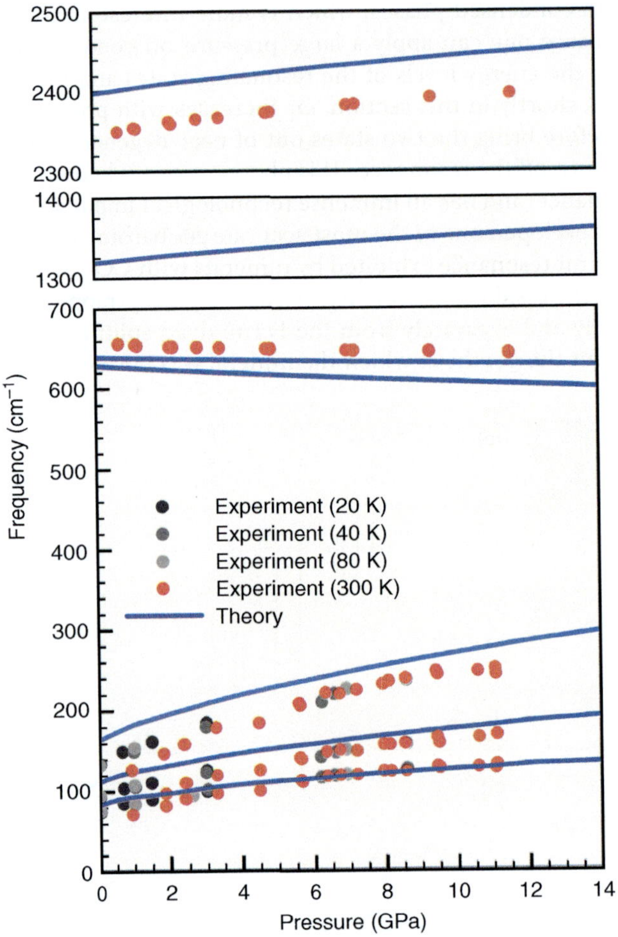

Figure 9.20 The pressure-dependence of the infrared and Raman band positions of CO_2-I. The experimental data were taken from Refs. [62 and 175]. *Source*: Sode *et al.* 2013 [169]. Reprinted with permission of AIP Publishing.

the host magma, and thus its formation history. Many Raman bands depend on pressure, but resonant bands and especially the Fermi dyads of CO_2 have greatly amplified pressure-dependence.

The BSSE-corrected MP2/aug-cc-pVDZ calculation [169] of dry ice (solid CO_2 phase I) predicts the pressure-dependence of the infrared and Raman band positions shown in Figure 9.20. The frequencies are obtained in the harmonic approximation. Both the symmetric stretching (ω_1) modes (1300–1400 cm^{-1}) and antisymmetric stretching (ω_3) modes (2300–2500 cm^{-1}) increase their frequencies, which is expected as molecules have greater difficulty pushing one another away in a more crowded environment. Librational modes below 300 cm^{-1} also rapidly increase their frequencies with pressure for the same reason. However, the bending (ω_2) modes soften slightly with pressure and thus have negative mode Grüneisen parameters. One may imagine that

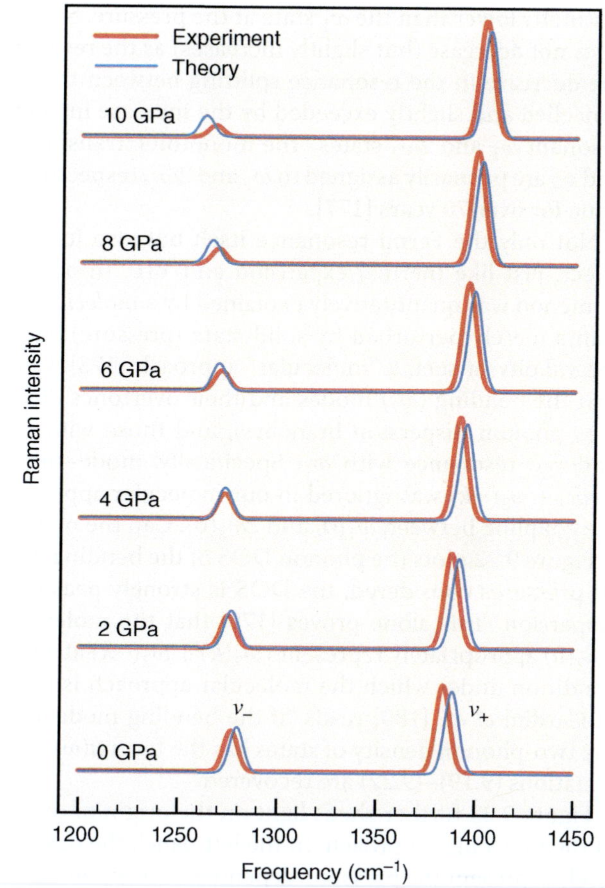

Figure 9.21 The pressure-dependence of Raman spectra of CO_2-I in the Fermi-resonant region. The experimental spectra were reconstructed from the data in Ref. [175]. *Source*: Sode *et al.* 2013 [169]. Reprinted with permission of AIP Publishing.

the bending of otherwise linear CO_2 contracts each molecular footprint and alleviates the crowdedness, not unlike the hydrogen-bond bending modes in ice-Ih.

Although the absolute values of the observed frequencies are not reproduced well by MP2 in some cases because of the lack of anharmonicity in the latter, the pressure-dependence is surprisingly accurately predicted. For instance, the average slopes of ω_3 and ω_2 are 4.2 and -1.5 cm^{-1} GPa^{-1}, respectively, as compared with the observed values of 4.2 and -1.3 cm^{-1} GPa^{-1}. Furthermore, the crystal-field splitting of ω_2 at 0 GPa is 5 cm^{-1} experimentally and 11 cm^{-1} computationally.

Figure 9.21 compares the simulated and observed Fermi dyad positions and relative intensities of CO_2-I as a function of pressure up to 10 GPa [169]. The simulation was based on the full vibrational CI (VCI) calculation of the CO_2 molecule whose force constants were corrected for the crystal-field (pressure) effects. Specifically, an accurate molecular quartic force field [176] obtained from CCSD(T)/cc-pVTZ was used as a basis and its quadratic force constants were adjusted with the average pressure-dependence of the corresponding force constants of solid CO_2 predicted by embedded-fragment BSSE-corrected MP2/aug-cc-pVDZ. The agreement between theory and experiment is excellent. At 10 GPa, one finds only one prominent peak in both the simulated and observed spectra, meaning that the resonance is essentially turned off because the $2\omega_2$ state lies distinctly lower than the ω_1 state at the pressure. Somewhat paradoxically, the splitting does not decrease (but slightly increases) as the resonance is turned off. This is because the decrease in the resonance splitting between the ν_+ and ν_- states with pressure is cancelled and slightly exceeded by the increase in energy difference between the non-resonant ω_1 and $2\omega_2$ states. The monotonic transition from ν_+ to ω_1 suggests that ν_+ and ν_- are primarily assigned to ω_1 and $2\omega_2$, respectively, which has been a controversial issue for over 70 years [177].

Not only the Fermi resonance itself but also its pressure tuning is an anharmonic effect, just like thermal expansion and VIE. In our work described above, this phenomenon was quantitatively explained by a *molecular* VCI calculation using force constants merely perturbed by solid-state (pressure) effect. A question was raised about the validity of such a "molecular" approach [178]. Cardino and Schettino [178] noted that the bending (ω_2) modes and their overtones ($2\omega_2$) formed continuous manifolds (i.e., phonon dispersion branches), and those with nonzero wave vectors could also undergo resonance with ω_1. Specifically, mode–mode coupling between $\omega_1(\mathbf{0})$ and $\omega_2(\boldsymbol{k}) + \omega_2(-\boldsymbol{k})$ was ignored in our molecular approach, which took into account only the coupling between $\omega_1(\mathbf{0})$ and $2\omega_2(\mathbf{0})$. Can the molecular approach be justified?

Figure 9.22 plots the phonon DOS of the bending modes, that is, the ω_2-manifold. At all pressures considered, the DOS is strongly peaked, indicating that $\omega_2(\boldsymbol{k})$ has small dispersion. This alone proves [179] that the molecular approach is justified because $2\omega_2(\mathbf{0})$ appropriately represents $\omega_2(\boldsymbol{k}) + \omega_2(-\boldsymbol{k})$ at all $\boldsymbol{k}$ in this case. In other words, the condition under which the molecular approach is justified is met, which in the words of Cardini *et al.* [180] reads "if the bending mode ω_2 is not dispersed and, therefore, the two-phonon density of states has the form $n(\omega) = \delta(\omega - 2\omega_2^0)$ the isolated molecule equations (9.19)–(9.22) are recovered."

Figure 9.23 further sheds light on the origin of diverging opinions about the validity of the molecular approach. In the left panel, the dashed and dotted blue curves respectively represent the pressure-dependence of ω_1 and $2\omega_2$ in the molecular approach. The latter corresponds to the peak in the $2\omega_2$-manifold (the phonon DOS due to the bending overtones indicated by the blue-shaded area). The anharmonic mode–mode coupling spawns the ν_+ and ν_- states, plotted as black curves with open circles, which agree accurately with the experimental data. The corresponding data of an empirical molecular model suggested by Hanson and Jones [175] is superimposed as purple dots. Their accurate agreement with our *ab initio* derived data [169, 179] highlights the impressive accuracy of the intuitions of Hanson and Jones. In contrast, an empirical "solid-state"

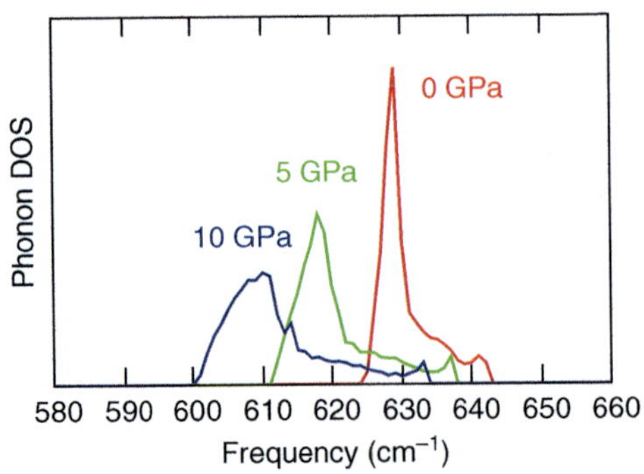

Figure 9.22 Phonon DOS of CO_2-I at 0, 5, and 10 GPa. *Source:* Hirata *et al.* 2014b [179]. Reprinted with permission of AIP Publishing.

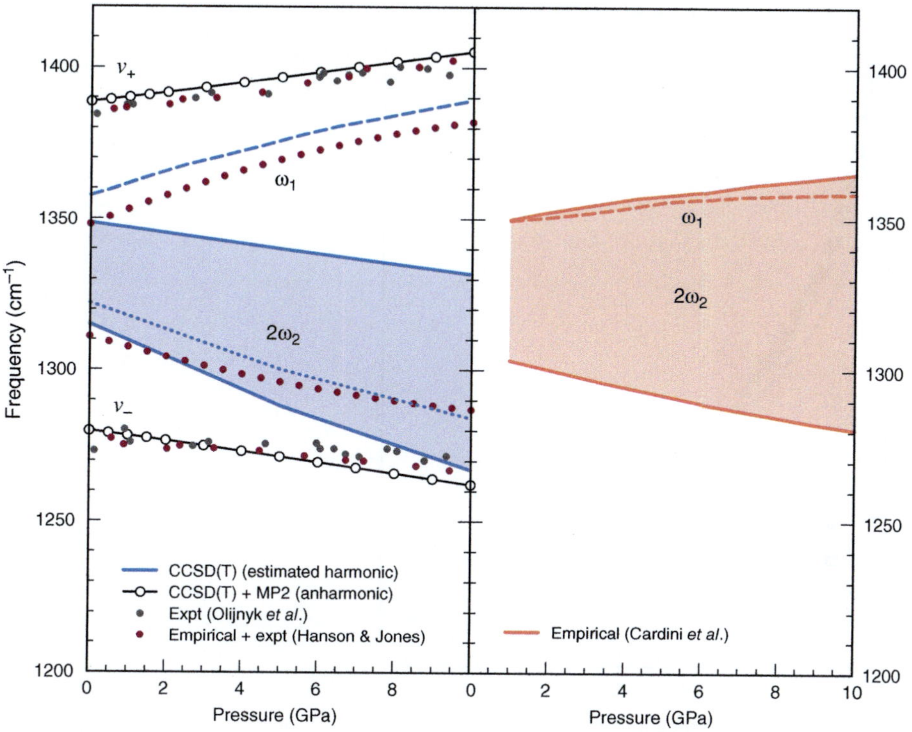

Figure 9.23 Comparison of the pressure-dependence of ω_1 and $2\omega_2$ of CO_2-I from the hybrid CCSD(T)-MP2 calculation and the empirical solid-state model of Cardini *et al.* [180]. The left panel also plots the calculated and observed anharmonic frequencies of the Fermi dyad (ν_+ and ν_-). The peaks in the phonon DOS of the $2\omega_2$-manifold of CCSD(T)-MP2 are indicated by blue dotted lines. The left panel also plots ω_1 and $2\omega_2$ in the empirical molecular model of Hanson and Jones [175]. *Source*: Hirata *et al.* 2014b [179]. Reprinted with permission of AIP Publishing.

model of Cardini *et al.* [180], shown in the right panel, seems somewhat unrealistic, as compared with our *ab initio* data. It is not that the solid-state approach of Cardini *et al.* [180] is inaccurate; it is indeed more complete and accurate than the molecular approach adopted by us and by Hanson and Jones. Rather, the unrealistic force field assumed by Cardini *et al.* [180] seems to have led to their wrong conclusion [178] that only the solid-state approach is valid.

9.3.6 CO_2-I and III: Solid–Solid Phase Transition

Figure 9.24 shows the equations of state of solid CO_2 [60, 181]. The dots are experimental data [182], and the blue curve is our MP2/aug-cc-pVDZ result for CO_2-I. MP2 overbinds as usual, predicting too dense and stiff a solid especially at low pressures (see Sections 9.3.1 and 9.3.4), but otherwise its curve is in reasonable agreement with the observed. The experimental results show a volume collapse of ≈ 2% at 10–13 GPa (shaded area), suggesting a solid–solid phase transition. At higher pressures, the equation of state, therefore, corresponds to a different phase, which turns out to be orthorhombic phase III

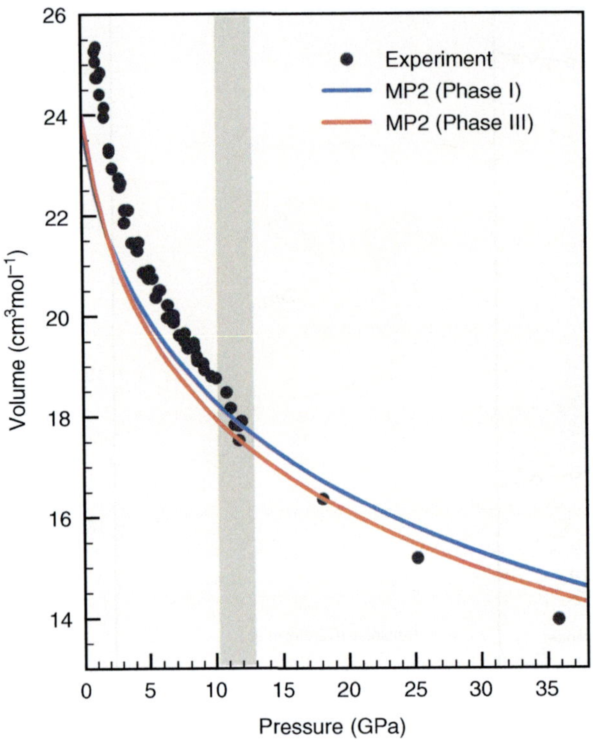

Figure 9.24 Equations of state of CO_2-I and III. *Source*: Li *et al.* 2013 [60]. Reproduced with permission of Nature.

(Figure 9.25), as mentioned in Introduction. The calculated equation of state of CO_2-III (red curve) lies lower than that of CO_2-I (blue curve) at most pressures (also by $\approx 2\%$ at around the transition pressure), which is consistent with the observed volume collapse. Le Châtelier's principle then predicts a transition from less dense CO_2-I to denser CO_2-III at a high enough pressure. The experimental values of the transition pressure range widely from 2.5 GPa at 80 K (Ref. 62) to above 18 GPa at room temperature [63], although many coalesce around 12 GPa [182, 183]. At what pressure does the transition occur computationally?

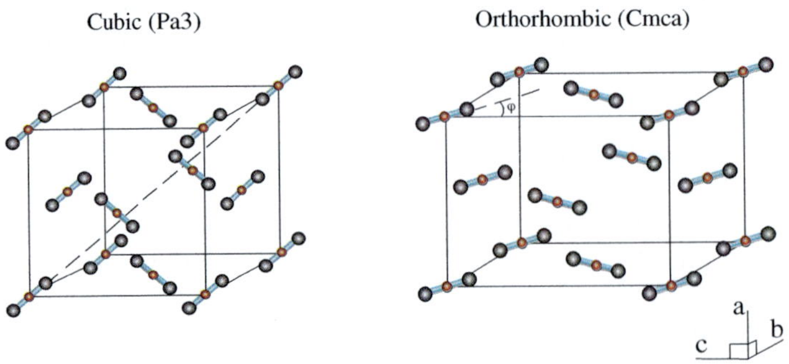

Figure 9.25 Structures of CO_2-I (*left*) and III (*right*). *Source*: Li *et al.* 2013 [60]. Reproduced with permission of Nature.

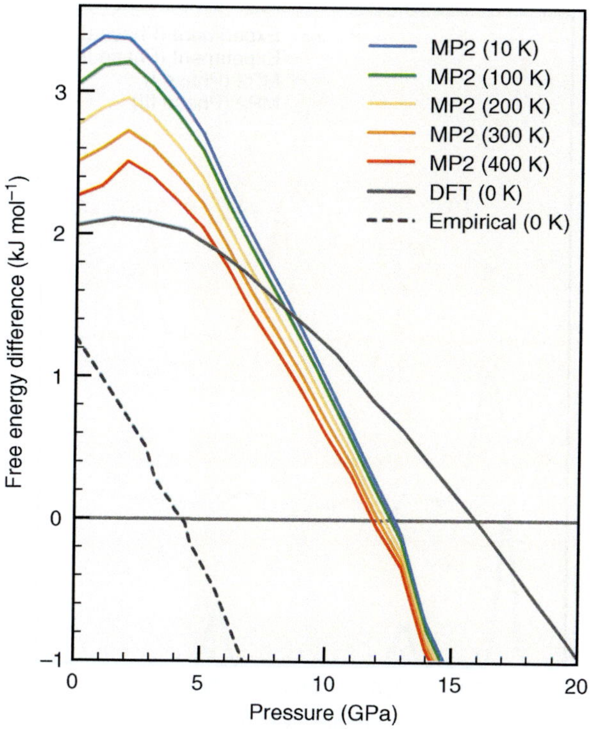

Figure 9.26 Gibbs energy differences between CO_2-I and III. The grey curve is from DFT calculations by Bonev *et al.* [184] and the broken curve from empirical-force-field calculations by Kuchta and Etters [185]. *Source*: Li *et al.* 2013 [60]. Reproduced with permission of Nature.

Figure 9.26 plots the Gibbs energy differences between CO_2-I and III. A positive value means that phase I is more stable than phase III. According to our MP2 calculations [60, 181], the Gibbs energy order flips at 13 GPa at 0 K, and the transition pressure decreases slightly with increasing temperature. This is in line with the majority of experimental data, which placed the transition pressure at 12 GPa with small temperature-dependence (see also Figure 9.1). This result is robust with respect to a basis-set extension to aug-cc-pVTZ, which lowers the transition pressures by a few GPa, but maintains the same overall picture [60, 181]. It also agrees qualitatively with a previous DFT study [184], although the latter predicted a higher transition pressure of 16 GPa at 0 K with much greater temperature-dependence. In contrast, an empirical calculation [185] proposed a exceedingly low transition pressure of 4.3 GPa.

We made certain that the phases we study were indeed the ones experimentally observed by comparing the calculated and observed high-pressure Raman spectra of the two phases. Figure 9.27 show those spectra in the liberational region. Our MP2 calculations [60, 181] reproduce the observed spectra of both phases with remarkable accuracy. Given that the emergence of three new peaks (at 245, 270, and 320 cm^{-1}) in the 18.0-GPa spectrum [63] was the most decisive proof of the phase transition, the quantitative agreement renders strong computational support for Hanson's assignment [63] of these peaks to phase III.

A still unsolved mystery is phase II [64–66]. Phase III is believed to be a metastable, kinetically trapped state, reachable only by the low-temperature compression of phase I through a Martensitic transition between the structurally similar phases (Figure 9.25). There should, therefore, be an underlying, stable phase with a lower Gibbs energy, which

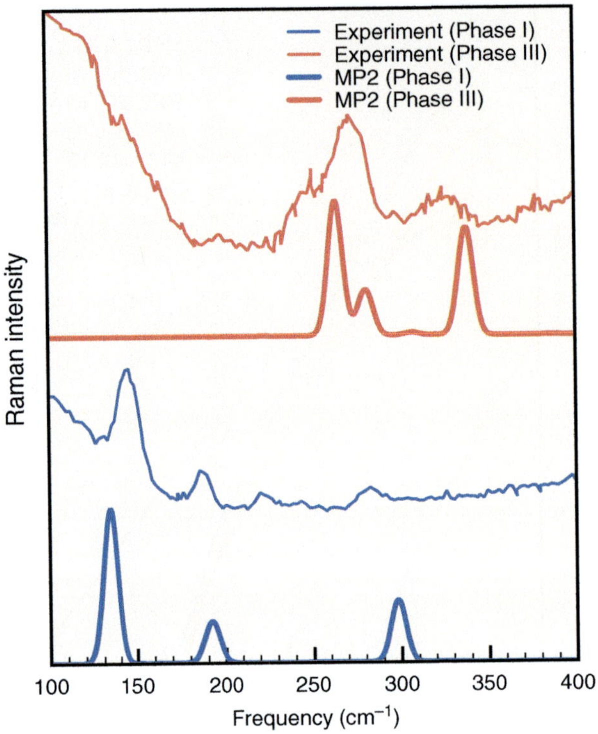

Figure 9.27 Calculated and observed [63] Raman spectra of CO_2-I (at 14.5 GPa) and III (at 18.0 GPa). *Source*: Li *et al.* 2013 [60]. Reproduced with permission of Nature.

is phase II (Figure 9.1). Complete *ab initio* characterization of the structural, thermodynamic and spectroscopic properties of the three phases (I, II, and III) is awaited.

9.3.7 CO_2-I: Thermal Expansion

This last subsection showcases a challenging example [186], where even BSSE-corrected MP2/aug-cc-pVTZ incurs a rather large error. Figure 9.28 compares the calculated and observed thermal expansion coefficients (α_V) of CO_2-I. The MP2/aug-cc-pVDZ calculation based on Grüneisen's theory (see Section 9.2.5) underestimates the experimental values [187–189] of α_V by more than a factor of 2 at all temperatures. The use of the aug-cc-pVTZ basis set with a BSSE correction improves the prediction, but only slightly; it leaves a large discrepancy between theory and experiment by still a factor of 2. Where does this large error come from?

Recall $\alpha_V = C_V\gamma/(BV)$ according to Grüneisen's theory (equation 9.27). Hence, the source of the error should be found in one or more of these factors. In the following, we examine them one by one, starting with C_V. Figure 9.29 attests to that the calculated heat capacity (C_V) cannot be a major source of the error, as it agrees with the experiment [187, 190] within $\approx 10\%$. It can only be a minor source.

Figure 9.30 testifies that the calculated value and temperature-dependence of the Grüneisen parameter (γ) is also reasonable. The calculated γ is nearly flat at ≈ 2, while the two experimental values also hover around 2. The γ curve proposed by Trusler

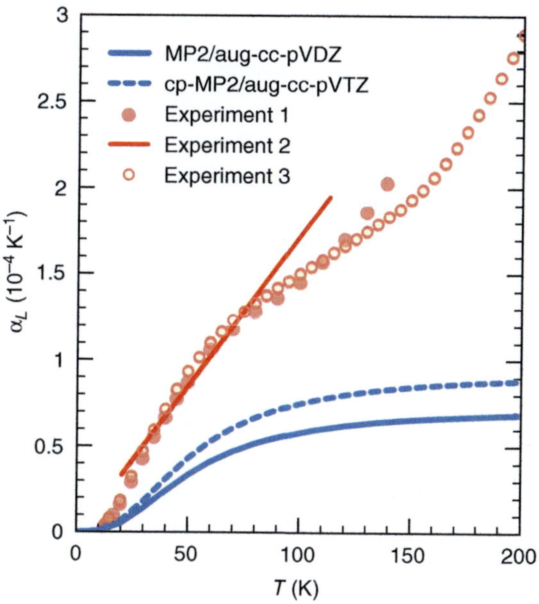

Figure 9.28 Linear thermal expansion coefficient ($\alpha_L = \alpha_V/3$) of CO_2-I as a function of temperature. Experimental data were taken from Refs. [187–189]. *Source*: Li *et al.* 2015b [186]. Reprinted with permission of American Chemical Society.

[191, 192] on the basis of an empirical fitting seems a little too large (≈ 2.6), whereas the one similarly obtained by Giauque and Egan seems far fetched.

Figure 9.31 gives a more detailed account of γ. MP2 accurately reproduces nearly every aspect of the acoustic branches (primarily responsible for thermal expansion) *if* the calculated frequencies are scaled by 0.8. The need for the scaling can be ascribed to the systematic overbinding of molecular crystals by this theory (See Sections 9.3.1 and 9.3.4).

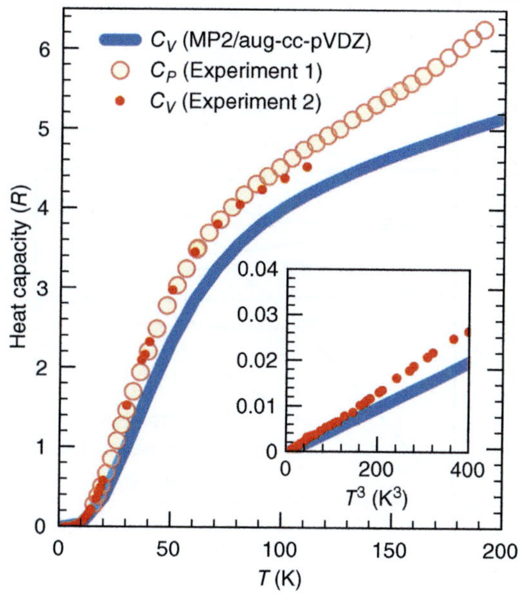

Figure 9.29 Isobaric and isochoric heat capacities (C_P and C_V) of CO_2-I as a function of temperature. Experiments 1 and 2 refer to Giauque and Egan [190] and Manzhelll *et al.* [187], respectively. *Source*: Li *et al.* 2015b [186]. Reprinted with permission of American Chemical Society.

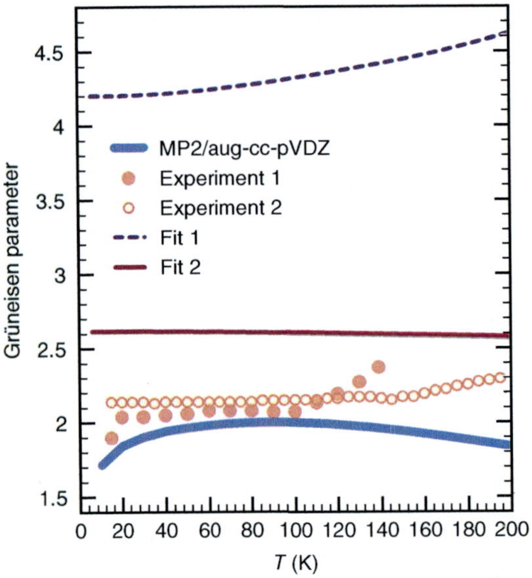

Figure 9.30 Grüneisen parameter (γ) of CO_2-I as a function of temperature. Experiments 1 and 2 refer to Manzhelii *et al.* [187] and Krupskii *et al.* [188], respectively. Fits 1 and 2 were obtained by Giordano *et al.* [193] and by Trusler [191, 192], respectively, by fitting an empirical equation of motion to experimental data. *Source*: Li *et al.* 2015b [186]. Reprinted with permission of American Chemical Society.

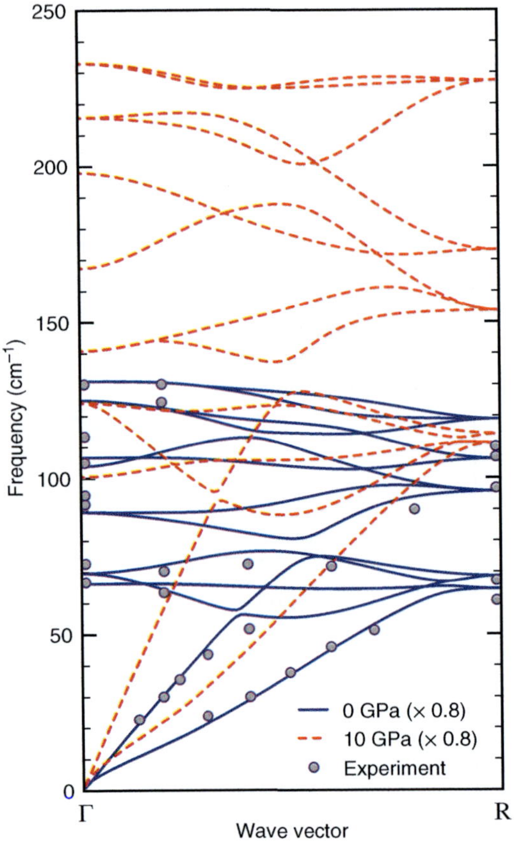

Figure 9.31 Phonon dispersions in CO_2-I at 0 and 10 GPa. The calculated frequencies are scaled by 0.8 just to make the comparison with the experimental data [194] easier. *Source*: Li *et al.* 2015b [186]. Reprinted with permission of American Chemical Society.

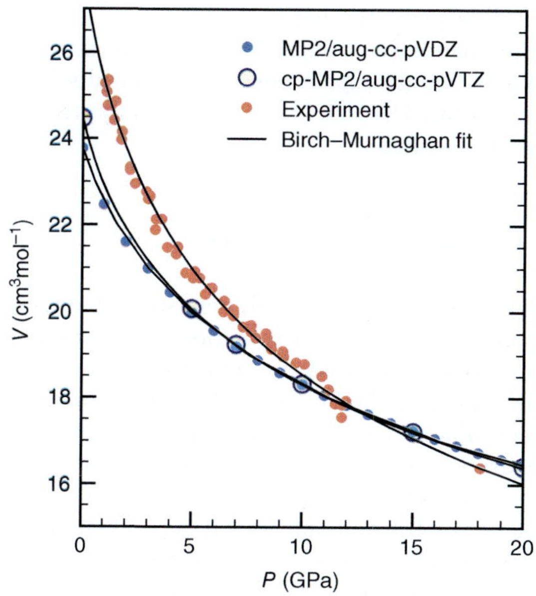

Figure 9.32 Equation of state of CO_2-I and the third-order Birch–Murnaghan equation fits. The experimental data at 296 ± 2 K were taken from Liu [182]. The volume collapse at about 12 GPa is caused by the solid–solid phase transition to CO_2-III discussed in Section 9.3.6. *Source*: Li *et al.* 2015b [186]. Reprinted with permission of American Chemical Society.

With a pressure loading of 10 GPa, the whole acoustic branches blue-shift, corresponding to positive mode Grüneisen parameters. Judging from the comparison in this figure, MP2-calculated γ is more likely to be slightly overestimated rather than underestimated. Again, γ cannot be a major source of the error in α_V.

Figure 9.32 plots the MP2-calculated and observed [182] equations of state (P-V curves), which finally exposes the major source of the error in α_V. While V is only somewhat underestimated by overbinding MP2, this underestimation has a greatly amplified effect on B because it is proportional to $\partial P/\partial V$ or the reciprocal of the slope of the P–V curve. Although the experimental values of B at 0 K and 0 GPa are widely scattered, the consensus value is around 8.0 GPa. In contrast, MP2/aug-cc-pVDZ predicts a value of 16.1 GPa, which is too large by a factor of two. This calculated value of B is lowered by BSSE-corrected MP2/aug-cc-pVTZ to 12.1 GPa, which is still greatly overestimated. These overestimated values of B (that enter Grüneisen's formula for α_V as a denominator factor) are primarily responsible for the gross underestimation of α_V. In other words, MP2 overbinds CO_2-I, which is thus computationally too stiff and cannot expand as much as it actually does in reality, just like MP2 ice-Ih (Section 9.3.1). In the SCS-MP2 simulation of liquid water also, a large negative pressure had to be applied for it to have the correct density of 1 g/cm^3 (Section 9.3.4).

Recently, Heit *et al.* [67] showed that the use of QHA, which includes the effects of phonon pressure and temperature-dependent B, can bring the MP2/aug-cc-pVDZ and experimental results for α_V of CO_2-I in near perfect agreement. However, we find it unlikely that this is the case with our method at the MP2/aug-cc-pVDZ level, given the consistency and severity of overbinding by this theoretical level across multiple molecular crystals and a molecular liquid. We note that the fragment method of Heit *et al.* [67] and ours differ in the embedding-field construction (Section 9.2.1). We believe that MP2/aug-cc-pVDZ may be accurate enough for structures, spectra, and heat capacities, as well as qualitative behavior of thermal expansion, but it fundamentally lacks the ability to quantitatively reproduce the stiffness of weakly bound solids such as molecular crystals. We do expect our *ab initio* results for α_V to improve systematically (as also demonstrated by Heit *et al.* [67]), but only gradually as the theoretical level and basis set size are elevated.

9.4 Concluding Remarks

Quantitative computational chemistry for condensed matter has entered a new phase, in which systematical approximations to the solutions of their electron Schrödinger equations converging toward exactitude (such as MP and CC with systematic GTO basis sets) can be applied for a variety of their properties. The embedded-fragment method that enables such applications is linear scaling of cost with respect to global system size, quadratic scaling (with a dimer-level truncation) with respect to unit cell size, easy to implement for a variety of properties and theoretical levels, massively parallel, strongly size-consistent, and only weakly dependent on perfect periodicity. The accessible properties include the structures and RDF, equations of state, heat capacities, bulk moduli, thermal expansion coefficients, volume isotope effects, phonon dispersion curves and DOS, infrared spectra (including band intensities), Raman spectra (including Fermi resonances), INS spectra, phase diagrams, self-diffusion coefficients, etc. We are hopeful that they replace empirical force-field or DFT calculations of molecular solids or Car–Parrinello MD for molecular liquids whenever predictive accuracy is sought. We also expect that they will be further extended to a broader class of solids and liquids such as covalently bonded solids, ionic solids and liquids, molecular solids with charge transfer including organic superconductors and organic optoelectronic materials, supramolecular assemblies, and more complex solutions. We therefore expect embedded-fragment *ab initio* WFT to become an essential partner to experiment in many fields ranging from solid-state physics [195, 196], high-pressure chemistry [195, 197, 198], geochemistry [197], planetary science [195], medicinal chemistry [94, 199], supramolecular chemistry [200, 201], materials science [196, 199], liquid theory [30, 115, 202], ion solvation, to chemical reactions in solutions [203].

9.5 Disclaimer

This is a summary of work published by the principal author's laboratory as journal articles that are cited in figure captions. It has some overlapping scope with S. Hirata *et al.* *Accounts of Chemical Research* **47**, 2721 (2014a).

Acknowledgments

The original studies described herein have been supported over the years by the U.S. National Science Foundation under awards CHE-1118616 and CHE-1361586, by the U.S. Department of Energy under awards DE-FG02-11ER16211 and DE-FG02-12ER46875, by the Camille Dreyfus Teacher-Scholar Award of the Camille & Henry

Dreyfus Foundation, Inc., and by the Scialog program of the Research Corporation for Science Advancement. Some of them are also part of the Blue Waters sustained-petascale computing project, which is supported by the U.S. National Science Foundation (awards OCI-0725070 and ACI-1238993) and the state of Illinois. Blue Waters is a joint effort of the University of Illinois at Urbana-Champaign and its National Center for Supercomputer Applications. S.H. has also been supported by the CREST program of the Japan Science and Technology Agency. We thank Drs. Kwang S. Kim, Gregory A. Voth, Soohaeng Y. Willow, Sotiris S. Xantheas, and Xiao Cheng Zeng for contributions to the original articles summarized in this chapter.

References

1 R. G. Parr and W. Yang, *Density-Functional Theory of Atoms and Molecules* (Oxford University Press, New York, 1989).

2 W. Koch and M. C. Holthausen, *A Chemist's Guide to Density Functional Theory*, 2nd ed. (Wiley-VCH, Weinheim, 2001).

3 A. Szabo and N. S. Ostlund, *Modern Quantum Chemistry: Introduction to Advanced Electronic Structure Theory* (MacMillan, New York, 1982).

4 T. Helgaker, P. Jørgensen, and J. Olsen, *Molecular Electronic-Structure Theory* (John Wiley & Sons, Chichester, 2000).

5 I. Shavitt and R. J. Bartlett, *Many-Body Methods in Chemistry and Physics: MBPT and Coupled-Cluster Theory* (Cambridge University Press, Cambridge, 2009).

6 T. H. Dunning, "Gaussian-basis sets for use in correlated molecular calculations. I. The atoms boron through neon and hydrogen," *Journal of Chemical Physics* **90**, 1007–1023 (1989).

7 P. O. Widmark, P. Å. Malmqvist, and B. O. Roos, "Density-matrix averaged atomic natural orbital (ANO) basis-sets for correlated molecular wave-functions. I. First row atoms," *Theoretica Chimica Acta* **77**, 291–306 (1990).

8 J. Almlöf and P. R. Taylor, "Atomic natural orbital (ANO) basis sets for quantum-chemical calculations," *Advances in Quantum Chemistry* **22**, 301–373 (1991).

9 W. Klopper, F. R. Manby, S. Ten-no, and E. F. Valeev, "R12 methods in explicitly correlated molecular electronic structure theory," *International Reviews in Physical Chemistry* **25**, 427–468 (2006).

10 T. Shiozaki, E. F. Valeev, and S. Hirata, "Explicitly correlated coupled-cluster methods," *Annual Reports in Computational Chemistry* **5**, 131–148 (2009).

11 L. G. Kong, F. A. Bischoff, and E. F. Valeev, "Explicitly correlated R12/F12 methods for electronic structure," *Chemical Reviews* **112**, 75–107 (2012).

12 P. R. Taylor, in *Lecture Notes in Quantum Chemistry II*, edited by B. O. Roos (Springer-Verlag, Berlin, 1994) pp. 125–202.

13 S. Hirata and K. Yagi, "Predictive electronic and vibrational many-body methods for molecules and macromolecules," *Chemical Physics Letters* **464**, 123–134 (2008).

14 T. Helgaker, W. Klopper, and D. P. Tew, "Quantitative quantum chemistry," *Molecular Physics* **106**, 2107–2143 (2008).

15 J. E. Del Bene, in *Pioneers of Quantum Chemistry*, edited by E. T. Strom and A. K. Wilson (American Chemical Society, Washington, DC, 2013) pp. 301–315.

16 M. Nooijen, K. R. Shamasundar, and D. Mukherjee, "Reflections on size-extensivity, size-consistency and generalized extensivity in many-body theory," *Molecular Physics* **103**, 2277–2298 (2005).
17 S. Hirata, "Thermodynamic limit and size-consistent design," *Theoretical Chemistry Accounts* **129**, 727–746 (2011).
18 W. M. C. Foulkes, L. Mitas, R. J. Needs, and G. Rajagopal, "Quantum Monte Carlo simulations of solids," *Reviews of Modern Physics* **73**, 33–83 (2001).
19 G. H. Booth, A. Grüneis, G. Kresse, and A. Alavi, "Towards an exact description of electronic wavefunctions in real solids," *Nature* **493**, 365–370 (2013).
20 S. Y. Willow, K. S. Kim, and S. Hirata, "Brueckner–Goldstone quantum Monte Carlo for correlation energies and quasiparticle energy bands of one-dimensional solids," *Physical Review B* **90**, 201110 (2014).
21 S. Hirata, "Quantum chemistry of macromolecules and solids," *Physical Chemistry Chemical Physics* **11**, 8397–8412 (2009).
22 M. S. Gordon, D. G. Fedorov, S. R. Pruitt, and L. V. Slipchenko, "Fragmentation methods: A route to accurate calculations on large systems," *Chemical Reviews* **112**, 632–672 (2012).
23 G. J. O. Beran and S. Hirata, "Fragment and localized orbital methods in electronic structure theory," *Physical Chemistry Chemical Physics* **14**, 7559–7561 (2012).
24 S. Hirata, K. Gilliard, X. He, J. J. Li, and O. Sode, "Ab initio molecular crystal structures, spectra, and phase diagrams," *Accounts of Chemical Research* **47**, 2721–2730 (2014a).
25 G. J. O. Beran, S. H. Wen, K. Nanda, Y. H. Huang, and Y. Heit, "Accurate and robust molecular crystal modeling using fragment-based electronic structure methods," *Prediction and Calculation of Crystal Structures: Methods and Applications* **345**, 59–93 (2014).
26 G. J. O. Beran, "A new era for ab initio molecular crystal lattice energy prediction," *Angewandte Chemie International Edition* **54**, 396–398 (2015).
27 G. J. O. Beran, "Modeling polymorphic molecular crystals with electronic structure theory," *Chemical Reviews* **116**, 5567–5613 (2016).
28 V. F. Petrenko and R. W. Whitworth, *Physics of Ice* (Oxford University Press, New York, 1999).
29 C. G. Salzmann, P. G. Radaelli, B. Slater, and J. L. Finney, "The polymorphism of ice: Five unresolved questions," *Physical Chemistry Chemical Physics* **13**, 18468–18480 (2011).
30 F. Frank, *Water: Comprehensive Treatease Vol. 1* (Plenum Press, New York, 1975).
31 P. Ball, "Water – an enduring mystery," *Nature* **452**, 291–292 (2008).
32 F. Paesani and G. A. Voth, "The properties of water: Insights from quantum simulations," *Journal of Physical Chemistry B* **113**, 5702–5719 (2009).
33 D. J. Lawrence, W. C. Feldman, J. O. Goldsten, S. Maurice, P. N. Peplowski, B. J. Anderson, D. Bazell, R. L. McNutt, L. R. Nittler, T. H. Prettyman, D. J. Rodgers, S. C. Solomon, and S. Z. Weider, "Evidence for water ice near Mercury's north pole from messenger neutron spectrometer measurements," *Science* **339**, 292–296 (2013).
34 A. T. Young, "Are clouds of venus sulfuric acid?" *Icarus* **18**, 564–582 (1973).
35 C. Sagan and S. F. Dermott, "The tide in the seas of Titan," *Nature* **300**, 731–733 (1982).
36 S. F. Dermott and C. Sagan, "Tidal effects of disconnected hydrocarbon seas on Titan," *Nature* **374**, 238–240 (1995).

37 T. Guillot, "Probing the giant planets," *Physics Today* **57**, 63–69 (2004).

38 M. C. Malin and K. S. Edgett, "Evidence for recent groundwater seepage and surface runoff on Mars," *Science* **288**, 2330–2335 (2000).

39 F. J. Martin-Torres, M. P. Zorzano, P. Valentin-Serrano, A. M. Harri, M. Genzer, O. Kemppinen, E. G. Rivera-Valentin, I. Jun, J. Wray, M. B. Madsen, W. Goetz, A. S. McEwen, C. Hardgrove, N. Renno, V. F. Chevrier, M. Mischna, R. Navarro-Gonzalez, J. Martinez-Frias, P. Conrad, T. McConnochie, C. Cockell, G. Berger, A. R. Vasavada, D. Sumner, and D. Vaniman, "Transient liquid water and water activity at Gale crater on Mars," *Nature Geoscience* **8**, 357–361 (2015).

40 C. Pilorget and F. Forget, "Formation of gullies on Mars by debris flows triggered by CO_2 sublimation," *Nature Geoscience* **9**, 65 (2016).

41 J. C. Li and D. K. Ross, "Evidence for two kinds of hydrogen bond in ice," *Nature* **365**, 327–329 (1993).

42 J. C. Li, S. M. Bennington, and D. K. Ross, "Further evidence for the existence of two kinds of hydrogen bonds in ice Ih," *Physics Letters A* **192**, 295–300 (1994).

43 P. Zhang, L. Tian, Z. P. Zhang, G. Shao, and J. C. Li, "Investigation of the hydrogen bonding in ice Ih by first-principles density function methods," *Journal of Chemical Physics* **137**, 044504 (2012).

44 X. He, O. Sode, S. S. Xantheas, and S. Hirata, "Second-order many-body perturbation study of ice Ih," *Journal of Chemical Physics* **137**, 204505 (2012).

45 K. Röttger, A. Endriss, J. Ihringer, S. Doyle, and W. F. Kuhs, "Lattice constants and thermal expansion of H_2O and D_2O ice Ih between 10 and 265 K," *Acta Crystallographica Section B: Structural Science* **50**, 644–648 (1994).

46 K. Röttger, A. Endriss, J. Ihringer, S. Doyle, and W. F. Kuhs, "Lattice constants and thermal expansion of H_2O and D_2O ice Ih between 10 and 265 K (vol 50, pg 644, 1994)," *Acta Crystallographica Section B: Structural Science* **68**, 91–91 (2012).

47 O. Mishima, L. D. Calvert, and E. Whalley, "'Melting ice' I at 77 K and 10 kbar: A new method of making amorphous solids," *Nature* **310**, 393–395 (1984).

48 P. H. Poole, F. Sciortino, U. Essmann, and H. E. Stanley, "Phase behavior of metastable water," *Nature* **360**, 324–328 (1992).

49 D. T. Limmer and D. Chandler, "The putative liquid-liquid transition is a liquid-solid transition in atomistic models of water," *Journal of Chemical Physics* **135**, 134503 (2011).

50 D. T. Limmer and D. Chandler, "The putative liquid-liquid transition is a liquid-solid transition in atomistic models of water. II," *Journal of Chemical Physics* **138**, 214504 (2013).

51 D. T. Limmer and D. Chandler, "Theory of amorphous ices," *Proceedings of the National Academy of Sciences of the United States of America* **111**, 9413–9418 (2014).

52 P. Wernet, D. Nordlund, U. Bergmann, M. Cavalleri, M. Odelius, H. Ogasawara, L. A. Naslund, T. K. Hirsch, L. Ojamae, P. Glatzel, L. G. M. Pettersson, and A. Nilsson, "The structure of the first coordination shell in liquid water," *Science* **304**, 995–999 (2004).

53 K. Laasonen, M. Sprik, M. Parrinello, and R. Car, "Ab-initio liquid water," *Journal of Chemical Physics* **99**, 9080–9089 (1993).

54 P. L. Silvestrelli, M. Bernasconi, and M. Parrinello, "Ab initio infrared spectrum of liquid water," *Chemical Physics Letters* **277**, 478–482 (1997).

55 P. L. Silvestrelli and M. Parrinello, "Water molecule dipole in the gas and in the liquid phase," *Physical Review Letters* **82**, 3308–3311 (1999a).

56 P. L. Silvestrelli and M. Parrinello, "Water molecule dipole in the gas and in the liquid phase (vol 82, pg 3308, 1999)," *Physical Review Letters* **82**, 5415–5415 (1999b).

57 Y. Q. Tu and A. Laaksonen, "The electronic properties of water molecules in water clusters and liquid water," *Chemical Physics Letters* **329**, 283–288 (2000).

58 Y. S. Badyal, M. L. Saboungi, D. L. Price, S. D. Shastri, D. R. Haeffner, and A. K. Soper, "Electron distribution in water," *Journal of Chemical Physics* **112**, 9206–9208 (2000).

59 A. Hassanali, F. Giberti, J. Cuny, T. D. Kühne, and M. Parrinello, "Proton transfer through the water gossamer," *Proceedings of the National Academy of Sciences of the United States of America* **110**, 13723–13728 (2013).

60 J. J. Li, O. Sode, G. A. Voth, and S. Hirata, "A solid-solid phase transition in carbon dioxide at high pressures and intermediate temperatures," *Nature Communications* **4**, 2647 (2013).

61 V. Iota and C. S. Yoo, "Phase diagram of carbon dioxide: Evidence for a new associated phase," *Physical Review Letters* **86**, 5922–5925 (2001).

62 H. Olijnyk, H. Däufer, H. J. Jodl, and H. D. Hochheimer, "Effect of pressure and temperature on the Raman spectra of solid CO_2," *Journal of Chemical Physics* **88**, 4204–4212 (1988).

63 R. C. Hanson, "A new high-pressure phase of solid CO_2," *Journal of Physical Chemistry* **89**, 4499–4501 (1985).

64 L. G. Liu, "Dry ice II, a new polymorph of CO_2," *Nature* **303**, 508–509 (1983).

65 V. Iota, C. S. Yoo, J. H. Klepeis, Z. Jenei, W. Evans, and H. Cynn, "Six-fold coordinated carbon dioxide VI," *Nature Materials* **6**, 34–38 (2007).

66 C. S. Yoo, H. Kohlmann, H. Cynn, M. F. Nicol, V. Iota, and T. LeBihan, "Crystal structure of pseudo-six-fold carbon dioxide phase II at high pressures and temperatures," *Physical Review B* **65** (2002).

67 Y. N. Heit, K. D. Nanda, and G. J. O. Beran, "Predicting finite-temperature properties of crystalline carbon dioxide from first principles with quantitative accuracy," *Chemical Science* **7**, 246–255 (2016).

68 S. Hirata, "Fast electron-correlation methods for molecular crystals: An application to the α, β_1, and β_2 modifications of solid formic acid," *Journal of Chemical Physics* **129**, 204104 (2008).

69 M. Kamiya, S. Hirata, and M. Valiev, "Fast electron correlation methods for molecular clusters without basis set superposition errors," *Journal of Chemical Physics* **128**, 074103 (2008).

70 S. Saebø, W. Tong, and P. Pulay, "Efficient elimination of basis set superposition errors by the local correlation method: Accurate ab initio studies of the water dimer," *Journal of Chemical Physics* **98**, 2170–2175 (1993).

71 S. Saebø and P. Pulay, "Local treatment of electron correlation," *Annual Review of Physical Chemistry* **44**, 213–236 (1993).

72 M. Schütz, G. Rauhut, and H. J. Werner, "Local treatment of electron correlation in molecular clusters: Structures and stabilities of $(H_2O)_n$, $n = 2 - 4$," *Journal of Physical Chemistry A* **102**, 5997–6003 (1998).

73 S. Hirata, M. Valiev, M. Dupuis, S. S. Xantheas, S. Sugiki, and H. Sekino, "Fast electron correlation methods for molecular clusters in the ground and excited states," *Molecular Physics* **103**, 2255–2265 (2005).

74 F. A. Momany, "Determination of partial atomic charges from ab initio molecular electrostatic potentials: Application to formamide, methanol, and formic acid," *Journal of Physical Chemistry* **82**, 592–601 (1978).

75 C. I. Bayly, P. Cieplak, W. D. Cornell, and P. A. Kollman, "A well-behaved electrostatic potential based method using charge restraints for deriving atomic charges: The RESP model," *Journal of Physical Chemistry* **97**, 10269–10280 (1993).

76 K. Kitaura, T. Sawai, T. Asada, T. Nakano, and M. Uebayasi, "Pair interaction molecular orbital method: An approximate computational method for molecular interactions," *Chemical Physics Letters* **312**, 319–324 (1999a).

77 K. Kitaura, E. Ikeo, T. Asada, T. Nakano, and M. Uebayasi, "Fragment molecular orbital method: An approximate computational method for large molecules," *Chemical Physics Letters* **313**, 701–706 (1999b).

78 D. G. Fedorov, N. Asada, I. Nakanishi, and K. Kitaura, "The use of many-body expansions and geometry optimizations in fragment-based methods," *Accounts of Chemical Research* **47**, 2846–2856 (2014).

79 E. E. Dahlke and D. G. Truhlar, "Electrostatically embedded many-body expansion for large systems, with applications to water clusters," *Journal of Chemical Theory and Computation* **3**, 46–53 (2007).

80 C. Muller and B. Paulus, "Wavefunction-based electron correlation methods for solids," *Physical Chemistry Chemical Physics* **14**, 7605–7614 (2012).

81 R. M. Richard, K. U. Lao, and J. M. Herbert, "Aiming for benchmark accuracy with the many-body expansion," *Accounts of Chemical Research* **47**, 2828–2836 (2014).

82 K. D. Nanda and G. J. O. Beran, "Prediction of organic molecular crystal geometries from MP2-level fragment quantum mechanical/molecular mechanical calculations," *Journal of Chemical Physics* **137**, 174106 (2012).

83 D. W. Zhang and J. Z. H. Zhang, "Molecular fractionation with conjugate caps for full quantum mechanical calculation of protein-molecule interaction energy," *Journal of Chemical Physics* **119**, 3599–3605 (2003).

84 K. Babu and S. R. Gadre, "Ab initio quality one-electron properties of large molecules: Development and testing of molecular tailoring approach," *Journal of Computational Chemistry* **24**, 484–495 (2003).

85 F. R. Manby, D. Alfè, and M. J. Gillan, "Extension of molecular electronic structure methods to the solid state: Computation of the cohesive energy of lithium hydride," *Physical Chemistry Chemical Physics* **8**, 5178–5180 (2006).

86 W. Li, S. H. Li, and Y. S. Jiang, "Generalized energy-based fragmentation approach for computing the ground-state energies and properties of large molecules," *Journal of Physical Chemistry A* **111**, 2193–2199 (2007).

87 A. L. Ringer and C. D. Sherrill, "First principles computation of lattice energies of organic solids: The benzene crystal," *Chemistry: A European Journal* **14**, 2542–2547 (2008).

88 R. Podeszwa, B. M. Rice, and K. Szalewicz, "Predicting structure of molecular crystals from first principles," *Physical Review Letters* **101**, 115503 (2008).

89 M. A. Collins, "Ab initio lattice dynamics of nonconducting crystals by systematic fragmentation," *Journal of Chemical Physics* **134**, 164110 (2011).

90 R. M. Richard and J. M. Herbert, "A generalized many-body expansion and a unified view of fragment-based methods in electronic structure theory," *Journal of Chemical Physics* **137**, 064113 (2012).

91 G. J. O. Beran and K. Nanda, "Predicting organic crystal lattice energies with chemical accuracy," *Journal of Physical Chemistry Letters* **1**, 3480–3487 (2010).
92 S. H. Wen and G. J. O. Beran, "Accurate molecular crystal lattice energies from a fragment QM/MM approach with on-the-fly ab initio force field parametrization," *Journal of Chemical Theory and Computation* **7**, 3733–3742 (2011).
93 S. H. Wen and G. J. O. Beran, "Crystal polymorphism in oxalyl dihydrazide: Is empirical DFT-D accurate enough?" *Journal of Chemical Theory and Computation* **8**, 2698–2705 (2012a).
94 S. H. Wen and G. J. O. Beran, "Accidental degeneracy in crystalline aspirin: New insights from high-level ab initio calculations," *Crystal Growth & Design* **12**, 2169–2172 (2012b).
95 S. H. Wen, K. Nanda, Y. H. Huang, and G. J. O. Beran, "Practical quantum mechanics-based fragment methods for predicting molecular crystal properties," *Physical Chemistry Chemical Physics* **14**, 7578–7590 (2012c).
96 Y. H. Huang, Y. H. Shao, and G. J. O. Beran, "Accelerating MP2C dispersion corrections for dimers and molecular crystals," *Journal of Chemical Physics* **138**, 224112 (2013).
97 K. D. Nanda and G. J. O. Beran, "What governs the proton ordering in ice XV?" *Journal of Physical Chemistry Letters* **4**, 3165–3169 (2013).
98 J. D. Hartman and G. J. O. Beran, "Fragment-based electronic structure approach for computing nuclear magnetic resonance chemical shifts in molecular crystals," *Journal of Chemical Theory and Computation* **10**, 4862–4872 (2014).
99 Y. Heit and G. J. O. Beran, "Exploiting space-group symmetry in fragment-based molecular crystal calculations," *Journal of Computational Chemistry* **35**, 2205–2214 (2014).
100 J. D. Hartman, S. Monaco, B. Schatschneider, and G. J. O. Beran, "Fragment-based ^{13}C nuclear magnetic resonance chemical shift predictions in molecular crystals: An alternative to planewave methods," *Journal of Chemical Physics* **143**, 102809 (2015).
101 S. Y. Willow, M. A. Salim, K. S. Kim, and S. Hirata, "Ab initio molecular dynamics of liquid water using embedded-fragment second-order many-body perturbation theory towards its accurate property prediction," *Scientific Reports* **5**, 14358 (2015).
102 O. Sode and S. Hirata, "Second-order many-body perturbation study of solid hydrogen fluoride under pressure," *Physical Chemistry Chemical Physics* **14**, 7765–7779 (2012).
103 W. Marshall and S. W. Lovesey, *Theory of Thermal Neutron Scattering* (Oxford University Press, London, 1971).
104 S. Hirata, H. Torii, Y. Furukawa, M. Tasumi, and J. Tomkinson, "Inelastic neutron scattering from trans-polyacetylene," *Chemical Physics Letters* **261**, 241–245 (1996).
105 E. I. Blount, "Formalisms of band theory," *Solid State Physics: Advances in Research and Applications* **13**, 305–373 (1962).
106 R. Resta, "Macroscopic polarization in crystalline dielectrics: The geometric phase approach," *Reviews of Modern Physics* **66**, 899–915 (1994).
107 B. Champagne and J. M. André, "Determination of ab initio polarizabilities of polymers: Application to polyethylene and polysilane," *International Journal of Quantum Chemistry* **42**, 1009–1024 (1992a).
108 B. Champagne, J. G. Fripiat, and J. M. André, "From uncoupled to coupled Hartree–Fock polarizabilities of infinite polymeric chains: Pariser–Parr–Pople

applications to the polyacetylene chains," *Journal of Chemical Physics* **96**, 8330–8337 (1992b).

109 N. W. Ashcroft and N. D. Mermin, *Solid State Physics* (Brooks Cole, Boston, 1976).

110 D. C. Wallace, *Thermodynamics of Crystals* (Wiley, New York, 1972).

111 R. Kubo, "Statistical theory of linear polymers. II. Elasticity of vulcanized rubber," *Journal of the Physical Society of Japan* **2**, 51–56 (1947).

112 P. B. Allen, "Anharmonic phonon quasiparticle theory of zero-point and thermal shifts in insulators: Heat capacity, bulk modulus, and thermal expansion," *Physical Review B* **92**, 064106 (2015).

113 M. J. Gillan, "The quantum simulation of hydrogen in metals," *Philosophical Magazine A: Physics of Condensed Matter Structure Defects and Mechanical Properties* **58**, 257–283 (1988).

114 D. M. Ceperley, "Path integrals in the theory of condensed helium," *Reviews of Modern Physics* **67**, 279–355 (1995).

115 M. P. Allen and D. J. Tildesley, *Computer Simulation of Liquids* (Oxford University Press, New York, 1989).

116 M. Bernasconi, P. L. Silvestrelli, and M. Parrinello, "Ab initio infrared absorption study of the hydrogen-bond symmetrization in ice," *Physical Review Letters* **81**, 1235–1238 (1998).

117 A. Putrino and M. Parrinello, "Anharmonic Raman spectra in high-pressure ice from ab initio simulations," *Physical Review Letters* **88**, 176401 (2002).

118 P. H. Berens and K. R. Wilson, "Molecular dynamics and spectra. I. Diatomic rotation and vibration," *Journal of Chemical Physics* **74**, 4872–4882 (1981).

119 T. Hasegawa and Y. Tanimura, "A polarizable water model for intramolecular and intermolecular vibrational spectroscopies," *Journal of Physical Chemistry B* **115**, 5545–5553 (2011).

120 P. H. Berens, S. R. White, and K. R. Wilson, "Molecular dynamics and spectra. II. Diatomic Raman," *Journal of Chemical Physics* **75**, 515–529 (1981).

121 G. Fini and P. Mirone, "Evidence for short-range orientation effects in dipolar aprotic liquids from vibrational spectroscopy. II. Carbonyl compounds," *Journal of the Chemical Society: Faraday Transactions II* **70**, 1776–1782 (1974).

122 J. L. McHale, "The influence of angular dependent intermolecular forces on vibrational spectra of solution phase molecules," *Journal of Chemical Physics* **75**, 30–35 (1981).

123 H. Torii and M. Tasumi, "Local order and transition dipole coupling in liquid methanol and acetone as the origin of the Raman noncoincidence effect," *Journal of Chemical Physics* **99**, 8459–8465 (1993).

124 S. Habershon, G. S. Fanourgakis, and D. E. Manolopoulos, "Comparison of path integral molecular dynamics methods for the infrared absorption spectrum of liquid water," *Journal of Chemical Physics* **129**, 074501 (2008).

125 F. Paesani and G. A. Voth, "A quantitative assessment of the accuracy of centroid molecular dynamics for the calculation of the infrared spectrum of liquid water," *Journal of Chemical Physics* **132**, 014105 (2010).

126 H. Tanaka, "Hydrogen bonds between water molecules: Thermal expansivity of ice and water," *Journal of Molecular Liquids* **90**, 323–332 (2001).

127 M. A. Salim, S. Y. Willow, and S. Hirata, "Ice Ih anomalies: Thermal contraction, anomalous volume isotope effect, and pressure-induced amorphization," *Journal of Chemical Physics* **144**, 204503 (2016).

128 J. S. Tse, D. D. Klug, C. A. Tulk, I. Swainson, E. C. Svensson, C. K. Loong, V. Shpakov, V. R. Belosludov, R. V. Belosludov, and Y. Kawazoe, "The mechanisms for pressure-induced amorphization of ice Ih," *Nature* **400**, 647–649 (1999).

129 O. Mishima, "Relationship between melting and amorphization of ice," Nature **384**, 546 (1996).

130 T. Strässle, S. Klotz, G. Hamel, M. M. Koza, and H. Schober, "Experimental evidence for a crossover between two distinct mechanisms of amorphization in ice Ih under pressure," *Physical Review Letters* **99**, 175501 (2007).

131 K. Gilliard, O. Sode, and S. Hirata, "Second-order many-body perturbation and coupled-cluster singles and doubles study of ice VIII," *Journal of Chemical Physics* **140**, 174507 (2014).

132 B. Kamb, "Structure of ice VI," *Science* **150**, 205–209 (1965).

133 E. Whalley, D. W. Davidson, and J. B. R. Heath, "Dielectric properties of ice VII and ice VIII: A new phase of ice," *Journal of Chemical Physics* **45**, 3976 (1966).

134 A. F. Goncharov, V. V. Struzhkin, M. S. Somayazulu, R. J. Hemley, and H. K. Mao, "Compression of ice to 210 gigapascals: Infrared evidence for a symmetric hydrogen-bonded phase," *Science* **273**, 218–220 (1996).

135 P. Pruzan, J. C. Chervin, E. Wolanin, B. Canny, M. Gauthier, and M. Hanfland, "Phase diagram of ice in the VII-VIII-X domain. Vibrational and structural data for strongly compressed ice VIII," *Journal of Raman Spectroscopy* **34**, 591–610 (2003).

136 A. N. Dunaeva, D. V. Antsyshkin, and O. L. Kuskov, "Phase diagram of H_2O: Thermodynamic functions of the phase transitions of high-pressure ices," *Solar System Research* **44**, 202–222 (2010).

137 T. L. Roush, "Physical state of ices in the outer solar system," *Journal of Geophysical Research: Planets* **106**, 33315–33323 (2001).

138 D. H. Green, "Experimental melting studies on a model upper mantle composition at high-pressure under water-saturated and water-undersaturated conditions," *Earth and Planetary Science Letters* **19**, 37–53 (1973).

139 J. M. Besson, S. Klotz, G. Hamel, W. G. Marshall, R. J. Nelmes, and J. S. Loveday, "Structural instability in ice VIII under pressure," *Physical Review Letters* **78**, 3141–3144 (1997).

140 K. R. Hirsch and W. B. Holzapfel, "Effect of high pressure on the Raman spectra of ice-VIII and evidence for ice-X," *Journal of Chemical Physics* **84**, 2771–2775 (1986).

141 M. Song, H. Yamawaki, H. Fujihisa, M. Sakashita, and K. Aoki, "Infrared investigation on ice VIII and the phase diagram of dense ices," *Physical Review B* **68**, 014106 (2003).

142 M. Kobayashi, T. Nanba, M. Kamada, and S. Endo, "Proton order-disorder transition of ice investigated by far-infrared spectroscopy under high pressure," *Journal of Physics: Condensed Matter* **10**, 11551–11555 (1998).

143 D. D. Klug, J. S. Tse, Z. X. Liu, X. Gonze, and R. J. Hemley, "Anomalous transformations in ice VIII," *Physical Review B* **70**, 144113 (2004).

144 J. S. Tse and D. D. Klug, "Anomalous isostructural transformation in ice VIII," *Physical Review Letters* **81**, 2466–2469 (1998).

145 Y. Yoshimura, S. T. Stewart, M. Somayazulu, H. Mao, and R. J. Hemley, "High-pressure x-ray diffraction and Raman spectroscopy of ice VIII," *Journal of Chemical Physics* **124**, 024502 (2006).

146 K. Umemoto and R. M. Wentzcovitch, "Theoretical study of the isostructural transformation in ice VIII," *Physical Review B* **71**, 012102 (2005).

147 S. Yoo, X. C. Zeng, and S. S. Xantheas, "On the phase diagram of water with density functional theory potentials: The melting temperature of ice Ih with the Perdew–Burke–Ernzerhof and Becke–Lee–Yang–Parr functionals," *Journal of Chemical Physics* **130**, 221102 (2009).
148 S. Yoo and S. S. Xantheas, "Communication: The effect of dispersion corrections on the melting temperature of liquid water," *Journal of Chemical Physics* **134**, 121105 (2011).
149 L. B. Skinner, C. C. Huang, D. Schlesinger, L. G. M. Pettersson, A. Nilsson, and C. J. Benmore, "Benchmark oxygen-oxygen pair-distribution function of ambient water from x-ray diffraction measurements with a wide Q-range," *Journal of Chemical Physics* **138**, 074506 (2013).
150 S. Y. Willow, X. C. Zeng, S. S. Xantheas, K. S. Kim, and S. Hirata, "Why is MP2-water 'cooler' and 'denser' than DFT-water?" *Journal of Physical Chemistry Letters* **7**, 680–684 (2016).
151 M. Del Ben, M. Schönherr, J. Hutter, and J. VandeVondele, "Bulk liquid water at ambient temperature and pressure from MP2 theory," *Journal of Physical Chemistry Letters* **4**, 3753–3759 (2013).
152 M. Del Ben, M. Schönherr, J. Hutter, and J. VandeVondele, "Bulk liquid water at ambient temperature and pressure from MP2 theory (vol 4, pg 3753, 2013)," *Journal of Physical Chemistry Letters* **5**, 3066–3067 (2014).
153 M. Del Ben, J. Hutter, and J. VandeVondele, "Probing the structural and dynamical properties of liquid water with models including non-local electron correlation," *Journal of Chemical Physics* **143**, 054506 (2015).
154 M. Del Ben, J. Hutter, and J. VandeVondele, "Second-order Møller–Plesset perturbation theory in the condensed phase: An efficient and massively parallel Gaussian and plane waves approach," *Journal of Chemical Theory and Computation* **8**, 4177–4188 (2012).
155 Y. Komeiji, T. Nakano, K. Fukuzawa, Y. Ueno, Y. Inadomi, T. Nemoto, M. Uebaysai, D. G. Fedorov, and K. Kitaura, "Fragment molecular orbital method: Application to molecular dynamics simulation, 'ab initio FMO-MD'", *Chemical Physics Letters* **372**, 342–347 (2003).
156 Y. Komeiji, Y. Mochizuki, and T. Nakano, "Three-body expansion and generalized dynamic fragmentation improve the fragment molecular orbital-based molecular dynamics (FMO-MD)," *Chemical Physics Letters* **484**, 380–386 (2010).
157 T. Fujita, T. Nakano, and S. Tanaka, "Fragment molecular orbital calculations under periodic boundary condition," *Chemical Physics Letters* **506**, 112–116 (2011).
158 Y. Mochizuki, T. Nakano, Y. Komeiji, K. Yamashita, Y. Okiyama, H. Yoshikawa, and H. Yamataka, "Fragment molecular orbital-based molecular dynamics (FMO-MD) method with MP2 gradient," *Chemical Physics Letters* **504**, 95–99 (2011).
159 K. R. Brorsen, N. Minezawa, F. Xu, T. L. Windus, and M. S. Gordon, "Fragment molecular orbital molecular dynamics with the fully analytic energy gradient," *Journal of Chemical Theory and Computation* **8**, 5008–5012 (2012).
160 K. R. Brorsen, F. Zahariev, H. Nakata, D. G. Fedorov, and M. S. Gordon, "Analytic gradient for density functional theory based on the fragment molecular orbital method," *Journal of Chemical Theory and Computation* **10**, 5297–5307 (2014).
161 M. Holz, S. R. Heil, and A. Sacco, "Temperature-dependent self-diffusion coefficients of water and six selected molecular liquids for calibration in accurate ^{1}H NMR PFG measurements," *Physical Chemistry Chemical Physics* **2**, 4740–4742 (2000).

162 J. R. Scherer, M. K. Go, and S. Kint, "Raman spectra and structure of water from −10 to 90°," *Journal of Physical Chemistry* **78**, 1304–1313 (1974).

163 J. E. Bertie and Z. D. Lan, "Infrared intensities of liquids. XX: The intensity of the OH stretching band of liquid water revisited, and the best current values of the optical constants of H_2O (l) at 25°C between 15,000 and 1 cm^{-1}," *Applied Spectroscopy* **50**, 1047–1057 (1996).

164 J. D. Eaves, A. Tokmakoff, and P. L. Geissler, "Electric field fluctuations drive vibrational dephasing in water," *Journal of Physical Chemistry A* **109**, 9424–9436 (2005).

165 M. Heyden, J. Sun, S. Funkner, G. Mathias, H. Forbert, M. Havenith, and D. Marx, "Dissecting the THz spectrum of liquid water from first principles via correlations in time and space," *Proceedings of the National Academy of Sciences of the United States of America* **107**, 12068–12073 (2010).

166 H. Torii, "Time-domain calculations of the polarized Raman spectra, the transient infrared absorption anisotropy, and the extent of delocalization of the OH stretching mode of liquid water," *Journal of Physical Chemistry A* **110**, 9469–9477 (2006).

167 S. Grimme, "Semiempirical GGA-type density functional constructed with a long-range dispersion correction," *Journal of Computational Chemistry* **27**, 1787–1799 (2006).

168 E. B. Wilson, J. C. Decius, and P. C. Cross, *Molecular Vibration* (Dover, New York, 1955).

169 O. Sode, M. Keçeli, K. Yagi, and S. Hirata, "Fermi resonance in solid CO_2 under pressure," *Journal of Chemical Physics* **138**, 074501 (2013).

170 E. Roedder, "Geobarometry of ultramafic xenoliths from Loihi seamount, Hawaii, on the basis of CO_2 inclusions in olivine," *Earth and Planetary Science Letters* **66**, 369–379 (1983).

171 O. Navon, "High internal-pressures in diamond fluid inclusions determined by infrared absorption," *Nature* **353**, 746–748 (1991).

172 M. Schrauder and O. Navon, "Solid carbon dioxide in a natural diamond," *Nature* **365**, 42–44 (1993).

173 Y. Kawakami, J. Yamamoto, and H. Kagi, "Micro-Raman densimeter for CO_2 inclusions in mantle-derived minerals," *Applied Spectroscopy* **57**, 1333–1339 (2003).

174 M. L. Frezzotti, F. Tecce, and A. Casagli, "Raman spectroscopy for fluid inclusion analysis," *Journal of Geochemical Exploration* **112**, 1–20 (2012).

175 R. C. Hanson and L. H. Jones, "Infrared and Raman studies of pressure effects on the vibrational modes of solid CO_2," *Journal of Chemical Physics* **75**, 1102–1112 (1981).

176 V. Rodriguez-Garcia, S. Hirata, K. Yagi, K. Hirao, T. Taketsugu, I. Schweigert, and M. Tasumi, "Fermi resonance in CO_2: A combined electronic coupled-cluster and vibrational configuration-interaction prediction," *Journal of Chemical Physics* **126**, 124303 (2007).

177 Y. Garrabos, M. A. Echargui, and F. Marsaultherail, "Comparison between the density effects on the levels of the Raman spectra of the Fermi resonance doublet of the $^{12}C^{16}O_2$ and $^{13}C^{16}O_2$ molecules," *Journal of Chemical Physics* **91**, 5869–5881 (1989).

178 G. Cardini and V. Schettino, "Comment on "Fermi resonance in solid CO_2 under pressure" [*Journal of Chemical Physics* **138**, 074501 (2013)]," *Journal of Chemical Physics* **140**, 177101 (2014).

179 S. Hirata, O. Sode, M. Keçeli, K. Yagi, and J. J. Li, "Response to "Comment on 'Fermi resonance in solid CO_2 under pressure'" [*Journal of Chemical Physics* **140**, 177101 (2014)]," *Journal of Chemical Physics* **140**, 177102 (2014b).

180 G. Cardini, P. R. Salvi, V. Schettino, and H. J. Jodl, "Pressure tuning of Fermi resonance in crystal CO_2," *Journal of Chemical Physics* **91**, 3869–3876 (1989).

181 J. J. Li, O. Sode, G. A. Voth, and S. Hirata, "A solid-solid phase transition in carbon dioxide at high pressures and intermediate temperatures (vol 4, 2647, 2013)," *Nature Communications* **6**, 8907 (2015a).

182 L. G. Liu, "Compression and phase behavior of solid CO_2 to half a megabar," *Earth and Planetary Science Letters* **71**, 104–110 (1984).

183 K. Aoki, H. Yamawaki, and M. Sakashita, "Phase study of solid CO_2 to 20 GPa by infrared-absorption spectroscopy," *Physical Review B* **48**, 9231–9234 (1993).

184 S. A. Bonev, F. Gygi, T. Ogitsu, and G. Galli, "High-pressure molecular phases of solid carbon dioxide," *Physical Review Letters* **91** (2003).

185 B. Kuchta and R. D. Etters, "Prediction of a high-pressure phase transition and other properties of solid CO_2 at low temperatures," *Physical Review B* **38**, 6265–6269 (1988).

186 J. J. Li, O. Sode, and S. Hirata, "Second-order many-body perturbation study on thermal expansion of solid carbon dioxide," *Journal of Chemical Theory and Computation* **11**, 224–229 (2015b).

187 V. G. Manzhelii, A. M. Tolkachev, M. I. Bagatskii, and E. I. Voitovich, "Thermal expansion, heat capacity, and compressibility of solid CO_2," *Physica Status Solidi B: Basic Research* **44**, 39 (1971).

188 I. N. Krupskii, A. I. Prokhvatilov, A. I. Érenburg, and A. S. Baryl'nik, "Thermal expansion x-ray studies of solid CO_2," *Fizika Nizkikh Temperatur* **8**, 533–541 (1982).

189 W. H. Keesom and J. W. L. Köhler, "The lattice constant and expansion coefficient of solid carbon dioxide," *Physica* **1**, 655–658 (1934).

190 W. F. Giauque and C. J. Egan, "Carbon dioxide. The heat capacity and vapor pressure of the solid. the heat of sublimation. Thermodynamic and spectroscopic values of the entropy," *Journal of Chemical Physics* **5**, 45–54 (1937).

191 J. P. M. Trusler, "Equation of state for solid phase I of carbon dioxide valid for temperatures up to 800 K and pressures up to 12 GPa," *Journal of Physical and Chemical Reference Data* **40** (2011).

192 J. P. M. Trusler, "Equation of state for solid phase I of carbon dioxide valid for temperatures up to 800 K and pressures up to 12 GPa (vol 40, 043105, 2011)," *Journal of Physical and Chemical Reference Data* **41** (2012).

193 V. M. Giordano, F. Datchi, F. A. Gorelli, and R. Bini, "Equation of state and anharmonicity of carbon dioxide phase I up to 12 GPa and 800 K," *Journal of Chemical Physics* **133**, 144501 (2010).

194 B. M. Powell, G. Dolling, L. Piseri, and P. Martel, *Neutron Inelastic Scattering* (International Atomic Energy Agency, Vienna, 1972).

195 J. M. McMahon, M. A. Morales, C. Pierleoni, and D. M. Ceperley, "The properties of hydrogen and helium under extreme conditions," *Reviews of Modern Physics* **84**, 1607–1653 (2012).

196 T. Ishiguro, K. Yamaji, and G. Saito, *Organic Superconductors*, 2nd ed. (Springer-Verlag, Berlin, 1998).

197 R. J. Hemley, "Effects of high pressure on molecules," *Annual Review of Physical Chemistry* **51**, 763 (2000).

198 D. D. Dlott, "Ultrafast spectroscopy of shock waves in molecular materials," *Annual Review of Physical Chemistry* **50**, 251–278 (1999).
199 J. Bernstein, *Polymorphism in Molecular Crystals* (Clarendon Press, Oxford, 2002).
200 L. Lieserowitz and J. F. Stoddart, "Molecular crystals," *Current Opinion in Solid State & Materials Science* **3**, 397–398 (1998).
201 Y. Inokuma, M. Kawano, and M. Fujita, "Crystalline molecular flasks," *Nature Chemistry* **3**, 349–358 (2011).
202 J.-P. Hansen and I. R. McDonald, *Theory of Simple Liquids*, 3rd ed. (Academic Press, London, 2006).
203 H. Nakata, D. G. Fedorov, T. Nagata, K. Kitaura, and S. Nakamura, "Simulations of chemical reactions with the frozen domain formulation of the fragment molecular orbital method," *Journal of Chemical Theory and Computation* **11**, 3053–3064 (2015).

10

A Linear-Scaling Divide-and-Conquer Quantum Chemical Method for Open-Shell Systems and Excited States

Takeshi Yoshikawa and Hiromi Nakai

Department of Chemistry and Biochemistry, School of Advanced Science and Engineering, Waseda University, Japan

10.1 Introduction

The acceleration of *ab initio* electronic structure calculations [1, 2] has been one of the most important themes in the field of quantum chemistry since the mid-1980s when a series of single-reference theories starting from the Hartree–Fock (HF) method were already mature. The standard single-reference quantum chemical calculations consist of three time-consuming steps, namely, the Fock-matrix construction, its diagonalization to obtain the molecular orbitals (MOs) and/or one-electron density matrix, and the post-HF correlation calculation that does not appear in the HF and Kohn–Sham (KS) density functional theory (DFT) calculations. Several strategies have focused on accelerating one of these steps. For example, linear-scaling Coulomb [3–8] and exchange [9–11] methods have succeeded in accelerating the Fock-matrix construction and have been implemented in many quantum chemistry packages. Several groups have proposed linear-scaling approaches that directly yield MOs or a one-electron density matrix without the explicit diagonalization of the Fock matrix [12–20]. The Laplace-transformed many-body perturbation approach [21–25] can accelerate the post-HF second-order Møller–Plesset perturbation (MP2) calculations. Although many of these methods use the so-called locality of the electronic interactions to achieve a low-scaling computational cost, the simplest way to achieve an entirely linear-scaling cost is by the spatial fragmentation of the entire system.

The local correlation method developed by Pulay with Saebø [26–29] can be considered as the first fragmentation-based low-scaling quantum chemical method. In their strategy, the correlating (virtual) orbital space for a given pair of localized occupied orbitals is spatially truncated. Although there have been many extensions made by several groups [30–34], the underlying concept has remained unchanged. However, the local correlation scheme only aims to accelerate the post-HF correlation calculation. The first fragmentation-based scheme that aims to accelerate the self-consistent field (SCF) calculation, which consists of the construction and diagonalization of the Fock matrix, is the divide-and-conquer (DC) method that was first proposed by Yang and

Fragmentation: Toward Accurate Calculations on Complex Molecular Systems, First Edition.
Edited by Mark S. Gordon.

coworkers [35–37]. In their first formalism, the local electronic density in each fragment was evaluated using MOs composed of atomic orbitals (AOs) of the fragment, and the density of the entire system was constructed as the weighted sum of the local densities with the assistance of a spatial partition function. To ensure that the number of electrons in the entire system remained constant, they used the finite-temperature formalism and introduced a Fermi level that was constant over all fragments. Because the method only provides the electronic density, this scheme could only be applied to DFT calculations.

In 1995, Yang and Lee [38] proposed a novel DC formalism that provides a one-electron density matrix instead of an electronic density distribution. By using this formalism, the DC method became applicable to the HF and semi-empirical MO calculations, as well as DFT. In their method, a simple partition matrix plays the same role as the partition function in the density-based formalism. The important works by Walker and Mezey [39, 40], the Lego or adjustable density matrix assembler (ADMA), are worth mentioning because they first introduced the partition matrix to enable large-scale quantum chemical calculations, although they did not use the Fermi level to keep the number of electrons. The density matrix-based DC method was soon applied to semi-empirical MO calculations by several research groups [41–46], as well as to DFT calculations [47–49]. Recently, Nakai and coworkers implemented and assessed the method applied to the HF and hybrid HF/DFT calculations [50–52], which was further examined by He and Merz [53]. Furthermore, they first applied it to the MP2 [54–56] and coupled cluster (CC) [57, 58] calculations, and succeeded in achieving an entirely linear-scaling computation of the correlated quantum chemical calculations. Further theoretical developments of the DC method were accomplished to enhance its applicability, such as open-shell [59, 60], dynamic polarizability [61–63], excited-state [64], and relativistic calculations [65]. Recently, the DC method was applied to the density-functional tight-binding (DFTB) theory to develop a massively parallel program that achieves on-the-fly molecular reaction dynamics simulations for more than one million-atom systems from scratch [66]. In this chapter, we review the linear-scaling quantum chemical calculation for the extension to open-shell and excited-state theories based on the DC method.

10.2 Theories for the Divide-and-Conquer Method

10.2.1 Review of DC-SCF and DC-Based Correlation Theories

10.2.1.1 DC-HF/DFT

In the DC method, the system under consideration is spatially divided into disjointed subsystems, where a set of AOs in a subsystem s is represented as $\mathbf{S}(s)$:

$$\mathbf{S}(s) \cap \mathbf{S}(s') = \emptyset \quad \forall s \neq s', \tag{10.1}$$

and the union of $\mathbf{S}(s)$ becomes a set of AOs in the entire system represented as $\boldsymbol{T}$:

$$\bigcup_{s} \mathbf{S}(s) = \boldsymbol{T}. \tag{10.2}$$

The disjointed subsystem is called the central region. To improve the description of the subsystem, the neighboring region from the central region, which is called the buffer region, is taken into consideration when expanding the subsystem MOs in the DC calculation. A set of AOs corresponding to the buffer region of subsystem s, which is denoted by $\mathbf{B}(s)$, is added to $\mathbf{S}(s)$ to construct a set of AOs in the localization region of subsystem α, $\mathbf{L}(s)$, as follows:

$$\mathbf{L}(s) \equiv \mathbf{S}(s) \cup \mathbf{B}(s). \tag{10.3}$$

In the DC-HF method, the density matrix of the total system $\boldsymbol{D}$ is given as the sum of local density matrices $\boldsymbol{D}^s$:

$$D_{\mu\nu} \approx D_{\mu\nu}^{\mathrm{DC}} = \sum_s P_{\mu\nu}^s D_{\mu\nu}^s = \sum_s D_{\mu\nu}^s, \tag{10.4}$$

with the following partition matrix $\boldsymbol{P}^s$:

$$P_{\mu\nu}^s = \begin{cases} 1 & [\mu \in \mathbf{S}(s) \wedge \nu \in \mathbf{S}(s)] \\ 1/2 & [\mu \in \mathbf{B}(s) \wedge \nu \in \mathbf{S}(s)] \vee [\mu \in \mathbf{S}(s) \wedge \nu \in \mathbf{B}(s)] \\ 0 & \text{otherwise.} \end{cases} \tag{10.5}$$

Hereafter, the indices $\{\mu, \nu, \ldots\}$ refer to AOs, $\{i, j, \ldots\}$ to occupied MOs, $\{a, b, \ldots\}$ to virtual MOs, and $\{p, q, \ldots\}$ to all MOs. When treating a system with n_e electrons, the local density matrix $\boldsymbol{D}^s$ is given by

$$D_{\mu\nu}^s = P_{\mu\nu}^s \sum_i^{n_e} C_{\mu i}^s C_{\nu i}^s = \sum_i \eta\left(\varepsilon_{\mathrm{F}} - \varepsilon_i^s\right) C_{\mu i}^s C_{\nu i}^s, \tag{10.6}$$

where $\eta(x)$ is the Heaviside step function,

$$\eta(x) = \begin{cases} 1 & (x \geq 0) \\ 0 & (x < 0) \end{cases}, \tag{10.7}$$

and ε_{F} is the Fermi level. In the DC approximation, the discrete step function $\eta(x)$ is substituted with the Fermi function,

$$f_\beta(x) = \frac{1}{1 + \exp(-\beta x)}, \tag{10.8}$$

with inverse temperature parameter β:

$$D_{\mu\nu}^s = P_{\mu\nu}^s \sum_q f_\beta\left(\varepsilon_{\mathrm{F}} - \varepsilon_q^s\right) C_{\mu q}^s C_{\nu q}^s. \tag{10.9}$$

Here, $\{C_{\mu q}^s\}$ and $\{\varepsilon_q^s\}$ are the subsystem MO coefficient and orbital energy, respectively, which are determined by solving the following Roothaan–Hall equation for subsystem s,

$$\boldsymbol{F}^s \boldsymbol{C}^s = \boldsymbol{\varepsilon}^s \boldsymbol{S}^s \boldsymbol{C}^s. \tag{10.10}$$

Here, $\boldsymbol{S}^s$ and $\boldsymbol{F}^s$ represent the local overlap and Fock matrices for subsystem s that are the submatrices of the entire overlap and Fock matrices in the basis of $\mathbf{L}(s)$. The Fermi level ε_{F} can be determined uniquely by the constraint of the total number of electrons n_e:

$$n_e = \mathrm{Tr}\left[\boldsymbol{D}^{\mathrm{DC}} \boldsymbol{S}\right] = \sum_s \sum_{\mu \in L(s)} (\boldsymbol{D}^s \boldsymbol{S}^s)_{\mu\mu}. \tag{10.11}$$

The entire Fock matrix $\boldsymbol{F}$ is constructed in the usual manner in the DC-HF method,

$$\begin{aligned} F_{\mu\nu} &= H_{\mu\nu}^{\text{core}} + \sum_{\lambda\rho} D_{\lambda\rho}^{\text{DC}} \left[\langle \mu\nu | \rho\lambda \rangle - \frac{1}{2} \langle \mu\lambda | \rho\nu \rangle \right] \\ &= H_{\mu\nu}^{\text{core}} + \sum_{s} \sum_{\lambda\rho \in L(s)} D_{\lambda\rho}^{s} \left[\langle \mu\nu | \rho\lambda \rangle - \frac{1}{2} \langle \mu\lambda | \rho\nu \rangle \right], \end{aligned} \tag{10.12}$$

with the two-electron integral notation of $\langle \mu\nu | \lambda\sigma \rangle = \iint \phi_{\mu}^{*}(\mathbf{r}_1)\phi_{\nu}^{*}(\mathbf{r}_2) r_{12}^{-1} \phi_{\lambda}(\mathbf{r}_1)\phi_{\sigma}(\mathbf{r}_2) d\mathbf{r}_1 d\mathbf{r}_2$ and the core Hamiltonian matrix $\boldsymbol{H}^{\text{core}}$. The Fock construction of equation (10.12) and the density matrix construction of equations (10.4) and (10.9) are iterated until they converge in the same manner as the standard SCF procedure. Finally, the DC-HF energy is given as

$$E_{\text{DC-HF}} = \frac{1}{2} \sum_{s} \sum_{\mu\nu \in L(s)} D_{\mu\nu}^{s} \left(H_{\nu\mu}^{\text{core}} + F_{\nu\mu} \right). \tag{10.13}$$

The same procedure can be adopted in the DC-DFT calculation by substituting the Fock matrix of equation (10.12) with the KS Hamiltonian, and the energy expression of equation (10.13) with the KS energy.

The DC-SCF procedure is separable into two parts: (i) solving the SCF equations (equation 10.10) for all subsystems and (ii) constructing the Fock matrix according to equation (10.12). The cost for step (i) is $(\sum_s |L(s)|^3)$, which can be reduced as $O(|T|)$ because only the number of subsystems is dependent on the entire system size that is proportional to $|\boldsymbol{T}|$. The cost for step (ii) is reduced from $O(|\boldsymbol{T}|^4)$ to $O(|\boldsymbol{T}|^2)$ because (μ, ν) and (λ, ρ) in equation (10.12) should be in the same subsystem bases. The standard integral screening and the fast Coulomb or exchange schemes will further diminish the computational cost.

10.2.1.2 DC-Based Correlation Theory

The DC-based correlation energy is estimated by summing the correlation energies that correspond to the individual subsystems as follows:

$$\Delta E_{\text{corr}}^{\text{DC}} = \sum_{s} \Delta E_{\text{corr}}^{s}. \tag{10.14}$$

The DC-HF calculation gives MOs in each subsystem (localization region). The correlation energy $\Delta E_{\text{corr}}^{s}[\boldsymbol{L}(s)]$ corresponding to localization region s will be obtained straightforwardly with the two-electron integral notation as

$$\Delta E_{\text{corr}}^{s}[\boldsymbol{L}(s)] = \sum_{ij}^{\text{occ}} \sum_{ab}^{\text{vir}} \langle i^s j^s | a^s b^s \rangle \left(2\tilde{t}_{ij,ab}^{s} - \tilde{t}_{ij,ab}^{s} \right) \neq \Delta E_{\text{corr}}^{s}, \tag{10.15}$$

where $\{i^s, j^s, \ldots\}$ and $\{a^s, b^s, \ldots\}$ refer to the occupied and virtual MOs of subsystem s that have orbital energies smaller than and greater than ε_{F} and $\tilde{t}_{ij,ab}^{s}$ represents an effective two-electron excitation coefficient for subsystem s. However, the sum of the correlation energies over all localization regions would provide an inappropriate correlation energy for the entire system because the localization regions overlap with each other owing to the existence of the buffer region. Therefore, the correlation energy that corresponds to the central region should be extracted. This task was accomplished by using the concept of the energy density analysis (EDA) [67], which was proposed and developed by Nakai.

The EDA partitions the energy into atomic contributions in an analogous manner as the Mulliken population analysis. By leaving the last integral transformation undone, one can rewrite the correlation energy as the sum of the atomic contributions:

$$\begin{aligned}\Delta E^s_{\text{corr}} &= \sum_{A \in \mathbf{S}(s)}^{\text{atom}} \left[\sum_{ij}^{\text{occ}} \sum_{ab}^{\text{vir}} \sum_{\mu \in A} C^*_{\mu\iota} \left\langle \mu j^s \middle| a^s b^s \right\rangle \left(2\tilde{t}^s_{ij,ab} - \tilde{t}^s_{ij,ab} \right) \right] \\ &= \sum_{ij}^{\text{occ}} \sum_{ab}^{\text{vir}} \sum_{\mu \in \mathbf{S}(s)} C^*_{\mu\iota} \left\langle \mu j^s \middle| a^s b^s \right\rangle \left(2\tilde{t}^s_{ij,ab} - \tilde{t}^s_{ij,ab} \right)\end{aligned} \tag{10.16}$$

In the MP2 case, $\tilde{t}^s_{ij,ab}$ is represented as

$$\tilde{t}^s_{ij,ab} = \frac{\langle a^s b^s | i^s j^s \rangle}{\varepsilon^s_a + \varepsilon^s_b - \varepsilon^s_i - \varepsilon^s_j}, \tag{10.17}$$

and in the coupled cluster theory with singles and doubles excitations (CCSD) case,

$$\tilde{t}^s_{ij,ab} = t^s_{i,a} t^s_{j,b} + t^s_{ij,ab}, \tag{10.18}$$

where the T_1 and T_2 amplitudes, $t^s_{i,a}$ and $t^s_{ij,ab}$, are determined by solving the CCSD equations for the localization region of subsystem s.

10.2.1.3 Dual-Buffer DC-Based Correlation Method

In DC-based correlation and excited-state calculations, the correlation-energy error converges to zero more rapidly with respect to the buffer size than the HF-energy error. Because the computational costs for the correlation energy calculations are generally higher than those for HF, the DC-based correlation calculation can be accelerated while keeping its accuracy by adopting a larger and smaller buffer for the HF and correlation calculations, respectively. The procedure of this dual-buffer treatment is as follows:

1. Iteratively solve the DC-HF equations by adopting a larger buffer region, and obtain a converged total density matrix and HF energy. This process can be substituted with any accelerated HF methodologies that can provide density matrix and HF energy, or even with standard HF.
2. Solve the DC-HF equations in subsystems with smaller buffer regions once for all subsystems and obtain the subsystem MOs. The Fermi level, which is required to separate the occupied and virtual MOs, may (or may not) be redefined from the constraint of equation (10.11).
3. Perform the DC-based correlation calculation using the subsystem MOs determined in step 2 and obtain the correlation energy.

10.2.2 Linear-Scaling Divide-and-Conquer Method for Open-Shell Systems

10.2.2.1 DC-USCF and DC-UMP2

In the DC unrestricted SCF method, the σ-spin ($\sigma =\uparrow$ or $\downarrow$) density matrix of the entire system, $\boldsymbol{D}^{\sigma}$, is approximated as the sum of the local density matrices, $\boldsymbol{D}^{\sigma s}$, as

$$D^{\sigma}_{\mu\nu} \approx D^{\sigma \mathrm{DC}}_{\mu\nu} = \sum_{s} P^{s}_{\mu\nu} D^{\sigma s}_{\mu\nu} = \sum_{s} D^{\sigma s}_{\mu\nu}. \tag{10.19}$$

The local density matrix is constructed from subsystem orbitals:

$$D^{\sigma s}_{\mu\nu} = P^{s}_{\mu\nu} \sum_{q} f_{\beta}\left(\varepsilon^{\sigma}_{\mathrm{F}} - \varepsilon^{\sigma s}_{q}\right) C^{\sigma s}_{\mu q} C^{\sigma s *}_{\nu q}, \tag{10.20}$$

where $\{C^{\sigma s}_{\mu q}\}$ and $\{\varepsilon^{\sigma s}_{q}\}$ are the subsystem MO coefficient and orbital energy, respectively, which are determined by solving the following Pople–Nesbet equation for subsystem s,

$$\boldsymbol{F}^{\sigma s} \boldsymbol{C}^{\sigma s} = \boldsymbol{\varepsilon}^{\sigma s} \boldsymbol{S}^{s} \boldsymbol{C}^{\sigma s}. \tag{10.21}$$

Here, $\boldsymbol{F}^{\sigma s}$ represents the σ-spin Fock matrices for subsystem s that are the submatrices of the entire overlap and Fock matrices in the basis of $\boldsymbol{L}(\alpha)$.

Each Fermi level $\varepsilon^{\sigma}_{\mathrm{F}}$ in equation (10.20) is determined independently and uniquely through the entire system by the following constraint of the total number of σ-spin electrons n^{σ}_{e}:

$$n^{\sigma}_{e} = \mathrm{Tr}[\boldsymbol{D}^{\sigma \mathrm{DC}} \boldsymbol{S}] = \sum_{s} \sum_{\mu \in L(s)} (\boldsymbol{D}^{\sigma s} \boldsymbol{S}^{s})_{\mu\mu}. \tag{10.22}$$

The entire Fock matrix $\boldsymbol{F}^{\sigma}$ is constructed in the usual manner,

$$F^{\sigma}_{\mu\nu} = H^{\mathrm{core}}_{\mu\nu} + \sum_{s} \sum_{\lambda\rho \in \boldsymbol{L}(s)} \left[\left(D^{s\uparrow}_{\lambda\rho} + D^{s\downarrow}_{\lambda\rho}\right)(\mu\nu\,|\,\rho\lambda) - D^{s\sigma}_{\lambda\rho}(\mu\lambda\,|\,\rho\nu)\right]. \tag{10.23}$$

Finally, the DC-UHF energy is given as

$$E_{\mathrm{DC\text{-}UHF}} = \frac{1}{2} \sum_{s} \sum_{\sigma}^{\uparrow,\downarrow} \sum_{\mu\nu \in \boldsymbol{L}(s)} D^{s\sigma}_{\mu\nu} \left(H^{\mathrm{core}}_{\nu\mu} + F^{\sigma}_{\nu\mu}\right). \tag{10.24}$$

The same procedure can be adopted in the DC-UDFT calculation by substituting the Fock matrix of equation (10.23) with the unrestricted KS Hamiltonian, and the energy expression of equation (10.24) with the KS energy.

In the unrestricted MP2 (UMP2) theory, the correlation energy can be rewritten with spatial orbitals as the sum of up-spin, down-spin, and cross terms as follows:

$$\begin{aligned}\Delta E_{\mathrm{UMP2}} = &\sum_{i^{\uparrow}<j^{\uparrow}}^{\mathrm{occ}} \sum_{a^{\uparrow}<b^{\uparrow}}^{\mathrm{vir}} \left\langle i^{\uparrow} j^{\uparrow} \middle| a^{\uparrow} b^{\uparrow} \right\rangle \left[\tilde{t}_{i^{\uparrow} j^{\uparrow}, a^{\uparrow} b^{\uparrow}} - \tilde{t}_{i^{\uparrow} j^{\uparrow}, b^{\uparrow} a^{\uparrow}}\right] \\ &+ \sum_{i^{\downarrow}<j^{\downarrow}}^{\mathrm{occ}} \sum_{a^{\downarrow}<b^{\downarrow}}^{\mathrm{vir}} \left\langle i^{\downarrow} j^{\downarrow} \middle| a^{\downarrow} b^{\downarrow} \right\rangle \left[\tilde{t}_{i^{\downarrow} j^{\downarrow}, a^{\downarrow} b^{\downarrow}} - \tilde{t}_{i^{\downarrow} j^{\downarrow}, b^{\downarrow} a^{\downarrow}}\right] \\ &+ \sum_{i^{\uparrow}}^{\mathrm{occ}} \sum_{j^{\downarrow}}^{\mathrm{occ}} \sum_{a^{\uparrow}}^{\mathrm{vir}} \sum_{b^{\downarrow}}^{\mathrm{vir}} \left\langle i^{\uparrow} j^{\downarrow} \middle| a^{\uparrow} b^{\downarrow} \right\rangle \tilde{t}_{i^{\uparrow} j^{\downarrow}, a^{\uparrow} b^{\downarrow}},\end{aligned} \tag{10.25}$$

where $\tilde{t}_{i^{\sigma} j^{\sigma'}, a^{\sigma} b^{\sigma'}}$ represents an effective two-electron excitation coefficient as follows:

$$\tilde{t}_{i^{\sigma} j^{\sigma'}, a^{\sigma} b^{\sigma'}} = -\frac{\left\langle a^{\sigma} b^{\sigma'} \middle| i^{\sigma} j^{\sigma'} \right\rangle}{\varepsilon_a^{\sigma} + \varepsilon_b^{\sigma'} - \varepsilon_i^{\sigma} - \varepsilon_j^{\sigma'}}. \tag{10.26}$$

In the DC-based correlation theory, the total correlation energy is estimated by summing the correlation energies that correspond to the individual subsystems. This strategy has been extended to the UMP2 theory:

$$\Delta E_{\mathrm{DC\text{-}UMP2}} = \sum_{s}^{\mathrm{subsystem}} \Delta E_{\mathrm{UMP2}}^{s}. \tag{10.27}$$

Here, the correlation energy of subsystem s, $\Delta E_{\mathrm{UMP2}}^{s}$, is estimated using subsystem orbitals, which are constructed in the localization region, containing not only the central region but also the buffer region. While the buffer regions overlap in several subsystems, the central ones have no overlap. To avoid double counting, the correlation energies corresponding to the central regions should be estimated. Thus, the EDA technique was applied to the UMP2 correlation energy representation as follows:

$$\begin{aligned}\Delta E_{\mathrm{UMP2}}^{s} = &\sum_{\sigma}^{\uparrow,\downarrow} \sum_{i^{\sigma s}<j^{\sigma s}}^{\mathrm{occ}} \sum_{a^{\sigma s}<b^{\sigma s}}^{\mathrm{vir}} \sum_{\mu \in \mathbf{S}(s)} C_{\mu i}^{\sigma s *} \left\langle \mu j^{\sigma s} \mid a^{\sigma s} b^{\sigma s} \right\rangle \left[\tilde{t}_{i^{\sigma} j^{\sigma}, a^{\sigma} b^{\sigma}}^{s} - \tilde{t}_{i^{\sigma} j^{\sigma}, b^{\sigma} a^{\sigma}}^{s}\right] \\ &+ \sum_{i^{\uparrow s}}^{\mathrm{occ}} \sum_{j^{\downarrow s}}^{\mathrm{occ}} \sum_{a^{\uparrow s}}^{\mathrm{vir}} \sum_{b^{\downarrow s}}^{\mathrm{vir}} \sum_{\mu \in \mathbf{S}(s)} \frac{1}{2} \left(C_{\mu i}^{\uparrow s *} \left\langle \mu j^{\downarrow s} \mid a^{\uparrow s} b^{\downarrow s} \right\rangle + C_{\mu j}^{\downarrow s *} \left\langle i^{\uparrow s} \mu \mid a^{\uparrow s} b^{\downarrow s} \right\rangle \right) \tilde{t}_{i^{\uparrow} j^{\downarrow}, a^{\uparrow} b^{\downarrow}}^{s},\end{aligned} \tag{10.28}$$

where $\tilde{t}_{i^{\sigma} j^{\sigma'}, a^{\sigma} b^{\sigma'}}^{s}$ represents an effective two-electron excitation coefficient for subsystem s as follows:

$$\tilde{t}_{i^{\sigma} j^{\sigma'}, a^{\sigma} b^{\sigma'}}^{s} = -\frac{\left\langle a^{\sigma s} b^{\sigma' s} \mid i^{\sigma s} j^{\sigma' s} \right\rangle}{\varepsilon_a^{\sigma s} + \varepsilon_b^{\sigma' s} - \varepsilon_i^{\sigma s} - \varepsilon_j^{\sigma' s}}. \tag{10.29}$$

10.2.2.2 Expected Value of the Squared Spin Operator $\hat{S}^2$

A quantity that is specific in the open-shell calculation is the expected value of the squared spin operator $\hat{S}^2$. The UHF wavefunction is the pure eigenfunction of the z-spin operator $\hat{S}_z$, but usually not of $\hat{S}^2$. The expected value of the $\hat{S}^2$ operator for the UHF wavefunction, which is given by

$$\langle \hat{S}^2 \rangle = S_z^2 + \frac{n_\uparrow + n_\downarrow}{2} - \sum_i^{n_\uparrow} \sum_j^{n_\downarrow} \left\langle \psi_i^\uparrow \mid \psi_j^\downarrow \right\rangle^2, \tag{10.30}$$

in the case of $n_\uparrow \geq n_\downarrow$ with $S_z = (n_\uparrow - n_\downarrow)/2$, is a good indicator to estimate the level of spin contamination. Since this value can be estimated in terms of the density and overlap matrices as

$$\sum_i^{n_\uparrow} \sum_j^{n_\downarrow} \left\langle \psi_i^\uparrow \mid \psi_j^\downarrow \right\rangle^2 = \mathrm{Tr}\left[\mathbf{D}^\uparrow \mathbf{S} \mathbf{D}^\downarrow \mathbf{S}\right], \tag{10.31}$$

the expected value for the UMP2 wavefunction is generally given with the reduced two-electron density matrix Γ as [68]:

$$\sum_i^{n_\uparrow} \sum_j^{n_\downarrow} \left\langle \psi_i^\uparrow \mid \psi_j^\downarrow \right\rangle^2 = \sum_{p^\uparrow q^\downarrow r^\downarrow s^\uparrow} \left\langle p^\uparrow \middle| r^\downarrow \right\rangle \left\langle q^\downarrow \middle| s^\uparrow \right\rangle \Gamma_{p^\uparrow q^\downarrow r^\downarrow s^\uparrow}. \tag{10.32}$$

These estimations in the framework of the DC method are possible by replacing the density matrix expression.

10.2.3 Linear-Scaling Divide-and-Conquer Method for Excited-State Calculations

The DC-based excited-state theory uses the subsystem MOs obtained from the preceding DC-HF or DC-DFT calculations. Following the ground-state DC calculation, the subsystem that includes the pigment is selected as the excitation subsystem (ES). The excited-state wavefunctions of the ES are constructed from its MOs, which can be classified as either occupied or virtual according to the Fermi level.

10.2.3.1 DC-CIS/TDDFT

In the DC-CIS method, the nth excited-state wavefunction in the ES is expressed as the linear combination of singly-excited determinants, $\left|\Phi_{i,a}^{\mathrm{ES}}\right\rangle$, from the reference determinant, $\left|\Phi_0^{\mathrm{ES}}\right\rangle$, as follows:

$$\left|\Psi_{\mathrm{CIS}}^{\mathrm{ES}(n)}\right\rangle = \sum_i^{\mathrm{occ(ES)}} \sum_a^{\mathrm{vir(ES)}} d_{i,a}^{\mathrm{ES}(n)} \left|\Phi_{i,a}^{\mathrm{ES}}\right\rangle. \tag{10.33}$$

Here, $\left|\Phi_0^{\mathrm{ES}}\right\rangle$ is defined as the Slater determinant of the occupied MOs in the ES that have lower energies than the Fermi level determined via the DC-HF calculation. Hereafter, $\boldsymbol{d}^{\mathrm{ES}(n)}$ is referred to as the configuration interaction (CI) coefficient that is deduced as the normalized eigenvector of the Hamiltonian matrix, $\boldsymbol{A}^{\mathrm{ES}}$, which is expressed using the two-electron integral notation, as follows:

$$A^{\mathrm{ES}}_{ia,jb} = \left\langle \Phi^{\mathrm{ES}}_{i,a} \middle| \hat{H} - E^{\mathrm{ES}}_0 \middle| \Phi^{\mathrm{ES}}_{j,b} \right\rangle = \left(\varepsilon^{\mathrm{ES}}_a - \varepsilon^{\mathrm{ES}}_i\right) \delta_{ij}\delta_{ab} - \left\langle j^{\mathrm{ES}} a^{\mathrm{ES}} \middle\| i^{\mathrm{ES}} b^{\mathrm{ES}} \right\rangle, \quad (10.34)$$

where $E^{\mathrm{ES}}_0 = \langle \Phi^{\mathrm{ES}}_0 | \hat{H} | \Phi^{\mathrm{ES}}_0 \rangle$. The excitation energies, $\omega^{\mathrm{ES}(n)}$, can be obtained from the eigenvalues of this matrix. This procedure can also be applied to TDDFT calculations by substituting the CIS Hamiltonian of equation (10.34) with that of TDDFT.

10.2.3.2 DC-SAC/SACCI

The symmetry-adapted cluster (SAC) expansion for a completely symmetric singlet ground state is expressed as

$$\left|\Psi_{\mathrm{SAC}}\right\rangle = \exp\left(\hat{S}\right) \left|\Phi_0\right\rangle, \quad (10.35)$$

where $\hat{S}$ represents the symmetry-adapted single and double excitation operators, as follows:

$$\hat{S} = \hat{S}_{\mathrm{S}} + \hat{S}_{\mathrm{D}} = \sum_i^{\mathrm{occ}} \sum_a^{\mathrm{vir}} c_{i,a} \hat{S}_{i,a} + \sum_{i,j}^{\mathrm{occ}} \sum_{a,b}^{\mathrm{vir}} \left(c_{ij,ab} \hat{S}_{ij,ab} + c'_{ij,ab} \hat{S}'_{ij,ab} \right), \quad (10.36)$$

where

$$\hat{S}_{i,a} = \left(a^{\dagger}_{a\uparrow} a_{i\uparrow} + a^{\dagger}_{a\downarrow} a_{i\downarrow} \right) \Big/ \sqrt{2}, \quad (10.37)$$

$$\hat{S}_{ij,ab} = \left(a^{\dagger}_{a\uparrow} a_{i\uparrow} a^{\dagger}_{b\downarrow} a_{j\downarrow} + a^{\dagger}_{a\downarrow} a_{i\downarrow} a^{\dagger}_{b\uparrow} a_{j\uparrow} + a^{\dagger}_{b\uparrow} a_{i\uparrow} a^{\dagger}_{a\downarrow} a_{j\downarrow} + a^{\dagger}_{b\downarrow} a_{i\downarrow} a^{\dagger}_{a\uparrow} a_{j\uparrow} \right) \Big/ 4, \quad (10.38)$$

$$\begin{aligned} \hat{S}'_{ij,ab} = \Big(& 2a^{\dagger}_{a\uparrow} a_{i\uparrow} a^{\dagger}_{b\uparrow} a_{j\uparrow} + 2a^{\dagger}_{a\downarrow} a_{i\downarrow} a^{\dagger}_{b\downarrow} a_{j\downarrow} + a^{\dagger}_{a\uparrow} a_{i\uparrow} a^{\dagger}_{b\downarrow} a_{j\downarrow} + a^{\dagger}_{a\downarrow} a_{i\downarrow} a^{\dagger}_{b\uparrow} a_{j\downarrow} \\ & - a^{\dagger}_{b\uparrow} a_{i\uparrow} a^{\dagger}_{a\downarrow} a_{j\downarrow} - a^{\dagger}_{b\downarrow} a_{i\downarrow} a^{\dagger}_{a\uparrow} a_{j\uparrow} \Big) \Big/ 4\sqrt{3}. \end{aligned} \quad (10.39)$$

Here, $a_{p\sigma}$ and $a^{\dagger}_{p\sigma}$ represent the annihilation and creation operators, respectively, for spatial orbital p with σ spin. The SAC expansion coefficients, that is, $c_{i,a}$, $c_{ij,ab}$, and $c'_{ij,ab}$, are determined by solving the following non-variational equations:

$$\left\langle \Phi_0 \middle| \hat{H} - E_{\mathrm{SAC}} \middle| \Psi_{\mathrm{SAC}} \right\rangle = 0, \quad (10.40)$$

$$\left\langle \Phi_0 \middle| \hat{S}_I \left(\hat{H} - E_{\mathrm{SAC}}\right) \middle| \Psi_{\mathrm{SAC}} \right\rangle = 0, \quad (10.41)$$

with $\hat{S}_I = \hat{S}_{i,a}, \hat{S}_{ij,ab}$, or $\hat{S}'_{ij,ab}$. Here, E_{SAC} is the SAC energy, which comprises the HF energy, E_{HF}, and the SAC correlation energy, ΔE_{SAC}. The SAC correlation energy can be calculated as

$$\Delta E_{\mathrm{SAC}} = \left\langle ij \middle| ab \right\rangle \left(2c_{i,a}c_{j,b} - c_{i,b}c_{j,a} + c_{ij,ab} + \sqrt{3}c'_{ij,ab} \right). \quad (10.42)$$

Based on the SAC wavefunction, the SAC configuration interaction (SACCI) expansion for the nth excited state is expressed as

$$\left|\Psi_{\text{SACCI}}^{(n)}\right\rangle = \sum_K d_K^{(n)} \hat{P}\hat{R}_K^\dagger \left|\Psi_{\text{SAC}}\right\rangle, \tag{10.43}$$

where $\{\hat{R}_K^\dagger\}$ represents a set of excitations, each of which generates a basis for the target state. $\hat{P} = 1 - |\Psi_{\text{SAC}}\rangle\langle\Psi_{\text{SAC}}|$ is the projection operator that projects out the ground-state wavefunction. The SACCI coefficients, $d_K^{(n)}$, and the excitation energies, $\omega^{(n)}$, are calculated by solving the non-variational equation, as follows:

$$\left\langle \Phi_0 \right| \hat{R}_K \left(\hat{H} - E_{\text{SAC}} - \omega^{(n)}\right) \left|\Psi_{\text{SACCI}}^{(n)}\right\rangle = 0. \tag{10.44}$$

In the DC-SAC/SACCI theory, the ground-state SAC wavefunction for the ES is

$$\left|\Psi_{\text{SAC}}^{\text{ES}}\right\rangle = \exp(\hat{S}^{\text{ES}}) \left|\Phi_0^{\text{ES}}\right\rangle, \tag{10.45}$$

where $\hat{S}^{\text{ES}}$ represents the symmetry-adapted single and double excitation operator from the reference determinant, $|\Phi_0^{\text{ES}}\rangle$, in the ES, as follows:

$$\hat{S}^{\text{ES}} = \sum_i^{\text{occ(ES)}} \sum_a^{\text{vir(ES)}} c_{i,a}^{\text{ES}} \hat{S}_{i,a}^{\text{ES}} + \sum_{i,j}^{\text{occ(ES)}} \sum_{a,b}^{\text{vir(ES)}} \left(c_{ij,ab}^{\text{ES}} \hat{S}_{ij,ab}^{\text{ES}} + c_{ij,ab}^{\prime\text{ES}} \hat{S}_{ij,ab}^{\prime\text{ES}} \right). \tag{10.46}$$

Here, $\hat{S}_{i,a}^{\text{ES}}$, $\hat{S}_{ij,ab}^{\text{ES}}$, and $\hat{S}_{ij,ab}^{\prime\text{ES}}$ are the same as in equations (10.37)–(10.39), but are defined using the annihilation and creation operators of the MOs in the ES, that is, $a_{p\sigma}^{\text{ES}}$ and $a_{p\sigma}^{\text{ES}\dagger}$. The expansion coefficients are determined by solving the non-variational equation for the ES, as follows:

$$\left\langle \Phi_0^{\text{ES}} \middle| \hat{H} - E_{\text{SAC}}^{\text{ES}} \middle| \Psi_{\text{SAC}}^{\text{ES}} \right\rangle = 0, \tag{10.47}$$

$$\left\langle \Phi_0^{\text{ES}} \middle| \hat{S}_I^{\text{ES}} (\hat{H} - E_{\text{SAC}}^{\text{ES}}) \middle| \Psi_{\text{SAC}}^{\text{ES}} \right\rangle = 0. \tag{10.48}$$

Here, $E_{\text{SAC}}^{\text{ES}}$ is the SAC energy of the ES. It is defined as

$$\begin{aligned} E_{\text{SAC}}^{\text{ES}} &= E_0^{\text{ES}} + \Delta E_{\text{SAC}}^{\text{ES}} \\ &= E_0^{\text{ES}} + \left\langle i^{\text{ES}} j^{\text{ES}} \middle| a^{\text{ES}} b^{\text{ES}} \right\rangle \left(2c_{i,a}^{\text{ES}} c_{j,b}^{\text{ES}} - c_{i,b}^{\text{ES}} c_{j,a}^{\text{ES}} + c_{ij,ab}^{\text{ES}} + \sqrt{3} c_{ij,ab}^{\prime\text{ES}} \right) \end{aligned}. \tag{10.49}$$

Based on the SAC wavefunction in the ES, the DC-SACCI expansion for the nth excited state is

$$\left|\Psi_{\text{SACCI}}^{\text{ES}(n)}\right\rangle = \sum_K d_K^{\text{ES}(n)} \hat{P}^{\text{ES}} \hat{R}_K^{\text{ES}\dagger} \left|\Psi_{\text{SAC}}^{\text{ES}}\right\rangle, \tag{10.50}$$

where $\{\hat{R}_K^{\text{ES}\dagger}\}$ represents a set of excitations, each of which generates a basis for the target state in the ES. $\hat{P}^{\text{ES}} = 1 - |\Psi_{\text{SAC}}^{\text{ES}}\rangle\langle\Psi_{\text{SAC}}^{\text{ES}}|$ is the projection operator that projects out the ground-state wavefunction in the ES. The DC-SACCI coefficients, $d_K^{\mathrm{ES}(n)}$, and excitation energies, $\omega^{\mathrm{ES}(n)}$, are calculated by solving the following non-variational equation:

$$\left\langle \Phi_0^{\mathrm{ES}} \right| \hat{R}_{\mathrm{K}}^{\mathrm{ES}} \left(\hat{H} - E_{\mathrm{SAC}}^{\mathrm{ES}} - \omega^{\mathrm{ES}(n)} \right) \left| \Psi_{\mathrm{SACCI}}^{\mathrm{ES}(n)} \right\rangle = 0. \tag{10.51}$$

10.3 Assessment of the Divide-and-Conquer Method

10.3.1 Divide-and-Conquer Calculations for Open-Shell Systems

10.3.1.1 DC-USCF and DC-UMP2

The DC-UHF, UDFT, and UMP2 methods were assessed in the calculations of the spin- and charge-delocalized oligothiophenes ($H(C_4H_2S)_nH$, with D_{2h} symmetry), one of which is depicted in Figure 10.1 for $n = 14$. All calculations were performed with the modified version of the GAMESS program package [69]. In the DC-SCF calculations, the inverse temperature parameter of the Fermi function β in equation (10.28) was fixed to 200 a.u. unless otherwise noted. One thiophene unit in the system was adopted as one subsystem, and adjacent left and right n_b units were adopted as the corresponding buffer region (see Figure 10.1 representing an example for $n_\mathrm{b} = 2$). Carbon atoms and subsystems were numbered from the edge, as shown in Figure 10.1. All of the following calculations were performed with the 6-31G** basis set [70–72].

Figure 10.2a shows the buffer-size dependence of the absolute HF and correlation energy errors in the DC-HF and MP2 calculations of oligothiophene ($n = 14$), respectively. Here, the open- and closed-shell singlet, triplet, and cation doublet states are calculated. In DC-HF, the absolute energy errors for a buffer size of $n_\mathrm{b} \geq 3$ achieved 1.2 mhartree or less for all four states, and those of $n_\mathrm{b} \geq 4$ achieved 0.07 mhartree or less error. Moreover, the correlation energy errors for a buffer size of $n_b \geq 1$ achieved 1.2 mhartree or less for all four states. As mentioned in Section 10.2.1.3, the correlation energy errors are much smaller than the HF energy errors in the DC-MP2 method. Figure 10.2b shows the buffer-size dependence of the expected values of the $\hat{S}^2$ operator, which presents a similar dependence on the buffer size to the energy. The deviations of $\langle \hat{S}^2 \rangle$ from the conventional UHF results are sufficiently small (< 0.03) for $n_\mathrm{b} \geq 3$, compared to those for $n_\mathrm{b} = 2$.

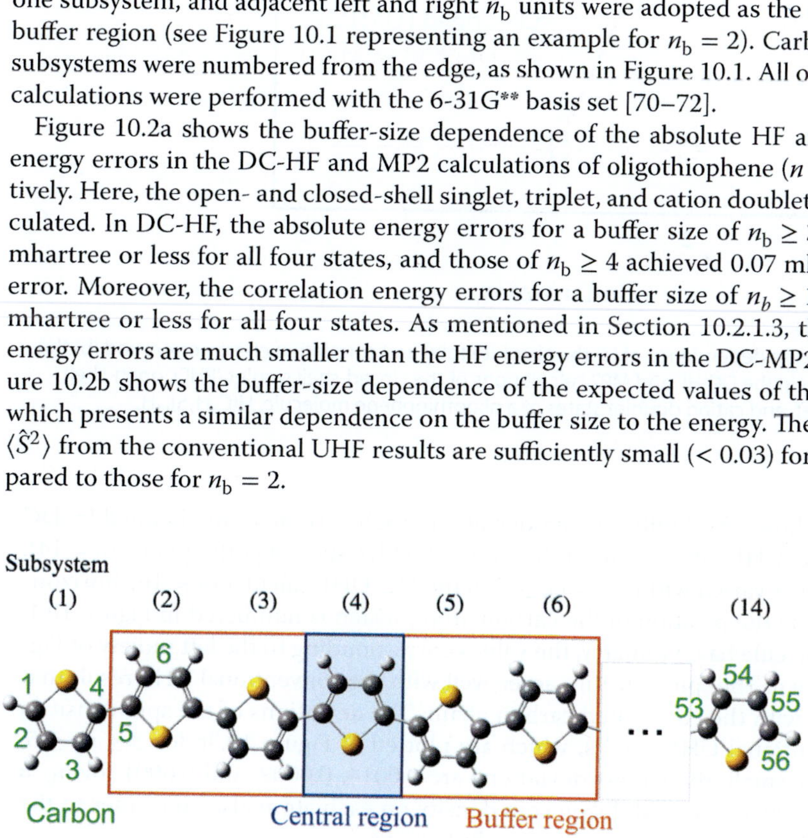

Figure 10.1 Structure of the oligothiophene, $H(C_4H_2S)_nH$ ($n = 14$), and the schematic of the central and buffer regions in the DC calculations with $n_\mathrm{b} = 2$.

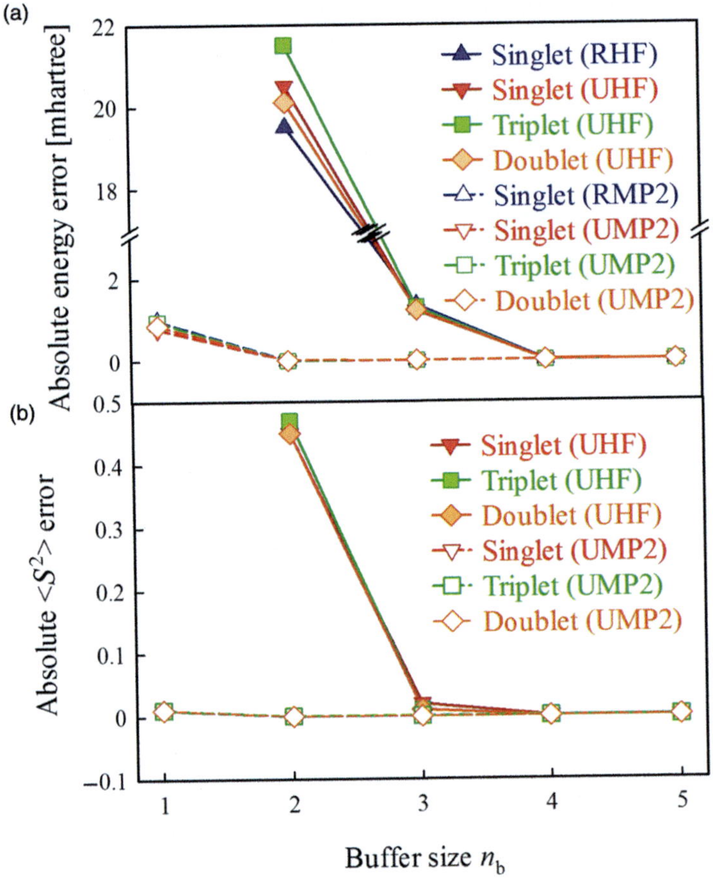

Figure 10.2 Buffer-size dependence of (a) the absolute total and correlation energy errors and (b) the absolute $\langle S^2 \rangle$ errors in the DC-HF and MP2 calculations of the closed-shell singlet (RHF), open-shell singlet (UHF), triplet, and cation doublet states of a oligothiophene molecule, $H(C_4H_2S)_{14}H$, respectively.

Figure 10.3a shows the Mulliken spin density on each carbon atom obtained by DC and conventional UHF calculations of the cation doublet-state oligothiophene ($n = 14$). The buffer size was varied within $3 \leq n_b \leq 5$ in the DC-UHF calculations. The horizontal axis represents the position of the carbon atom, which is numbered in Figure 10.1. Because the molecule has symmetry, the values corresponding to the left moiety of Figure 10.1 are shown. The spin densities agree well with the conventional UHF result that oscillates along with the conjugated carbon chain. The deviations of the spin densities from the conventional UHF results, which are plotted in Figure 10.3b for $3 \leq n_b \leq 5$, are considerably small; the largest deviations are 0.0014, 0.0002, and 0.0001 for n_b = 3, 4, and 5, respectively. The Mulliken spin density on each atom also converges to the conventional UHF result with respect to the buffer size.

Table 10.1 shows the Mulliken charge and spin densities corresponding to each subsystem obtained by DC and conventional UHF calculations of the cation doublet-state

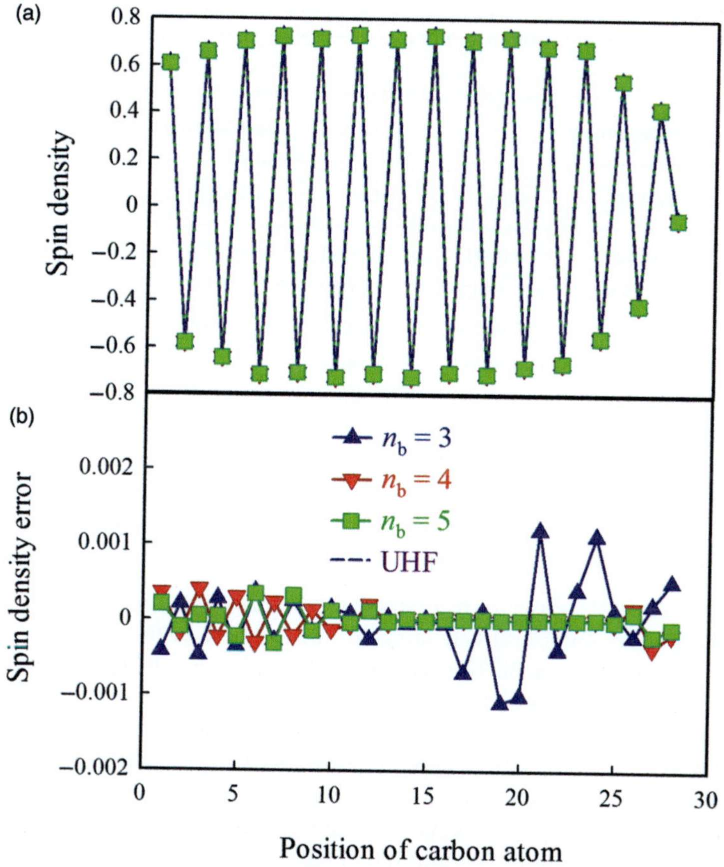

Figure 10.3 (a) Mulliken spin density on each carbon atom obtained by DC and conventional UHF calculations of the cation doublet-state oligothiophene $H(C_4H_2S)_{14}H$ with the 6-31G** basis set. The values are plotted for each buffer size of $3 \leq n_b \leq 5$. (b) The deviations of the DC-UHF spin densities from the conventional UHF results for $3 \leq n_b \leq 5$.

oligothiophene ($n = 14$) with $n_b = 4$. The differences of the densities between the DC and conventional UHF results, which are given in parentheses, are sufficiently small (≤ 0.0004). Because the excess charge and spin are delocalized, each subsystem has fractional charge and spin densities. Note again that the number of charges or spins in each subsystem is determined from equation (10.22), and one need not specify the value before the DC-UHF/UDFT calculation. Despite this simple scheme, the DC-UHF/UDFT method can treat charge- and spin-delocalized open-shell systems with reasonable accuracy.

Next, we show the assessment on the accuracy of the DC-UDFT energies with typical functionals (i.e., BLYP [73, 74], B3LYP [75], BHHLYP [76], and long-range corrected (LC) BOP [77, 78]). Table 10.2 shows the adopted functional dependence of the DC and conventional UDFT energies of triplet oligothiophene ($n = 14$) with $n_b = 4$. Here, the results for $\beta = 200$ and 400 a.u. (the inverse temperature parameter that appeared in equation 10.20) are listed together. The energy errors from the conventional UDFT

Table 10.1 Mulliken charge and spin densities corresponding to each subsystem obtained by DC and conventional UHF calculations of the doublet (cation) oligothiophene, $H(C_4H_2S)_{14}H$, with the 6-31G** basis set.

	Charge density			Spin density		
Subsystem	**UHF**	**DC-UHF**	**(diff.)**	**UHF**	**DC-UHF**	**(diff.)**
(1), (*14*)	+0.0129	+0.0128	(−0.0001)	+0.0052	+0.0520	(−0.0000)
(2), (13)	+0.0068	+0.0068	(+0.0000)	+0.0038	+0.0038	(+0.0000)
(3), (12)	+0.0093	+0.0093	(+0.0000)	+0.0007	+0.0007	(−0.0000)
(4), (11)	+0.0196	+0.0197	(+0.0001)	+0.0025	+0.0024	(−0.0001)
(5), (10)	+0.0441	+0.0442	(+0.0001)	+0.0138	+0.0139	(+0.0001)
(6), (9)	+0.1146	+0.1148	(+0.0002)	+0.0852	+0.0853	(+0.0001)
(7), (8)	+0.2924	+0.2921	(−0.0003)	+0.3887	+0.3886	(−0.0001)

A buffer size of $n_b = 4$ was adopted in the DC-UHF calculation.

results are given in parentheses. Attending to the results with $\beta = 200$ a.u., the errors for the BLYP and B3LYP energies are comparatively large, although those of the BHHLYP and LC-BOP energies are sufficiently small. The error introduced by the DC treatment can be roughly classified into two: the cut-off error by constructing subsystem orbitals in reduced AO space, $\boldsymbol{L}(\alpha)$, and the finite temperature error by the Fermi distribution of the occupation number in equation (10.22). In general, a smaller band gap (the energy difference between the highest occupied and lowest unoccupied MOs) leads to a larger temperature error. Because the band gaps of this system evaluated from the conventional BLYP and B3LYP calculations are small (0.007 and 0.033 hartree, compared to that of HF of 0.272 hartree), large temperature errors for the DC-BLYP and DC-B3LYP energies are prospective. Actually, the energy errors of DC-BLYP and DC-B3LYP can be reduced by using a lower electronic temperature ($\beta = 400$ a.u.), as shown in Table 10.2.

Table 10.2 Adopted DFT functional dependence of the total energy obtained by DC and conventional UDFT calculations of triplet oligothiophene, $H(C_4H_2S)_{14}H$, with the 6-31G** basis set.

		DC-UDFT ($\beta = 200$ a.u.)		DC-UDFT ($\beta = 400$ a.u.)	
Functional	**UDFT energy [hartree]**	**Energy [hartree]**	**(diff.) [mhartree]**	**Energy [hartree]**	**(diff.) [mhartree]**
HF	−7702.945769	−7702.945713	(−0.056)	−7702.945713	(−0.056)
BLYP	−7725.223192	−7725.230293	(+7.101)	−7725.228066	(+4.874)
B3LYP	−7726.362603	−7726.369123	(+6.520)	−7726.364156	(+1.553)
BHHLYP	−7724.787869	−7724.788023	(+0.154)	−7724.787987	(+0.118)
LC-BOP	−7716.662273	−7716.662276	(+0.003)	−7716.662273	(+0.000)

A buffer size of $n_b = 4$ was adopted in the DC calculations.

Finally, we describe the efficiencies of the DC-UHF and UMP2 methods in measuring the central processing unit (CPU) time. An Intel Xeon X5470 (3.33 GHz) processor was used on a single core. Triplet oligothiophenes (n = 4–22) were calculated with the buffer size of $n_b = 3$ and correlation buffer-size $n_b = 1$ in DC-UHF and DC-UMP2 calculations, respectively. Figure 10.4a shows the CPU times for solving the SCF equations for the first SCF iteration of the conventional and DC-UHF calculations. The horizontal axis represents the number of units, n. The DC-UHF method drastically reduces the computational time from the conventional UHF time. Although the larger system improves the efficiency of the DC-UHF method, the DC-UHF time becomes shorter than the conventional UHF time, even for n = 14. According to the scaling analysis by the double-logarithmic plot, the times scale with $O(n^{3.56})$ and $O(n^{1.33})$ for the conventional and DC-UHF calculations, which reasonably agree with the theoretical asymptotic values of $O(n^3)$ and $O(n^1)$, respectively. Figure 10.4b shows the times for

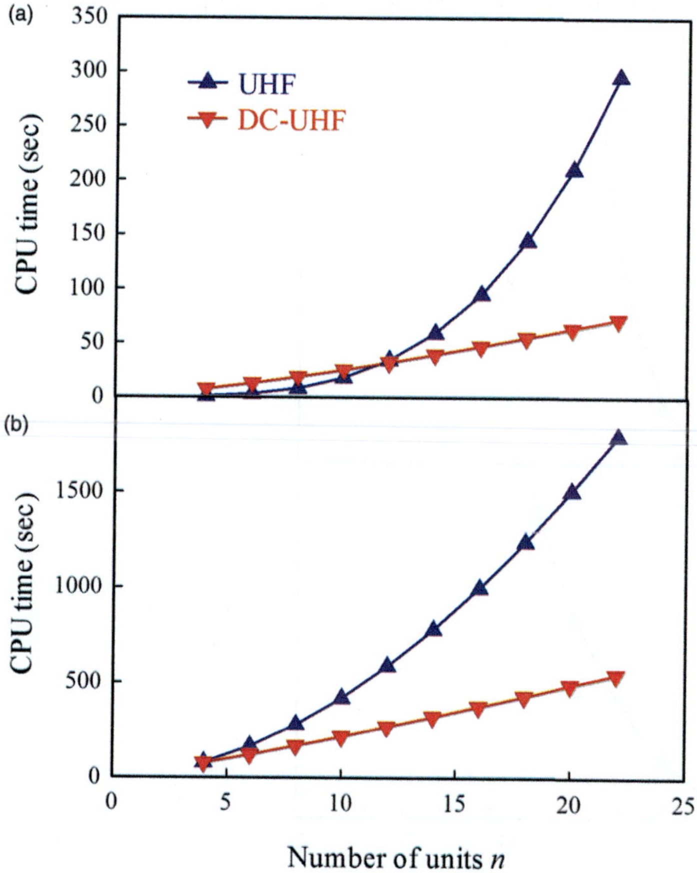

Figure 10.4 System-size dependence of the CPU times for (a) solving the SCF equations and (b) constructing the Fock matrix in the first SCF iteration of the DC and conventional UHF calculations of triplet oligothiophenes, $H(C_4H_2S)_nH$ (n = 4–22), with the 6-31G** basis set. An Intel Xeon X5470 (3.33 GHz) processor was used on a single core. The buffer region was fixed at $n_b = 3$ in the DC-UHF calculations. The FMM option implemented in the GAMESS program was switched on.

constructing the Fock matrix in the first SCF iteration. In these calculations, the fast multipole method (FMM) implemented in the GAMESS program was switched on. Although the Fock matrix construction scheme for the DC calculations is the same as that for the conventional calculations, the DC-UHF times are always shorter than the conventional UHF times. Furthermore, the DC-UHF method reduces the order of this step; the time scale with $O(n^{1.85})$ and $O(n^{1.15})$ for conventional and DC-UHF calculations, respectively. The combination of the DC method with FMM works considerably well and achieves a linear-scaling computational time because the sparse density matrix of the DC-UHF calculation makes the screening more efficient than that of conventional UHF.

The CPU times and required memory sizes of the conventional and DC-UMP2 calculations are plotted in Figures 10.5a and 10.5b, respectively. The DC-UMP2 method drastically reduces the CPU time from the conventional UMP2 time. According to the scaling analysis by the double-logarithmic plot, the CPU times scale with $O(n^{5.56})$ and $O(n^{1.33})$ for the conventional and DC-UMP calculations, respectively, which

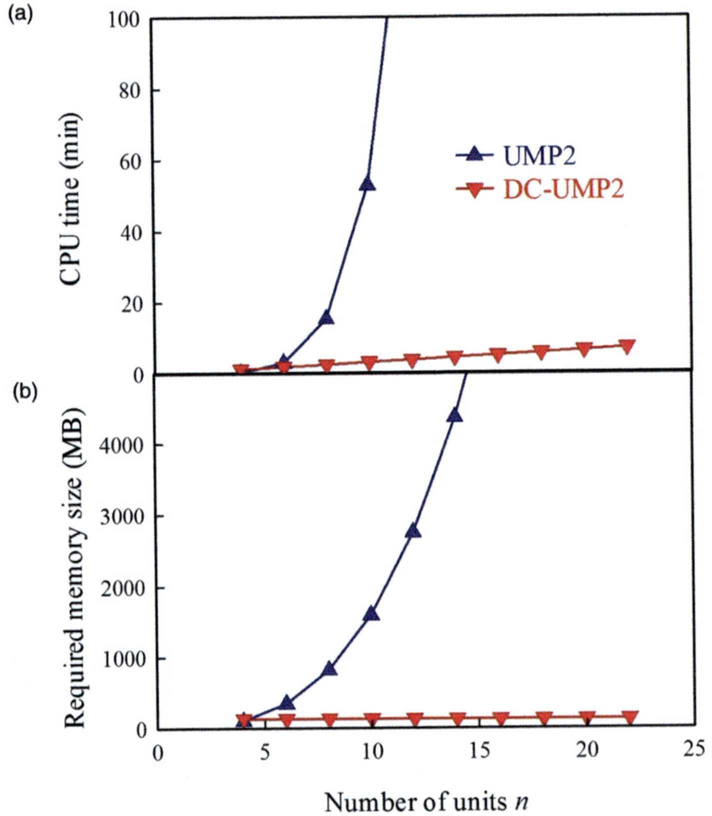

Figure 10.5 System-size dependence of (a) the CPU times and (b) the required memory size of the DC and conventional UMP2 calculations of triplet oligothiophenes, $H(C_4H_2S)_nH$ ($n = 4$–22), with the 6-31G** basis set. An Intel Xeon X5470 (3.33 GHz) processor was used on a single core. The buffer region was fixed at $n_b = 1$ in the DC-UMP2 calculations.

reasonably agree with the theoretical asymptotic values of $O(n^5)$ and $O(n^1)$. For $n = 4$, the required memory size for the DC-UMP2 calculation is larger than that for the conventional UMP2 calculation because the evaluation of the correlation energy corresponding to the central region requires additional memories than the conventional UMP2 calculation. However, the memory size required for the DC-UMP2 calculation is constant with respect to the system size n, while that for the conventional calculation increases roughly proportionally to n^3.

10.3.2 Excited-State Calculations based on the Divide-and-Conquer Method

In this section, the performance of the present DC-CIS, DC-TDDFT, and DC-SACCI methods is numerically assessed by comparing the results with those of conventional CIS, TDDFT, and SACCI calculations. All CIS and TDDFT calculations were performed using a modified version of the GAMESS program, while SACCI calculations were performed using a modified version of the Gaussian09 program [79]. In this section, only the excited-state calculations were performed using the DC scheme following a conventional calculation of the ground states. All chemical core orbitals were frozen during the excited-state calculations. For the TDDFT calculations, the long-range corrected Becke–Lee–Yang–Parr (LCBLYP) exchange-correlation functional was adopted, unless otherwise noted. The perturbation selection with a LevelThree threshold in the Gaussian program was used for the SACCI calculations.

10.3.2.1 Conjugated Aldehyde

Here, we explain the application of the DC-based excited-state theory to the n–π^* excited-state calculations of a conjugated aldehyde, $C_{16}H_{17}CHO$, which is depicted in Figure 10.6. The C=C, C-C, C=O, and C-H lengths were fixed at 1.36, 1.46, 1.22, and 1.10 Å, respectively, and all atoms were in the same plane. The ES consisted of the terminal CHO group and several C_2H_2 units, the number of which was denoted by n_{ex} (see Figure 10.6, which shows an example for $n_{ex} = 2$). Figure 10.6 shows the ES-size (n_{ex}) dependence of the n–π^* excitation energies obtained from the DC-CIS, TDDFT, and SACCI calculations of $C_{16}H_{17}CHO$ using the 6-31G** basis set. The dashed lines in Figure 10.7 represent the conventional CIS, TDDFT, and SACCI excitation energies. For all three methods, the results of the DC-based excited-state calculations show a smooth convergence with respect to the ES size and deviate by an absolute value of 0.12 eV or less for the ES size of $n_{ex} \geq 3$.

The efficiencies of the DC-CIS and TDDFT methods were determined by measuring the CPU time required using an Intel Xeon X5470 (3.33 GHz) processor on a single

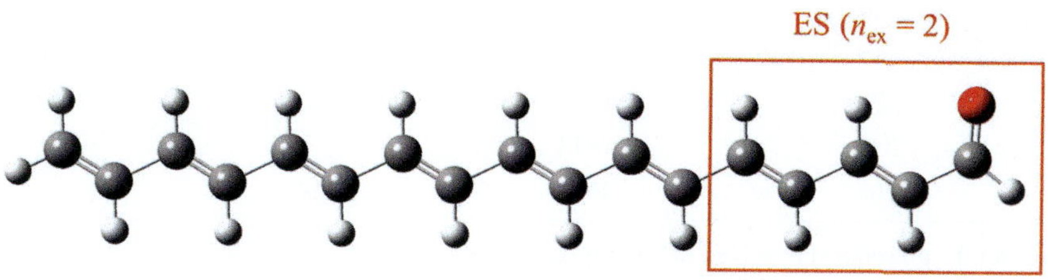

Figure 10.6 Structure of the $C_{16}H_{17}CHO$ molecule and schematic of the ES in the DC calculation.

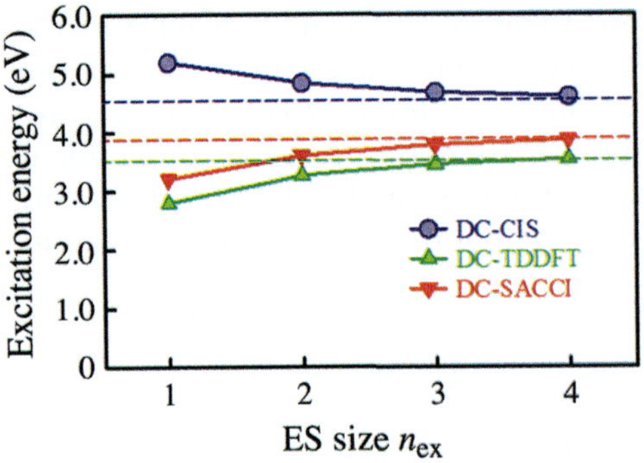

Figure 10.7 ES-size (n_{ex}) dependence of the excitation energies obtained using DC-CIS, TDDFT, and SACCI calculations of $C_{16}H_{17}CHO$ with the 6-31G** basis set. The conventional CIS, TDDFT, and SACCI excitation energies are indicated by dashed lines.

core. The Davidson algorithm was adopted to diagonalize the Hamiltonian matrix; the maximum number of expansion vectors used by the solver's iterations was set at 50. The lowest 10 singlet excited states were obtained. Table 10.3 shows the CPU times for the DC-CIS and TDDFT calculations for $C_{16}H_{17}CHO$ using the 6-31G** basis set. The times required for the preceding HF or DFT calculation are not included. The DC method drastically reduced the required computational time with only a slight energy error.

Table 10.4 shows the basis-set dependence of the n–π^* excitation energies obtained from the DC-CIS, TDDFT, and SACCI calculations. The 6-31G, 6-311G [80], 6-311G** [80], cc-pVDZ [81], and cc-pVTZ [81] basis sets were adopted in addition to the 6-31G** set; note that conventional SACCI with the cc-pVTZ basis set could not be performed due to the large computational cost. The ES size was fixed at $n_{ex} = 3$. In Table 10.4, the deviations of the DC excitation energies from those obtained using conventional methods are presented in parentheses. The energy deviations do not show a significant dependence on the adopted basis set and are quite small at a maximum of 0.15 eV. Therefore, the use of a larger basis set does not reduce the effectiveness of the present method.

Finally, the accuracies of the DC-TDDFT excitation energies obtained using typical functionals, that is, SVWN, BLYP, B3LYP, and LCBLYP, were examined. Table 10.5 shows the functional dependence of the n–π^* excitation energies obtained from DC (with $n_{ex} = 3$) and conventional TDDFT methods. The excitation energies obtained using the pure DFT functionals (i.e., SVWN and BLYP) were lower than those obtained using

Table 10.3 CPU times (in seconds) of the DC-CIS and TDDFT calculations of $C_{16}H_{17}CHO$ using the 6-31G** basis set.

n_{ex}	CIS	TDDFT
1	14	114
2	49	402
3	179	1290
4	242	1944
Conv.	708	4530

The times required for the preceding HF or DFT calculations are not included. An Intel Xeon X5470 (3.33 GHz) processor was used on a single core.

Table 10.4 Basis-set dependence of the n–π^* excitation energy (in eV) of $C_{16}H_{17}CHO$ obtained using DC and conventional CIS, TDDFT, and SACCI calculations.

Basis set	CIS	DC-CIS	(diff.)	TDDFT	DC-TDDFT	(diff.)	SACCI	DC-SACCI	(diff.)
6-31G	4.41	4.34	(−0.07)	3.48	3.39	(−0.09)	3.85	3.82	(−0.03)
6-31G**	4.64	4.58	(−0.06)	3.53	3.50	(−0.03)	3.87	3.76	(−0.11)
6-311G	4.41	4.31	(−0.10)	3.48	3.63	(+0.15)	3.84	3.80	(−0.04)
6-311G**	4.61	4.49	(−0.12)	3.51	3.49	(−0.02)	3.79	3.69	(−0.10)
cc-pVDZ	4.59	4.51	(−0.08)	3.49	3.46	(−0.02)	3.84	3.76	(−0.08)
cc-pVTZ	4.63	4.58	(−0.06)	3.50	3.46	(−0.04)	—	3.78	—

The energy deviations from the conventional results are shown in parentheses. The ES size of $n_{ex} = 3$ was adopted for the DC calculations.

wavefunction-based correlated SACCI, that is, 3.87 and 3.76 eV by the conventional and DC-SACCI methods, respectively (Table 10.4). The hybrid DFT functionals (i.e., B3LYP and LCBLYP) provided closer results. The results from TDHF, which are listed in Table 10.5, overestimated the excitation energy compared to the SACCI results. For all of the functionals, the DC-TDDFT calculations reproduced the excitation energies obtained using the conventional method within 0.15 eV.

10.3.2.2 Photoactive Yellow Protein

The DC-based excited-state theory was applied to the photoactive yellow protein (PYP); the crystal structure, which is depicted in Figure 10.8a, is registered under the accession code 2PHY in the Protein Data Bank [82]. PYP is a photoreceptor protein that comprises 125 amino acid residues and a pigment, which is deprotonated 4-hydroxycinnamic acid (HC4) linked with Cys69 by a thioester bond. PYP shows an absorption maximum at 446 nm (2.78 eV), which corresponds to the lowest π–π^* excitation energy from HOMO to LUMO, while the absorption maximum of HC4 is at 284 nm (4.37 eV) in water [83]. Thus, a large red shift of 1.59 eV is induced via binding of the pigment to the apoprotein.

Table 10.5 DFT functional dependence of the n–π^* excitation energy (in eV) of a conjugated aldehyde $C_{16}H_{17}CHO$ obtained using DC and conventional TDDFT methods.

Functional	Conv.	DC	(diff.)
SVWN	2.15	2.30	(+0.15)
BLYP	2.31	2.45	(+0.14)
B3LYP	3.17	3.15	(−0.03)
LCBLYP	3.53	3.50	(−0.03)
HF	4.48	4.42	(−0.06)

The deviations of the energies from the conventional results are shown in parentheses. An ES size of $n_{ex} = 3$ was adopted for the DC calculations.

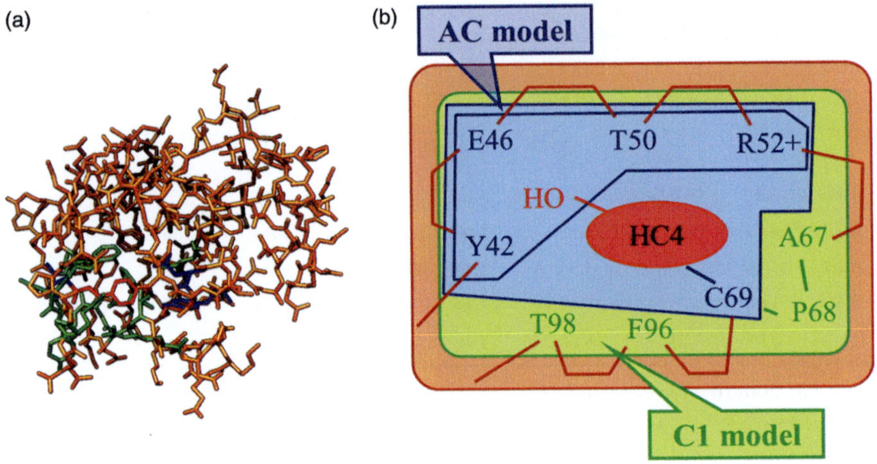

Figure 10.8 (a) Structure of PYP. (b) Schematic of the three-model systems: (i) active center (AC) model [HC4, Tyr42, Glu46, Thr50, Arg52, and Cys69] (blue), (ii) first coordination sphere (C1) model [AC, Ala67, Pro68, Phe96, and Thr98] (green), and (iii) the entire PYP protein (orange).

It is possible to separately quantify the effects of the pigment on the absorption maximum using quantum chemical calculations. The red shift in PYP was investigated using DC-CIS, TDDFT, and SACCI calculations. For the TDDFT calculations, the LCBLYP exchange-correlation functional was adopted. Accordingly, three models of this system, of which the schematic is depicted in Figure 10.8b, were constructed: (i) The active center (AC) model, which comprises HC4, Tyr42, Glu46, Thr50, Arg52, and Cys69, and is depicted in blue, (ii) the first coordination sphere (C1) model, which comprises AC, Ala67, Pro68, Phe96, and Thr98, and is depicted in green, and (iii) the entire PYP protein, which is depicted in orange. The 6-31G basis set was adopted for the following calculations.

First, the $\pi-\pi^*$ excitation energy of a HC4 molecule in water solvent was evaluated. The solvent effect was included using the integral equation formulation of the polarizable continuum model [84]. The excitation energies calculated using the conventional CIS, TDDFT, and SACCI methods are given in Table 10.6 together with the

Table 10.6 Excitation energies (in eV) of the HC4, AC, and PYP models.

Method	HC4	(diff.)	AC	(diff.)	PYP	(diff.)
CIS	4.64	(+0.27)	4.40	(+1.61)	4.33	(+1.55)
TDDFT	4.60	(+0.23)	3.83	(+1.06)	3.38	(+0.60)
SACCI	4.52	(+0.15)	3.59	(+0.81)	—	—
Exptl.	4.37		2.78		2.78	

The HC4 and AC models were calculated using the conventional method, and the PYP model was calculated using the ONIOM scheme. The deviations of the excitation energies from the experimental values are shown in parentheses.

experimental results. The differences between the calculated and experimental excitation energies are presented in parentheses. Although the SACCI method gives the best result, with 0.15 eV deviation from the experimental value, the deviations of the CIS and TDDFT results are also small, that is, <0.3 eV.

Table 10.6 shows the excitation energies for the AC model obtained using conventional CIS, TDDFT, and SACCI methods. In contrast to HC4, the excitation-energy results significantly depend on the methodology. Note that all the excitation energies exhibit red shifts from those of HC4, which are caused by the formation of the thioester bond with Cys69 and deprotonation of the OH group in HC4. The SACCI method, which gives the lowest excitation energy, still has an excitation-energy deviation of 0.81 eV. Therefore, it is evident that the simple AC model could not sufficiently reproduce the experimental shift; this is due to the lack of interactions with the outside residues.

The effects of the outside residues were included using the ONIOM method [85] with the Gaussian09 program; this method has been widely used for calculating biomolecules in quantum chemistry. The AC model was adopted as the high-level layer where the CIS and TDDFT methods were applied (note that ONIOM-SACCI is not implemented in Gaussian). The other atoms in the PYP system were treated using the universal force field [86]. The calculated and experimental excitation energies are shown in Table 10.6. Although the excitation energies obtained using the ONIOM method are red-shifted from the results of the AC model, they still did not reach the experimental value. The ONIOM method with a classical electrostatic field for the low-layer region also could not reproduce the experimental value.

Next, DC-CIS, TDDFT, and SACCI calculations were performed on the C1 and PYP model systems; the results are given in Table 10.7. The AC region is adopted as the ES. Although the excitation energies of the C1 model are red-shifted from the results for the AC model (Table 10.6), they still do not match the experimental value. However, the results for the whole PYP model with TDDFT and SACCI are close to the experimental results; in particular, the DC-SACCI method reproduces the excitation energy within 0.11 eV. Comparing the DC result with the ONIOM results given in Table 10.6, the DC-TDDFT treatment of the entire PYP system leads to an additional 0.23 eV shift from the ONIOM-TDDFT excitation energy. Therefore, the inclusion of the classical electrostatic effects from the outside residues is not enough to reproduce the biochemical excitation shift, but the quantum-mechanical treatment of the whole PYP system using a highly accurate excited-state method is required.

Table 10.7 Excitation energies (in eV) of the C1 and PYP models calculated using the DC scheme.

Method	C1	(diff.)	PYP	(diff.)
CIS	4.22	(+1.44)	4.23	(+1.46)
TDDFT	3.52	(+0.74)	3.15	(+0.37)
SACCI	3.29	(+0.52)	2.89	(+0.11)
Exptl.	2.78		2.78	

The deviations of the excitation energy from the experimental values are shown in parentheses.

Table 10.8 CPU times (in minutes) for conventional and DC-SACCI calculations of the AC, C1, and PYP models.

Model	CPU time	N_{gr}	N_{ex}
AC	897.0	340013	444369
C1	852.6	318392	326815
PYP	662.4	305349	274343

An Intel Xeon X5470 (3.33 GHz) processor was used on a single core. The numbers of the configurations selected in the ground- and excited-state, N_{gr} and N_{ex}, respectively, are listed together.

The efficiency of the DC-SACCI method was examined by measuring the required CPU time using an Intel Xeon X5470 (3.33 GHz) processor on a single core. Table 10.8 shows the CPU times for a conventional SACCI calculation of the AC model and for DC-SACCI calculations of the C1 and PYP models. The CPU time remained nearly constant because the SACCI calculation was only performed on the same ES region, that is, the AC region. The SAC/SACCI method introduces the perturbative configuration selection for reducing the computational time. The number of configurations selected in the ground- and excited-state calculations, N_{gr} and N_{ex}, respectively, are listed in Table 10.8. The number of configurations selected in the calculation of the AC model is larger than that of the C1 and PYP models because 10 hydrogen atoms were added at the cleavages of the covalent bonds that formed when constructing the active-site model. The DC-SACCI methodology enabled to accelerate the excited-state calculations while maintaining a high accuracy.

10.4 Conclusion

In this chapter, we reported a theoretical extension of the DC method for accurate and practical calculations for nanomaterials and biomolecules. The DC-SCF method was extended to spin-unrestricted open-shell calculations. Numerical assessments revealed that the accuracy of the DC-UHF method was comparable to that of the closed-shell DC-HF method: the energy and spin-density errors were reduced along with the buffer size. Non-integer charge and spin densities in each subsystem were also confirmed. Furthermore, the CPU time for the DC-UHF calculation scaled linearly with respect to the system size by combining with the FMM. The correlated open-shell treatment was enabled in the framework of the DC-based correlation method. The present DC-UMP2 method was assessed in calculations of the spin- and charge-delocalized polyene cation systems. Numerical assessments revealed that the DC-UMP2 method has the advantageous features of the closed-shell DC-MP2 method. The correlation energy errors are generally small, achieving chemical accuracy, and are controllable with the buffer size; the CPU time scales quasi-linearly with respect to the system size and the required memory size becomes constant.

Furthermore, we explained a theoretical extension for excited-state calculations of the DC method. The DC-based SAC method was developed as a good starting point for excited-state calculations. In the numerical assessments of the present method, the local orbitals constructed in the DC method improved the performance of the configuration selection in the SAC calculations, and the computational time scaled quasi-linearly with respect to the system size. A DC-based SACCI theory in conjunction with the DC-CIS theory and DC-TDDFT was developed. The excitation energy errors are generally small, indicating that a high chemical accuracy was achieved. The present scheme enabled efficient excited-state calculations of large systems.

References

1 *Linear-Scaling Techniques in Computational Chemistry and Physics: Methods and Applications*, ed. R. Zaleśny, M. G. Papadopoulos, P. G. Mezey, J. Leszczynski, Springer, Dordrecht, 2011.
2 *Computational Methods for Large Systems: Electronic Structure Approaches for Biotechnology and Nanotechnology*, ed. J. R. Reimers, John Wiley & Sons, Hoboken, 2011.
3 C. A. White, M. Head-Gordon, *J. Chem. Phys.* 1994, **101**, 6593.
4 M. C. Strain, G. E. Scuseria, M. J. Frisch, *Science* 1996, **271**, 51.
5 C. H. Choi, K. Ruedenberg, M. S. Gordon, *J. Comput. Chem.* 2001, **22**, 1484.
6 L. Füsti-Molnár, P. Pulay, *J. Chem. Phys.* 2002, **117**, 7827.
7 Y. Kurashige, T. Nakajima, K. Hirao, *J. Chem. Phys.* 2007, **126**, 144106.
8 M. A. Watson, Y. Kurashige, T. Nakajima, K. Hirao, *J. Chem. Phys.* 2008, **128**, 054105.
9 E. Schwegler, M. Challacombe, *J. Chem. Phys.* 1996, **105**, 2726.
10 J. C. Burant, G. E. Scuseria, M. J. Frisch, *J. Chem. Phys.* 1996, **105**, 8969.
11 C. Ochsenfeld, C. A. White, M. Head-Gordon, *J. Chem. Phys.* 1998, **109**, 1663.
12 S. Goedecker, *Rev. Mod. Phys.* 1999, **71**, 1085.
13 S. Y. Wu, C. S. Jayanthi, *Phys. Rep.* 2002, **358**, 1.
14 F. Mauri, G. Galli, R. Car, *Phys. Rev. B: Condens. Matter Mater. Phys.* 1993, **47**, 9973.
15 P. Ordejon, D. A. Drabold, M. P. Grumbach, R. M. Martin, *Phys. Rev. B: Condens. Matter Mater. Phys.* 1993, **48**, 14646.
16 J. Kim, F. Mauri, G. Galli, *Phys. Rev. B: Condens. Matter Mater. Phys.* 1995, **52**, 1640.
17 G. Galli, M. Parrinello, *Phys. Rev. Lett.* 1992, **69**, 3547.
18 X.-P. Li, R. W. Nunes, D. Vanderbilt, *Phys. Rev. B: Condens. Matter Mater. Phys.* 1993, **47**, 10891.
19 G. E. Scuseria, *J. Phys. Chem. A* 1999, **103**, 4782.
20 A. M. N. Niklasson, M. Challacombe, *Phys. Rev. Lett.* 2004, **92**, 193001.
21 J. Almlöf, *Chem. Phys. Lett.* 1991, **181**, 319.
22 M. Häser, *Theor. Chim. Acta* 1993, **87**, 147.
23 P. Y. Ayala, G. E. Scuseria, *J. Chem. Phys.* 1999, **110**, 3660.
24 P. R. Surján, *Chem. Phys. Lett.* 2005, **406**, 318.
25 M. Kobayashi, H. Nakai, *Chem. Phys. Lett.* 2006, **420**, 250.
26 P. Pulay, *Chem. Phys. Lett.* 1983, **100**, 151.
27 P. Pulay, S. Saebø, *Theor. Chim. Acta* 1986, **69**, 357.
28 S. Saebø, P. Pulay, *Annu. Rev. Phys. Chem.* 1993, **44**, 213.
29 S. Saebø, P. Pulay, *J. Chem. Phys.* 2001, **115**, 3975.
30 M. Schütz, G. Hetzer, H.-J. Werner, *J. Chem. Phys.* 1999, **111**, 5691.
31 P. E. Maslen, M. Head-Gordon, *Chem. Phys. Lett.* 1998, **283**, 102.

32 H.-J. Werner, F. R. Manby, P. J. Knowles, *J. Chem. Phys.* 2003, **118**, 8149.
33 C. Hampel, H.-J. Werner, *J. Chem. Phys.* 1996, **104**, 6286.
34 J. E. Subotnik, M. Head-Gordon, *J. Phys.: Condens. Matter* 2008, **20**, 294211.
35 W. Yang, *Phys. Rev. Lett.* 1991, **66**, 1438.
36 W. Yang, *Phys. Rev. A: At., Mol., Opt. Phys.* 1991, **44**, 7823.
37 Q. Zhao, W. Yang, *J. Chem. Phys.* 1995, **102**, 9598.
38 W. Yang, T.-S. Lee, *J. Chem. Phys.* 1995, **103**, 5674.
39 P. D. Walker, P. G. Mezey, *J. Am. Chem. Soc.* 1993, **115**, 12423.
40 P. G. Mezey, *J. Math. Chem.* 1995, **18**, 141.
41 T.-S. Lee, D. M. York, W. Yang, *J. Chem. Phys.* 1996, **105**, 2744.
42 D. M. York, T.-S. Lee, W. Yang, *J. Am. Chem. Soc.* 1996, **118**, 10940.
43 J. Khandogin, D. M. York, *Proteins* 2004, **56**, 724.
44 S. L. Dixon, K. M. Merz, Jr., *J. Chem. Phys.* 1996, **104**, 6643.
45 A. van der Vaart, V. Gogonea, S. L. Dixon, K. M. Merz, Jr., *J. Comput. Chem.* 2000, **21**, 1494.
46 J. M. Cabrera-Trujillo, J. Robles, *Phys. Rev. B: Condens. Matter Mater. Phys.* 2001, **64**, 165408.
47 S. K. Goh, A. St-Amant, *Chem. Phys. Lett.* 1997, **274**, 429.
48 S. K. Goh, R. T. Gallant, A. St-Amant, *Int. J. Quantum Chem.* 1998, **69**, 405.
49 D. M. Shaw, A. St-Amant, *J. Theor. Comput. Chem.* 2004, **3**, 419.
50 T. Akama, M. Kobayashi, H. Nakai, *J. Comput. Chem.* 2007, **28**, 2003.
51 T. Akama, A. Fujii, M. Kobayashi, H. Nakai, *Mol. Phys.* 2007, **105**, 2799.
52 T. Akama, M. Kobayashi, H. Nakai, *Int. J. Quantum Chem.* 2009, **109**, 2706.
53 X. He, K. M. Merz, Jr., *J. Chem. Theory Comput.* 2010, **6**, 405.
54 M. Kobayashi, T. Akama, H. Nakai, *J. Chem. Phys.* 2006, **125**, 204106.
55 M. Kobayashi, Y. Imamura, H. Nakai, *J. Chem. Phys.* 2007, **127**, 074103.
56 M. Kobayashi, H. Nakai, *Int. J. Quantum Chem.* 2009, **109**, 2227.
57 M. Kobayashi, H. Nakai, *J. Chem. Phys.* 2008, **129**, 044103.
58 M. Kobayashi, H. Nakai, *J. Chem. Phys.* 2009, **131**, 114108.
59 M. Kobayashi, T. Yoshikawa, H. Nakai, *Chem. Phys. Lett.* 2010, **500**, 172.
60 T. Yoshikawa, M. Kobayashi, H. Nakai, *Theor. Chem. Acc.* 2011, **130**, 411.
61 T. Touma, M. Kobayashi, H. Nakai, *Chem. Phys. Lett.* 2010, **485**, 247.
62 T. Touma, M. Kobayashi, H. Nakai, *Theor. Chem. Acc.* 2011, **130**, 701.
63 M. Kobayashi, T. Touma, H. Nakai, *J. Chem. Phys.* 2012, **136**, 084108.
64 T. Yoshikawa, M. Kobayashi, A. Fujii, H. Nakai, *J. Phys. Chem. B* 2013, **117**, 5565.
65 J. Seino, H. Nakai, *J. Chem. Phys.* 2013, **139**, 034109.
66 H. Nishizawa, Y. Nishimura, M. Kobayashi, S. Irle, H. Nakai, *J. Comput. Chem.* 2016, **37**, 1983.
67 H. Nakai, *Chem. Phys. Lett.* 2002, **363**, 73.
68 A. I. Krylov, *J. Chem. Phys.* 2000, **113**, 6052.
69 M. W. Schmidt, K. K. Baldridge, J. A. Boatz, S. T. Elbert, M. S. Gordon, J. H. Jensen, S. Koseki, N. Matsunaga, K. A. Nguyen, S. Su, T. L. Windus, M. Dupuis, J. A. Montgomery Jr., *J. Comput. Chem.* 1993, **14**, 1347.
70 W. J. Hehre, R. Ditchfield, J. A. Pople, *J. Chem. Phys.* 1972, **56**, 2257.
71 P. C. Hariharan, J. A. Pople, *Theor. Chim. Acta* 1973, **28**, 213.
72 M. M. Francl, W. J. Petro, W. J. Hehre, J. S. Binkley, M. S. Gordon, D. J. DeFrees, J. A. Pople, *J. Chem. Phys.* 1982, 77, 3654.

73 A. D. Becke, *Phys. Rev. A* 1988, **38**, 3098.
74 C. Lee, W. Yang, R. G. Parr, *Phys. Rev. B* 1988, **37**, 785.
75 P. J. Stephens, F. J. Devlin, C. F. Chabalowski, M. J. Frisch, *J. Phys. Chem.* 1994, **98**, 11623.
76 A. D. Becke, *J. Chem. Phys.* 1993, **98**, 1372.
77 T. Tsuneda, T. Suzumura, K. Hirao, *J. Chem. Phys.* 1999, **110**, 10664.
78 H. Iikura, T. Tsuneda, T. Yanai, K. Hirao, *J. Chem. Phys.* 2001, **115**, 3540.
79 M. J. Frisch, G. W. Trucks, H. B. Schlegel, G. E. Scuseria, M. A. Robb, J. R. Cheeseman, G. Scalmani, V. Barone, B. Mennucci, G. A. Petersson, H. Nakatsuji, M. Caricato, X. Li, H. P. Hratchian, A. F. Izmaylov, J. Bloino, G. Zheng, J. L. Sonnenberg, M. Hada, M. Ehara, K. Toyota, R. Fukuda, J. Hasegawa, M. Ishida, T. Nakajima, Y. Honda, O. Kitao, H. Nakai, T. Vreven, J. A. Montgomery, Jr. J. E. Peralta, F. Ogliaro, M. Bearpark, J. J. Heyd, E. Brothers, K. N. Kudin, V. N. Staroverov, R. Kobayashi, J. Normand, K. Raghavachari, A. Rendell, J. C. Burant, S. S. Iyengar, J. Tomasi, M. Cossi, N. Rega, J. M. Millam, M. Klene, J. E. Knox, J. B. Cross, V. Bakken, C. Adamo, J. Jaramillo, R. Gomperts, R. E. Stratmann, O. Yazyev, O. A. J. Austin, R. Cammi, C. Pomelli, J. W. Ochterski, R. L. Martin, K. Morokuma, V. G. Zakrzewski, G. A. Voth, P. Salvador, J. J. Dannenberg, S. Dapprich, A. D. Daniels, O. Farkas, J. B. Foresman, J. V. Ortiz, J. Cioslowski, D. J. Fox, Gaussian 09, Revision A.02; Gaussian, Inc: Wallingford, CT, USA, 2009.
80 R. Krishnan, J. S. Binkley, R. Seeger, J. A. Pople, *J. Chem. Phys.* 1980, **72**, 650.
81 T. H. Dunning, Jr., *J. Chem. Phys.* 1989, **90**, 1007.
82 http://www.rcsb.org/
83 M. Yoda, H. Houjou, Y. Inoue, M. Sakurai, *J. Phys. Chem. B* 2001, **105**, 9887.
84 G. Scalmani. M. J. Frisch, *J. Chem. Phys.* 2006, **124**, 094107.
85 T. Vreven, K. Morokuma, *Theor. Chem. Acc.* 2003, **109**, 125.
86 A. K. Pappe, C. J. Casewit, K. S. Colwell, W. A. G. III, W. M. Skiff, *J. Am. Chem. Soc.* 1992, **114**, 10024.

11

MFCC-Based Fragmentation Methods for Biomolecules

Jinfeng Liu,[1] Tong Zhu,[1,2] Xiao He,[1,2] and John Z. H. Zhang[1,2,3]

[1] *School of Chemistry and Molecular Engineering, East China Normal University, China*
[2] *NYU-ECNU Center for Computational Chemistry, NYU Shanghai, China*
[3] *Department of Chemistry, New York University, USA*

11.1 Introduction

At present, it is still impractical to apply conventional full system quantum mechanical (QM) methods to biomolecules due to computational scaling problem. Accurate and efficient quantum calculations for biomolecules present a grand challenge to theoretical and computational chemists [1]. Over the past two decades, much effort has been devoted to the development of various linear-scaling and fragmentation methods to extend the applicability of rigorous QM methods to large systems [2–8]. Owing to the computational efficiency and easy implementation, the development of fragmentation methods has attracted significant interest over the past decade, and these methods have become highly efficient for linear-scaling QM calculations of large molecular systems [9, 10].

The fragmentation approach is based on the chemical locality of macromolecular systems, which assumes that the local electronic property of a macromolecule is only weakly influenced by atoms that are far from the region. Based on this chemically intuitive approach, a macromolecule can be divided into several subsystems (fragments) according to the specific fragmentation scheme and energies of the fragments are separately computed. The total energy or other molecular properties of the entire system is then approximately obtained by taking proper linear combinations of the corresponding terms of individual fragments.

The fragmentation approach is more attractive as a practical tool for electronic structure calculation of large systems than conventional linear-scaling methods in several aspects, such as easy implementation of parallelization without extensively modifying the existing QM programs and straightforward application at all levels of *ab initio* electronic structure theories. A range of fragmentation QM methods for large systems have been proposed, including the fragment molecular orbital (FMO) method [11–13], the molecular fractionation with conjugate caps (MFCC) approach [14–20], the systematic fragmentation method (SFM) [21–23], the adjustable density matrix assembler (ADMA) approach [24–26], the molecular tailoring approach (MTA) [27–29], the generalized energy-based fragmentation (GEBF) method [30, 31], the

Fragmentation: Toward Accurate Calculations on Complex Molecular Systems, First Edition.
Edited by Mark S. Gordon.

electrostatically embedded many-body (EE-MB) expansion approach [32–34], the explicit polarization (X-Pol) potential [35, 36], the combined fragmentation method (CFM) [37], the molecules-in-molecules (MIM) approach [38], many-overlapping-body (MOB) expansion approach [39], the generalized many body expansion (GMBE) approach [40, 41], and the automated fragmentation quantum mechanics/molecular mechanics (AF-QM/MM) method [42–44]. A comprehensive review of the fragmentation QM methods can be found in two recent review articles written by Gordon *et al.* [10], and Collins and Bettens [45], respectively.

In this chapter, we describe the MFCC-based fragmentation methods and their applications to biological systems. The MFCC approach was originally proposed by Zhang and Zhang in 2003 and it was initially used to compute protein–ligand interaction energy in gas phase [14]. Subsequently, the MFCC approach was further expanded to compute the full electron density of the system and the total energy of protein at the DFT level [17, 18]. To make the MFCC approach more accurate and efficient for applications in geometry optimization and molecular dynamics simulation of proteins, the generalized molecular fractionation with conjugate caps/molecular mechanics (GMFCC/MM) method was developed for calculating total energies of biomolecules. In the GMFCC/MM method, molecular force field interactions were introduced to calculate long-range interaction between distant non-neighboring residues (the two residues are not sequentially connected to each other), whereas the short-range non-neighboring fragment interactions are calculated by quantum mechanics [16].

To account for the environmental polarization effect for each fragment calculation, electrostatic embedding was introduced into the GMFCC/MM method which became the latest electrostatically embedded generalized molecular fractionation with conjugate caps (EE-GMFCC) method for more accurate calculation of the protein energy. The EE-GMFCC method calculates the total energy of the protein system by taking a linear combination of fragment-based energies of neighboring residues and interaction energies between non-neighboring residues that are spatially in close contact, with a key element being that all of the calculated fragments are embedded in a field of point charges to mimic the electrostatic environment of the remaining system [19]. The EE-GMFCC method is computationally efficient and linear-scaling with a low prefactor, and has been successfully applied to protein geometry optimization [46], molecular dynamics simulation [47], protein–ligand binding affinity calculation [48], and protein vibrational spectrum calculation at QM levels [46].

11.2 Theory and Applications

11.2.1 The MFCC Approach

The MFCC approach was initially developed to provide efficient and linear-scaling QM calculations of protein–ligand interaction energies. The main idea of the MFCC approach is to divide a protein molecule into amino acid fragments by cutting off covalent bonds connecting amino acids (see Figure 11.1) [14, 49–51]. A pair of conjugate caps (concaps) at the cutting location are introduced to saturate the dangling bonds and also to mimic the local chemical environment of the original protein to the cutoff fragments. Using this fragmentation scheme, the interaction energy between the protein and ligand

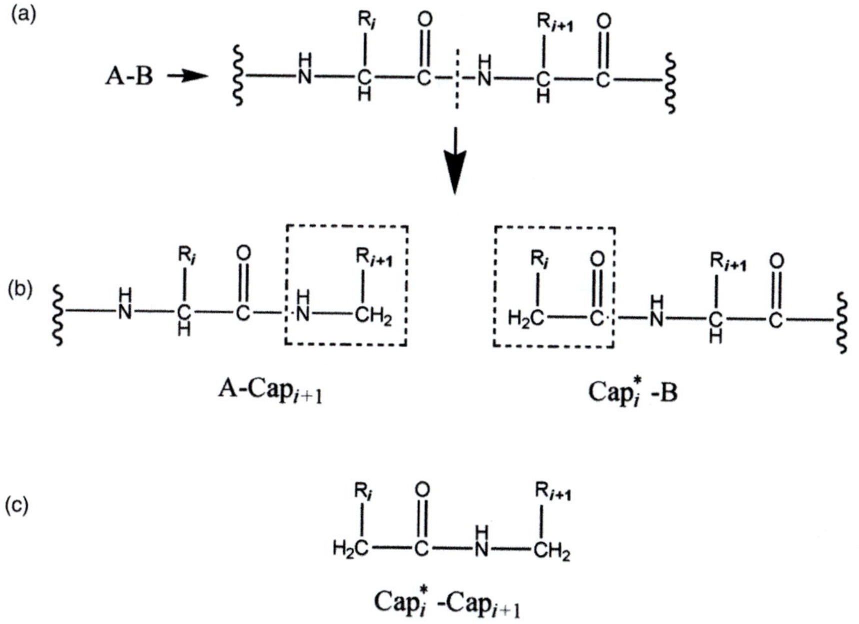

Figure 11.1 The MFCC scheme in which the peptide bond is cut in (a) and the fragments are capped with Cap_{i+1} and the conjugate Cap_i^* in (b), where subscript *i* represents the index of the *i*th amino acid in the given protein; the atomic structure of concap is shown in (c). The concap is defined as the fused molecular species of $Cap_i^* - Cap_{i+1}$. *Source*: He *et al.* 2014 [20]. Reprinted with permission of American Chemical Society.

can be calculated by separate calculations of individual fragments interacting with the ligand. Furthermore, the pair of concaps is fused to form proper molecular species such that the doubly counted interaction energy between caps of the fragments (also called as overlapping fragments) and the ligand can be thoroughly subtracted. Hydrogen atoms are added to terminate the molecular caps to prevent dangling bonds.

In the MFCC scheme, the peptide bonds of the protein are cut to generate residue-based fragments and pairs of concaps, Cap_i^* (CH_2R_iCO-) and Cap_{i+1} ($-NHCH_2R_{i+1}$) (*i* denotes the index of the *i*th residue), are inserted at the cutting locations to cap the fragments. In this scheme, there are $N-2$ capped fragments and $N-3$ concaps for a protein containing N amino acids. The interaction energy for the protein–ligand binding system (E_{P-L}) is given by the following expression [14],

$$E_{P-L} = \sum_{k=1}^{N-2} E_{F_k-L} - \sum_{k=1}^{N-3} E_{CC_k-L} - \sum_{k=1}^{N-2} E_{F_k} + \sum_{k=1}^{N-3} E_{CC_k} - E_L, \tag{11.1}$$

where E_{F_k-L} and E_{CC_k-E} represent the total energy of the *k*th capped fragment with the ligand and the total energy of the *k*th concap with the ligand, respectively. E_{F_k} and E_{CC_k} are the self-energies of the *k*th capped fragment and *k*th concap, respectively. E_L is the energy of the ligand. In situations where the protein has additional chemical bonds-between non-neighboring residues such as disulfide bonds, additional cutting of these bonds is needed as described in Ref. [51]. The MFCC method can accurately reproduce

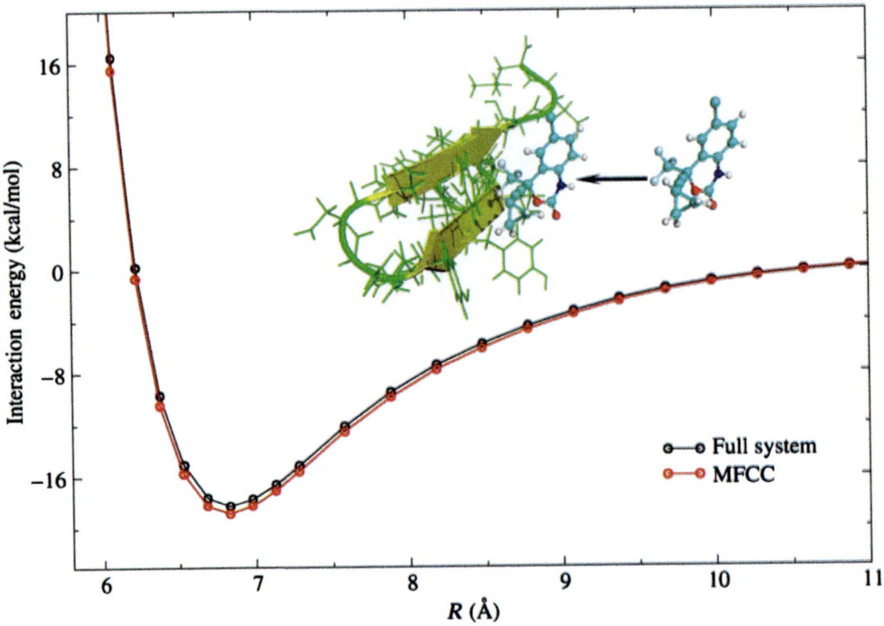

Figure 11.2 Interaction potential energy curves for Efavirenz and a fraction of HIV-1 reverse transcriptase (Asn175 to Leu193 of chain A, PDB id: 1FKO). The interaction energy is calculated at the M062X/6-311G** level along the direction from the geometric center of Efavirenz to that of the polypeptide. *Source*: He *et al.* 2014 [20]. Reprinted with permission of American Chemical Society.

the QM interaction energy of protein–ligand binding complex with low computational cost, which has been validated in previous works [52, 53]. Here, the comparison between the full system and MFCC results for calculation of the interaction energies between Efavirenz and a polypeptide which contains 19 residues (a model system extracted from HIV-1 reverse transcriptase) is shown in Figure 11.2.

11.2.2 Electron Density and Total Energy

Accurate determination of electron density and electrostatic potential of molecular systems is of both fundamental importance and practical utility. For example, the spatial distribution of density or electrostatic potential is often used to understand chemical structure, reaction, binding, catalysis, and solvation [54, 55]. The electrostatic potential is of particular interest in rational drug design for optimization of lead drug candidates and pharmacophore search [56]. It is also useful in areas such as force field parameterization [57, 58]. Furthermore, accurate determination of total electron density of proteins provides the basis for accurate QM calculation of total energy of proteins.

Based on the MFCC ansatz, the electron density ($\rho(\mathbf{r})$) of the protein at position $\mathbf{r}$ can be obtained by linear combination of individual electron densities of capped fragments [15, 59, 60],

$$\rho(\mathbf{r}) = \sum_{k=1}^{N-2} \rho_{F_k}(\mathbf{r}) - \sum_{k=1}^{N-3} \rho_{CC_k}(\mathbf{r}), \tag{11.2}$$

where ρ_{F_k} is the electron density of the kth capped fragment and ρ_{CC_k} is the electron density of the kth concap. It can be proved that the total number of electrons in equation (11.2) is conserved by integration.

$$\int \rho(\mathbf{r})d\mathbf{r} = \sum_{k=1}^{N-2} \int \rho_{F_k}(\mathbf{r})d\mathbf{r} - \sum_{k=1}^{N-3} \int \rho_{CC_k}(\mathbf{r})d\mathbf{r} = \sum_{k=1}^{N-2} N_{F_k} - \sum_{k=1}^{N-3} N_{CC_k} = N_{\text{total}} \tag{11.3}$$

However, it should be noted that equation (11.3) does not guarantee that the electron density is positive everywhere due to approximations in calculated electron densities.

After the electron density of the protein is obtained from the MFCC calculation, one can employ DFT to compute total energy of the protein using the following equation:

$$E(\rho) = T[\rho] - \sum_{\alpha} Z_\alpha \int \frac{\rho(\mathbf{r})}{|\mathbf{R}_\alpha - \mathbf{r}|} d\mathbf{r} + \frac{1}{2} \int \varphi(\mathbf{r})\rho(\mathbf{r})d\mathbf{r} + \sum_{\alpha,\beta} \frac{Z_\alpha Z_\beta}{R_{\alpha\beta}} + E_{XC}[\rho], \tag{11.4}$$

where $T[\rho]$ is the kinetic energy, $\phi(\mathbf{r})$ is the electrostatic potential (electron contribution only), and $E_{XC}[\rho]$ is the exchange-correlation energy. The kinetic energy can also be approximately obtained using the MFCC ansatz [15],

$$T[\rho] = \sum_{k=1}^{N-2} T_{F_k}[\rho_{F_k}] - \sum_{k=1}^{N-3} T_{CC_k}[\rho_{CC_k}] \tag{11.5}$$

Another approach to computing the protein energy is to construct the full density matrix (DM) of the system based on fragment density matrices. The density matrix of the molecular system can be obtained using the MFCC-DM scheme [17]:

$$P^{\mu\nu} = \sum_{k=1}^{N-2} P^{\mu\nu}_{F_k} - \sum_{k=1}^{N-3} P^{\mu\nu}_{CC_k}, \tag{11.6}$$

where $P^{\mu\nu}_{F_k}$ and $P^{\mu\nu}_{CC_k}$ are the density matrix element of the kth fragment and kth concap, respectively. After the density matrix of the full protein system is obtained, the total HF or DFT energy of the protein can be calculated directly from the full density matrix [17].

To include off-diagonal density matrix elements for the close contact interactions from non-neighboring residues, pairwise interaction correction (PIC) was introduced to MFCC-DM [18]. In the MFCC-DM-PIC approach, two non-neighboring residues which are in close contact are paired, termed as interacting unit. The PIC on the density matrix element associated with the interacting units (residue i and j) are obtained by the following relation:

$$P^{\mu\nu}_{\text{PIC}}(i-j) = P^{\mu\nu}_{ij} - P^{\mu\nu}_{i} - P^{\mu\nu}_{j} \tag{11.7}$$

and they are added to the total density matrix of the full system from the MFCC-DM calculation. The PIC corrected density matrix is then used to calculate molecular properties and total energy of the protein [18].

11.2.3 The EE-GMFCC Method for Energy Calculation

In the framework of MFCC approach [14], a given protein with N amino acids (defined as $A_1A_2A_3 \ldots A_N$) is decomposed into $N-2$ individual amino acid units by cutting through the peptide bond as illustrated in Figure 11.1. A pair of conjugate caps is inserted to preserve the local chemical environment of the remaining fragment for each separated amino acid. The MFCC energy of the protein can be approximately expressed as the sum of the capped fragments minus that of the conjugate caps [17],

$$E_{\text{MFCC}} = \sum_{i=2}^{N-1} E(\text{Cap}_{i-1}^{*}A_i\text{Cap}_{i+1}) - \sum_{i=2}^{N-2} E(\text{Cap}_i^{*}\text{Cap}_{i+1}) \tag{11.8}$$

where i denotes the index for ith residue, $E(\text{Cap}_{i-1}^{*}A_i\text{Cap}_{i+1})$ is the self-energy of the ith residue (A_i) capped with a left cap Cap_{i-1}^{*} and a right cap Cap_{i+1}, and $E(\text{Cap}_i^{*}\text{Cap}_{i+1})$ is the energy of the ith concap, respectively. The energy of concap is deducted to avoid double counting of the self-energy of the amino acids.

The MFCC approach only includes the self-energy of individual residue and interaction energy between sequentially connected tripeptides. To include the interaction energy between non-neighboring residues, a mixed quantum-classical GMFCC/MM approach for protein total energy calculations is proposed [16]. In the GMFCC/MM scheme, if the distance of any two atoms between two residues is less than a predefined distance threshold, the interaction energy between these two fragments are treated by QM, otherwise, the interaction energy is calculated using MM. As shown in Figure 11.3, we introduce a generalized concap (Gconcap) to account for the two-body QM interaction energies between sequentially non-neighboring fragments which are spatially in close contact. The energy expression of the GMFCC/MM method is given as follows [16]:

$$\begin{aligned} E_{\text{GMFCC/MM}} = E_{\text{MFCC}} &+ \sum_{\substack{i,j>i+2,\\ |\text{R}_i-\text{R}_j|\leq\lambda}} (E_{ij} - E_i - E_j)_{\text{QM}} \\ &+ \left\{ \sum_{k,l} (E_{kl} - E_k - E_l)_{\text{MM}} - \sum_{\substack{i,j>i+2,\\ |R_i-R_j|\leq\lambda}} (E_{ij} - E_i - E_j)_{\text{MM}} \right\}, \end{aligned} \tag{11.9}$$

where E_i and E_j denote the energies of the residues i and j capped with hydrogen atoms on both sides, respectively. E_{ij} is the energy of the dimer consisting of fragments i and j. λ is the distance threshold. If the distance of any two atoms between fragments i and j is less than or equal to λ ($|R_i - R_j| \leq \lambda$), the interaction energy between these two fragments are treated by QM (see Figure 11.3b), otherwise, the interaction energy is calculated by MM. k and l in equation (11.9) represent all the pairs whose interaction energies have not been included in the QM calculation of E_{MFCC}. As the conjugate caps are defined in Figure 11.1, it can be proved that $k = (\text{H})\text{Cap}_{i-1}^{*} - \text{NH}$ (where (H) denotes that the added hydrogen atom on the left is excluded) and $l = \text{CO} - A_{i+2}A_{i+3} \cdots A_N$, ($i = 2, 3, \ldots, N-2$).

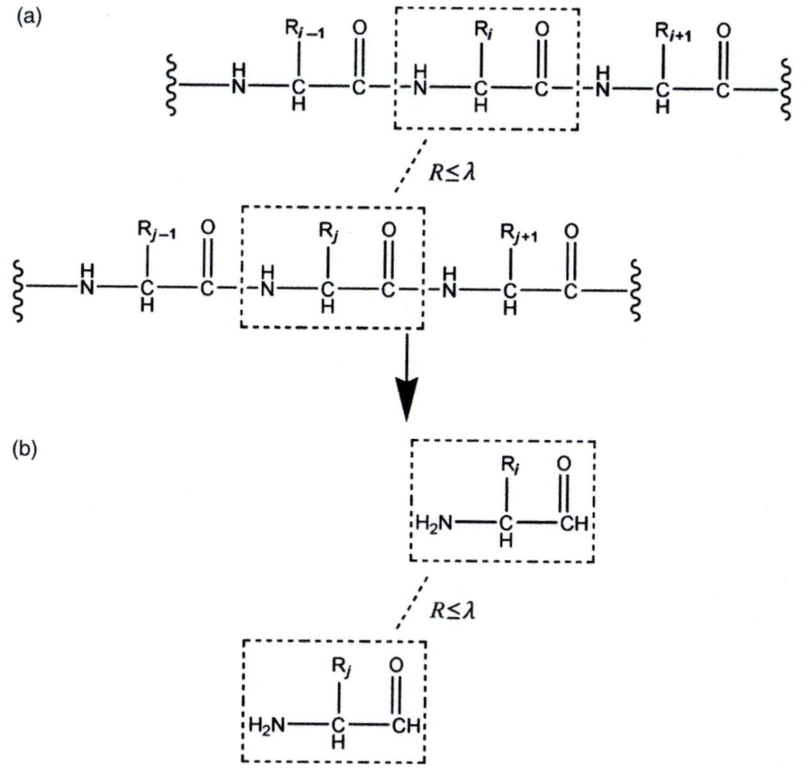

Figure 11.3 (a) The generalized concap (Gconcap) scheme, when the distance (R) between two non-neighboring residue i and j is within a distance threshold λ ($R \leq \lambda$). (b) The atomic structure of Gconcap. *Source*: He *et al.* 2014 [20]. Reprinted with permission of American Chemical Society.

However, the major drawback of the GMFCC/MM approach is that the many-body contribution to the total energy from multi-fragment interactions was neglected. To account for the electronic polarization effect arising from the remaining part of the protein for each fragment calculation, the EE-GMFCC method using electrostatic embedding was proposed, in which all the fragment calculations in GMFCC are embedded in the electrostatic field of the point charges representing the remaining fragments in the protein [19]. The total energy of protein using the EE-GMFCC approach is calculated as follows [19]:

$$E_{\mathrm{EE-GMFCC}} = \sum_{i=2}^{N-1} \tilde{E}(\mathrm{Cap}_{i-1}^{*} A_i \mathrm{Cap}_{i+1}) - \sum_{i=2}^{N-2} \tilde{E}(\mathrm{Cap}_{i}^{*} \mathrm{Cap}_{i+1}) + \sum_{\substack{i,j>i+2,\\ |R_i-R_j|\leq\lambda}} (\tilde{E}_{ij} - \tilde{E}_i - \tilde{E}_j)_{\mathrm{QM}}$$
$$- \left\{ \sum_{k,l} \sum_{m,n} \frac{q_{m(k)} q_{n(l)}}{R_{m(k)n(l)}} - \sum_{\substack{i,j>i+2,\\ |R_i-R_j|\leq\lambda}} \sum_{m',n'} \frac{q_{m'(i)} q_{n'(j)}}{R_{m'(i)n'(j)}} \right\}, \quad (11.10)$$

where $\tilde{E}$ represents the summation of the self-energy of fragment and the interaction energy between fragment and background charges of the remaining system. The other notations are similar to equation (11.9). The definition of concap and Gconcap is same as GMFCC/MM as shown in Figures 11.1 and 11.3. The term $\tilde{E}_{ij} - \tilde{E}_i - \tilde{E}_j$ represents the two-body QM interaction energy between residues i and j whose closest distance is less than a predefined threshold λ. The interaction energy doubly counted in the first three terms of equation (11.10) is deducted by the pairwise charge-charge interactions approximately in the last term, where k = (H)Cap$^*_{i-1}$ – NH((H) denotes that the added hydrogen atom on the left is excluded) and $l = \mathrm{CO} - \mathrm{A}_{i+2}\mathrm{A}_{i+3} \cdots \mathrm{A}_N$, ($i = 2, 3, \ldots, N-2$, A_i denotes the ith residue), and $q_{n(l)}$ represents the point charge of the nth atom in the fragment l.

The embedding scheme ensures the electronic polarization effect is taken into account. Through the capped fragment and Gconcap calculations, one- and two-body kinetic, Coulomb, exchange, and correlation interactions are included nearly exactly. Meanwhile, by using the embedding scheme, three- and all higher-order Coulomb effects are also included at the HF or DFT level. In the current treatment, each MFCC fragment typically has three residues, and each of the neighboring concap shown in Figure 11.1 and non-neighboring Gconcap in Figure 11.3 contains two residues. Although the total number of fragment pairs is proportional to the square of the number of residues, the number of non-neighboring residues in close contact increases linearly because each residue can only have limited number of residues in its vicinity. Because the largest fragment has just three residues and contains less than 70 atoms consisting of C, H, O, N, and S, this method can be applied to high-level *ab initio* calculation such as MP2.

Numerical studies have been carried out to test the accuracy and efficiency of the EE-GMFCC approach. As shown in Table 11.1, the total energies calculated by the EE-GMFCC approach are in excellent agreement with the full system results. For 18 globular proteins, the deviations of total energies calculated using EE-GMFCC at the HF/6-31G* level are all within 0.01 a.u. (as compared to corresponding full system calculations) when the distance threshold λ is equal to or greater than 4.0 Å. Furthermore, we varied the distance threshold to test the convergence of the EE-GMFCC method. The mean unsigned error (MUE) dropped from 3.79 to 2.39 kcal/mol when λ is increased from 3.0 to 4.0 Å; however, the MUE was little changed when λ is increased from 4.0 to 4.5 Å (2.39 vs. 2.35 kcal/mol). Figure 11.4 shows the total energies of four proteins calculated by EE-GMFCC as a function of the distance threshold. One can see from the figure that the total energies are all close to convergence at $\lambda = 4$ Å. Figure 11.5 shows the relative energy profile of 19 conformers for one globular protein (PDB id: 2KCF, 576 atoms) calculated at the HF/6-31G* level when λ was set to 4.0 Å. The conformations were extracted from molecular dynamics (MD) simulation. As shown in Figure 11.5, the relative EE-GMFCC energies agree well with the full system results.

We also applied the EE-GMFCC approach for protein energy calculations at DFT (B3LYP) and MP2 levels of theory with the 6-31G* basis set. The overall agreement between EE-GMFCC and full system calculations is also very good. The deviations of EE-GMFCC from full system results are mostly below 0.01 a.u. for diverse proteins when the distance threshold λ was set to 4.0 Å [19]. Figure 11.6 shows the comparison of CPU time between the conventional MP2 and EE-GMFCC method on the 12-core Intel Xeon E5645 2.4 GHz processor. As can be seen from the figure, the

Table 11.1 Total energies of 18 three-dimensional globular proteins obtained using standard full system HF/6-31G* calculations and EE-GMFCC HF/6-31G* calculations with different distance threshold λ.

				Energy deviation from full system calculation (kcal/mol)				
				EE-GMFCC				
System	No. of atoms	No. of basis functions	Full system (a.u.)	$\lambda = 3.0$ Å	$\lambda = 3.5$ Å	$\lambda = 4.0$ Å	$\lambda = 4.5$ Å	Max no. of atoms in fragments
1AML	598	5178	−15141.90793	4.28	0.85	0.14	0.57	59
2XL1	243	2137	−6183.35015	−0.94	−1.29	−1.34	−1.08	65
1BZG	573	4851	−13680.77639	6.77	3.35	2.76	2.78	70
1BBA	582	5033	−15103.38688	1.48	0.63	0.25	0.09	67
2JPK	589	5000	−13855.22544	5.52	4.27	3.34	2.99	69
2PPZ	608	5111	−14958.00905	3.15	2.14	1.71	1.59	63
1VTP	396	3418	−10015.00448	0.97	0.94	0.47	0.81	56
2YSC	578	5108	−14634.63499	5.56	4.09	3.66	3.02	68
2RLK	588	5089	−14589.93489	9.60	7.12	5.66	5.68	64
1WN8	354	2913	−8878.73074	5.02	3.88	3.70	3.46	59
2I9M	246	2069	−6267.19552	4.11	3.41	3.47	3.34	63
1LE1	218	1944	−5470.03139	1.97	1.16	0.49	0.36	60
2L8E	760	6493	−20270.54997	2.36	1.63	0.45	−0.35	65
3V1E	706	6061	−18015.49410	−0.31	−1.11	−1.66	−2.21	56
3T97 (B)	999	8302	−24598.98868	8.80	6.58	5.62	6.10	66
2LH0 (A)	1142	9702	−28500.53341	−0.01	−4.91	−5.92	−5.45	63
3L32 (A)	738	6264	−18054.99086	2.57	0.57	−0.45	−0.77	63
3NTW (A)	960	8016	−23683.45062	4.81	3.45	1.94	1.61	62
MUE				**3.79**	**2.85**	**2.39**	**2.35**	
RMSE				**4.66**	**3.47**	**3.06**	**2.99**	

Source: Wang *et al.* 2013 [19]. Reprinted with permission of American Chemical Society.
The point changes are taken from the Amber94 force field.

computational scale of the EE-GMFCC MP2 calculation is linear-scaling only with a low prefactor, whereas the computational cost for traditional MP2 calculation on the entire system shows $O(N^4 \sim N^5)$ as a function of the number of basis functions. Furthermore, the EE-GMFCC method for MP2 calculations has more computational advantages over conventional ones in terms of disk space requirement and memory usage.

11.2.4 The EE-GMFCC-CPCM Method for Protein Solvation Energy

Incorporating the fragment QM methods with solvation models made an important pace toward rigorous study of protein properties in solution [58, 61]. Wave function

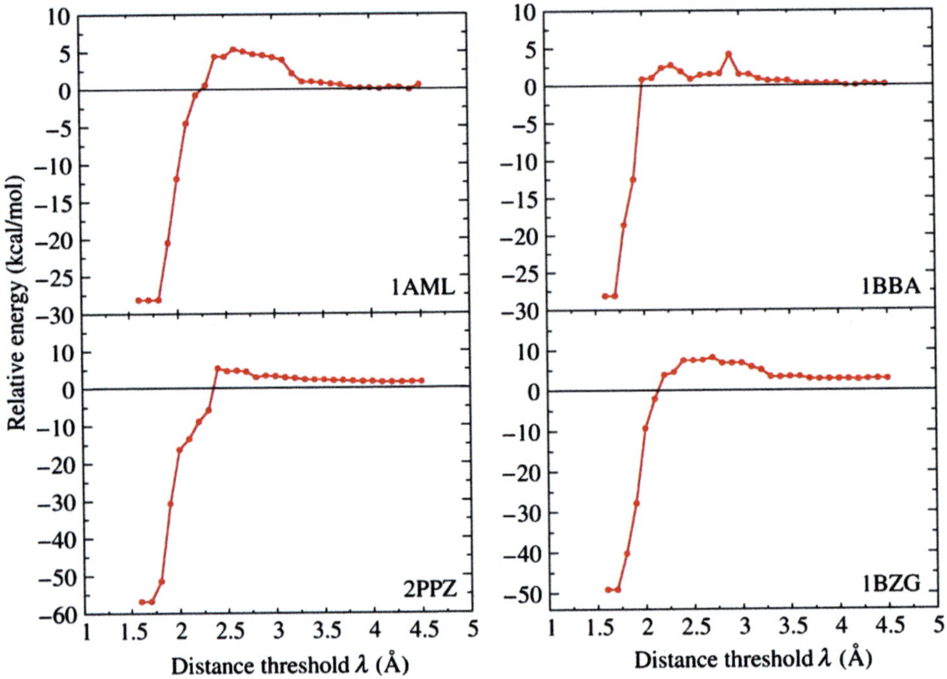

Figure 11.4 The relative energies (in kcal/mol, red line) as a function of the distance threshold λ for four proteins. The total energy from full system calculation (black line) is regarded as the reference. *Source*: Wang *et al.* 2013 [19]. Reprinted with permission of American Chemical Society.

distortion energy, which is missing in classical treatment, can be accurately captured from QM calculations. In 2006, Mei *et al.* proposed a practical MFCC-CPCM method which combined the MFCC approach with the conductor-like PCM model to calculate the solvation energies of proteins [58]. More recently, we combined the more accurate EE-GMFCC method with the CPCM model, denoted as EE-GMFCC-CPCM [62], for accurate calculation of protein solvation energy. As compared to standard full system QM/CPCM calculation, EE-GMFCC-CPCM shows clear improvement over the MFCC-CPCM approach for solvation energy calculations on proteins.

The Hamiltonian of a solute molecule in solution can be expressed as

$$H = H_0 + H', \tag{11.11}$$

where H_0 is the gas phase Hamiltonian of the solute molecule and H' denotes the perturbation from the solvent, which is written as

$$H' = \sum_{\mu,\alpha} \frac{q_\mu Z_\alpha}{\left|\mathbf{r}_\mu - \mathbf{R}_\alpha\right|} - \sum_{\mu,i} \frac{q_\mu}{\left|\mathbf{r}_\mu - \mathbf{r}_i\right|}, \tag{11.12}$$

where q_μ and $\mathbf{r}_\mu$ represent the surface charge and its position, respectively, which are determined through the CPCM method [63]. Z_α and $\mathbf{R}_\alpha$ are the nuclear charge and its

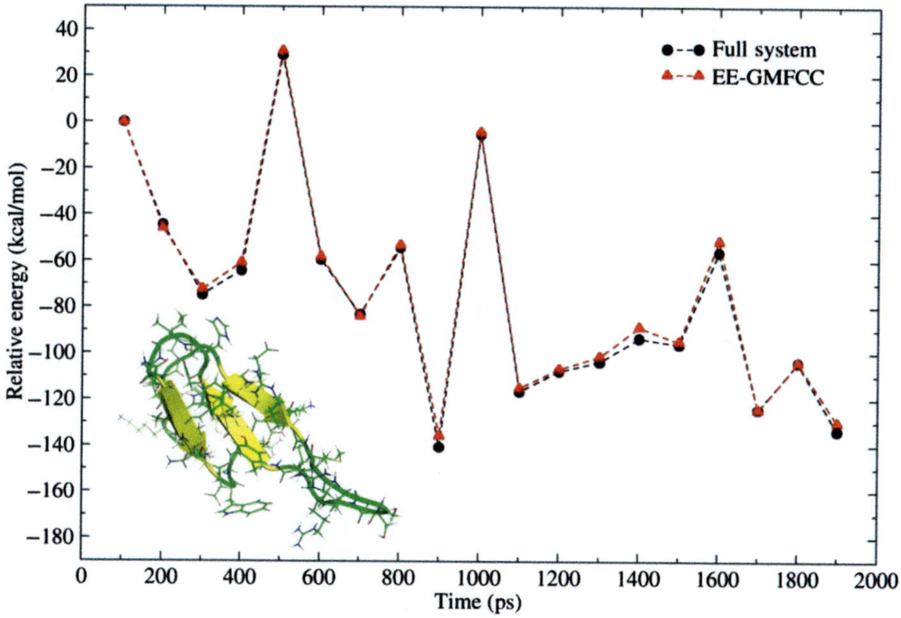

Figure 11.5 Comparison of the relative energies of 19 conformers (PDB id: 2KCF, 576 atoms) selected from 2 ns MD simulation between the standard full system HF/6-31G* calculations (black dots) and EE-GMFCC results (red triangles). *Source*: He *et al.* 2014 [20]. Reprinted with permission of American Chemical Society.

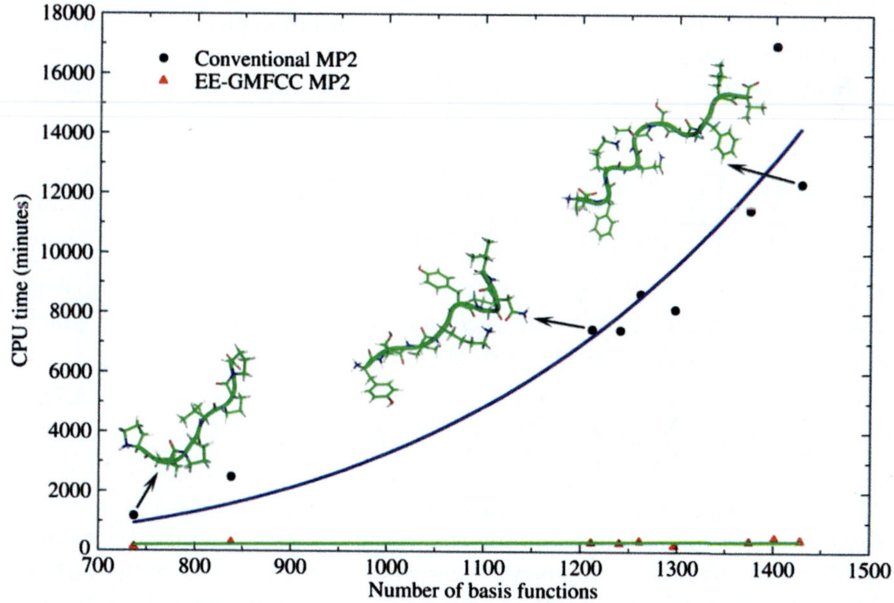

Figure 11.6 CPU time for the full system and EE-GMFCC calculations as a function of the number of basis functions at the MP2/6-31G* level. *Source*: He *et al.* 2014 [20]. Reprinted with permission of American Chemical Society.

corresponding coordinate, and $\mathbf{r}_i$ represents the position of electron i. The electrostatic solvation energy (G(ele)) is obtained by

$$G(\text{ele}) = \langle \Psi | H_0 | \Psi \rangle - \langle \Psi_0 | H_0 | \Psi_0 \rangle + \frac{1}{2} \langle \Psi | H' | \Psi \rangle = E(\text{wfd}) + G(\text{es}), \quad (11.13)$$

where Ψ_0 and Ψ represent the solute's wave functions in gas phase and in solution, respectively. The wave function distortion energy E(wfd) is expressed as

$$E(\text{wfd}) = \langle \Psi | H_0 | \Psi \rangle - \langle \Psi_0 | H_0 | \Psi_0 \rangle \quad (11.14)$$

and the electrostatic solute-solvent interaction energy is given by

$$G(\text{es}) = \frac{1}{2} \langle \Psi | H' | \Psi \rangle \quad (11.15)$$

In the EE-GMFCC-CPCM approach, the wave function distortion energy of equation (11.14) can be approximately obtained through

$$E(\text{wfd}) = \sum_{k=1}^{N_F} \Delta \tilde{E}_k - \sum_{k=1}^{N_C} \Delta \tilde{E}_k^c - \sum_{k=1}^{N_{GC}} \Delta \left(\tilde{E}_k^{ij} - \tilde{E}_k^i - \tilde{E}_k^j \right), \quad (11.16)$$

where $\Delta\tilde{E}$ represents the difference of the fragment energy between gas phase and solution phase, which is the sum of the self-energy of each fragment and the interaction energy between the fragment and background charges of the remaining protein. N_F, N_C, and N_{GC} denote the number of capped individual residues, concaps, and Gconcaps, respectively. We normally use the fixed Amber ff99SB charge model for each point charge $q_{n(l)}$ in equation (11.10). Therefore, the doubly counted charge-charge interaction energy is exactly canceled out between the gas phase and solution phase in the current EE-GMFCC-CPCM calculations.

On the other hand, the electrostatic solute-solvent interaction energy G(es) is calculated by

$$G(\text{es}) = \frac{1}{2} \sum_{\mu} q_\mu \left[\sum_{\alpha} \frac{Z_\alpha}{|\mathbf{r}_\mu - \mathbf{R}_\alpha|} - \int \frac{\rho(\mathbf{r})}{|\mathbf{r}_\mu - \mathbf{r}|} d\mathbf{r} \right] = \frac{1}{2} \sum_{\mu} q_\mu \phi(\mathbf{r}_\mu), \quad (11.17)$$

where $\rho(\mathbf{r})$ is the electron density at position $\mathbf{r}$ and $\phi(\mathbf{r}_\mu)$ is the electrostatic potential on cavity surface site μ generated by both the nuclei and the electrons of the solute, which can be approximately obtained from the EE-GMFCC method

$$\phi(\mathbf{r}_\mu) = \sum_{k=1}^{N_F} \phi_k(\mathbf{r}_\mu) - \sum_{k=1}^{N_C} \phi_k^C(\mathbf{r}_\mu) - \sum_{k=1}^{N_{GC}} \left[\phi_k^{ij}(\mathbf{r}_\mu) - \phi_k^i(\mathbf{r}_\mu) - \phi_k^j(\mathbf{r}_\mu) \right] \quad (11.18)$$

In the EE-GMFCC-CPCM method, the two-body correction was added to both the total energy expression of the solute and the electrostatic potential on the cavity surface. The QM calculation of each fragment is carried out in the presence of the surface charges. The updated wave function is then employed to determine the surface charges of the CPCM equation. The procedure is repeated in a self-consistent fashion until the total electrostatic solvation energy and the surface charges reach convergence, as in the standard self-consistent reaction field calculation.

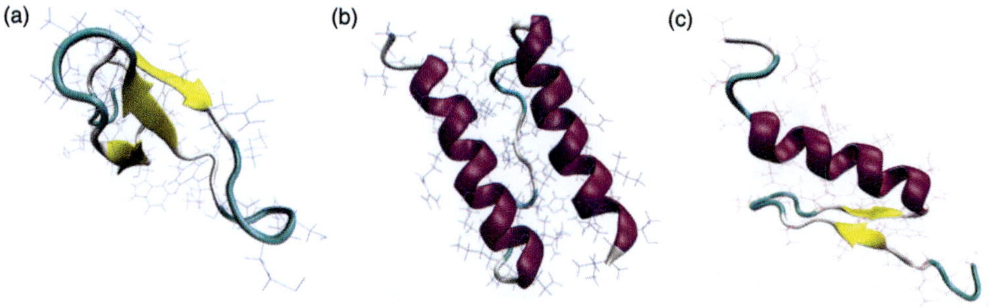

Figure 11.7 Three representative three-dimensional protein structures studied in this study. PDB IDs are 2KCF (a), 2Y0Q (b), and 1BHI (c), respectively. *Source*: Jia *et al.* 2013 [62]. Reprinted with permission of AIP Publishing.

We applied the EE-GMFCC-CPCM method for protein solvation energy calculations at the HF/6-31G* level. The proteins we studied have diverse tertiary structures composed of β-sheet, α-helix or a mixture of them (see Figure 11.7). The calculated electrostatic solvation energies are listed in Table 11.2. Except for 2XL1, the EE-GMFCC-CPCM calculated electrostatic solvation energies are in excellent agreement with the corresponding full system calculations with all deviations less than 0.01 Hartree. For electrostatic solvation energy, the RMSD of the EE-GMFCC-CPCM calculations from full system results is 3.93 kcal/mol. Therefore, the EE-GMFCC-CPCM method efficiently gives accurate electrostatic solvation energy for proteins.

Table 11.2 Electrostatic solvation energies in kcal/mol from the full system CPCM, EE-GMFCC-CPCM calculations, respectively, at the HF/6-31G* level.

Protein (pdb id)	No. of atoms	Full system (kcal/mol)	EE-GMFCC-CPCM[a] (kcal/mol)
1LE1	218	−166.73	−171.14 (−4.41)
2I9M	246	−237.15	−238.29 (−1.14)
2GB1	247	−439.70	−442.94 (−3.24)
1L2Y	304	−324.78	−328.88 (−4.10)
1WN8	354	−257.99	−258.99 (−1.00)
2XL1	243	−177.02	−184.90 (−7.88)
1VTP	396	−781.62	−786.24 (−4.62)
1BBA	582	−623.25	−622.11 (1.14)
2KCF	576	−685.46	−686.40 (−0.94)
2F21	603	−624.79	−619.79 (5.00)
1BHI	591	−528.10	−525.61 (2.49)
2Y0Q	803	−503.47	−499.04 (4.43)
RMSD			3.93
MUE			3.36

[a] Data in parentheses represent the errors with respect to the full system calculations.

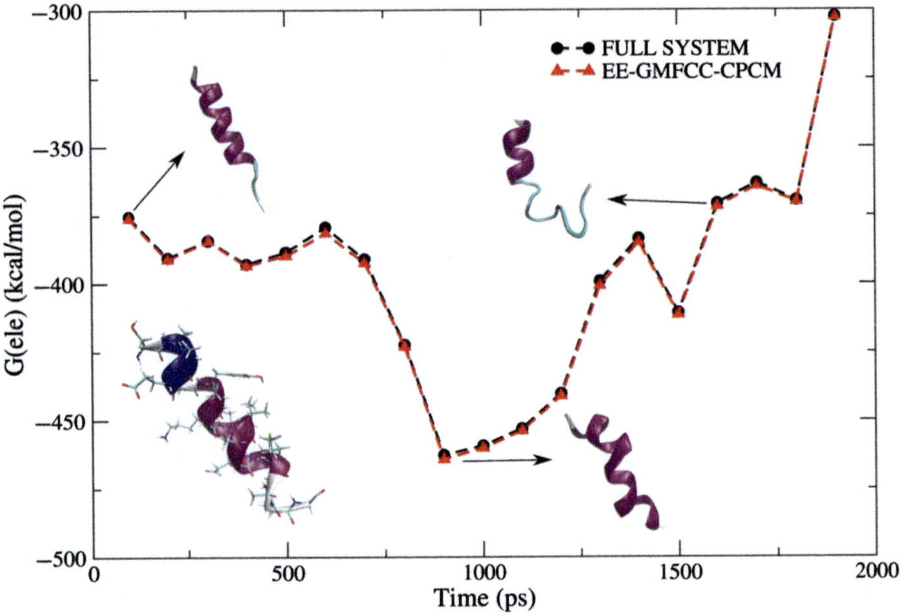

Figure 11.8 Variations of calculated electrostatic solvation energies over 19 different conformations selected from 2 ns MD simulation for the protein 2I9M. The calculations are performed using the EE-GMFCC-CPCM and standard full system HF/6-31G* methods, respectively. *Source*: He *et al.* 2014 [20]. Reprinted with permission of American Chemical Society.

During protein folding process, protein undergoes large conformational change towards a relatively compact structure. The interaction between the protein and solvent molecules has a significant impact on thermodynamic properties of the protein. The relative electrostatic solvation energies of different conformers of protein 2I9M (pdb id) are calculated using the EE-GMFCC-CPCM method and compared with corresponding full system results. Previous studies have shown that the Amber ff99SB force field tends to disfavor the α-helical structure, the native conformation of 2I9M is not stable and unfolding occurs during MD simulation using Amber ff99SB force field [64]. Therefore, it is worth studying the free energy profile of protein 2I9M in water solution at the *ab initio* level. 19 conformations are selected at an interval of 100 ps from 2 ns MD simulation. The electrostatic solvation energies calculated for the 19 conformers are shown in Figure 11.8. The total electrostatic solvation energy undergoes large fluctuation from −470 to −300 kcal/mol. Results from the EE-GMFCC-CPCM calculations are in excellent agreement with corresponding full system calculations. The RMSD of solvation energy from EE-GMFCC-CPCM calculation is only 1.14 kcal/mol with reference to full system calculations.

The EE-GMFCC-CPCM approach is an efficient linear-scaling QM method to treat biological macromolecules in solution. It can also be applied with density functional theory or other high-level electron correlation methods. Moreover, other implicit solvent models (such as SMD) can be readily combined with the EE-GMFCC approach for further improving the accuracy of theoretical calculation on solvation energies of proteins.

11.2.5 The EE-GMFCC-CPCM Method for Protein–Ligand Binding Energy

The EE-GMFCC-CPCM method was applied for calculation of binding affinities (ΔG_{bind}) of 14 avidin-biotin analogues based on the thermodynamics cycle [65],

$$\Delta G_{bind} = G(PL) - G(P) - G(L) \tag{11.19}$$

where P, L, and PL stand for the receptor, ligand, and their complex form. The free energy of each component is estimated as

$$\Delta G = \Delta E_{ele} + \Delta E_{vdw} + \Delta G_{solv} + \Delta G_{np} \tag{11.20}$$

where E_{ele} is the molecular electronic energy calculated by EE-GMFCC (at HF or DFT levels). E_{vdw} is the dispersion correction energy obtained from the density functional dispersion correction (DFT-D3) method [66]. G_{solv} is the electrostatic solvation energy, computed by the EE-GMFCC-CPCM method, and G_{np} is the non-electrostatic solvation energy contribution, which is calculated by the non-electrostatic term in the CPCM model [63]. The free energy contribution from conformational entropy change upon ligand binding was not included in our study.

The binding affinity of each ligand bound to avidin was calculated at the B3LYP/6-31G* level. The correlation coefficient (R) between the calculated binding energies and the experimental values is 0.85 based on single complex structure calculation. The correlation coefficient can be further improved to 0.88 when multiple snapshots are used for binding free energy calculation (see Figure 11.9). Our study demonstrates that the EE-GMFCC-CPCM method is capable of providing reliable prediction of binding affinities for various ligands bound to the same target.

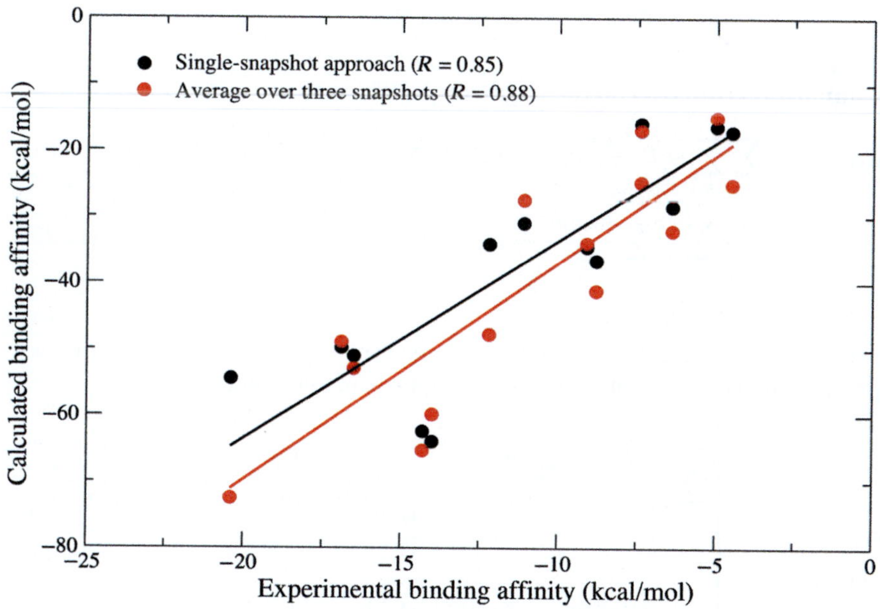

Figure 11.9 The correlation between the experimental and calculated binding affinities at the B3LYP/6-31G*-D level. *Source*: Liu *et al.* 2015 [48]. Reproduced with permission of The Royal Society of Chemistry.

11.2.6 The EE-GMFCC Method for Geometry Optimization and Vibrational Spectrum of Proteins

From the EE-GMFCC scheme, we have obtained the total energy expression for proteins. To derive the force on each atom, we need to compute the derivative of $E_{\text{EE-GMFCC}}$ with respect to the nuclear coordinate $\mathbf{r}_{\text{m}}$. There are both real atoms and background charges in each fragment. The mth atom, which appears as a real atom in fragment k, may become a background charge in other fragments. Hence, the force on the mth atom in fragment k can be computed as below [46].

$$\boldsymbol{f}(\mathbf{r}_m) = \begin{cases} -\dfrac{\partial \tilde{E}_k}{\partial \mathbf{r}_m} & \text{(If } m \text{ is a real atom)} \\ q(\mathbf{r}_m)^* EF_k^{bc}(\mathbf{r}_m) & \text{(If } m \text{ is a background charge)} \end{cases} \tag{11.21}$$

where $\tilde{E}_k$ represents the self-energy of fragment k and the interaction energy between fragment k and background charges, and $q(\mathbf{r}_m)$ stands for the point charge of mth atom. In the case where the atom m appears as the background charge, we calculate the electric field $EF_k^{bc}(\mathbf{r}_m)$ at the position $\mathbf{r}_m$.

Based on the above equations, the overall force on the mth atom can be obtained analytically through the EE-GMFCC approach using the following expression.

$$\begin{aligned} \mathrm{F}_{\text{EE-GMFCC}}(\mathrm{r}_m) = {} & \sum_{i=1}^{N-2} f_{F_i}(\mathrm{r}_m) - \sum_{i=1}^{N-3} f_{\mathrm{CC}_i}(\mathrm{r}_m) \\ & + \sum_{\substack{i,j>i+2, \\ |\mathrm{r}_i-\mathrm{r}_j|\le\lambda}} \left[f_{ij}(\mathrm{r}_m) - f_i(\mathrm{r}_m) - f_j(\mathrm{r}_m)\right] - \mathrm{F}_{\mathrm{MM}}(\mathrm{r}_m) \end{aligned} \tag{11.22}$$

where the summation runs over all the fragments, and

$$\begin{aligned} \mathrm{F}_{\mathrm{MM}}(\mathrm{r}_{m,s}) = {} & \sum_{l}\sum_{n} \frac{q_{m(k)}q_{n(l)}}{R^3_{m(k)n(l)}}(\mathrm{r}_{m,s} - \mathrm{r}_{n,s}) + \sum_{k,l}\sum_{n} \frac{q_{n(k)}q_{m(l)}}{R^3_{n(k)m(l)}}(\mathrm{r}_{m,s} - \mathrm{r}_{n,s}) \\ & - \sum_{\substack{j>i+2 \\ |r_i-r_j|\le\lambda}} \sum_{n'} \frac{q_{m(i)}q_{n'(j)}}{R^3_{m(i)n'(j)}}(\mathrm{r}_{m,s} - \mathrm{r}_{n',s}) \; (s = x, y, z) \end{aligned} \tag{11.23}$$

Furthermore, the second derivative of the total energy with respect to nuclear coordinates $\mathbf{r}_{\text{I}}$ and $\mathbf{r}_{\text{J}}$ (the Hessian matrix element) is approximately calculated as

$$\frac{\partial^2 E_{\text{EE-GMFCC}}}{\partial \mathrm{r}_I \partial \mathrm{r}_J} \approx \begin{cases} \displaystyle\sum_{i=1}^{N-2} \frac{\partial^2 \tilde{E}_{F_i}}{\partial \mathrm{r}_I \partial \mathrm{r}_J} - \sum_{i=1}^{N-3} \frac{\partial^2 \tilde{E}_{\mathrm{CC}_i}}{\partial \mathrm{r}_I \partial \mathrm{r}_J} + \sum_{\substack{i,j>i+2 \\ |\mathrm{r}_i-\mathrm{r}_j|\le\lambda}} \left(\frac{\partial^2 \tilde{E}_{ij}}{\partial \mathrm{r}_I \partial \mathrm{r}_J} - \frac{\partial^2 \tilde{E}_i}{\partial \mathrm{r}_I \partial \mathrm{r}_J} - \frac{\partial^2 \tilde{E}_j}{\partial \mathrm{r}_I \partial \mathrm{r}_J} \right)_{\mathrm{QM}} & (11.24) \\ q_I q_J \left[\dfrac{(-3)(\mathrm{r}_{I,s} - \mathrm{r}_{J,s})(\mathrm{r}_{I,t} - \mathrm{r}_{J,t})}{R^2_{IJ}} + \delta_{s,t} \right] R_{IJ}^{-3} \quad (s, t = x,\ y,\ z) & (11.25) \end{cases}$$

Equation (11.24) is applied to atoms I and J as coexisting atoms in each fragment calculation (neighboring residues and sequentially non-neighboring fragments that are spatially in close contact), whereas equation (11.25) is employed for atoms I and J, which are not coexisting in any fragment calculations.

Vibrational spectroscopy has been widely used for studying the structure and dynamics of proteins. After the optimized structures were obtained from the EE-GMFCC calculation, one can calculate the Hessian matrix of the whole system based on equations (11.24) and (11.25). Then, vibrational frequencies and their corresponding normal modes can be obtained by diagonalizing the mass-weighted Hessian matrix of the protein. The dipole moment μ and polarizability α_{ij} can also be derived using the EE-GMFCC approach

$$\Omega_{\text{EE-GMFCC}} = \sum_{i=1}^{N-2} \Omega_{F_i} - \sum_{i=1}^{N-3} \Omega_{CC_i} + \sum_{\substack{i,j>i+2 \\ |\mathbf{r}_i - \mathbf{r}_j| \le \lambda}} (\Omega_{ij} - \Omega_i - \Omega_j)_{\text{QM}} (\Omega = \mu, \alpha_{ij}) \quad (11.26)$$

By applying the chain rule, the derivatives of the dipole moment and polarizability with respect to the normal coordinates Q_k can be obtained using the equation below.

$$\frac{\partial \Omega}{\partial Q_k} = \sum_{i=1}^{N_{atom}} \sum_{j=x,y,z} \frac{\partial \Omega}{\partial \xi_{ij}} \frac{\partial \xi_{ij}}{\partial Q_k} (k = 1, 2, \ldots, N_f) \quad (11.27)$$

where N_{atom} is the total number of atoms, N_f is the number of vibrational degrees of freedom, and ξ_{ij} is the mass-weighted Cartesian coordinate defined as $\xi_{ij} = \sqrt{m_i} r_{ij}$ $(j = x, y, z)$.

Using these derivatives, the infrared (IR) intensity I_k and Raman intensities R_k can be calculated with the expressions below,

$$I_k \propto \left| \frac{\partial \mu_{\text{EE-GMFCC}}}{\partial Q_k} \right|^2 \quad (11.28)$$

$$R_k \propto \frac{3}{2} \left(\sum_{i=x,y,z} \frac{\partial \alpha^{ii}_{\text{EE-GMFCC}}}{\partial Q_k} \right)^2 + \frac{21}{2} \sum_{i,j=x,y,z} \left(\frac{\partial \alpha^{ij}_{\text{EE-GMFCC}}}{\partial Q_k} \right)^2 \quad (11.29)$$

The EE-GMFCC calculated IR and Raman vibrational spectra for proteins Trpzip2 and AKA (see Figure 11.10) are shown in Figure 11.11 as reference to the full system calculations. The half-width at half-maximum (HWHM) used for Lorentz broadenings

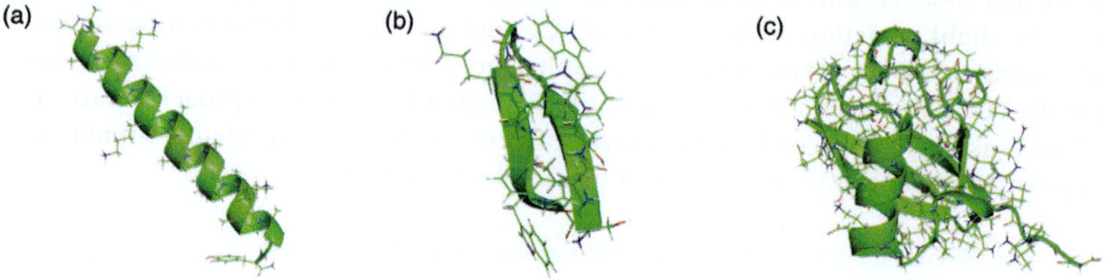

Figure 11.10 Three representative three-dimensional protein structures: (a) AKA, (b) Trpzip2, and (c) Ubiquitin. *Source*: Liu *et al.* 2016 [46]. Reproduced with permission of PCCP Owner Societies.

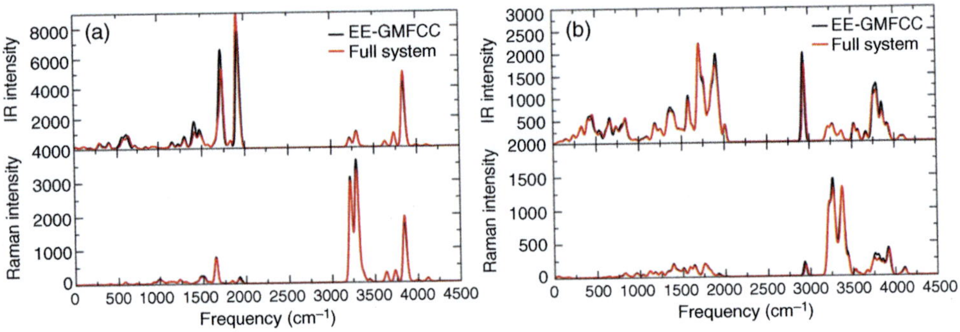

Figure 11.11 The IR and Raman spectra of AKA (a) and Trpzip2 (b) obtained from EE-GMFCC and full system calculations at the HF/6-31G* level. *Source*: Liu *et al.* 2016 [46]. Reproduced with permission of PCCP Owner Societies.

of the spectrum lines was set to 30 cm^{-1}. We can see that the spectrum shapes from EE-GMFCC calculations are in excellent agreement with those from full system calculations for both the IR and Raman spectra.

In IR spectroscopy, the amide I vibrational mode, primarily associated with the peptide bond carbonyl stretch, is often utilized for structural analysis [67]. The frequencies at which the amide I band occur depend on different patterns of intra- and intermolecular couplings. Thus, different protein secondary structures have characteristic spectral features in the amide I region (1600–1700 cm^{-1}) [67–69]. For example, β-sheets have a strong absorption band near 1630–1640 cm^{-1} and a weaker band at high frequencies (about 1680 cm^{-1}), while the peak for α-helices is located at 1640–1660 cm^{-1} [68]. We also investigated the amide I vibrational band of protein systems using EE-GMFCC approach at the M05-2X/6-31+G* level, which is capable of describing the N–H stretch region of the amide I vibrational mode [70]. Owing to their characteristic secondary structures, AKA, Trpzip2, and Ubiquitin (as shown in Figure 11.10) were used for investigating structure-spectroscopy relationships and validating the reliability of the EE-GMFCC method. The calculated vibrational frequencies were multiplied by a scale factor 0.9312 to compare with the experimental spectra [46]. The EE-GMFCC calculated and experimental amide I vibrational spectra for these three systems are shown in Figure 11.12. As one can see from the figure, the α-helical AKA has only one maximal absorption peak, while β-hairpin Trpzip2 has a distinguishing two-peak pattern, as compared to AKA. Ubiquitin has a complex secondary structure with mixed α-helix and β-sheet, and it shows a similar feature as that of Trpzip2. In these three cases, the slight deviation between EE-GMFCC and experimental observation is most likely owing to the approximation that the anharmonicity and solvation effect were not explicitly taken into account in our study. Overall, the EE-GMFCC approach is capable of distinguishing the secondary structures of proteins from IR spectrum simulation and providing quantitative prediction of major vibrational bands.

11.2.7 The EE-GMFCC-Based *Ab Initio* Molecular Dynamics for Proteins

Quantum mechanics, in principle, can provide accurate potential energy function for biomolecules, including all important quantum effects. Following the pioneering work of Car and Parrinello [71–73], the so-called *ab initio* molecular dynamics (AIMD)

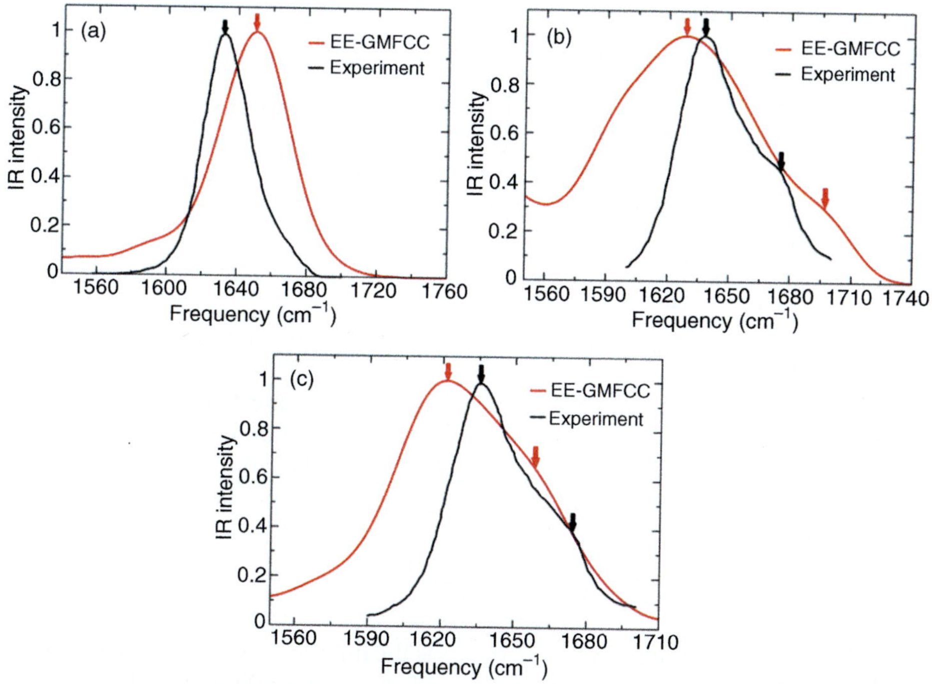

Figure 11.12 Theoretical (at the M05-2X/6-31+G* level) and experimental amide I vibrational spectra for (a) AKA, (b) Trpzip2, and (c) Ubiquitin. *Source*: Liu *et al.* 2016 [46]. Reproduced with permission of PCCP Owner Societies.

method was developed to address these problems. In the AIMD approach, atomic forces are calculated by QM methods such as HF and DFT, whereas the nuclear dynamics of the system is described by classical mechanics. AIMD is currently a popular and expanding computational tool employed to study physical, chemical, and biological phenomena.

As the application of QM methods has been limited by its large computational cost, most of the previous AIMD studies focused on relatively small systems such as polypeptides. To extend the applicability of AIMD methods to large systems such as proteins, we employed the EE-GMFCC method for quantum calculation of protein potential energy. Based on the potential energy of the protein computed by the EE-GMFCC method, the atomic forces are derived using equation (11.22) and used in MD simulation of protein dynamics.

It has been long recognized that water plays an important role in protein structure and dynamics. The EE-GMFCC approach can also be applied in explicit solvent environment. To reduce the computational cost, we use the mechanical embedded (ME)-QM/MM approach to describe the protein dynamics in explicit solvent, in which the protein is treated by the EE-GMFCC method, while the water molecules are described by mechanical mechanics. Therefore, the total energy of protein–solvent system under this scheme can be expressed as the summation of protein energy, protein-solvent interaction energy, and solvent energy as follows:

$$E_{\text{total}} = E_{\text{protein}}^{\text{EE-GMFCC}} + E_{\text{water}}^{\text{MM}} + E_{\text{protein-water}}^{\text{QM/MM}} \tag{11.30}$$

The water molecules and their interactions are described by the classical force field method,

$$E_{\text{water}}^{\text{MM}} = \sum_{i}^{N_{\text{water}}} E_{\text{water}}^{\text{intra}} + \sum_{\substack{\text{non-bonded} \\ \text{atom pairs } m,n}} \left\{ E_{m,n}^{VDW} + \frac{1}{4\pi\varepsilon_0} \frac{q_m q_n}{r_{m,n}} \right\} \tag{11.31}$$

$$E_{\text{water}}^{\text{intra}} = k_{\text{OH}}(r - r_0)^2 + k_{\text{HOH}}(\theta - \theta_0)^2 \tag{11.32}$$

$$E_{m,n}^{\text{VDW}} = \frac{A_{m,n}}{r_{m,n}^{12}} - \frac{B_{m,n}}{r_{m,n}^{6}} \tag{11.33}$$

q_m, q_n, r_0, θ_0, k_{OH}, k_{HOH}, $A_{m,n}$, and $B_{m,n}$ are force field parameters taken from Amber ff99SB force field.

The coupling between QM and MM parts was treated by mechanical embedding, which means that the EE-GMFCC calculation of protein is performed in gas phase (i.e., without background charges of water molecules in the QM calculation of each fragment) and the interaction between protein and water is described by the classical force field:

$$E_{\text{protein-water}}^{\text{QM/MM}} = \sum_{m}^{N_{\text{protein}}} \sum_{n}^{3\times N_{\text{water}}} \left(E_{m,n}^{\text{VDW}} + \frac{1}{4\pi\varepsilon_0} \frac{q_m q_n}{r_{m,n}} \right) \tag{11.34}$$

N_{protein} is the number of atoms in protein, N_{water} is the number of water molecules. The atomic charges of protein and water are taken from the Amber ff99SB force field. Then, the atomic force of a given atom m in the system can be expressed as

$$\mathbf{F}_m = -\nabla_m E_{\text{total}} = -\nabla_m E_{\text{protein}}^{\text{EE-GMFCC}} - \nabla_m E_{\text{water}}^{\text{MM}} - \nabla_m E_{\text{protein-water}}^{\text{QM/MM}} \tag{11.35}$$

where $\nabla_m E_{\text{protein}}^{\text{EE-GMFCC}}$ is the same as that in equation (11.22). $\nabla_m E_{\text{water}}^{\text{MM}}$ and $\nabla_m E_{\text{protein-water}}^{\text{QM/MM}}$ are the standard gradient expressions of the classical force field.

We have performed AIMD simulation for a small protein Trpcage (PDBID: 1L2Y, 20 residues with 304 atoms) solvated in explicit water molecules. For every step of simulation, atomic forces of the protein were calculated using the EE-GMFCC approach and then passed to the MD engine (the Sander module) of Amber. A water cap composed of TIP3P water molecules were added to account for the solvation effect. To avoid boiling off water molecules in MD simulation, a confining potential was added. Prior to AIMD simulation, the whole system was optimized and then heated up to 298 K. The resulting geometry was used as the initial structure for AIMD simulation, the integration time step of AIMD was 1 fs. All of the QM calculations were carried out at the M06-2X/6-31G* level using the Gaussian09 package. For comparison, we also performed classical MD simulation for this protein in solution with the same time scale using the Amber ff99SB force field.

The results are shown in Figures 11.13 and 11.14. The overall energy curve of the protein during AIMD simulation is stable, the standard deviations of the potential energy and total energy are 0.1018 and 0.1066 hartree, respectively. Furthermore, it can be seen from Figure 11.14 that the protein structure under AIMD simulation is more stable than that under the Amber ff99SB force field. One possible reason causing this could be that Trpcage is mainly composed of helical structures, and the Amber ff99SB force field tends to disfavor helical conformation. In addition, due to highly empirical nature of torsion

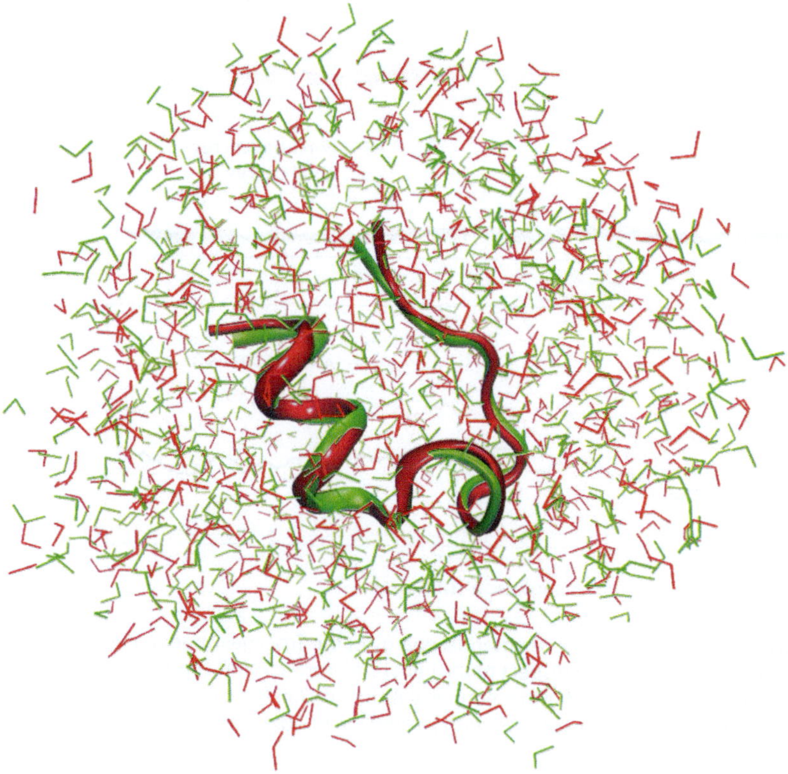

Figure 11.13 Comparison of the initial structure (red) and that from the last snapshot (green) of Trpcage from AIMD simulation with explicit solvent. *Source*: Liu *et al.* 2015 [47]. Reprinted with permission of American Chemical Society.

parameters, different classical force fields tend to have different bias towards secondary structure propensity. In contrast, the AIMD approach is unbiased in this respect and more balanced in terms of secondary structure formation.

As compared to classical force fields, one of the most important advantages of AIMD simulation is the inclusion of QM effects in the system, such as electrostatic polarization and charge transfer. In the EE-GMFCC approach, interactions of sequentially connected tripeptides and residues that are spatially in close contact are treated by QM. Therefore, the local QM effect such as charge transfer can be readily included. To illustrate the electrostatic polarization and charge transfer effects in the protein, we calculated the atomic charges by Mulliken population analysis when the QM calculation was performed for each fragment. In this scheme, the atomic charge of atom k in the protein can be calculated as follows:

$$q_k = \sum_{i=2}^{N-1} q_k\left(\text{Cap}_{i-1}^{*}A_i\text{Cap}_{i+1}\right) - \sum_{i=2}^{N-2} q_k\left(\text{Cap}_i^{*}\text{Cap}_{i+1}\right) + \sum_{\substack{i,j>i+2,\\ |R_i-R_j|\le\lambda}} (q_k(A_{ij}) - q_k(A_i) - q_k(A_j)) \tag{11.36}$$

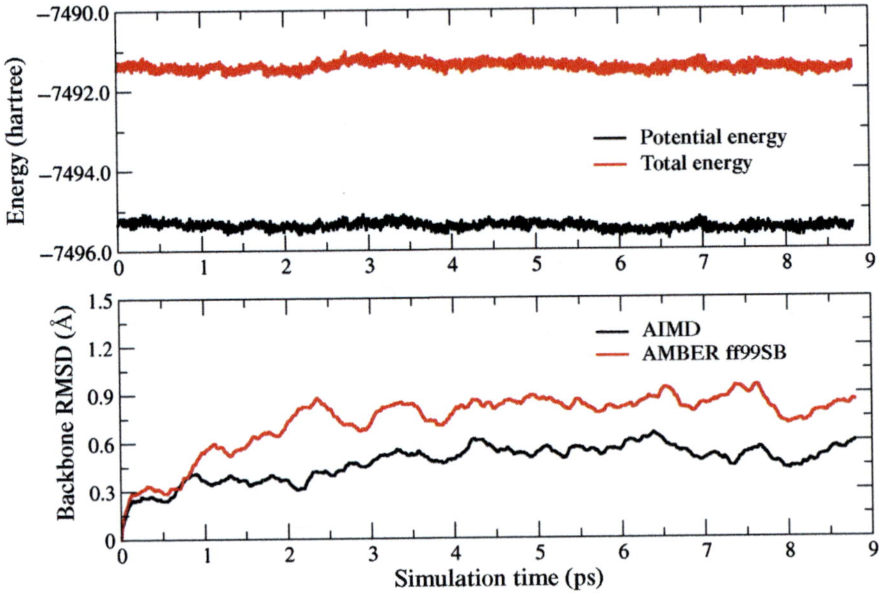

Figure 11.14 Time evolution of the total energy, potential energy of the Trpcage (upper panel) and protein backbone RMSD (lower panel) with respect to the initial structure in AIMD simulation in explicit solvent. Protein backbone RMSD computed by the empirical Amber ff99SB force field was also shown for comparison. *Source*: Liu *et al.* 2015 [47]. Reprinted with permission of American Chemical Society.

where $q_k(\text{Cap}^*_{i-1}A_i\text{Cap}_{i+1})$ denotes the Mulliken charge of atom k that belongs to the fragment $\text{Cap}^*_{i-1}A_i\text{Cap}_{i+1}$. To avoid double counting, the atom k's charge in the concap $\text{Cap}^*_i\text{Cap}_{i+1}$ needs to be deducted. The third term in equation (11.36) includes the two-body effect between non-sequentially connected residues that are spatially in close contact. Figure 11.15 plots the atomic charges, obtained from both EE-GMFCC and full system QM calculations, from the last snapshot of AIMD simulation in solution. As shown, the calculated results from EE-GMFCC calculation are very close to those from full system QM calculation, which demonstrates that the polarization and charge transfer effect are correctly captured in the EE-GMFCC method. Other atomic charge models, such as natural bond orbital (NBO) analysis and restrained electrostatic potential charge fitting (RESP) can also be combined with EE-GMFCC for other biological applications.

The EE-GMFCC method is trivially parallel since each fragment can be calculated independently on different processors. For this particular system, it took about 7 minutes to perform one step MD simulation using 96 2.66 GHz processers (12 processers for the QM calculation of each fragment), this time can be reduced to approximately 2.8 minutes if 240 processers were used. In fact, if the number of processors increases linearly with the size of the system, the time of performing each MD step will approximately keep constant.

Further development of this approach will focus on accurate treatment of protein–solvent interaction. A problem with mechanical embedding approach is that it ignores the electrostatic polarization effect of the QM system by the MM region. To fix this

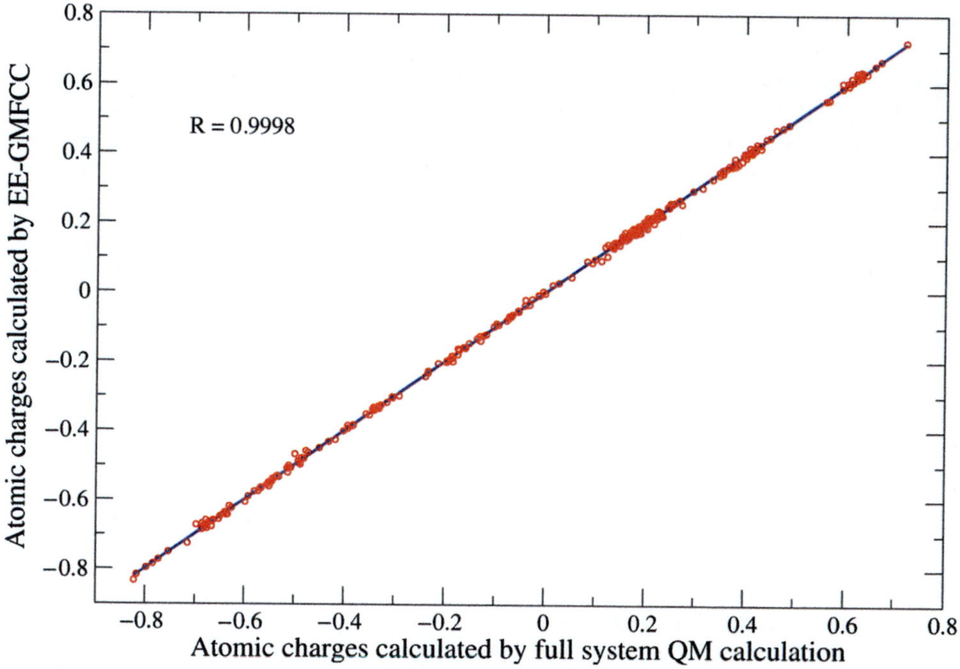

Figure 11.15 Comparison of atomic charges of the last snapshot of AIMD simulation from both EE-GMFCC and full system QM calculations. *Source*: Liu *et al.* 2015 [47]. Reprinted with permission of American Chemical Society.

problem, one can apply more rigorous methods to calculate protein–water interactions, such as using full QM methods to describe water molecules in the framework of EE-GMFCC approach, or one can modify the classical force field parameters of the water model to make them more suitable for electrostatic embedding. In addition, the EE-GMFCC method itself can be optimized to increase its computational efficiency. The last one is how to deal with the periodic boundary condition in AIMD simulation. This is still a developing field and has recently attracted increased attention.

11.3 Conclusion

In this chapter, we reviewed the MFCC-based fragmentation methods and their applications in biological systems. By introducing the electrostatic embedding field and two-body QM interaction correction on top of the original MFCC approach, the EE-GMFCC method is an accurate and efficient approach for QM calculation of protein molecular properties, such as total energy, electrostatic solvation energy, protein–ligand binding affinity, and vibrational spectra. The current EE-GMFCC method is linear-scaling with a small prefactor, trivially parallel, and can be readily applied in performing high-level *ab initio* molecular dynamics for protein in aqueous solution. Further development of this approach is underway in our laboratory, including improving the efficiency of EE-GMFCC and more accurate description of the interaction between the protein and aqueous environment in EE-GMFCC-based AIMD simulation.

Acknowledgments

This work was supported by the National Natural Science Foundation of China (Grants No. 21303057, 21673074, 21433004 and 21403068), Ministry of Science and Technology of China (Grant no. 2016YFA0501700), Specialized Research Fund for the Doctoral Program of Higher Education (Grant No. 20130076120019), Shanghai Putuo District (Grant 2014-A-02), the NYU-ECNU Center for Computational Chemistry at NYU Shanghai, and NYU Global Seed Grant for Collaborative Research. We also thank the Supercomputer Center of East China Normal University for providing us with computational time.

References

1 Szabo, A.; Ostlund, N. S. McGraw-Hill: New York **1989**.
2 Goedecker, S. *Rev. Mod. Phys.* **1999**, *71*, 1085.
3 Friesner, R. A.; Murphy, R. B.; Beachy, M. D.; Ringnalda, M. N.; Pollard, W. T.; Dunietz, B. D.; Cao, Y. X. *J. Phys. Chem. A* **1999**, *103*, 1913.
4 Strain, M. C.; Scuseria, G. E.; Frisch, M. J. *Science* **1996**, *271*, 51.
5 Yang, W. T. *Phys. Rev. Lett.* **1991**, *66*, 1438.
6 Hampel, C.; Werner, H. J. *J. Chem. Phys.* **1996**, *104*, 6286.
7 Schutz, M.; Rauhut, G.; Werner, H. J. *J. Phys. Chem. A* **1998**, *102*, 5997.
8 He, X.; Merz, K. M. *J. Chem. Theory Comput.* **2010**, *6*, 405.
9 He, X.; Sode, O.; Xantheas, S. S.; Hirata, S. *J. Chem. Phys.* **2012**, *137*, 204505.
10 Gordon, M. S.; Fedorov, D. G.; Pruitt, S. R.; Slipchenko, L. V. *Chem. Rev.* **2012**, *112*, 632.
11 Kitaura, K.; Ikeo, E.; Asada, T.; Nakano, T.; Uebayasi, M. *Chem. Phys. Lett.* **1999**, *313*, 701.
12 Nakano, T.; Kaminuma, T.; Sato, T.; Fukuzawa, K.; Akiyama, Y.; Uebayasi, M.; Kitaura, K. *Chem. Phys. Lett.* **2002**, *351*, 475.
13 Fedorov, D. G.; Kitaura, K. *J. Phys. Chem. A* **2007**, *111*, 6904.
14 Zhang, D. W.; Zhang, J. Z. H. *J. Chem. Phys.* **2003**, *119*, 3599.
15 He, X.; Zhang, J. Z. H. *J. Chem. Phys.* **2005**, *122*, 031103.
16 He, X.; Zhang, J. Z. H. *J. Chem. Phys.* **2006**, *124*, 184703.
17 Chen, X. H.; Zhang, Y. K.; Zhang, J. Z. H. *J. Chem. Phys.* **2005**, *122*, 184105.
18 Chen, X. H.; Zhang, J. Z. H. *J. Chem. Phys.* **2006**, *125*, 044903.
19 Wang, X. W.; Liu, J. F.; Zhang, J. Z. H.; He, X. *J. Phys. Chem. A* **2013**, *117*, 7149.
20 He, X.; Zhu, T.; Wang, X.; Liu, J.; Zhang, J. Z. H. *Acc. Chem. Res.* **2014**, *47*, 2748.
21 Deev, V.; Collins, M. A. *J. Chem. Phys.* **2005**, *122*, 154102.
22 Collins, M. A.; Deev, V. A. *J. Chem. Phys.* **2006**, *125*, 104104.
23 Mullin, J. M.; Roskop, L. B.; Pruitt, S. R.; Collins, M. A.; Gordon, M. S. *J. Phys. Chem. A* **2009**, *113*, 10040.
24 Exner, T. E.; Mezey, P. G. *J. Phys. Chem. A* **2002**, *106*, 11791.
25 Exner, T. E.; Mezey, P. G. *J. Comput. Chem.* **2003**, *24*, 1980.
26 Exner, T. E.; Mezey, P. G. *J. Phys. Chem. A* **2004**, *108*, 4301.
27 Babu, K.; Gadre, S. R. *J. Comput. Chem.* **2003**, *24*, 484.
28 Ganesh, V.; Dongare, R. K.; Balanarayan, P.; Gadre, S. R. *J. Chem. Phys.* **2006**, *125*, 104109.

29 Isegawa, M.; Wang, B.; Truhlar, D. G. *J. Chem. Theory Comput.* **2013**, *9*, 1381.
30 Li, W.; Li, S. H.; Jiang, Y. S. *J. Phys. Chem. A* **2007**, *111*, 2193.
31 Li, S. H.; Li, W.; Ma, J. *Acc. Chem. Res.* **2014**, *47*, 2712.
32 Richard, R. M.; Herbert, J. M. *J. Chem. Phys.* **2012**, *137*, 064113
33 Dahlke, E. E.; Truhlar, D. G. *J. Chem. Theory Comput.* **2007**, *3*, 46.
34 Dahlke, E. E.; Truhlar, D. G. *J. Chem. Theory Comput.* **2007**, *3*, 1342.
35 Xie, W. S.; Gao, J. L. *J. Chem. Theory Comput.* **2007**, *3*, 1890.
36 Xie, W. S.; Song, L. C.; Truhlar, D. G.; Gao, J. L. *J. Chem. Phys.* **2008**, *128*, 234108.
37 Le, H. A.; Tan, H. J.; Ouyang, J. F.; Bettens, R. P. A. *J. Chem. Theory Comput.* **2012**, *8*, 469.
38 Mayhall, N. J.; Raghavachari, K. *J. Chem. Theory Comput.* **2011**, *7*, 1336.
39 Mayhall, N. J.; Raghavachari, K. *J. Chem. Theory Comput.* **2012**, *8*, 2669.
40 Richard, R. M.; Herbert, J. M. *J. Chem. Phys.* **2012**, *137*, 064113.
41 Richard, R. M.; Herbert, J. M. *J. Chem. Theory Comput.* **2013**, *9*, 1408.
42 He, X.; Wang, B.; Merz, K. M. *J. Phys. Chem. B* **2009**, *113*, 10380.
43 Zhu, T.; He, X.; Zhang, J. Z. H. *Phys. Chem. Chem. Phys.* **2012**, *14*, 7837.
44 Zhu, T.; Zhang, J. Z. H.; He, X. *J. Chem. Theory Comput.* **2013**, *9*, 2104.
45 Collins, M. A.; Bettens, R. P. A. *Chem. Rev.* **2015**, *115*, 5607.
46 Liu, J. F.; Zhang, J. Z. H.; He, X. *Phys. Chem. Chem. Phys.* **2016**, *18*, 1864.
47 Liu, J. F.; Zhu, T.; Wang, X. W.; He, X.; Zhang, J. Z. H. *J. Chem. Theory Comput.* **2015**, *11*, 5897.
48 Liu, J. F.; Wang, X. W.; Zhang, J. Z. H.; He, X. *Rsc Advances* **2015**, *5*, 107020.
49 Zhang, D. W.; Zhang, J. Z. H. *J. Theor. Comput. Chem.* **2004**, *3*, 43.
50 Xiang, Y.; Zhang, D. W.; Zhang, J. Z. H. *J. Comput. Chem.* **2004**, *25*, 1431.
51 Chen, X. H.; Zhang, D. W.; Zhang, J. Z. H. *J. Chem. Phys.* **2004**, *120*, 839.
52 He, X.; Mei, Y.; Xiang, Y.; Zhang, D. W.; Zhang, J. Z. H. *Proteins: Struct. Funct. Bioinform.* **2005**, *61*, 423.
53 Mei, Y.; He, X.; Xiang, Y.; Zhang, D. W.; Zhang, J. Z. H. *Proteins: Struct. Funct. Bioinform.* **2005**, *59*, 489.
54 Hummer, G.; Pratt, L. R.; García, A. E.; Berne, B. J.; Rick, S. W. *J. Phys. Chem. B* **1997**, *101*, 3017.
55 Tobias, D. J. *Curr. Opin. Struct. Biol.* **2001**, *11*, 253.
56 Besler, B. H.; Merz, K. M., Jr., Kollman, P. A. *J. Comput. Chem.* **1990**, *11*, 431.
57 Cornell, W. D.; Cieplak, P.; Bayly, C. I.; Kollmann, P. A. *J. Am. Chem. Soc.* **1993**, *115*, 9620.
58 Mei, Y.; Ji, C. G.; Zhang, J. Z. H. *J. Chem. Phys.* **2006**, *125*, 094906.
59 Gao, A. M.; Zhang, D. W.; Zhang, J. Z. H.; Zhang, Y. K. *Chem. Phys. Lett.* **2004**, *394*, 293.
60 Mei, Y.; Zhang, D. W.; Zhang, J. Z. H. *J. Phys. Chem. A* **2005**, *109*, 2.
61 Fedorov, D. G.; Kitaura, K.; Li, H.; Jensen, J. H.; Gordon, M. S. *J. Comput. Chem.* **2006**, *27*, 976.
62 Jia, X.; Wang, X.; Liu, J.; Zhang, J. Z. H.; Mei, Y.; He, X. *J. Chem. Phys.* **2013**, *139*, 214104
63 Barone, V.; Cossi, M. *J. Phys. Chem. A* **1998**, *102*, 1995.
64 Best, R. B.; Hummer, G. *J. Phys. Chem. B* **2009**, *113*, 9004.
65 Kollman, P. A.; Massova, I.; Reyes, C.; Kuhn, B.; Huo, S. H.; Chong, L.; Lee, M.; Lee, T.; Duan, Y.; Wang, W.; Donini, O.; Cieplak, P.; Srinivasan, J.; Case, D. A.; Cheatham, T. E. *Acc. Chem. Res.* **2000**, *33*, 889.
66 Grimme, S.; Antony, J.; Ehrlich, S.; Krieg, H. *J. Chem. Phys.* **2010**, *132*, 154104.
67 Wang, L.; Middleton, C. T.; Zanni, M. T.; Skinner, J. L. *J. Phys. Chem. B* **2011**, *115*, 3713.

68 Ganim, Z.; Chung, H. S.; Smith, A. W.; Deflores, L. P.; Jones, K. C.; Tokmakoff, A. *Acc. Chem. Res.* **2008**, *41*, 432.

69 Surewicz, W. K.; Mantsch, H. H.; Chapman, D. *Biochemistry* **1993**, *32*, 389.

70 Buchanan, E. G.; James, W. H.; Choi, S. H.; Guo, L.; Gellman, S. H.; Muller, C. W.; Zwier, T. S. *J. Chem. Phys.* **2012**, *137*, 094301

71 Laasonen, K.; Pasquarello, A.; Car, R.; Lee, C.; Vanderbilt, D. *Phys. Rev. B: Condens. Matter* **1993**, *47*, 10142.

72 Remler, D. K.; Madden, P. A. *Mol. Phys.* **1990**, *70*, 921.

73 White, J. A.; Bird, D. M. *Phys. Rev. B: Condens. Matter* **1994**, *50*, 4954.

Index

Fragmentation: Toward Accurate Calculations on Complex Molecular Systems, First Edition.
Edited by Mark S. Gordon.

n

o

p

q

Printed and bound by CPI Group (UK) Ltd, Croydon, CR0 4YY
17/09/2026
14964303-0003